Energy, Environment, and Sustainability

AIMS AND SCOPE

This books series publishes cutting edge monographs and professional books focused on all aspects of energy and environmental sustainability, especially as it relates to energy concerns. The Series is published in partnership with the International Society for Energy, Environment, and Sustainability. The books in these series are edited or authored by top researchers and professional across the globe. The series aims at publishing state-of-the-art research and development in areas including, but not limited to:

- Renewable Energy
- Alternative Fuels
- Engines and Locomotives
- Combustion and Propulsion
- Fossil Fuels
- Carbon Capture
- Control and Automation for Energy
- Environmental Pollution
- Waste Management
- Transportation Sustainability

Review Process

The proposal for each volume is reviewed by the main editor and/or the advisory board. The chapters in each volume are individually reviewed single blind by expert reviewers (at least four reviews per chapter) and the main editor.

Ethics Statement for this series can be found in the Springer standard guidelines here https://www.springer.com/us/authors-editors/journal-author/journal-author-helpdesk/before-you-start/before-you-start/1330#c14214

Sunil Kumar Sharma · Ram Krishna Upadhyay ·
Vikram Kumar
Editors

Dynamics of Transportation Ecosystem, Modeling, and Control

Springer

Editors
Sunil Kumar Sharma
School of Technology
Gati Shakti Vishwavidyalaya
Vadodara, Gujarat, India

Ram Krishna Upadhyay
School of Technology
Gati Shakti Vishwavidyalaya
Vadodara, Gujarat, India

Vikram Kumar
Engine Research Laboratory
Department of Mechanical Engineering
Indian Institute of Technology Kanpur
Kanpur, Uttar Pradesh, India

ISSN 2522-8366 ISSN 2522-8374 (electronic)
Energy, Environment, and Sustainability
ISBN 978-981-97-0436-1 ISBN 978-981-97-0437-8 (eBook)
https://doi.org/10.1007/978-981-97-0437-8

This Springer imprint is published by the registered company Springer Nature Singapore Pte Ltd.
The registered company address is: 152 Beach Road, #21-01/04 Gateway East, Singapore 189721, Singapore

Preface

Dynamics of Transportation Ecosystem, Modeling, and Control offers a comprehensive exploration of the multifaceted transportation industry, encompassing an array of topics that span from the intricacies of hydraulic system modeling to the transformative potential of blockchain technology in distribution. In a rapidly changing global landscape, where transportation plays a pivotal role in connecting communities and facilitating commerce, this collection of chapters provides a valuable lens through which to examine the challenges, innovations, and sustainable solutions shaping the present and future of transportation.

The International Society for Energy, Environment and Sustainability (ISEES) was founded at the Indian Institute of Technology Kanpur (IIT Kanpur), India, in January 2014 to spread knowledge and awareness and to catalyze research activities in the fields of Energy, Environment, Sustainability, and Combustion. Society's goal is to contribute to the development of clean, affordable, and secure energy resources and a sustainable environment for society, spread knowledge in the areas mentioned above and create awareness about the environmental challenges the world is facing today. The unique way adopted by ISEES was to break the conventional silos of specializations (engineering, science, environment, agriculture, biotechnology, materials, fuels, etc.) to tackle the problems related to energy, environment, and sustainability holistically. This is evident by the participation of experts from all fields to resolve these issues. The ISEES is involved in various activities, such as conducting workshops, seminars, conferences, etc., in the domains of its interests. The society also recognizes the outstanding works of young scientists, professionals, and engineers for their contributions by conferring awards in various categories.

The Seventh International Conference on "Sustainable Energy and Environmental Challenges" (VII-SEEC) was organized under the auspices of ISEES from December 16 to 18, 2022, at the Indian Institute of Technology, Banaras Hindu University, Varanasi (IIT-BHU), India. This conference provided a platform for discussions between eminent scientists and engineers from several countries, including India, the USA, Spain, Poland, Austria, the Czech Republic, and Korea. At this conference, eminent international speakers presented their views on energy, combustion, emissions, and alternative energy resources for sustainable development and a cleaner environment. The conference presented two high-voltage plenary talks by Dr. Ajay Kumar, Former Union Defense Secretary, GoI and Dr. S. S. V. Ramakumar, Director (R&D), Indian Oil.

The conference included 29 technical sessions on energy and environmental sustainability topics, including two plenary talks, 15 keynote talks, 80 invited talks from prominent scientists, and 118 contributed talks by students and researchers. The conference included technical sessions on advanced engine technologies, air pollution monitoring, anaerobic digestion, combustion and flames, air pollution control, biodegradation of toxic chemicals, energy and exergy, desalination and wastewater treatment, environmental bioengineering, pollution and climate change: challenges and priorities, alternative transportation fuels and materials, emerging environmental contaminants, sustainable processing of biomass, human health and environmental sustainability, sprays and atomization, solid waste: challenges and mitigation, solid waste management, sustainable food and agri biotechnology, modeling and simulations, renewable energy technologies, bioremediation, biofuels and biorefineries, engine emissions and control, cleaner technologies for pollution mitigation, microbial processes, energy and environment, coal and biomass gasification, and environmental challenges mitigation.

More than 250 participants and speakers attended the conference, where 14 ISEES books published by Springer, Singapore, under a dedicated series, *Energy, Environment and Sustainability*, were released. This conference laid the roadmap for technology development, opportunities, and challenges in the Energy, Environment and Sustainability domain. These topics are relevant for the country and the world in the present context. We acknowledge the support from various funding agencies and organizations for the successful conduct of the VII-SEEC, where the idea of these books germinated. We would therefore gratefully like to acknowledge IIT BHU (special thanks to Prof. Akhilendra Pratap Singh); SERB, Government of India; Department of Scientific and Industrial Research (DSIR) (special thanks to Dr. Vipin Shukla) and our publishing partner Springer (special thanks to Swati Mehershi).

The editors would like to express their sincere gratitude to a large number of authors from all over the world for submitting their high-quality work on time and revising it appropriately at short notice. We want to express our special gratitude to our prolific reviewers: We want to express our special gratitude to our prolific set of reviewers, Dr. Prakash Kumar, Dr. Chandan Pandey, Dr. Jyoti Sharma, Dr. Swati Mohapatra, Dr. Kartik Selva Kumar, Dr. Pradeep Saroj, Dr. Dipen Rajak, Dr. Navneet Sharma, Mr. Surya Pratap Singh, Mr. Abhay Upadhyay, and Ms. Kiran Sharma, who reviewed various chapters of this monograph and provided their valuable suggestions to improve the manuscripts. Ms. Kiran Sharma played an invaluable role in the refinement of the manuscript through her expertise in editorial correction, proofreading, grammar, fragmentation, and manuscript flow. Her meticulous attention to detail ensured that the document was free of errors and maintained a coherent and logical structure, significantly improving the manuscript's readability and coherence.

This book aims to provide a holistic perspective on the intricate web of transportation systems, emphasizing their dynamic nature and the critical role they play in our interconnected world. By bringing together diverse topics such as hydraulic system modeling, noise and vibration analysis in railways, blockchain solutions for supply chain efficiency, and sustainable supply chain management, our goal is to offer readers a comprehensive understanding of the contemporary challenges and innovative solutions within the transportation industry. Through the collective wisdom of experts and researchers, this book not only sheds light on the current state of transportation but also strives to inspire progress and transformation in the field. It serves as a valuable resource for professionals, academics, and anyone seeking to contribute to the continuous evolution and improvement of transportation systems worldwide. Chapters include recent results and are focused on current trends in the railway, air, and automotive sectors. In this book, readers will gain insight into a wide range of topics within the transportation industry, from hydraulic systems and noise reduction in railways to blockchain applications in food distribution and sustainable supply chain management. These chapters provide a diverse and comprehensive overview of the evolving transportation landscape, offering valuable knowledge and inspiration for professionals and researchers alike. We hope the book greatly interests the professionals and post-graduate students involved in rail transportation, road transportation, marine transportation, air transportation, pipeline transportation, energy, and environmental research.

Vadodara, India — Sunil Kumar Sharma
Vadodara, India — Ram Krishna Upadhyay
Kanpur, India — Vikram Kumar

Contents

Editors and Contributors

About the Editors

Dr. Sunil Kumar Sharma is Assistant Professor at the Gati Shakti Vishwavidyalaya, Vadodara, a Central University, established by the Ministry of Railway, Government of India. He has received his Ph.D. from the Indian Institute of Technology, Roorkee. He worked at the Non-destructive Evaluation and Structural Health Monitoring Laboratory at C. N. University, South Korea. His research interests are vehicle dynamics, contact mechanics, mechatronics, and real-time software-enabled control systems for high-speed rail vehicles. He published several research articles in a national and international journal, book chapters, and edited a book. Dr. Sharma is also featured among the top 2% of scientists in a global list compiled by Stanford University, USA.

Dr. Ram Krishna Upadhyay is Assistant Professor at the Gati Shakti Vishwavidyalaya, Vadodara, a Central University, established by the Ministry of Railway, Government of India. He has received his Ph.D. and M.Tech. from the Indian Institute of Technology (ISM) Dhanbad in Mechanical Engineering with a broad specialization in Surface Engineering and Tribology. Before joining Gati Shakti Vishwavidyalaya, Dr. Ram worked as Postdoctoral Fellow at the Indian Institute of Technology Kanpur. His research interests include lubrication and materials wear for industrial application, tribology of additive manufactured parts, and nanocomposites. He is a recipient of the SERB-ACS NPDF best poster competition award by the Science and Engineering Research Board, New Delhi, and the American Chemical Society, USA. He published several journal papers, book chapters, edited a book, and completed a project funded by the Science and Engineering Research Board, New Delhi.

Dr. Vikram Kumar is currently at IIT Kanpur. He received his Ph.D. in Mechanical Engineering from the Indian Institute of Technology Kanpur, India, in 2018. His areas of research include polymer and composite coating; wear, friction, and lubrication; IC engine tribology; alternative fuels; advanced low-temperature combustion; engine emissions measurement; particulate characterization. Dr. Kumar has edited four books and authored 10 book chapters and 19 research articles in international journals and conferences. He has been awarded with ISEES Best Ph.D. Thesis Award (2018), Senior Research Associateship under SCIR-POOL Scientist (2018–2021). He is a lifetime member of ISEES.

Contributors

Samson Abiodun Aasa Department of Mechanical Engineering, Olabisi Onabanjo University, Ibogun Campus, Nigeria

Abiodun Oluwatosin Adeoye Department of Chemistry, Federal University Oye-Ekiti, Oye-Ekiti, Ekiti State, Nigeria

Nuhu Ahmad Department of Civil Engineering, Bayero University, Kano, Kano-State, Nigeria

Aditi Aiswarya School of Technology, Gati Shakti Vishwavidyalaya, Vadodara, India

Mohd Avesh Mazedan International Research Academy, Okhla, Delhi, New Delhi, India

Sarfraz Ahmad Awan Department of Computer Science and Information Technology, Virtual University of Pakistan, Lahore, Pakistan

Chintada Vinod Babu GMR Institute of Technology, Rajam, Andhra Pradesh, India

Lakshya Bhardwaj School of Technology, Gati Shakti Vishwavidyalaya, Vadodara, India

Ramajogi Naidu Doddi Aditya Institute of Technology and Management, Tekkali, Andhra Pradesh, India

Ruth Kooaka Dorgbaa Department of Metallurgical and Materials Engineering, University of Nigeria, Nsukka, Nigeria

Azad Duppala Aditya Institute of Technology and Management, Tekkali, Andhra Pradesh, India

Husaina Anchau Garba National Research Institute for Chemical Technology, Zaria, Kaduna State, Nigeria

Keerat Kumar Gupta Department of Civil Engineering, Graphic Era (Deemed to Be University), Dehradun, India

Ismail Hossain Department of Nuclear and Renewable Energy, School of Natural Sciences and Mathematics, Ural Federal University, Yekaterinburg, Russia

Naziru Imam Department of Chemistry, Federal University Oye-Ekiti, Oye-Ekiti, Ekiti State, Nigeria;
Analytical Chemistry Division, Department of Chemistry, University of Ibadan, Ibadan, Nigeria

Joseph T. Jose Indian Institute of Technology (Indian School of Mines), Dhanbad, India

Johan W. Joubert Department of Industrial Systems Engineering, University of Pretoria, Pretoria, South Africa

Jerzy Kisilowski Kazimierz Pulaski University of Technology and Humanities in Radom, Radom, Poland

Wladyslaw Koc Gdansk University of Technology, Gdansk, Poland

Ramji Koona Andhra University College of Engineering, Visakhapatnam, Andhra Pradesh, India

Rafał Kowalik Polish Air Force University, Deblin, Poland

Amit Kumar SSRAS safety and risks services Pvt. Ltd, Roorkee, India

Rakesh Kumar Department of Mechanical Engineering, School of Engineering, P P Savni University, Kosamba, Surat, India

Sanjeev Kumar Department of Civil Engineering, Graphic Era (Deemed to Be University), Dehradun, India

Vikram Kumar Department of Mechanical Engineering, Indian Institute of Technology Kanpur, Kanpur, India

Olayide Samuel Lawal Department of Chemistry, Federal University Oye-Ekiti, Oye-Ekiti, Ekiti State, Nigeria

Dosu Malomo Department of Chemistry, Federal University Oye-Ekiti, Oye-Ekiti, Ekiti State, Nigeria

Obiora A. Nnene Department of Civil Engineering, University of Cape Town, Cape Town, South Africa

Busuyi Patrick Omoniyi WATTCCON Chemical and Equipment Marketing Company Nig. LTD Commercial, Kaduna, Nigeria

Augustine Eyikwuojo Onakpa Green Chemistry and Processes for Biomass (Green CAP), Toulouse INP-ENSIACET, Toulouse, France

Srihari Palli Aditya Institute of Technology and Management, Tekkali, Andhra Pradesh, India

Rohit Patel Department of Environmental Science and Engineering, Indian Institute of Technology (Indian School of Mines), Dhanbad, India

Rukayat Oluwatobiloba Quadri Department of Chemistry, Federal University Oye-Ekiti, Oye-Ekiti, Ekiti State, Nigeria

Prabhat Ranjan Smt. Kashibai Navale College of Engineering, Pune, India

Emil Sadowski P.U.H. HESTA, Radom, Poland

Abhilasha Saksena School of Technology, Gati Shakti Vishwavidyalaya, Vadodara, India

Shivam Saw Department of Environmental Science and Engineering, Indian Institute of Technology (Indian School of Mines), Dhanbad, India

K. Karthik Selva Kumar Department of Green Energy Technology, Madanjeet School of Green Energy Technologies, Pondicherry University, Puducherry, India

Rakesh Chandmal Sharma Mechanical Engineering Department, Graphic Era (Deemed to Be University), Dehradun, India;
Mechanical Engineering Department, Graphic Era Hill University, Dehradun, India

Sunil Kumar Sharma School of Technology, Gati Shakti Vishwavidyalaya, Vadodara, Gujarat, India

Ashwani Shukla School of Technology, Gati Shakti Vishwavidyalaya, Vadodara, Gujarat, India

Abhinav Singh Department of Civil Engineering, Graphic Era (Deemed to Be University), Dehradun, India

Prasoon Kumar Singh Department of Environmental Science and Engineering, Indian Institute of Technology (Indian School of Mines), Dhanbad, India

Rahul Sinha Department of Mechanical Engineering, School of Engineering, P P Savni University, Kosamba, India;
Department of Mechanical Engineering, Budge Budge Institute of Technology, Kolkata, India

Ravi Shankar Sinha Department of Mechanical Engineering, Budge Budge Institute of Technology, Kolkata, India

Shashi Kant Tiwari Shri Balaji Institute of Modern Management, Sri Balaji University, Pune, India

Ram Krishna Upadhyay School of Technology, Gati Shakti Vishwavidyalaya, Vadodara, Gujarat, India

Gyan Wrat Energy Department, Aalborg University, Aalborg, Denmark

Abhinav Yash School of Technology, Gati Shakti Vishwavidyalaya, Vadodara, Gujarat, India

Jibrin Muhammad Yelwa Department of Scientific and Industrial Research, National Research Institute for Chemical Technology, Zaria, Kaduna-State, Nigeria

Mark H. P. Zuidgeest Department of Civil Engineering, University of Cape Town, Cape Town, South Africa

Part I
General

Chapter 1
Introduction to Dynamics of Transportation Ecosystem, Modeling, and Control

Sunil Kumar Sharma, Ram Krishna Upadhyay, and Vikram Kumar

The dynamics of the transportation ecosystem play a pivotal role in shaping the way people, goods, and information move within and between regions. In today's interconnected world, efficient and sustainable transportation systems are essential for economic growth, social well-being, and environmental sustainability. This introduction provides an overview of the key concepts associated with the dynamics of transportation, including modeling and control. The transportation ecosystem encompasses a wide range of interconnected elements, including vehicles, infrastructure, logistics networks, regulations, and technology. It involves various modes of transportation, with its unique characteristics and challenges. Understanding the dynamics of this complex system is crucial for optimizing its efficiency, safety, and sustainability.

Modeling is a fundamental tool for studying and analyzing transportation systems. Models are simplified representations of real-world systems that help researchers, policymakers, and practitioners gain insights into how transportation components interact and how they respond to various factors. Different types of transportation models, including traffic flow models, supply chain models, and urban transportation models, are designed to address specific aspects of the ecosystem.

Control theory plays a crucial role in ensuring the efficient operation of transportation systems. Control mechanisms are used to manage traffic flow, optimize routing and scheduling, and enhance safety. For example, traffic signals, adaptive cruise control, and route optimization algorithms are all examples of control systems used in transportation. These systems aim to regulate and improve various aspects of

S. K. Sharma (✉) · R. K. Upadhyay
School of Technology, Gati Shakti Vishwavidyalaya, Vadodara, Gujarat, India
e-mail: sk.shama@gsv.ac.in

V. Kumar
Department of Mechanical Engineering, Indian Institute of Technology Kanpur, Kanpur, India

S. K. Sharma et al. (eds.), *Dynamics of Transportation Ecosystem, Modeling, and Control*, Energy, Environment, and Sustainability,
https://doi.org/10.1007/978-981-97-0437-8_1

the transportation ecosystem, such as reducing congestion, minimizing travel times, and enhancing safety.

In the transportation dynamics, there exist several noteworthy challenges that warrant attention and action. One of the foremost challenges is congestion, a persistent issue exacerbated by the increasing urbanization and growing populations in many regions. As roadways and public transportation systems become more crowded, delays and reduced efficiency plague daily commutes and freight movements. Another pressing concern is the environmental impact of transportation. This sector remains a substantial contributor to greenhouse gas emissions, a matter of utmost significance in the context of climate change. Thus, developing sustainable transportation solutions takes on heightened importance, involving shifts toward cleaner energy sources and improved fuel efficiency.

Safety, too, remains a paramount challenge. Road accidents and transportation-related injuries continue to be a significant public health concern. Addressing this issue requires advancements in safety measures, including the adoption of advanced control and automation systems aimed at reducing accidents and protecting lives. Lastly, the rapid pace of technological advancements presents both opportunities and challenges. Innovations such as autonomous vehicles and smart transportation systems are transforming the landscape of transportation. Integrating these technologies effectively into existing systems while addressing privacy and security concerns is a complex task that requires careful consideration.

Transportation is the lifeblood of modern society, serving as the arteries that sustain our economies and our way of life. It encompasses an array of modes, from the bustling networks of highways and railways to the intricate supply chains that ensure products reach our doorsteps. Moreover, transportation is not merely about the movement of people and goods; it's about connecting people and ideas, fostering economic growth, and shaping the very fabric of our communities.

This book is structured into four distinct parts, each delving into various facets of the dynamic transportation ecosystem. Our aim is to provide readers with a comprehensive understanding of transportation, blending expertise, innovation, and practicality. Here is an overview of each part:

Part I: General

In this section, we provide an introduction to the dynamics of the transportation ecosystem, covering modeling and control.

Part II: Transportation System Modeling and Analysis

This part delves into the modeling and analysis of transportation systems, including topics such as hydraulic system parameters under varying loading conditions, railway-induced noise and vibration, determining the curvature of railway track axes, deterministic and probabilistic forced response analysis of a railway coach, and the reliability model of the railroad traffic control system equipped with a Peltier cell.

Part III: Innovative Technologies in Transportation

This section focuses on innovative technologies shaping transportation, including blockchain solutions for cost-efficiency and traceability in Indian food distribution, streamlining aerodynamic optimization with generalized ANN support, predicting train delays with machine learning techniques, enhancing transparency and efficiency in the fruits and vegetables cold chain through blockchain integration, the advanced chemistry and technical approach to pyrolysis of plastic waste toward achieving a circular economy, the future impact of green energy in revolutionizing transportation, and a comparison of meta-heuristic algorithms for transit network design.

Part IV: Sustainable Supply Chain and Environmental Impact

The final part addresses sustainable supply chain management and environmental considerations, covering topics such as challenges in sustainable supply chain management in petroleum transportation, assessing and mapping traffic noise levels in the coal capital of India, Dhanbad, designing a framework for the sustainable supply chain management of coal transportation, assessing environmental impacts and mitigation strategies for the DFCCIL rail projects, conducting a comparative analysis of weir plan forms and their discharge characteristics, and addressing warehousing and supply chain management of dangerous goods.

This monograph presents various aspects of transportation, management, and advanced techniques, offering a comprehensive perspective on sustainable transportation through Dynamics of Transportation Ecosystem, Modeling, and Control. The particular topics covered in this book include.

- Introduction to Dynamics of Transportation Ecosystem, Modeling, and Control.
- Modelling and Analysis of the Hydraulic System Parameters Under Different Loading Conditions.
- Railway-Induced Noise and Vibration.
- Theoretical and Application Aspects of Determining the Curvature of the Railway Track Axis by the Moving Chord Method.
- Deterministic and Probabilistic Forced Response Analysis of a Railway Coach.
- Reliability Model of Railroad Traffic Control System Equipped with Peltier Cell.
- Blockchain Solutions for Cost-Efficiency and Traceability in Indian Food Distribution.
- Streamlining Aerodynamic Optimization with Generalized ANN Support.
- Improving On-Time Performance: Predicting Train Delays with Machine Learning Techniques.
- Enhancing Transparency and Efficiency in the Fruits and Vegetables Cold Chain through Blockchain Integration.
- Pyrolysis of Plastic Wastes Towards Achieving a Circular Economy: An Advanced Chemistry and Technical Approach.
- Revolutionizing Transportation: The Future Impact of Green Energy.
- Comparing Meta-Heuristic Algorithms for Transit Network Design.
- Assessment and Mapping of Traffic Noise Levels in Coal Capital of India, Dhanbad: Implications for Noise Pollution Management.

- Assessing Environmental Impacts and Mitigation Strategies for the DFCCIL Rail Projects.
- Comparative Analysis of Weir Plan Forms and Their Discharge Characteristics.
- Warehousing and Supply Chain Management of Dangerous Goods.

The dynamics of transportation ecosystem, modeling, and control is a comprehensive compass that guides readers through the intricate world of transportation. It offers a roadmap to understand, model, and control transportation systems in an era marked by technological advancements and sustainability imperatives. This book is not just a collection of insights; it's a guide to navigating the dynamic and ever-evolving world of transportation, where every decision has far-reaching implications for our interconnected global society. It is our hope that readers will find this book to be a valuable resource, inspiring them to contribute to the continued evolution of transportation toward a more efficient, sustainable, and interconnected future.

Part II
Transportation System Modeling and Analysis

Chapter 2
Modelling and Analysis of the Hydraulic System Parameters Under Different Loading Conditions

Gyan Wrat, Prabhat Ranjan, Joseph T. Jose, and Rakesh Kumar

2.1 Introduction

Hydraulic systems are extensively employed in industries and construction machinery due to their high power-to-size ratio and quick response for applying substantial forces and torques. They are utilized in various domains, such as industrial hydraulic systems (Yao et al. 1998), electro-hydraulic positioning systems (FitzSimons and Palazzolo 1996; Silva and Lawrence 1998), active suspension control (Alleyne 1996; Alleyne and Karl Hedrick 1995), material testing (Lee and Srinivasan 1990), and hydraulic braking systems (Isermann 1984). Hydraulic systems are commonly utilized in heavy machinery such as excavators, bulldozers, cranes, backhoes, loaders, dump trucks, pavers, drilling rigs, forest machinery, mining equipment, agricultural machinery, and construction equipment. Ensuring the safety and reliability is of utmost importance to ensure continuous and efficient operation of hydraulic systems. The efficiency of hydraulic systems declines when their components become worn out or malfunction. Common faults in hydraulic systems include changes in actuator friction characteristics due to seal failure or component wear (Crowther et al. 1998; Skormin et al. 1994), variation in effective fluid bulk modulus caused by fluid contamination (Skormin et al. 1994; Tan and Sepehri 2002), abnormal

G. Wrat (✉)
Energy Department, Aalborg University, Aalborg, Denmark
e-mail: gyan007wart@gmail.com

P. Ranjan
Smt. Kashibai Navale College of Engineering, Pune, India

J. T. Jose
Indian Institute of Technology (Indian School of Mines), Dhanbad 826004, India

R. Kumar
Department of Mechanical Engineering, School of Engineering, P P Savni University, Kosamba 394125, India

S. K. Sharma et al. (eds.), *Dynamics of Transportation Ecosystem, Modeling, and Control*, Energy, Environment, and Sustainability,
https://doi.org/10.1007/978-981-97-0437-8_2

operation of servo-valves (Khoshzaban-Zavar-ehi 1997), and hydraulic internal and external leakage (Crowther et al. 1998; An and Sepehri 2005). These faults have the potential to diminish equipment performance and compromise the reliability of hydraulic systems. It is crucial to address and rectify these issues to maintain the continuous and efficient operation of hydraulic systems, ensuring optimal performance and minimizing downtime. To ensure the proper functioning and longevity of hydraulic systems, it is crucial to maintain low friction in the actuators. High friction can lead to seal failure, wear and tear (Agarwal and Lawson 2017) of components, and negatively impact the dynamic behaviour of the system. In applications requiring high dynamics, low friction aids in maximizing the acceleration of the actuator piston (Heipl and Murrenhoff 2015). The presence of Coulomb friction in hydraulic systems has been found to cause undesirable stick–slip motion (Tafazoli et al. 1998) and limit cycle oscillations (Merritt 1967). Implementing friction system compensation techniques can enhance system performance, but the effectiveness depends on the specific system being studied and its suitability for friction compensation. Therefore, minimizing friction is essential for preventing seal failure, reducing wear and tear, optimizing acceleration in dynamic applications, and mitigating adverse effects on system dynamics, and compensation techniques (Karpenko 2009) can be considered based on the specific system requirements. The effective bulk modulus of the fluid in the hydraulic system chamber is influenced by several factors, including working pressure, entrapped air, and temperature variations. These variations in the effective bulk modulus of the hydraulic oil can result in fluid contamination. Moreover, the dynamic and stiffness performance of components such as cylinders, hoses, and pipes in the hydraulic system are directly affected by the effective bulk modulus (Yang et al. 2011). Among the crucial components of the hydraulic system, servo-valves play a vital role as they act as the primary executor of the actuator. However, the degradation of hydraulic system components can cause changes in the effective orifice area, leading to faults in the servo-valve. Such faults can subsequently result in a drop in pump pressure, which impairs the movement of the actuator against the applied load (An 2007). Therefore, it is important to consider the combined effect of working pressure, entrapped air, and temperature variation on the effective bulk modulus of the hydraulic fluid. This helps prevent fluid contamination and ensures optimal performance of the cylinders, hoses, pipes, and servo-valves within the system. Maintaining the integrity of these components is crucial for avoiding faults in the servo-valve and subsequent pressure drops, enabling the actuator to function efficiently and effectively against the applied load. One of the primary issues encountered in hydraulic systems is hydraulic leakage, which occurs as a result of seal failure, clearance between mating parts, and component wear and tear. Leakage can be classified into two main types: internal leakage and external leakage, depending on the location within the hydraulic system. Internal leakage refers to the leakage of hydraulic fluid within the hydraulic circuit (Kulkarni 2011). In practical systems, a small degree of internal leakage is often deliberately incorporated to lubricate the contacting surfaces of mating parts and regulate their temperature. On the other hand, when hydraulic fluid leaks outside the hydraulic circuit and accumulates in the reservoir, it is termed as external leakage. External leakage is easily noticeable as it can be observed with the naked eye. Both internal and external leakage contribute to a pressure drop in the hydraulic system and reduce the velocity of the output shaft. To address these types of faults, various control techniques such as PID (proportional-integral-derivative) (Wrat et al. 2019), fuzzy logic (Wrat et al. 2020), and SVM (support vector machines) (Jose et al. 2021) can be employed. These control methods are utilized to compensate for the effects of leakage, restore system performance, and maintain the desired output. By implementing appropriate control strategies, the adverse impacts of hydraulic leakage can be mitigated, ensuring the efficient operation of the system. It is important to promptly address leakage issues and employ suitable control techniques to prevent pressure drops and velocity reduction in the output shaft, thus maintaining the overall functionality and effectiveness of the hydraulic system. Effective leakage prevention in hydraulic systems holds significant practical implications, ranging from cost savings and increased efficiency to environmental benefits. Effective leakage prevention contributes to cost savings through decreased maintenance costs and extended equipment life; to increase hydraulic efficiency through optimal performance and energy conservation; to environmental benefits through fluid conservation and contamination prevention; to operational reliability through consistent performance; and to safety enhancement through decreased slip hazards and minimized fire hazards.

This article utilizes automation studio software to simulate a hydraulic system and analyse various parameters under different loading conditions. The study involves comparing different loading conditions in terms of displacement, speed, energy, and other factors by introducing artificial leakage in the hydraulic cylinder. The system description section provides an explanation of the hydraulic system, while Fig. 2.2 illustrates the hydraulic circuit. The simulated results are analysed by plotting characteristic graphs in the result and discussion section.

2.2 System Description

The hydraulic system used for analysis at different load conditions is depicted in Fig. 2.1, which presents the circuit diagram. This oil hydraulic system features a double-acting cylinder connected to a simple helical spring, allowing for the application of various loads on the cylinder. Four pressure gauges are integrated into the circuit at different points to measure the pressure, and the system pressure can be adjusted using a relief valve. The system flow can be regulated by a variable pump, enabling control over the flow, pressure, and load. Artificial leakage is introduced into the hydraulic cylinder by incorporating an orifice. The variable area orifice in the system serves the purpose of introducing artificial leakage into the circuit. This enables the generation of different acceleration, flow, speed, and pressure curves under varying loading conditions and cross-sectional areas. A gear pump is responsible for supplying fluid from the reservoir to the pressure line. The fluid then passes through a direction control valve and flows into the cylinder through the orifice.

Finally, the fluid returns to the reservoir through the return line. A pictorial representation of the test setup can be seen in Fig. 2.2, providing a visual overview of the experimental arrangement. To facilitate understanding, Table 2.1 provides the nomenclature of the components used in the hydraulic system, allowing for easy reference throughout the analysis and discussion. The circuit diagram, along with the associated components and control mechanisms, provides a comprehensive setup for studying the behaviour of the hydraulic system under different load conditions. By incorporating artificial leakage and employing various measurement points, the system's performance can be assessed and analysed in terms of pressure, flow, speed, and other relevant parameters.

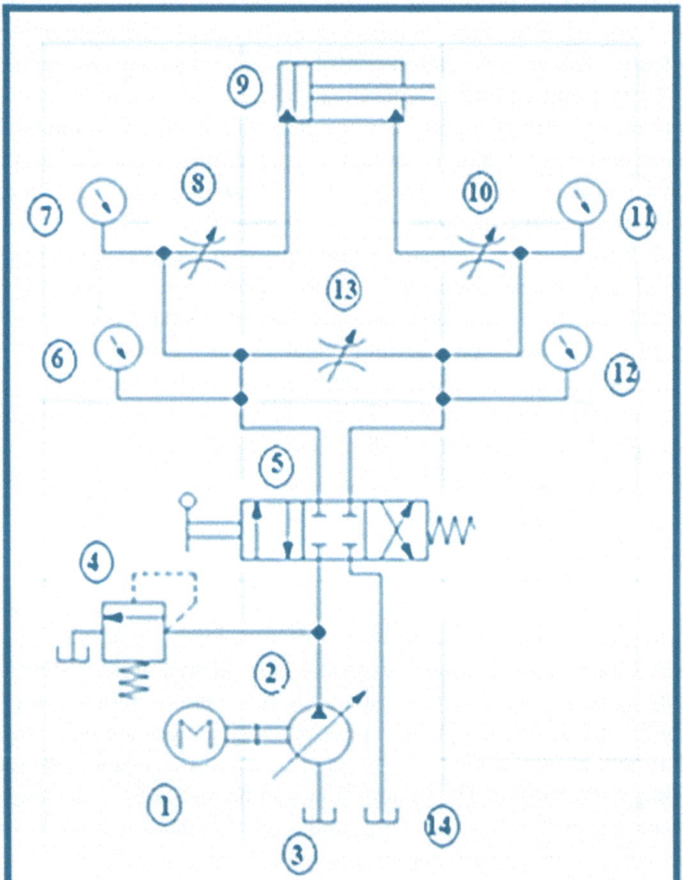

Fig. 2.1 Set-up circuit diagram

Fig. 2.2 Pictorial view of set-up

Table 2.1 Component nomenclature

Component number	Component name
1	Electric motor
2	Variable displacement pump
3,14	Oil reservoir
4	Pressure relief valve
5	Directional control valve
6,7,11,12	Pressure gauge
8,10,13	Flow control valve
9	Hydraulic actuator

2.3 System Modelling

The hydraulic system has been modelled using MATLAB-Simulink for simulation purposes. The simulation involves varying the orifice area and different loads on the cylinder to analyse their effects on the system's behaviour. The MATLAB simulation circuit is illustrated in Fig. 2.3, providing a visual representation of the model. Additionally, Table 2.2 presents the nomenclature and symbols of the parameter values used in the modelling.

Several assumptions have been made during the modelling process to simplify the analysis. First, the compressibility, thermal, and pressure aspects of the fluid are not taken into consideration. This assumption assumes that the fluid behaves as an incompressible and non-thermal medium. Furthermore, the detailed dynamics

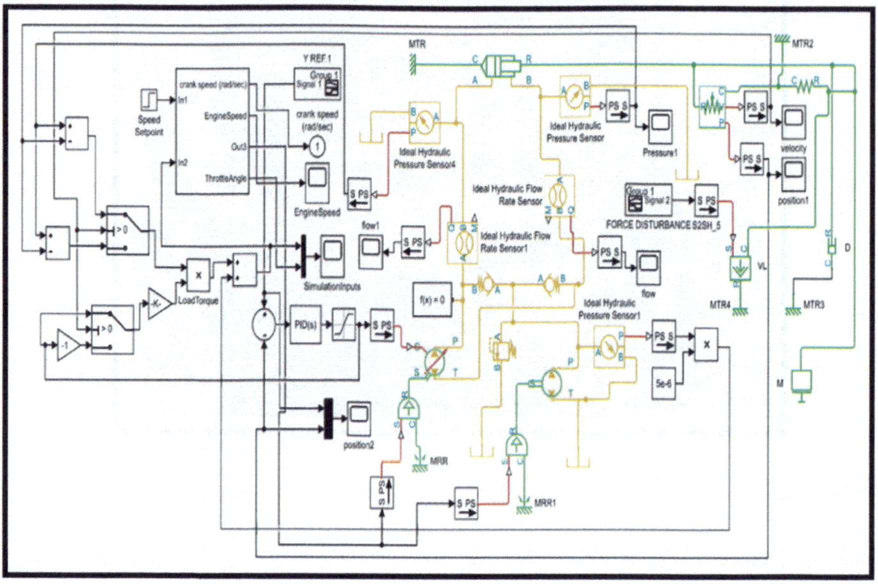

Fig. 2.3 MATLAB Simulink model

Table 2.2 Parameter values used in simulation

Description	Values
Displacement rate of the pump, Dp	0.45 × 10–6 m^3/rad
Leakage resistance of the pump, Rlkgp	0.1 × 10–13 Ns/m^5
Speed of pump, ωp	157 rad/s
Bulk compressibility of fluid, K	1 × 10–13 m^5/N
Viscous co-efficient of fluid, β	2.6037 × 10^7 Ns/m
Synchronous speed of the electric motor, ns	1500 rpm
Number of poles, p	4
Input frequency, f	50 Hz
Number of phases, Ø	3 Ø
Input voltage, V	410–440 V
Power of electric motor, P	7.5 k

of components such as slipper pads and the piston assembly are not considered, simplifying the model by neglecting their intricate behaviours. In terms of the valve, it is assumed that its opening varies linearly with respect to the pressure difference across it. This assumption provides a straightforward relationship between the valve opening and the pressure difference, allowing for simplified modelling and analysis. The model also neglects the inertial effects, frictional loads, and spring loads acting on the valve. By disregarding these factors, the model focuses on the primary dynamics of the system. Another assumption made is the neglect of the flow force. This means that the effect of flow force, which arises due to the fluid flow, is not taken into account in the model. This simplification allows for a more manageable and focused analysis of the system's behaviour. By considering these assumptions, the MATLAB-Simulink model provides a valuable tool for studying and simulating the hydraulic system's performance. The model enables the investigation of various scenarios by manipulating the orifice area and different loads on the cylinder, allowing for a comprehensive analysis of the system's dynamics and behaviour under different conditions.

2.4 Result and Discussion

The simulation analysis investigates the characteristics of different parameters in the hydraulic circuit by varying the orifice area under different loads. The initial comparison focuses on changing the orifice at a no-load condition to evaluate the speed, acceleration, flow, and pressure of the cylinder. This analysis provides insights into how the orifice area affects these parameters, allowing for a better understanding of the system's behaviour and performance.

The speed characteristic curve of a cylinder piston under a no-load condition, as depicted in Fig. 2.4, illustrates variations observed at different orifice openings. It is evident that the piston speed at a 3 mm orifice opening is comparatively lower than that at 0, 1, and 2 mm orifice openings. Conversely, the speeds at 0, 1, and 2 mm orifice openings appear to be quite similar. Notably, when the orifice diameter reaches around 3 mm, the speed of the piston rod experiences a sudden decrease. This decline can be attributed to the increased orifice area leading to greater leakage, consequently resulting in a decrease in piston speed. In summary, the speed characteristic curve demonstrates a direct relationship between orifice opening and piston speed, with a notable decrease occurring at the 3 mm orifice opening due to increased leakage resulting from a larger orifice area.

Figure 2.5 presents the flow characteristics curve of the cylinder piston side under a no-load condition, highlighting various orifice openings. At a zero-opening orifice, there is no flow, resulting in a flow rate of zero. However, as the orifice area increases, the flow rate also increases correspondingly. The maximum flow rate is achieved with a 3 mm orifice diameter. This implies that as the orifice size expands, more fluid can pass through, leading to a higher flow rate. Consequently, the flow characteristic

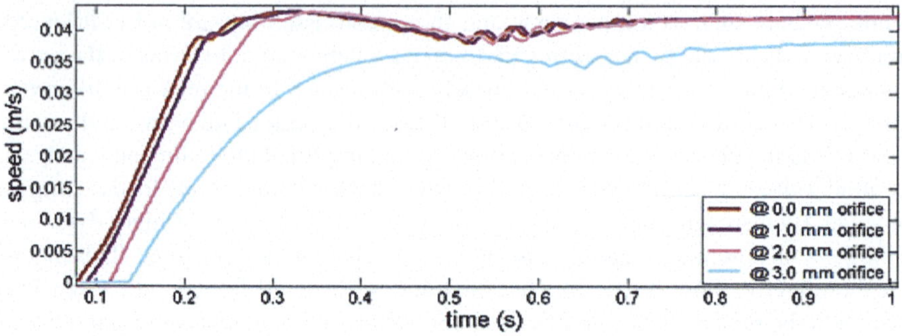

Fig. 2.4 Characteristics of piston speed at no loading condition at different orifice areas

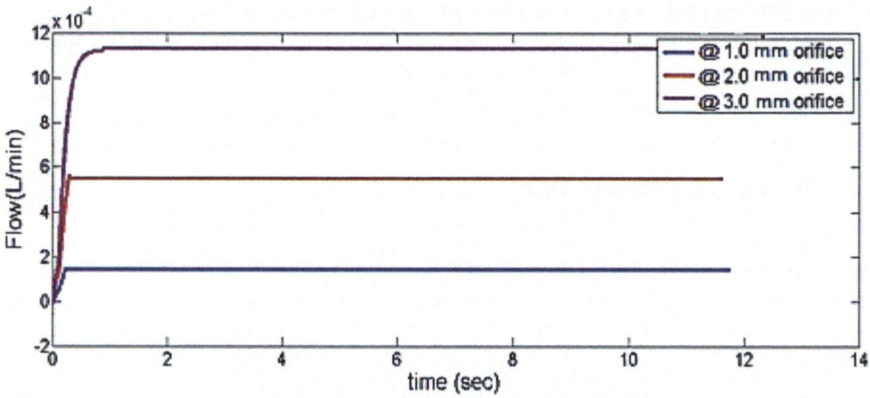

Fig. 2.5 Characteristics of piston side flow at no loading condition at different orifice areas

curve demonstrates a direct correlation between orifice opening and flow rate, with the largest orifice (3 mm diameter) yielding the highest flow rate.

Figure 2.6 displays the pressure characteristics curve of the cylinder piston under a no-load condition, presenting different orifice areas. In the absence of any load, the cylinder exhibits maximum pressure when there is no orifice (zero orifice condition). However, as the orifice area increases, the pressure within the cylinder decreases. This decrease in pressure can be attributed to an increase in system leakage resulting from the larger orifice area. It is worth noting that the pressure remains relatively constant up to a 2 mm orifice diameter, indicating little variation. However, beyond this point, there is a noticeable pressure drop around the 3 mm diameter. Consequently, the pressure characteristic curve demonstrates a clear relationship between orifice area and cylinder pressure, with an increase in orifice size leading to a decrease in pressure, particularly evident beyond a 2 mm orifice diameter.

Once the characteristic curves of various parameters were determined under a no-load condition while altering orifice areas, the system was subsequently subjected

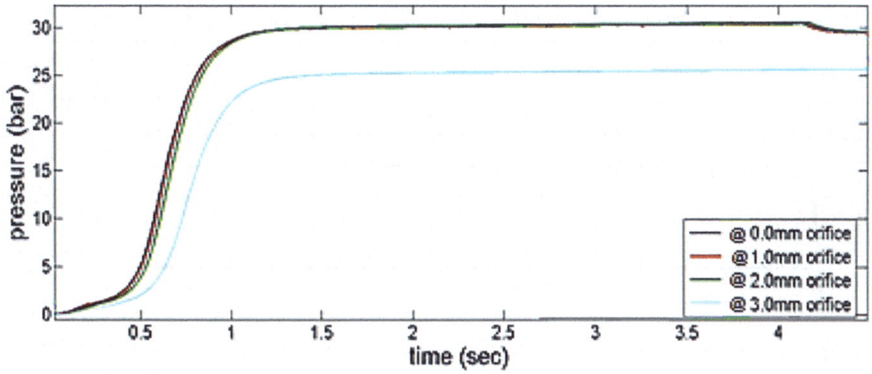

Fig. 2.6 Characteristics of cylinder pressure at no loading condition at different orifice areas

to a constant load of 10N. Figures 2.7 and 2.8 illustrate the speed and acceleration characteristics curves of the cylinder piston, respectively, with changing orifice areas under the constant loading condition. Upon applying the constant load, it becomes evident that the speed eventually stabilizes after a certain period of time. The piston cylinder speed exhibits a decrease as the orifice area increases. Notably, the speeds at 0, 1, and 2 mm orifice openings are relatively similar. This implies that the orifice area has a more significant impact on the speed than the specific opening size within this range. In summary, under the constant loading condition, the speed characteristics curve illustrates a decrease in piston cylinder speed with increasing orifice areas, while the speeds at 0, 1, and 2 mm orifice openings remain quite comparable.

Figure 2.8 provides a comparison of the flow characteristic curves at various orifice openings. It is observed that as the orifice area increases, the flow also increases. This relationship demonstrates that a larger orifice allows for a greater volume of fluid to pass through. Similarly, Fig. 2.9 shows the pressure characteristic curve

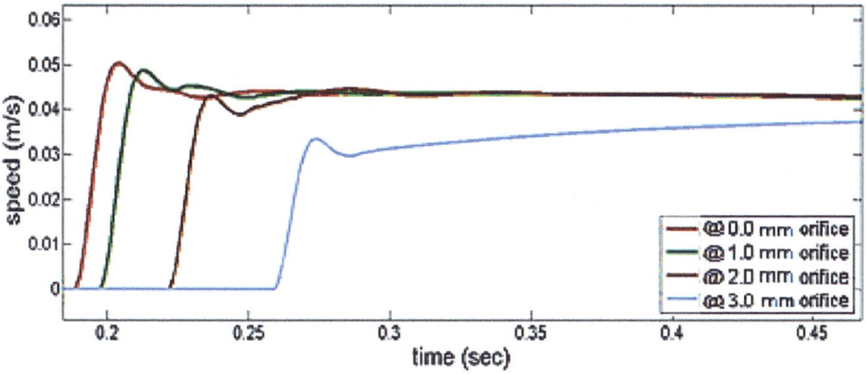

Fig. 2.7 Characteristics of piston speed at 10 N loading condition at different orifice areas

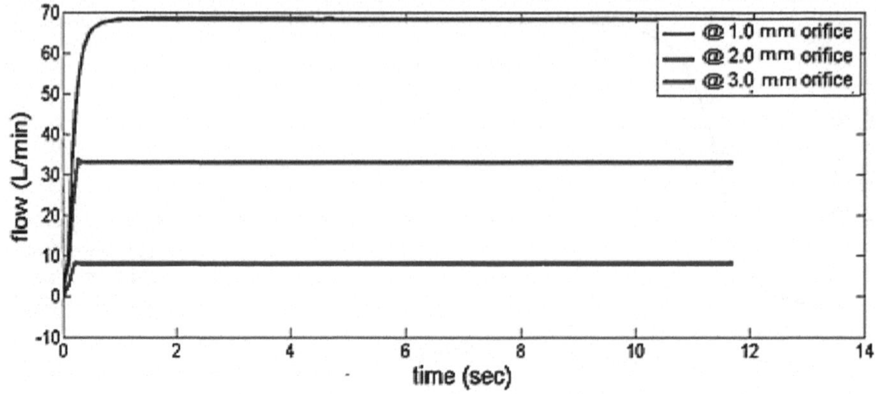

Fig. 2.8 Characteristics of piston side flow at 10 N loading condition at different orifice areas

of the cylinder piston at different orifice openings under a constant load of 10 N. The results are consistent with the findings of the no-load condition, as depicted in a previous figure. Notably, at a zero-orifice opening, the pressure is recorded as 35 bar, whereas the maximum pressure under the no-load condition was 30 bar. This suggests that the presence of the constant load leads to an increase in pressure at the zero-orifice opening. Furthermore, simulation analysis was conducted under various load conditions, including incremental and triangular loads, while considering the changing orifice areas. Remarkably, these analyses yielded similar results to those obtained under the 10 N load and no-load conditions. This reinforces the reliability and consistency of the observed trends in the characteristics of the cylinder piston, reaffirming the relationship between orifice area, load, and various parameters such as flow and pressure. In summary, the flow characteristic curve demonstrates that increasing the orifice area results in increased flow, while the pressure curve indicates similarities between the 10 N load and no-load conditions. Moreover, the simulation analyses conducted under different load conditions corroborate the observed trends, consolidating the findings.

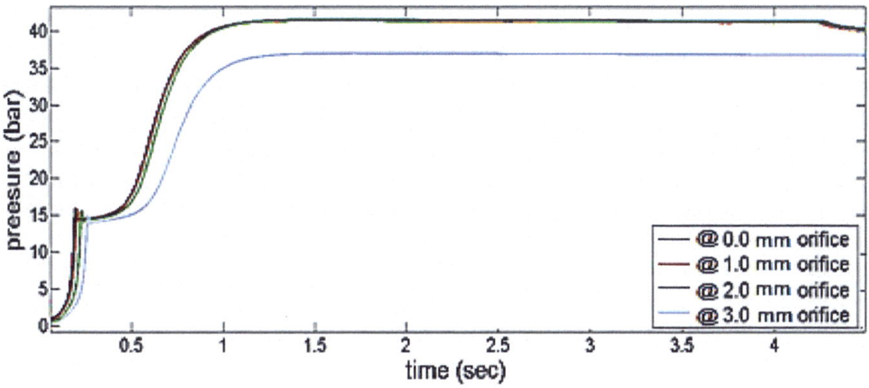

Fig. 2.9 Characteristics of cylinder pressure at 10 N loading condition at different orifice areas

2.5 Conclusion

This significant results of effect of loading in hydraulic circuit have been obtained by modelling and simulating a hydraulic system in MATLAB software. The artificial leakage created between the hydraulic cylinder by providing the variable orifice. Based on the results, it is observed that when the resistive force is zero, constant or triangular load along with increasing orifices areas the acceleration almost same and overlapped. In case of cylinder piston speed and pressure, as the orifice area increases, the leakage in system increases and for that the speed decreases, whereas for piston side flow as the orifice area increases the flow increases. Further analyses can be done considering the leakage of direction control valve, flow control valve, hydraulic pump or motor, etc.

The effects of loading on a hydraulic circuit were investigated through the development and simulation of a hydraulic system using MATLAB software. To introduce artificial leakage in the hydraulic cylinder, variable orifices were implemented. The obtained results provide valuable insights into the behaviour of the system under different loading conditions. Notably, when the resistive force is either zero or subjected to a constant or triangular load, the acceleration curves were found to be quite similar and overlapping. This suggests that the acceleration remains relatively consistent regardless of the specific loading condition and the varying orifice areas. In contrast, the cylinder piston speed and pressure exhibit distinct trends in response to changes in orifice area. As the orifice area increases, the leakage within the system also increases, resulting in a decrease in piston speed. This can be attributed to the reduced efficiency caused by the greater fluid loss. On the other hand, the flow rate on the piston side increases with larger orifice areas, indicating that a larger opening allows for more fluid to pass through, leading to an increase in flow. Further analyses can be conducted to explore the impact of leakage in other components of the hydraulic system, such as the direction control valve, flow control valve, hydraulic pump, or motor. By considering these additional factors, a more comprehensive understanding of the system's behaviour can be achieved. In summary, the modelling and simulation of the hydraulic system in MATLAB software provided significant insights into the effects of loading. The acceleration was found to be consistent across different loading conditions and orifice areas. However, the piston speed and pressure exhibited variations due to changes in orifice area, primarily caused by increased leakage. Future investigations may include the analysis of leakage in various other components to gain a comprehensive understanding of the hydraulic system's behaviour.

Acknowledgements In this study, we acknowledge the work conducted during the Ph.D. period, which laid the groundwork for the research presented here.

References

Agarwal NK, Lawson CP (2017) A practical method to account for seal friction in aircraft hydraulic actuator preliminary design. Proc Inst Mech Eng Part G J Aerosp Eng 231(5):941–950

Alleyne A (1996) Nonlinear force control of an electrohydraulic actuator. In: Proceedings of the JAPAN/USA symposium on flexible automation, pp 193–200

Alleyne A, Karl Hedrick J (1995) Nonlinear adaptive control of active suspensions. IEEE Trans Control Syst Technol 3(1):94–101

An L (2007) Actuator leakage fault detection and isolation based on extended Kalman filtering scheme

An, Sepehri N (2005) Hydraulic actuator circuit fault detection using extended Kalman filter. Int J Fluid Power 6(1):41–51

Crowther WJ, Edge KA, Burrows CR, Atkinson RM, Wollons DJ (1998) Fault diagnosis of a hydraulic actuator circuit using neural networks-An output vector space classification approach. Proc Inst Mech Eng Part I J Syst Control Eng 272(11):57–68

FitzSimons PM, Palazzolo JJ (1996) Part i modeling of a one-degree-of-freedom active hydraulic mount; part ii control of a one-degree-of-freedom hydraulic mount. ASME J Dyn Syst 118(4):439–448

Heipl O, Murrenhoff H (2015) Friction of hydraulic rod seals at high velocities. Tribol Int 85:66–73

Isermann R (1984) Process fault detection based on modeling and estimation methods—survey. Automatica 20:387–404

Jose JT et al (2021) Early detection and classification of internal leakage in boom actuator of mobile hydraulic machines using SVM. Eng Appl Artif Intell 106:104492

Karpenko M (2009) Quantitative fault tolerant control design for a hydraulic actuator with a leaking piston seal

Khoshzaban-Zavar-ehi M (1997) On-line condition monitoring and fault diagnosis in hydraulic system components using parameter estimation and pattern classification. Ph.D. dissertation, University of British Columbia

Kulkarni MAK (2011) Application of fuzzy logic to detection of internal leakage fault in hydraulic systems. Diss, University of Minnesota

Lee SR, Srinivasan K (1990) Self-tuning control application to closed loop servo-hydraulic material testing. ASME J Dyn Syst Measure Control 112:680–689

Merritt HB (1967) Hydrauli, C control systems. Wiley, New York

Silva CW, Lawrence PD (1998) Tracking control of an electrohydraulic manipulator in the presence of friction. IEEE Trans Control Syst Technol 6(3):401–411

Skormin VA, Apone J, Dunphy JJ (1994) Online diagnostics of a self-contained flight actuator. IEEE Trans Aerosp Electron Syst 30(1):186–195
Tafazoli S, de Silva CW, Lawrence PD (1998) Tracking control of an electrohydraulic manipulator in the presence of friction. IEEE Trans Control Syst Technol 6(3):401–411
Tan H, Sepehri N (2002) Parametric fault diagnosis for electrohydraulic cylinder drive units. IEEE Trans Industr Electron 49(1):96–106
Wrat G et al (2019) Position control and performance analysis of hydraulic system using two pump-controlling strategies. Proc Inst Mech Eng Part I J Syst Control Eng 233(9):1093–1105
Wrat G et al (2020) Energy saving and Fuzzy-PID position control of electro-hydraulic system by leakage compensation through proportional flow control valve. ISA Trans 101:269–280
Yang H, Feng B, Gong G (2011) Measurement of effective fluid bulk modulus in hydraulic system. J Dyn Syst Measure Control 133(6):061021
Yao B, Zhang J, Koehler D, Litherland J (1998) High performance swing velocity tracking control of hydraulic excavators. In: Proceedings of the American control conference, pp 818–822

Chapter 3
Railway-Induced Noise and Vibration

Rakesh Kumar, K. Karthik Selva Kumar, Shashi Kant Tiwari, Ravi Shankar Sinha, and Rahul Sinha

3.1 Introduction

Vibration is a repetitive or oscillating motion of an object or a system around a fixed point or equilibrium position. It involves the back-and-forth movement of the object or system from its rest position. Vibration is a common physical phenomenon observed in various natural and manufactured systems and can be annoying sometimes. It can occur in a wide range of systems, from macroscopic objects like bridges, buildings, and musical instruments, to microscopic systems like atoms and molecules. They can be beneficial or undesirable, depending on the context. For example, vibrations are essential for sound production in musical instruments but can be problematic in machinery and structures, leading to fatigue and failure over time. Similarly, metro rail can lead to noise and vibration, which can be an environmental concern for people living near the vicinity of the metro systems. The railway-induced vibration is mainly caused due to wheel-rail interaction (as shown in Fig. 3.1), while the other assisting factors include operating speed, geological characteristics, and bridge/tunnel effects.

R. Kumar · R. Sinha (✉)
Department of Mechanical Engineering, School of Engineering, P P Savni University, Kosamba 394125, India
e-mail: rks55900@gmail.com

K. K. Selva Kumar
Department of Green Energy Technology, Madanjeet School of Green Energy Technologies, Pondicherry University, Puducherry 605014, India

S. K. Tiwari
Shri Balaji Institute of Modern Management, Pune 411033, India

R. S. Sinha · R. Sinha
Department of Mechanical Engineering, Budge Budge Institute of Technology, Kolkata 700137, India

S. K. Sharma et al. (eds.), *Dynamics of Transportation Ecosystem, Modeling, and Control*, Energy, Environment, and Sustainability,
https://doi.org/10.1007/978-981-97-0437-8_3

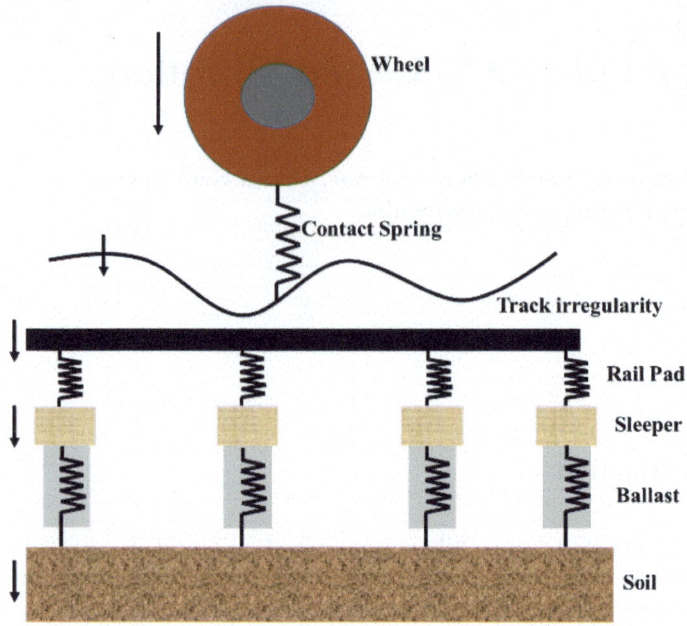

Fig. 3.1 Interaction of wheel with rail and ground (Ntotsios et al. 2017)

3.1.1 Need for Railway Noise and Vibration

With an increased demand for sustainable and climate-friendly means of transport, railway line projects have been proposed in most developed and developing countries such as Europe, the USA, India, and China, as shown in Fig. 3.2. Railways are the most common and affordable means of transport for most of the population residing in urban and metropolitan areas.

According to the UN report, the urban/metropolitan population will reach 66% around 2050 (Nations 2014; Lam 2023). Hence, with the population growth in young countries like India, railways will have a significant role as an efficient and eco-friendly means of mass transportation (Jebin et al. 2023; Zeybek et al. 2023). The development of railway tracks or metro stations will be an unavoidable reality comprising of the construction of new networks or amplification of the existing ones (Selvakumar et al. 2023; Bharule et al. 2019).

However, despite being a choice and ease of mass transportation, the railway also influences the surrounding environment. Railway-induced vibration and noise is still a significant concern for the scientific communities (Peplow et al. 2021; Amiri et al. 2016; Kumar et al. 2022; Ma et al. 2022; Seidler et al. 2023). The vibration is induced due to rail/wheel unevenness, parametric excitation due to sleeper periodicity, and impact excitation due to rail joints and wheel flats.

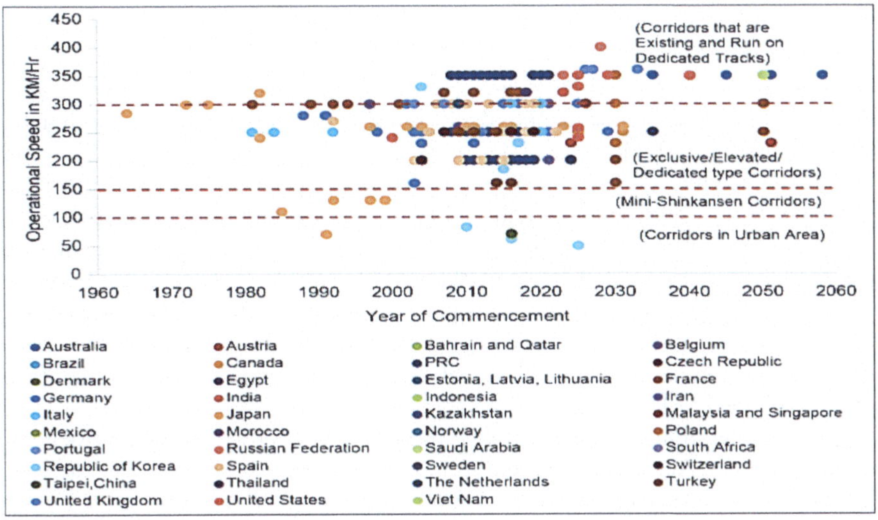

Fig. 3.2 Development of railways during the last few years (Bharule et al. 2019)

3.2 Background

Railway-induced vibration propagations occur through soil layers and excite nearby buildings or monuments through the foundation, resulting in structural vibration and noise. The report from WHO suggested that long-term exposure to railway-induced vibration and noise leads to adverse health effects (Organization and Burden 2011). Vibrations in the frequency range of 1–80 Hz are generally annoying and disturbing, while at frequencies 16–250 Hz can lead to vibration of walls and floors, resulting in structural borne noise (Cao et al. 2018; Xin et al. 2020). Several researchers have studied railway-induced vibration and noise, demonstrating its significance analytically, numerically, and experimentally. Cao et al. (2018) conducted a vibration test in a residential building near a metro depot and studied the dynamic interaction between rail and wheel and its effect on the nearby constructions (Cao et al. 2018). He et al. (2020) used analytical methods to predict the train-induced vibration. The study used the efficient 3D method, involving tunnel-ground-building dynamic interaction to predict vibration generated by the train running in the tunnel. It showed that building vibrations are significantly affected by the dynamic properties of the building and the ground-building interaction. Forrest and Hunt (2006a, b) described Pipe-in-Pipe (PiP) model for ground vibration prediction from a tunnel in homogenous full-space (Hussein et al. 2014) and further extended the PiP model to calculate vibrations from underground railway by incorporating multi-layered half-space geometry. However, the PiP model accuracy deteriorates when a tunnel is close to the layer interface or ground surface. Yuan et al. (2017), He et al. (2018, 2019), Zhou et al. (2020) and Heidary et al. (2023) proposed the analytical method to predict vibration from single or two parallel tunnels in multi-layered half-space. Apart from analytical studies, researchers have also made an effort to conduct studies based on empirical and hybrid methods. Various empirical models were developed to predict the ground vibration and noise induced by trains considering parameters such as train speed, rail/wheel interaction, tunnel shape, the distance between the building and tunnel, the foundation of the building, and time intervals of the train (Hao et al. 2022; Yang et al. 2023; Quagliata et al. 2018; Bahrekazemi 2004; Wang et al. 2023). Numerical approaches using finite element method (FE), boundary element method (BE) were also proposed to account for vibration induced due to railway traffic considering track-tunnel ground system (Chen and Zhou 2020; Kouroussis et al. 2019; He et al. 2017). The 2.5D numerical method was proposed based on finite element-perfectly matched layers FE-PML method and FE-BE method to predict vibration induced due to underground railway traffic (Coulier et al. 2014; Lopes et al. 2016; López-Mendoza et al. 2020) (Table 3.1).

A_{RT} = acceleration measurement on rail track (m/s^2), V_F = vibration velocity at the building floor, A_F = Acceleration of floor system (m/s^2), A_R = Acceleration of Rail (m/s^2), A_T = Acceleration of track Slab (m/s^2), A_{TW} = Acceleration of Tunnel

Table 3.1 Summary of experimental work related to railway-induced noise and vibration

Author	Methods	Input/measured parameter	Output	Limitation
Lopes et al. (2016)	Experimental and numerical approach	A_{RT} V_F,	Vibration data obtained were compared to numerical models based on three autonomous models	Absence of information related to track unevenness
Cao et al. (2018)	Experimental	A_F	Corresponding frequency spectra as well as 1/3 octave band RMS spectra obtained	Wheel/rail interaction
Xin et al. (2020)	Experimental	A_R, A_T, A_{TW}, A_{OG}, A_{IG}, SPL	Peak values, FFT	A study was carried out for a particular sensitive receptor
Ma et al. (2022)	Experimental	Geological parameters, (depth, ρ, Poisson's ratio, C_s, C_p), O_m	VAL (f_i), VL_Z, RMS	Influence of wheel/rail wear uncertainty
Luo et al. (2023)	Experimental	S, Mb, Mw, Kps, Lb, W, ρ_w, Y, Ar, ρ_r,	SPL	Conflict between numerical and experimental results was observed
He et al. (2020)	Analytical	C_s, C_p, $\beta_t \rho$ Poisson's ratio	SPL, RMS (velocity)	Comparison with experimental study needs to be done to validate the 3D method

wall (m/s^2), A_{OG} = Acceleration of outdoor ground (m/s^2), A_{IG} = Acceleration of Indoor Ground (m/s^2), SPL = sound pressure level (dB(A)), ρ = Density (kg/m^3), C_s = S-wave speed (m/s), C_p = P-Wave speed (m/s), O_m = net operating mileage (× 10^4 km), $VAL(f_i)$ = vibration acceleration level, VL_Z = vertical weighted vibration acceleration level, RMS = Root mean square (m/s^2), S = speed in (km/hr), Mb = Mass of bogie, Mw = Mass of wheelset, Kps = Stiffness of primary suspension, Lb = Length between bogie centers, W Wheelbase, $\rho_{w=}$ Wheel density, Y = Young's modulus of wheel, Ar = Cross-section area of rail, ρ_r = Rail density, β_t = Hysteretic material damping ratio.

Railway-induced vibration and noise is a major thrust area of research whose majority is still left untouched due to its complexity related to dynamics behavior and acoustic mechanism among rail, track, wheel/track interaction, bridge, and subgrade (Thompson and Jones 2000; Luo et al. 2022a, b). It is also an environmental issue as it has adverse health effects such as annoyance, sleep disturbance, heart rate, and cortisol levels in the population near the railway traffic (ISO14837–1 2005). The trend in the study showed an increased level of annoyance and disturbance with the magnitude of noise and vibration (Waddington et al. 2014; Maclachlan et al. 2018). Therefore, it is essential to compare the railway-induced vibration and noise with national and international standards such as British BS 6472 (2008): 1992, ISO 2631–2 (1997), DIN 4150–2 (1999) and DIN 4150–3 (1999) and thus to suggest proper measures to control and mitigate railway induced vibration and noise.

3.3 Mechanism of Railway Vibration Generation

Railway-induced vibration is one of the significant issues that can be categorized depending on the different types of railway applications. Moreover, railway-induced vibration is a complex interactive system in which it is difficult to predict the parameters causing the vibration with certainty. The mechanism of railway vibration is generally classified based on quasi-static and dynamic load (Ma et al. 2023; Liu et al. 2023). Quasi-static excitation, often called the "moving load," pertains to the static force generated by the train's mass as it moves along the track at its operational speed. This motion induces slight ground movement, resulting in an extremely low-frequency vibration that can be detected at some distance from the track. Generally, this quasi-static excitation has minimal impact on perceptible vibrations under normal conditions. The static load applied by each wheel on the track causes a localized deflection around the wheel. To describe this deflection, we consider a beam with bending stiffness EI supported by a stiffness 'k' per unit length. The deflection at a distance x from the applied static load F_0 can be expressed as (Thompson et al. 2019a):

$$u(x) = \frac{F_0\beta}{2k} e^{-\beta|x|}(\cos\beta|x| + \sin\beta|x|)$$

$u(x)$ = Vibration Displacement; $\beta = \left(\frac{k}{4EI}\right)^{\frac{1}{4}}$ and β = wave number of static beam deflection (Thompson et al. 2019b). The configuration of vibration spectra is influenced by the spacing between wheels within vehicles. During the train movement on a rail track, rolling noise arises from the interaction between the wheel and rail at their contact area, and both elements contribute significantly to the noise radiation. Neither the wheel nor the rail running surfaces are perfectly smooth, and their irregularities cause relative movement between them. When a wave with a wavelength "λ" is present on the rail (or wheel) and is traversed at speed v, it generates sinusoidal vibrations with a frequency proportional to:

$$f = \frac{v}{\lambda}$$

These high-frequency vibrations are transmitted into the wheel and the rail, regardless of which surface has more significant irregularities. As a result, sound is radiated into the air from these vibrating structures. Dynamic forces are generated at the wheel/rail contacts by the irregular profile of the track and wheel running surfaces, which can lead to wave propagation in the ground (Torstensson et al. 2011). Furthermore, dynamic forces arise due to impacts when the wheels pass over switches, crossings, or poorly maintained rail joints. The presence of uneven track support, such as sleeper pitch or longer wavelengths, can also induce dynamic displacements under the vehicle's loads (Bezin et al. 2009). These dynamic excitation mechanisms are primarily responsible for ground-borne noise and contribute significantly to perceptible vibrations (Küsel and Brommundt 2003).

3.3.1 Causes of Railway Vibration

Wheel/Rail Interaction and Excitation

Wheel-rail interaction is a complex mechanism that involves various processes and phenomena occurring at the contact between the train's wheels and the rails on which they run, as shown in Fig. 3.3.

The wheel and rail interact in a small area known as the contact patch. This is where the forces and stresses are concentrated during train operation. These include vertical, lateral, and longitudinal forces. The magnitude and distribution of these forces affect the train's stability, track wear, and overall ride quality. The interaction between the wheel and the rail induces dynamic effects due to irregularities on the rail surface and the wheel itself. These dynamic effects lead to vibrations in both the wheel and the rail system. The vibrations induced at the wheel-rail contact propagate through the wheel and track structures. These vibrations can also travel through the ground, leading to ground-borne noise and perceptible vibrations.

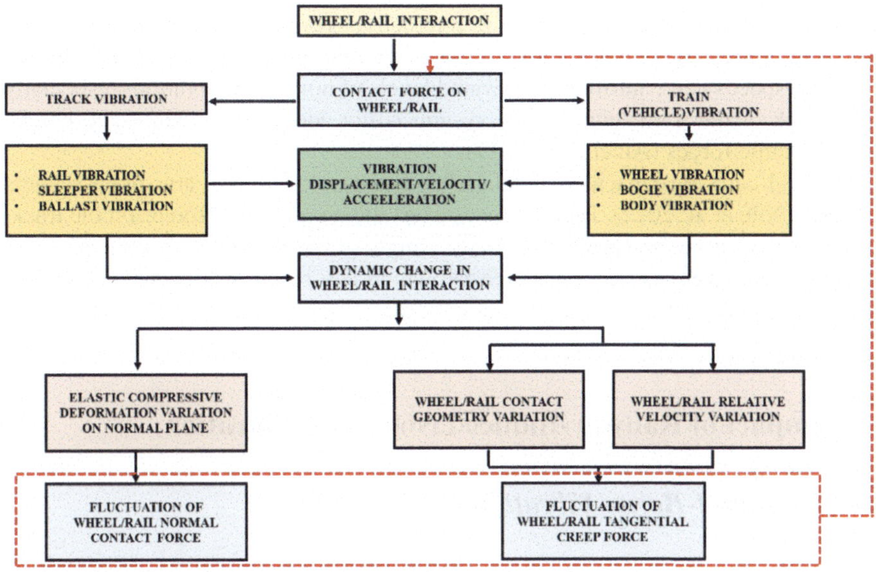

Fig. 3.3 General mechanism of wheel/rail interaction

Effect of Parameters Used for Track

The railway is one of the crucial modes of transport for the modern society. Hence, in this context, it is also essential to consider the parameters which can lead to vibration and noise generation. Various parameters can significantly influence the performance and behavior of the track system, including its ability to mitigate noise and vibrations (Morais et al. 2023). The geometric characteristics of the track, such as alignment, curvature, and vertical profile, play a crucial role in determining the dynamic behavior of the trains and the track. Proper track geometry ensures smooth and stable train operations, reducing the generation of noise and vibrations. The stiffness of the track components, including the rails, sleepers, and ballast, affects how they respond to dynamic loads from passing trains. An optimally designed track with appropriate stiffness can attenuate vibrations and minimize their propagation. The ballast layer beneath the sleepers provides support and distributes the load from the trains to the underlying substructure. High-quality ballast with adequate particle size and gradation can contribute to better vibration attenuation. The overall design and construction of the track structure, including the type of ballast, sleepers, and rail fastenings, impact the track's resilience to dynamic loads and its ability to mitigate vibrations (Charoenwong et al. 2023).

Operating Speed and Load-Carrying Capacity

The operating speed and load-carrying capacity of trains are significant factors that directly influence rail vibrations. When the train moves at higher speeds, it encounters rail irregularities more frequently, leading to a higher frequency of wheel-rail interactions. The increased frequency of interactions results in more vibrations being generated at the contact patch between the wheel and the rail. Consequently, higher operating speeds can contribute to elevated levels of noise and vibrations, especially if the track or the train's suspension system is not adequately designed to handle these dynamic forces (Ataei et al. 2017).

The load-carrying capacity of a train refers to the weight of the cargo or passengers it carries (Xue et al. 2022). Heavier loads exert more significant forces on the track, particularly at curves and gradients. As the weight of the train increases, the pressure on the wheel-rail interface also increases, which can result in higher levels of rail vibrations and potentially increasing noise emissions.

3.4 Impact of Railway-Induced Noise and Vibration

3.4.1 Ground-Borne Vibration

Railway operations generate ground-borne vibrations due to dynamic interactions between trains and track infrastructure. As trains move along the tracks, the wheels interact with the railhead, causing repetitive loading and unloading of the track. This interaction generates vibrations propagating through the track infrastructure (including ballast rails and sleepers) and into the nearby ground. The ground acts as a medium for the propagation of vibrations. Imperfections in the track, such as irregularities in the rail profile or unevenness in the ballast, can amplify the generation of the vibration caused due to motion of trains. These irregularities lead to variations in the applied forces and contribute to increased ground-borne vibrations. The vibrations generated in the track structure transfer into the ground through the contact points between the track and the underlying soil/rock. Once the vibrations are transmitted to the ground, they propagate outward as seismic waves (i.e., P-waves, S-waves, or Rayleigh waves), as shown in Fig. 3.4. When ground-borne vibrations encounter buildings or other structures, they can cause the structures to resonate or vibrate, potentially leading to structural damage or discomfort for occupants. The characteristics of ground-borne vibration, such as frequency, amplitude, and duration, depend on various factors, including train speed, track condition, soil properties, and the distance from the source. Additionally, factors like soft soil or irregularities in the track can influence the intensity and propagation of vibrations. If the frequency of the passing train aligns with the natural frequency of the track or ground, resonance can occur, amplifying the vibrations further.

Ground-borne vibration resulting from metro railway operations is a significant concern due to its potential impact on nearby communities and infrastructure. Scientific studies have extensively examined the effects of these vibrations on the built environment and human well-being. Research has shown that ground-borne vibrations from passing trains can cause structural damage to buildings, leading to safety risks and increased maintenance costs. The impact of ground-borne vibrations on

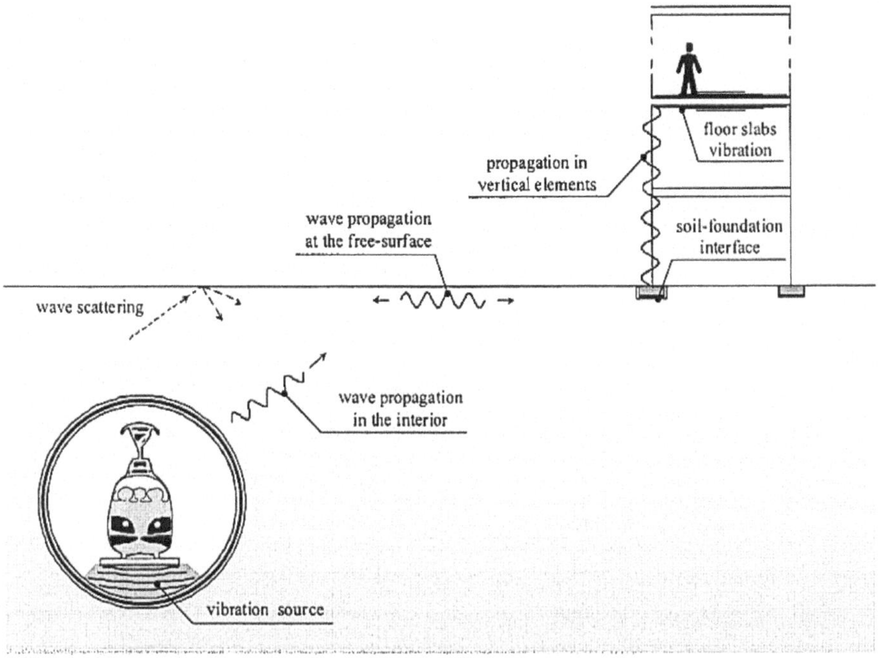

Fig. 3.4 Schematic representation of ground-borne vibration (Lai et al. 2005; Woods 1968; Connolly et al. 2015; Khan and Burdzik 2023)

buildings near metro railway lines, revealing that vibrations can lead to cracks and deterioration of structures (Khan and Burdzik 2023). Additionally, prolonged exposure to ground-borne vibrations has been linked to annoyance and sleep disturbances among residents living near rail tracks (Elias and Villot 2012). Such disturbances can adversely affect individuals' quality of life and overall well-being. Researchers and authorities are actively exploring mitigation measures to address these concerns, such as using resilient track systems and implementing vibration isolation technologies to reduce ground-borne vibration levels and protect urban environments (Javaid et al. 2023; Haladin et al. 2021; Kouroussis et al. 2014, 2015). The summary of various standards governing the assessment of ground-borne vibration is listed in Table 3.2.

3.4.2 Annoyance

Metro railway vibrations lead to annoyance. It has been extensively studied, highlighting the adverse effects on residents and workers (He et al. 2020; Luo et al. 2022a, 2022b; ISO14837–1 2005; Waddington et al. 2014; Maclachlan et al. 2018). For instance, a comprehensive study by Wong-McSweeney et al. examined the annoyance responses of individuals living near metro railway lines in urban areas

Table 3.2 Ground-borne vibration assessments based on various standards (Elias and Villot 2012)

Parameters	Limits (lower/upper levels)	Application	Standards
Effect on structures			
Peak particle velocity (PPV)	3.00 mm/s	Susceptible building	DIN4150-3 (Javaid et al. 2023)
	5.00 mm/s	Residential place	
Peak particle velocity (PPV)	3.00 mm/s	Sensitive building	SN 640 312 a (Haladin et al. 2021)
	6.00 mm/s	Typical building	
Effect on persons in the building			
Weighted RMS acceleration	0.32 m/s^2	Uncomfortable	ISO 2631–2 (Kouroussis et al. 2014)
	2.00 m/s^2	Extremely uncomfortable	
RMS Velocity	0.10 mm/s	Residential area frequent (>70 passbys per day)	USDT (Kouroussis et al. 2015)
	0.26 mm/s	Residential area infrequent (>70 passbys per day)	
Weighted dose $KB_{F, max}$	0.15 mm/s	Residential area	DIN 4150–2 (Faizan et al. 2022)
	0.10 mm/s	Sensitive area	

(Institution 2008; Vibration 2003). The study found a correlation between vibration exposure and annoyance in the affected population. Furthermore, investigations were carried out based on metro railway vibrations' physiological and psychological impact on residents' well-being (Normung D.I.f. 1999; Part 1999; Ma et al. 2023). The results demonstrated that prolonged exposure to vibrations significantly contributed to elevated stress levels, reduced sleep quality, and increased annoyance among the study participants.

Moreover, various research analyzed the impact of vibration frequencies generated by metro railways on human sensitivity and annoyance perception (Ma et al. 2023; Liu et al. 2023; Thompson et al. 2019a). The study identified low-frequency vibrations, typical of train operations, tend to be more annoying and disruptive to daily activities than higher-frequency vibrations. The research showed that annoyance levels increased with increased vibration generation due to the railway. Similarly, the noise generated due to passing trains propagates through the air and impacts the receiving building. Specific frequency response functions are devised and denoted as A, B, and C weighted curves to account for human perception at various frequency levels. Among these, the A-weighted decibel (dBA) is most suitable for representing the human response to environmental noise, which typically falls within the "normal" frequency range. When conducting measurements, noise-monitoring equipment directly reads the dBA value, achieved by setting the "weighting switch" to "A".

3.4.3 Human Vibration

Human exposure to metro railway vibration has become a growing concern in urban areas, as these vibrations can potentially impact the health and well-being of nearby residents and workers. Scientific studies have shown that prolonged exposure to train-induced vibrations can lead to physiological and psychological effects on individuals. The International Organization for Standardization (ISO) has developed guidelines, such as ISO 2631–1, to assess human exposure to whole-body vibration, providing a standardized framework for evaluating the impacts of metro railway vibrations on individuals (Thompson et al. 2019b; Torstensson et al. 2011; Bezin et al. 2009; Küsel and Brommundt 2003). Research has indicated that vibrations from metro railways can induce stress responses, affect cardiovascular health, and disrupt sleep patterns among affected populations. Moreover, studies have explored the relationship between vibration frequencies and human annoyance perception, revealing that low-frequency vibrations can be exceptionally bothersome. To address these concerns, various mitigation measures, such as resilient track systems and vibration isolation devices, are being explored to reduce vibration propagation (Morais et al. 2023; Charoenwong et al. 2023; Ataei et al. 2017).

3.4.4 Standards for Noise and Vibration Measurements

While standardized A-weighting is commonly used to evaluate human perception of airborne noise, the evaluation of vibration is a more complex process. According to ISO 2631–1, the frequency range considered for assessing the potential effects of vibration on health, comfort, perception, and motion sickness is as follows:

- 0.5–80 Hz for health, comfort, and perception and
- Hz to 0.5 Hz for motion sickness.

Vibration can be perceived in different forms, either as body vibration in the frequency range of 1–80 Hz or through tactile senses with a higher frequency range. In typical situations, frequencies relevant to ground-borne noise can range from as low as 16 Hz to as high as 500 Hz. However, dominant frequencies are generally below 100 Hz since they correspond to the response of building elements (Vibration 2003; Normung, D.I.f. 1999; Part 1999; Ma et al. 2023; Liu et al. 2023; Thompson et al. 2019a, 2019b; Torstensson et al. 2011; Bezin et al. 2009; Küsel and Brommundt 2003; Morais et al. 2023; Charoenwong et al. 2023; Ataei et al. 2017; Xue et al. 2022; Lai et al. 2005; Woods 1968; Connolly et al. 2015; Khan and Burdzik 2023; Elias and Villot 2012; Javaid et al. 2023; Haladin et al. 2021; Kouroussis et al. 2014, 2015; Faizan et al. 2022; ISO 1997). More guidance in German (1999) and British standards (1987) is mentioned, which involves the measurement criteria of measuring vibration dose value (VDV) and KB values for assessment of vibration level. VDV is defined to quantify the intermittent vibration

$$VDV = \left[\int_0^T a^4(t)dt \right]^{0.25} \quad m/s^{1.75}$$

where, $a(t)$ = frequency weighted (filtered) acceleration as a function of time; T = event duration.

$$KB_{FTr} = \sqrt{\frac{1}{T_r} \sum T_{i,j} KB_{FTm,j}{}^2}$$

where, T_r = evaluation period (day or night); $T_{e,r}$ = exposure period of each jth event; $KB_{FTm,j}{}^2$ = average of maximum filtered r.m.s signal value during each 30-s interval of the whole event.

3.5 Mitigation Measures

The primary source of noise in steel-wheel/steel-rail systems is the wheel/rail interaction, which consists of three distinct components: roar, impact, and squeal. Roar originates from rolling noise caused by minor roughness on the wheel tread and rail running surface. Impacts are produced due to irregularities on the rail's running surface or flat spots on the wheels. On the other hand, squeal occurs when the steel-wheel tread or its flange rubs against the rail, generating resonant vibrations within the wheel and leading to a screeching sound. This can be minimized by improving wheel roundness and suspension of the drive system.

Track defect and malfunctioning ballast are the source of vibration, hence regular maintenance of track is essential to avoid to rail-induced vibration. Track abnormalities such as "suspended sleepers" and malfunctioning ballast have the potential to induce vibrations, underscoring the significance of maintaining optimal track conditions to manage vibrations. Attaining proper track alignment can result in a notable 10 dB reduction in ground-borne noise, especially when trains are operating at speeds of 320 kph or below. It is advisable to position track discontinuities at a distance from areas sensitive to vibrations. Consistent tamping and leveling of tracks prove advantageous, particularly in areas with softer ground. While additional measures might yield only marginal enhancements in vibration reduction, they can still effectively address issues related to ground-borne noise (Bewes et al. 2013; Guo et al. 2021).

Resilient and vibration isolation rail fasteners can be utilized to lower the vibration effect (Bewes et al. 2013). Under sleeper pads are another option that can be utilized to protect ballast and thus increases the time for track maintenance. EN 16730:2016–09 defines requirements for under-sleeper pads in railway applications, specifically for concrete sleepers and bearers with under-sleeper pads (Standards 2016). The use of ballast mats, floating slabs, construction of trenches, change in operational speed and buffer zones are some of the measures that can be beneficial in control vibration (Jiang et al. 2013). Ballasted track systems in tunnels aim to minimize the impact of transitioning from open to enclosed environments by addressing vibration, stability, and load distribution issues. Under-ballast mats, placed between the ballast and the formation, protect the ballast from excessive pulverization and reduce vibration transfer into the tunnel walls. These mats, designed with specific stiffness specifications, absorb and dampen vibrations more effectively than the ballast alone, mitigating potential impacts on the tunnel structure and surrounding surroundings.

Slab track is a railway track system used in tunnels and underground environments, offering improved stability, reduced maintenance, and better control over track dynamics. It's designed to be mounted directly on a solid concrete or composite slab, eliminating the need for ballast. The slab track's suspension system can be tuned to maintain a desired frequency range, preventing resonances that could cause vibrations, noise, and structural damage.

3.6 Case Studies, Emerging Trends, and Challenges

3.6.1 Case Study 1—Physiological Effect of Residents Due to Train-Induced Vibration

This research study investigated the impacts of railway-induced noise and vibrations on residents in districts 17 and 18 of Tehran, Iran, with an estimated population of approximately 700,000. The researchers initially measured the noise levels at various points and found that they exceeded standard levels. They then surveyed using a questionnaire, gathering responses from 376 residents. The study's findings highlighted that resident living adjacent to the railway line experienced the adverse effects of noise and vibrations. Interestingly, the researchers observed that the age variable impacted vibration dissatisfaction but not noise dissatisfaction.

Additionally, the distance from the railway influenced both noise and vibration dissatisfaction. This study sheds light on the significant concerns of residents living near railway tracks in Tehran regarding the adversarial effects of noise and vibrations. It emphasizes the importance of intervention by railway authorities to mitigate these impacts. It highlights the influence of factors like age and distance on residents' dissatisfaction with noise and vibrations.

3.6.2 Case Study 2—Effect of Train-Induced Vibration Due to Intersecting Metro Lines at Bell Tower in Xi'an, China

This research paper evaluates the long-term consequences of vibration generation caused due to metro rails running on the two intersecting tunnels at the Bell Tower in Xi'an, China. It examined the two metro lines, one operational and the other under construction. Three-dimensional finite element model involving parameters such as ground type, track, tunnel, and train was developed to analyze and predict the vibration generation caused due to train speed. The study suggested that peak particle velocity (PPV) can be significantly reduced by reducing train speed and installing steel spring floating slabs. The research indicated that the proper design implementation could reduce ground vibration, causing damage to historic buildings.

3.6.3 Case Study 3—Study on Concrete-Rubber Composite Barriers' Performance in Mitigating Ground Vibration Caused by Metro Rails

Previous studies based on the isolation of low frequency surface waves were mainly focused on constructing periodic structures like trenches and piles. These structures necessitate large-scale deliveries or on-site casting, resulting in challenges with transportation and prolonged construction durations. Hence, the investigation focused on periodic barriers for isolating vibration. Periodic barriers based on composite concrete rubbers were proposed to surmount these challenges. Lab tests were undertaken to confirm the calculated frequency band gap and damping regions of optimized concrete-rubber barrier ground vibrations and shocks.

Additionally, a numerical framework was established to assess the vibration attenuation impact of the optimized concrete rubber barrier when subjected to metro rail loads. The Insertion Loss (IL) increased with an increased ratio of buried depth to tunnel depth, but its value remained below 2 dB. Therefore, a burial depth-to-tunnel base depth ratio of 1 is recommended for vibration isolation projects of underground metro trains. Overall, the research presents a promising solution using the novel concrete-rubber barrier efficiently to mitigate environmental vibrations caused by underground metro trains.

3.7 Concluding Remark

Railway-induced noise and vibration is a thrust area of research, and still, the most part, remained untouched by the research community. The area is broad and complex, and there is far more to explore related to noise and vibration. Various researchers have attempted through analytical, numerical, and experimental approaches to minimize the effect of vibration. The current chapter describes the need for railway transportation and its environmental impact. Various standards have been proposed nationally and internationally for noise and vibration assessment. Based on the assessment, the researchers suggested various mitigation measures for railway-induced noise and vibration.

References

Amiri M et al (2016) A new combination of artificial neural network and K-nearest neighbors models to predict blast-induced ground vibration and air-overpressure. Eng Comput 32(4):631–644

Ataei S, Miri A, Jahangiri M (2017) Assessment of load carrying capacity enhancement of an open spandrel masonry arch bridge by dynamic load testing. Int J Arch Herit 11(8):1086–1100

Bahrekazemi M (2004) Train-induced ground vibration and its prediction. Byggvetenskap

Bewes O, Jakielaszek L, Richardson M (2015) An assessment of the effectiveness of replacing slab track to control groundborne noise and vibration in buildings above an existing railway tunnel. In: Noise and vibration mitigation for rail transportation systems: proceedings of the 11th international workshop on railway noise, Uddevalla, Sweden, 9–13 September 2013. Springer

Bezin Y et al (2009) An investigation of sleeper voids using a flexible track model integrated with railway multi-body dynamics. Proc Inst Mech Eng, Part F: J Rail Rapid Transit 223(6):597–607

Bharule S, Kidokoro T, Seta F (2019) Evolution of high-speed rail and its development effects: stylized facts and review of relationships

Cao Z et al (2018) Measurement and analysis of vibrations in a residential building constructed on an elevated metro depot. Measurement 125:394–405

Charoenwong C et al (2023) Railway slab vs ballasted track: a comparison of track geometry degradation. Constr Build Mater 378:131121

Chen J, Zhou Y (2020) Dynamic vertical displacement for ballastless track-subgrade system under high-speed train moving loads. Soil Dyn Earthq Eng 129:105911

Connolly DP et al (2015) Benchmarking railway vibrations–track, vehicle, ground and building effects. Constr Build Mater 92:64–81

Coulier P, Lombaert G, Degrande G (2014) The influence of source–receiver interaction on the numerical prediction of railway induced vibrations. J Sound Vib 333(12):2520–2538

Elias P, Villot M (2012) Review of existing standards, regulations and guidelines, as well as laboratory and field studies concerning human exposures to vibration

Faizan AA et al (2022) Evaluation of human exposure and building damage to high-speed train-induced ground-borne vibration based on numerical studies: a comparison with international standards. Arab J Geosci 15(8):779

Forrest J, Hunt H (2006a) A three-dimensional tunnel model for calculation of train-induced ground vibration. J Sound Vib 294(4–5):678–705

Forrest J, Hunt H (2006b) Ground vibration generated by trains in underground tunnels. J Sound Vib 294(4–5):706–736

Guo Y, Markine V, Jing G (2021) Review of ballast track tamping: Mechanism, challenges and solutions. Constr Build Mater 300:123940

Haladin I, Bogut M, Lakušić S (2021) Analysis of tram traffic-induced vibration influence on earthquake damaged buildings. Buildings 11(12):590

Hao Y et al (2022) Study of noise and vibration impacts to buildings due to urban rail transit and mitigation measures. Sustainability 14(5):3119

He C et al (2018) Analytical method for calculation of ground vibration from a tunnel embedded in a multi-layered half-space. Comput Geotech 99:149–164

He C et al (2019) Theoretical modelling of the dynamic interaction between twin tunnels in a multi-layered half-space. J Sound Vib 456:65–85
He C et al (2017) Dynamic 2.5-D Green's function for a point load or a point fluid source in a layered poroelastic half-space. Eng Anal Bound Elem 77: 123–137
He et al (2020) An efficient three-dimensional method for the prediction of building vibrations from underground railway networks. Soil Dyn Earthq Eng 139:106269
Heidary R, Esmaeili M, Nik MG (2023) Experimental and numerical analysis of vibrations induced by a twin tunnel, underground railway. Acta Polytechnica Hungarica 20(1)
Hussein M et al (2014) The fictitious force method for efficient calculation of vibration from a tunnel embedded in a multi-layered half-space. J Sound Vib 333(25):6996–7018
Institution BS (2008) Guide to evaluation of human exposure to vibration in buildings: vibration sources other than blasting. British Standards Institution
ISO I 2631–1 (1997) Mechanical vibration and shock-evaluation of human exposure to whole-body vibration-Part 1: General requirements 42, p 43–44. ISO, Geneva, Switzerland
ISO 14837–1 (2005) Mechanical vibration–ground-borne noise and vibration arising from rail systems–Part 1: General guidance. ISO, Switzerland
Javaid MF, Azam R, Riaz MR (2023) Decision tree regression algorithm-based model for prediction of traffic-induced vibrations: development and application for parametric investigation. Innov Infrastruct Solut 8(3):103
Jebin N, Vivek V, Bindu CA (2023) Review of urban transport initiatives in India with a focus on urban transport planning scheme (UTPS). Available at SSRN 4488321
Jiang J et al (2015) Reducing railway induced ground-borne vibration by using trenches and buried soft barriers. In: Noise and vibration mitigation for rail transportation systems: proceedings of the 11th international workshop on railway noise, Uddevalla, Sweden, 9–13 September 2013. Springer
Khan D, Burdzik R (2023) Measurement and analysis of transport noise and vibration: a review of techniques, case studies, and future directions. Measurement, 113354
Kouroussis G et al (2015) The effect of railway local irregularities on ground vibration. Transp Res Part D: Transp Environ 39:17–30
Kouroussis G et al (2019) Urban railway ground vibrations induced by localized defects: using dynamic vibration absorbers as a mitigation solution. J Zhejiang Univ-Sci A (Appl Phys Eng) 20(2):83–97
Kouroussis G et al (2014) Building vibrations induced by railways: an analysis of commonly used evaluation standards. In: 21st International congress on sound and vibration
Kumar K, Bhartia A, Mishra R (2022) Monitoring, modeling, and mapping of rail-induced noise at selected stations in megacity Delhi. Int J Environ Sci Technol, 1–10
Küsel M, Brommundt E (2003) Wavy wear pattern on the tread of railway wheels. In: System dynamics and long-term behaviour of railway vehicles, track and subgrade. Springer, 121–132
Lai CG et al (2005) Prediction of railway-induced ground vibrations in tunnels. J Vib Acoust 127(5):503–514
Lam D (2023) Has the world survived the population bomb? A 10-year update. Popul Environ 45(2):10
Liu W et al (2023) A frequency-domain formulation for predicting ground-borne vibration induced by underground train on curved track. J Sound Vib 549:117578
Lopes P et al (2016) Vibrations inside buildings due to subway railway traffic. Experimental validation of a comprehensive prediction model. Sci Total Environ 568, 1333–1343
López-Mendoza D et al (2020) A transfer function method to predict building vibration and its application to railway defects. Constr Build Mater 232:117217
Luo Y-K et al (2022a) Evaluating railway noise sources using distributed microphone array and graph neural networks. Transp Res Part D: Transp Environ 107:103315
Luo Y-K et al (2022b) Towards the understanding of wheel-rail flange squeal: in-situ experiment and genuine 3D profile-enhanced transient modelling. Mech Syst Signal Process 180:109455

Luo Y-K et al (2023) Experimental evaluation and numerical interpretation of various noise mitigation strategies for in-service elevated suburban rail. Measurement, 113276
Ma M et al (2022) Effect of passing metro trains on uncertainty of vibration source intensity: monitoring tests. Measurement 193:110992
Ma C et al (2023) Initiation mechanism analysis of wheel polygonal wear on high-speed railway based on refined vibration model for ballastless track system. J Sound Vib 559:117782
Maclachlan L et al (2018) Annoyance in response to vibrations from railways. Int J Environ Res Public Health 15(9):1887
Morais J et al (2023) Railway track support condition assessment—methodology validation using numerical simulations. Eng Fail Anal, 107483
Nations U (2014) World urbanization prospects: The 2014 revision, highlights. Department of economic and social affairs. Population Division, United Nations, 32
Normung D.I.f. (1999) DIN 4150–2: Structural vibrations–Part 2: Human exposure to vibration in buildings. German Standards Organization (GSO), Berlin
Ntotsios E, Thompson D, Hussein M (2017) The effect of track load correlation on ground-borne vibration from railways. J Sound Vib 402:142–163
Organization WH (2011) Burden of disease from environmental noise: quantification of healthy life years lost in Europe. World Health Organization. Regional Office for Europe
Part SV (1999) 3: Effects of vibration on Structures. DIN Germany Institute, p 4150–4153
Peplow A, Persson P, Andersen LV (2021) Evaluating annoyance mitigation in the screening of train-induced noise and ground vibrations using a single-leaf traffic barrier. Sci Total Environ 790:147877
Quagliata A et al (2018) Transit noise and vibration impact assessment manual
Seidler A et al (2023) Health effects of railway-induced vibration combined with railway noise–a systematic review with exposure-effect curves. Environmental Research, p 116480
Selvakumar M, Ramulu DS, Sankar K (2023) A unique metro choice behaviour of suburban rail passengers in India. Urban Rail Transit 9(1):31–41
Standard B 6841 (1987) Measurement and evaluation of human exposure to whole-body mechanical vibration and repeated shock. British Standard 6841
Standards E (2016) Railway applications—track—concrete sleepers and bearers with under sleeper pads. European Standards, Dublin, Ireland
Thompson D, Jones C (2000) A review of the modelling of wheel/rail noise generation. J Sound Vib 231(3):519–536
Thompson D et al (2019a) Noise and vibration from railway vehicles. Handbook of railway vehicle dynamics, 2nd edn. CRC Press, pp 521–578
Thompson DJ, Kouroussis G, Ntotsios E (2019b) Modelling, simulation and evaluation of ground vibration caused by rail vehicles. Veh Syst Dyn 57(7):936–983
Torstensson P, Nielsen JC, Baeza L (2011) Dynamic train–track interaction at high vehicle speeds—Modelling of wheelset dynamics and wheel rotation. J Sound Vib 330(22):5309–5321
Vibration B (2003) Mechanical vibration and shock—evaluation of human exposure to whole-body vibration—Part 2: vibration in buildings (1 hz to 80 hz). ISO
Waddington DC et al (2014) Human response to vibration in residential environments. J Acoust Soc Am 135(1):182–193
Wang C et al (2023) Prediction method of ground vibration caused by high-speed railway based on the formula from the federal transportation administration. J Transp Eng, Part A: Syst 149(9):04023090
Woods RD (1968) Screening of surface wave in soils. J Soil Mech Found Div 94(4):951–979
Xin T et al (2020) Field measurement of rail corrugation influence on environmental noise and vibration: a case study in China. Measurement 164:108084
Xue R et al (2022) Vertical vibration analysis of a coupled vehicle-container model of a high-speed freight EMU. Veh Syst Dyn 60(4):1228–1252

Yang Y et al (2023) The evolution of China's environmental impact assessment system: retrospect and prospect from the perspective of effectiveness evaluation. Environ Impact Assess Rev 101:107122

Yuan Z, Boström A, Cai Y (2017) Benchmark solution for vibrations from a moving point source in a tunnel embedded in a half-space. J Sound Vib 387:177–193

Zeybek H, Roy S, Mitra S (2023) Transport as a driver of sustainable urban growth: evidence from Ankara, Turkey and Kolkata, India. In: Urban environment and smart cities in Asian countries: insights for social, ecological, and technological sustainability. Springer, pp 477–497

Zhou S et al (2020) Three-dimensional analytical model for coupled track-tunnel-soil system in a multilayered half-space. J Eng Mech 146(2):04019121

Chapter 4
Theoretical and Application Aspects of Determining the Curvature of the Railway Track Axis by the Moving Chord Method

Wladyslaw Koc

4.1 Introduction

The shape of the existing railway lines undergoes gradual changes during many years of operation and after some time begins to deviate from its original geometric layout. Due to passenger comfort and operational safety, deviations of the track axis in the horizontal plane may prove particularly unfavorable. Therefore, it is necessary to know the current geometric shape of the route so that any irregularities can be eliminated and possible corrections made. Appropriate measurement systems and calculation algorithms serve this purpose.

The currently used measurement methods are similar in different railway administrations (Committee for Standardization (CEN) 2010; Deutsche Bahn 2016; Federal Railroad Administration 2008; Network Rail 2008; Polish State Railways 2018). In addition to classical geodetic techniques, stationary satellite measurements based on the Global Navigation Satellite System (GNSS) technique are used, which use the so-called active geodetic networks (Szwilski et al. 2005). Methods of mobile satellite measurements (Specht and Koc 2016) are also introduced, in which, apart from GNSS receivers, Inertial Navigation System (INS) (Starlino Electronics Homepage 2023) devices are often used as supporting devices, and vision methods such as Terrestrial Laser Scanning (TLS) (Guimarães-Steinicke et al. 2019). Research is underway on the possibility of using systems consisting of satellite receivers mounted on various types of vehicles (Quan and Lau 2019; Wu et al. 2020).

The purpose of the measurements is to determine the geometric parameters (i.e., identification) of the measured route. This is done on the basis of the determined

W. Koc (✉)
Gdansk University of Technology, 80-180 Gdansk, Poland
e-mail: kocwl@pg.edu.pl

S. K. Sharma et al. (eds.), *Dynamics of Transportation Ecosystem, Modeling, and Control*, Energy, Environment, and Sustainability,
https://doi.org/10.1007/978-981-97-0437-8_4

Cartesian coordinates of the track axis points in the appropriate national spatial reference system. In Poland, for plane coordinates, the PL-2000 system (Regulation of the Council of Ministers 2012) is in force, created on the basis of a mathematically unambiguous assignment of points on the GRS 80 reference ellipsoid (Moritz 2000) to appropriate points on the plane according to the Gauss-Krüger mapping theory (Turiño 2008).

Determining the coordinates of the track axis makes it possible to visualize a given railway route, giving a general orientation about its location. Identification of the measured route requires the use of appropriate calculation algorithms. When considering the horizontal plane, the basis for the analysis are the determined values of the plane eastern Y_i coordinates and the northern X_i coordinates of a given measurement point in the PL-2000 system. However, the proper identification of the track axis is provided by appropriate graphs referring to the length parameter L. Therefore, in order to create the possibility of further analysis, it is necessary to go to the linear system, i.e., to determine the distances L_i of individual measurement points from the selected starting point O (Y_0, X_0).

The current approach to the problem of identifying the measured railway route consists of generating an optimized geometrical layout in a given area by minimizing the deviations of this layout from the measurement points (while meeting the appropriate maintenance and operational requirements). The traditional solution to a design problem is largely based on human experience (Lai and Schonfeld 2016; Zhang et al. 2020). In many works, attempts are made to automate the discussed process. There are two variants of the procedure: geometric identification and alignment optimization. These variants may be implemented separately or simultaneously.

When it comes to geometric identification, an elementary approach is possible, such as the use of discrete approximation methods using splines (Lenda 2014) to determining the geometric extent of the various components of the layout. There are also a number of other methods here, which have been presented in works (Gikas and Stratakos 2012; Luo et al. 2018; Othman et al. 2012). The research methods used to optimize the alignment are very diverse. Many of them concern transition curves (Cellmer et al. 2016; Skala-Szymanska et al. 2014; Song et al. 2020). Other applied alignment optimization methods are presented in Dong et al. (2007), Easa and Wang (2011) and Song et al. (2021). In works (Camacho-Torregrosa et al. 2015; Li et al. 2019, 2022), several methods of combining these variants into an iterative process of reproducing the track axis were proposed.

As it turns out, the problem of geometric identification of the railway track can be approached in a completely different way, departing from matching the hypothetical model layout to the points of the track axis, determined by direct measurements. The basis for identification may be the curvature of the geometric layout, it would only be necessary to develop an appropriate method for its determination. Paper (Koc 2020) presents a proposal for a new method for determining the curvature of the track axis, referred to as the "moving chord method".

4.2 Theoretical Basis of the Moving Chord Method

From the definition of curvature, it is necessary to operate with the angles of inclination of the tangent to the geometric layout. The measure of road curvature is the ratio of the angle $\Delta\Theta$, by which the direction of the longitudinal axis of the vehicle changes after passing a certain arc, to the length Δl of this arc (Fig. 4.1).

The curvature of the curve K at point M is the limit approached by the ratio of the acute angle $\Delta\Theta$ included between the tangents to the curve K at points M and M_1 to the length Δl of the arc, when point M_1 follows the curve K to point M.

$$\kappa = \lim_{\Delta l \to 0} \left| \frac{\Delta\Theta}{\Delta l} \right| = \frac{d\Theta}{dl} \tag{4.1}$$

If you have an analytical record of a given curve, determining the function $\Theta(l)$ and applying formula (4.1) is not a major problem. However, in a real, operated railway track, there is not such possibility. Therefore, in practice—assuming small values of Δl—a simplified formula can be used

$$\kappa = \frac{\Delta\Theta}{\Delta l} \tag{4.2}$$

The method of determining the angles of inclination of the tangents (and consequently the value of $\Delta\Theta$) has not yet been fully resolved.

Due to the difficulties in determining the difference of angles $\Delta\Theta$, a concept emerged that when determining the curvature of the track, not tangential but corresponding chords should be used. It should be noted that the use of chords of a certain length, derived from a given point forwards and backwards, has already taken place, among others in works (Fu 2016; Guo and Zhong 2014; Zhong 2010). However, this was not used to determine the value of the curvature, but to construct a geometric shape consisting of several curves (with clear segmentation points) using the CAGD (Computer Aided Geometric Design) software.

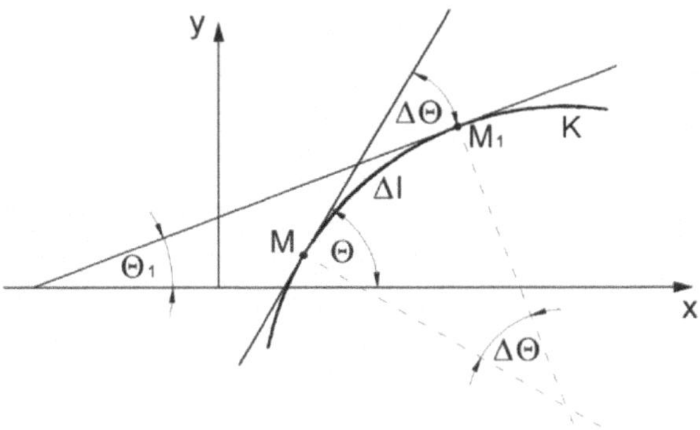

Fig. 4.1 Schematic diagram to explain the concept of curvature of the road route

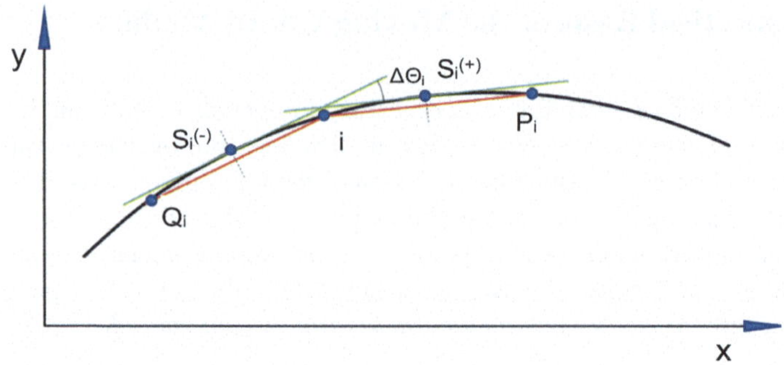

Fig. 4.2 Schematic diagram of curvature determination using the moving chord method

In the work (Koc 2020), a new method of determining the curvature was presented for the first time, based on the use of the difference in the angles of the chord of a fixed length occurring at a given point. This method is based on two assumptions:

(a) the tangent lines and the corresponding chords are parallel to each other,
(b) the points of contact project perpendicularly to the center of a given chord.

The above assumptions are met for a circular arc. However, this is not the case along the length of the transition curve, and the inconsistency resulting from a strict failure to meet these conditions is relatively small; it decreases as one passes to the initial region of the curve. Figure 4.2 shows a schematic diagram of curvature determination using the proposed method (Koc 2023a). In this case, it concerns a section of a circular arc.

To determine the value of curvature at point i, with known Cartesian coordinates, draw a chord of length l_c from this point pointing first backwards and then forwards. The end of the backward chord is at the Q_i point. Determining the coordinates of this point makes it possible to determine the angle of inclination of the chord $\Theta_i^{(-)}$. The same angle of inclination has the tangent to the circular arc at the point $S_i^{(-)}$, which is projected perpendicularly to the center of the chord. On the other hand, the end of the forward chord is at P_i point. The coordinates of this point (and the coordinates of point i) determine the angle of inclination of the chord $\Theta_i^{(+)}$, and thus also the angle of inclination of the tangent to the circular arc at the point $S_i^{(+)}$, which is projected perpendicularly to the center of the chord. In this situation, the difference in the angles of inclination of the tangent is

$$\Delta\Theta_i = \Theta_i^{(+)} - \Theta_i^{(-)} \tag{4.3}$$

and the angle of inclination of the tangent at point i (determining the direction angle of the route) is equal

$$\Theta_i = \frac{\Theta_i^{(-)} + \Theta_i^{(+)}}{2} \tag{4.4}$$

The existing curvature should be determined from the formula

$$\bar{\kappa}_i = \frac{\Delta\Theta_i}{\widehat{S_i^{(-)}S_i^{(+)}}} \tag{4.5}$$

where $\widehat{S_i^{(-)}S_i^{(+)}}$ is the distance between the points of contact $S_i^{(-)}$ and $S_i^{(+)}$ measured along the arc.

While the knowledge of Cartesian coordinates allows the determination of the angle difference $\Delta\Theta_i$, this does not apply to the determination of distance $\widehat{S_i^{(-)}S_i^{(+)}}$. In the case of a circular arc, it can be determined from the formula

$$\widehat{S_i^{(-)}S_i^{(+)}} = 2 \cdot R \cdot \arcsin\frac{l_c}{2R} \tag{4.6}$$

However, from a practical point of view, this formula (4.6) is of little use, since the value of the arc radius R is not known. In this situation, the using of moving chord method requires one more assumption:

(c) the curvature κ_i at a given point i is given by the formula

$$\kappa_i = \frac{\Delta\Theta_i}{l_c} \tag{4.7}$$

There is no doubt that in the case of arc sections, the curvature values obtained using formulas (4.5) and (4.7) must be different because condition $\widehat{S_i^{(-)}S_i^{(+)}} > l_c$ applies in each case. Assumption (c) therefore limits the applicability range of the discussed method of determining the curvature; the assumptions made in it concern—generally speaking – only a certain, limited range of geometric parameters.

4.3 Applicability of the Moving Chord Method

To determine the applicability range of the moving chord method, the differences between the distance $\widehat{S_i^{(-)}S_i^{(+)}}$ in formula (4.5) and the chord length l_c used in formula (4.7) were determined. This was possible for a circular arc, where the difference results from the following formula:

$$\Delta l_c = \widehat{S_i^{(-)}S_i^{(+)}} - l_c = 2 \cdot R \cdot \arcsin\frac{l_c}{2R} - l_c \tag{4.8}$$

Table 4.1 presents the values of the differences Δl_c calculated using the formula (4.8) for the radii of circular arcs in the range of $R = 50$–600 m, with the lengths of the moving chord $l_c = 10$–50 m. The graphs of the dependence $\Delta l_c(R)$ for different values of l_c are shown in Fig. 4.3.

Table 4.1 Comparison of the values of differences Δl_c for different arc radii R and chord lengths l_c

Arc radius	Differences Δl_c (mm) a>				
R (m)	$l_c = 10$m	$l_c = 20$m	$l_c = 30$ m	$l_c = 40$ m	$l_c = 50$ m
50	16.742	135.792	469.265	1151.685	2359.878
75	7.422	59.738	203.688	489.919	975.536
100	4.171	33.484	113.655	271.584	536.051
125	2.669	21.395	72.471	172.663	339.480
150	1.853	14.845	50.226	119.477	234.424
175	1.361	10.900	36.857	87.591	171.649
200	1.042	8.343	28.196	66.968	131.133
225	0.823	6.590	22.267	52.863	103.457
250	0.667	5.337	18.029	42.790	83.711
275	0.551	4.410	14.896	35.346	69.128
300	0.463	3.706	12.514	29.689	58.052
350	0.340	2.722	9.191	21.801	42.615
400	0.260	2.084	7.036	16.685	32.609
450	0.206	1.646	5.558	13.130	25.756
500	0.167	1.334	4.502	10.674	20.857
550	0.138	1.102	3.720	8.821	17.234
600	0.116	0.926	3.126	7.411	14.479

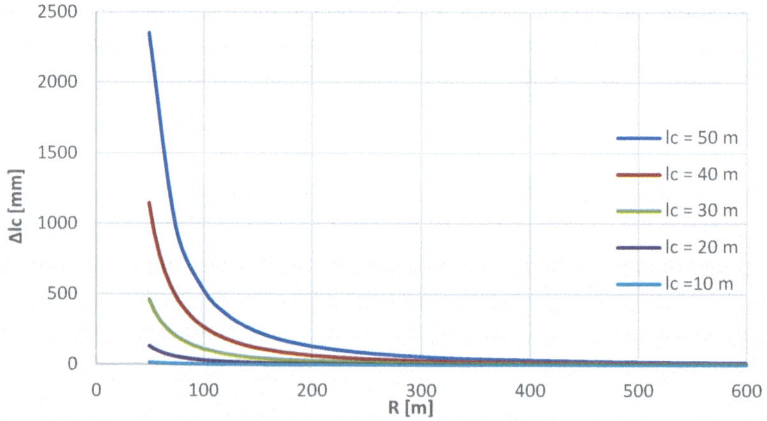

Fig. 4.3 Graphs of $\Delta l_c(R)$ for different l_c values (at $R = 50$–600 m)

The numerical values in Table 4.1 and the graphs in Fig. 4.3 fully explain the applicability range of the moving chord method. As can be seen, the values of Δl_c are the greater the smaller the radius of the circular arc and the greater the length of the chord used. Assuming—in a conventional way—the permissible value of Δl_c equal to about 3 mm, we limit the applicability of the discussed method to the radius $R = 120$ m at $l_c = 10$ m, $R = 350$ m at $l_c = 20$ m and $R = 600$ m at $l_c = 30$ m; for the chord $l_c = 40$ m and 50 m, the arc radii must be even larger.

The above statements clearly indicate that the moving chord method corresponds to the conditions on railway roads. The radii of circular curves $R \leq 300$ m are used here in principle only in turnouts, possibly on mountain lines. In typical situations, they are much larger so as not to limit the speed of trains, depending on the category of a given line.

It should be noted, however, that in the calculation practice for the railway track in operation, the length of the moving chord cannot be assumed arbitrarily. When selecting the chord length, the recommendations formulated in Koc (2022a) should be followed. Based on the analyses carried out there, it was clearly demonstrated that the length of the chord, used to determine the curvature in the operated track, should depend on the value of the radius of the circular arc. The following approximate lengths of l_c have been proposed, depending on a range of R values:

- for $R \leq 600$ m $\quad l_c = 20$ m,
- for $600 < R \leq 1000$ m $\quad l_c = 30$ m,
- for $1000 < R \leq 1400$ m $\quad l_c = 40$ m,
- for $R > 1400$ m $\quad l_c = 50$ m.

Figure 4.4 shows the graphs of the dependence $\Delta l_c(R)$ for a typical range of radii found on railway roads, with the relevant fragments referring to the required length of the moving chord in bold.

As it turns out, the adoption of assumptions from work (Koc 2022a) means that in no case the values of Δl_c exceed 4 mm, and in the vast majority of case—even 3 mm (Koc 2023a). The curvature values for a circular arc obtained using the formula (4.7) will therefore be slightly overestimated, and the radius of the arc—slightly underestimated. From a practical point of view, it shouldn't matter much. On the length of the transition curves, the impact of failure to meet the adopted assumptions turns out to be even less significant.

The above considerations largely explain why the cases of determining the curvature using the moving chord method in the operated track, presented in studies (Koc 2022b, c, 2023b), were fully successful. As you can see, the correctness of the proposed method also has a strong theoretical basis.

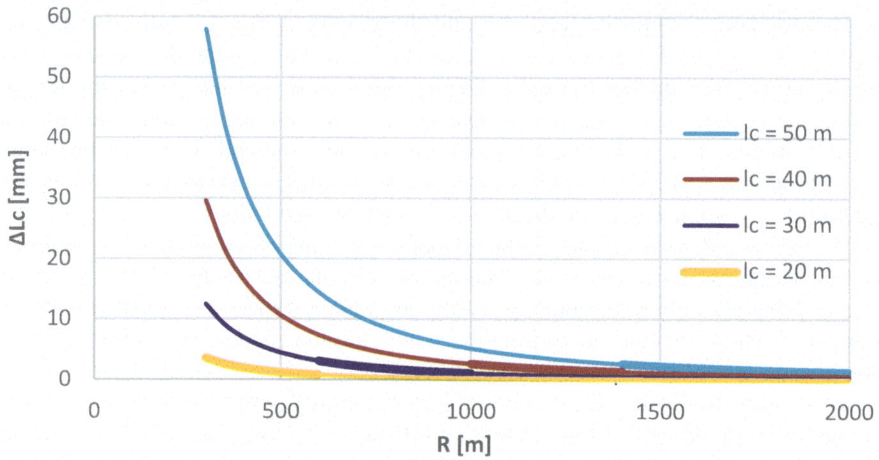

Fig. 4.4 Graphs of the dependence $\Delta l_c(R)$ for different values of l_c (at $R = 300$–2000 m) with bold fragments referring to the required length of the moving chord

4.4 Algorithm for Determining the Curvature of the Track Axis Using the Moving Chord Method

From a practical point of view, it will be beneficial to transfer the measurement data to the local x, y coordinate system. In most cases, this operation will consist of shifting the origin of the PL-2000 system to the selected point O (Y_0, X_0) and rotating it by the angle β. The following transformation formulas are then used:

$$x_i = (Y_i - Y_0)\cos\beta + (X_i - X_0)\sin\beta \tag{4.9}$$

$$y_i = -(Y_i - Y_0)\sin\beta + (X_i - X_0)\cos\beta \tag{4.10}$$

The positive value of the angle β occurs when the system is rotated to the left.

The methodology for determining the curvature of the track axis has been explained in detail in Koc (2022c). The sequence of actions to determine the curvature value at any measurement point is shown in Fig. 4.5.

We start determining the curvature κ_i from the measurement point i, which is located in such a way that it allows the projection of a virtual chord length l_c backwards; the end of calculations must take place at a point from which a virtual chord of the same length can still be placed forward. The basic operation that must be carried out first is determining the numbering of the points defining the intervals in which the ends of the virtual chords drawn from point i are located.

For a chord drawn from the point i forward, the interval in which the end of the chord occurs is determined by the points p_{i-1} and p_i (Fig. 4.5). After determining them, it is possible to analytically write the equation of a straight line passing through

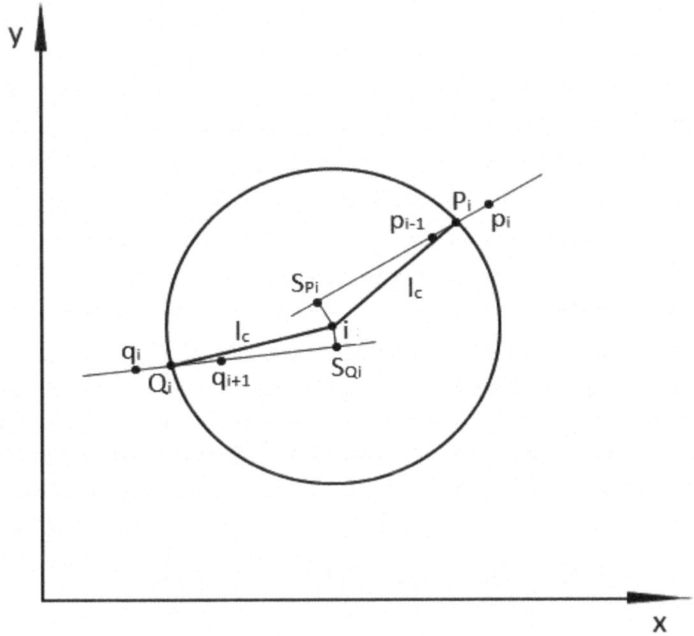

Fig. 4.5 Explanation of the method of determining the curvature value at any measurement point using the moving chord method

these points. This equation has the following form:

$$y = a_{pi} + b_{pi}x \tag{4.11}$$

As can be seen in Fig. 4.5, the end of the front chord (i.e., point P_i) lies on the straight line described by Eq. 4.11, at a distance l_c from point i. It is therefore the point of intersection of the circle with radius l_c and center at point i with the straight line (4.11). The coordinates of the P_i point are determined from the following formulas:

$$x_{Pi} = \frac{-B_{Pi} \pm \sqrt{B_{Pi}^2 - 4A_{Pi}C_{Pi}}}{2A_{Pi}} \tag{4.12}$$

$$y_{Pi} = a_{pi} + b_{pi}\frac{-B_{Pi} \pm \sqrt{B_{Pi}^2 - 4A_{Pi}C_{Pi}}}{2A_{Pi}} \tag{4.13}$$

where

$$A_{Pi} = 1 + b_{pi}^2,$$

$$B_{Pi} = -2(x_{Spi} + b_{pi}y_{Spi} - a_{pi}b_{pi})$$

$$C_{Pi} = x_{Spi}^2 + y_{Spi}^2 - 2a_{pi}y_{Spi} + a_{pi}^2 - l_c^2 + \left[(x_i - x_{Spi})^2 + (y_i - y_{Spi})^2\right]$$

$$x_{Spi} = \frac{b_{pi}}{1 + b_{pi}^2}\left(y_i + \frac{1}{b_{pi}}x_i - a_{pi}\right)$$

$$y_{Spi} = \frac{1}{1 + b_{pi}^2}\left(b_{pi}^2 y_i + b_{pi}x_i + a_{pi}\right)$$

The " + " sign in formulas (4.12) and (4.13) occurs when the abscissas values of the measured route points are increasing, while the "–" sign is valid for decreasing abscissas.

For a chord drawn from point i backwards, the interval in which the end of the chord occurs is determined by the points q_i and q_{i+1} (Fig. 4.5). We determine it in the same way as in the case of the forward chord. If the coordinates of points q_i and q_{i+1} are already known, it is possible to write the equation of a straight line that goes through these points analytically. This equation is as follows:

$$y = a_{qi} + b_{qi}x \tag{4.14}$$

As can be seen from Fig. 4.5, the sought end of the back chord (i.e., point Q_i) lies on the straight line described by Eq. 4.14 at a distance l_c from point i. Thus, it is the point of intersection of the circle with radius l_c and center at point i with the straight line (4.14). The coordinates of the Q_i point are determined from the formulas:

$$x_{Qi} = \frac{-B_{Qi} \pm \sqrt{B_{Qi}^2 - 4A_{Qi}C_{Qi}}}{2A_{Qi}} \tag{4.15}$$

$$y_{Qi} = a_{qi} + b_{qi}\frac{-B_{Qi} \pm \sqrt{B_{Qi}^2 - 4A_{Qi}C_{Qi}}}{2A_{Qi}} \tag{4.16}$$

where

$$A_{Qi} = 1 + b_{qi}^2,$$

$$B_{Qi} = -2(x_{Sqi} + b_{qi}y_{Sqi} - a_{qi}b_{qi})$$

$$C_{Qi} = x_{Sqi}^2 + y_{Sqi}^2 - 2a_{qi}y_{Sqi} + a_{qi}^2 - l_c^2 + \left[(x_i - x_{Sqi})^2 + (y_i - y_{Sqi})^2\right]$$

$$x_{Sqi} = \frac{b_{qi}}{1 + b_{qi}^2}\left(y_i + \frac{1}{b_{qi}}x_i - a_{qi}\right)$$

$$y_{Sqi} = \frac{1}{1 + b_{qi}^2}\left(b_{qi}^2 y_i + b_{qi} x_i + a_{qi}\right)$$

The "–" sign in formulas (4.15) and (4.16) occurs when the abscissas values of the measured route points are increasing, while the "+" sign is valid for decreasing abscissas. Thus, what we are dealing with here is the opposite situation to the case of a forward chord.

Having the Cartesian coordinates of point i (obtained from measurements) and the coordinates of the ends of virtual chords drawn forward and backward, we are able to determine the curvature values at a given measurement point. A forward chord connects point i with point P_i, and its coordinates are given by formulas (4.12) and (4.13). Its angle of inclination is

$$\Theta_{i \div Pi} = \Theta_i^{(+)} = \arctan\frac{y_{Pi} - y_i}{x_{Pi} - x_i} \tag{4.17}$$

The backward chord connects point i with point Q_i, and its coordinates are given by formulas (4.15) and (4.16). Its angle of inclination is

$$\Theta_{Qi \div i} = \Theta_i^{(-)} = \arctan\frac{y_i - y_{Qi}}{x_i - x_{Qi}} \tag{4.18}$$

In this situation, the curvature value at a given measurement point is determined using the formula

$$\kappa_i = \frac{\Theta_i^{(+)} - \Theta_i^{(-)}}{l_c} \tag{4.19}$$

A positive value of the curvature determined by the formula (4.19) corresponds to a curve with convexity directed downwards, and a negative value—a curve with convexity directed upwards.

The presented procedure is sequential in nature and consists of using the given calculation formulas. Determination of the curvature value does not require the development of special computer programs, and the entire operation can be carried out, for example, in a spreadsheet.

4.5 Creation of Test Geometric Layout

After presenting in paper (Koc 2020) the general assumptions of the moving chord method, many aspects of determining the curvature of the track axis using this method were discussed in further works. Papers (Koc 2020, 2021a, b) present relevant analyses on the application of the moving chord method for model geometric layouts (described with mathematical equations). They showed complete compliance of the obtained curvature diagrams with the expected values, which are the basis of the given geometric solution. This clearly confirmed the correctness of the method from the substantive point of view. Papers (Koc 2022b, c, 2023b) address the issue of its use for estimating the horizontal curvature of the axis of the operated railway track on the basis of Cartesian coordinates obtained by direct measurements. All works showed high efficiency of the moving chord method in determining the curvature of the railway track.

It seems that the main advantage of the discussed method is its unambiquity, resulting from a direct reference to the definition of curvature. Used in the process of geometric identification of the railway route, it allows for direct determination of the location of the so-called segmentation points, i.e., places where individual geometric elements connect with each other (e.g., straight sections with transition curves, transition curves with circular arcs, etc.). This is illustrated in Figs. 4.9, 4.10, 4.11, 4.12, 4.13 and 4.14 in Sect. 4.6 In other identification methods, this is the biggest problem with it.

There is no doubt that the discussed method will improve the process of identifying track geometrical layouts and increase its accuracy. This will enable a more precise design of the track geometric layout and its ongoing maintenance, which in turn will lead to increased safety and overall efficiency of rail operations.

What remains to be clarified is the compact of deviations from the model (theoretical) shape of the track on the obtained curvature values for the real (i.e., in operation) railway track. Since the moving chord method uses the determined Cartesian coordinates of the track axis, deviations from the model shape are the result not only of the existing deformations of the track, but—mainly—of the measurement error.

It was decided to design the geometric layout of the track, specifying the coordinates of the track axis at 5 m intervals; the method described in Koc (2022d) was used. Then, the layout was virtually deformed by correcting the coordinates in a random way for two variants that differed in the value of the maximum error. The obtained results made it possible to answer the question posed earlier.

The model geometrical layout is shown in Fig. 4.6. It is a symmetrical layout with a length of about 1.100 m, the turning angle of the route $\alpha = \pi/3$ rad, adapted to the speed of trains $V = 120$ km/h. It consists of a circular arc with a radius of 740 m, two transition curves in the form of a clothoid with a length of 120 m and two relatively short (approx. 100 m) straight sections.

Due to the symmetry, the analysis did not take into account the entire layout, but only its initial fragment with a length of 370 m, containing a straight track section, the first transition curve and a part of a circular arc. Figure 4.7 shows this fragment

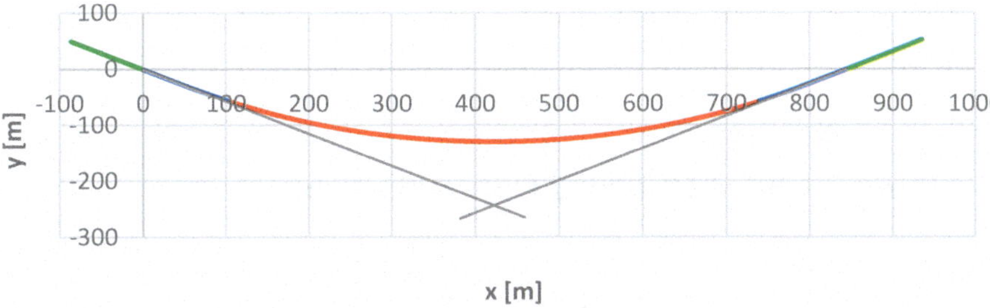

Fig. 4.6 The assumed model geometric layout; red color—circular arc, color blue—transition curves, green—straight sections

of the track, with the beginning of the straight section on the left side of the layout as its starting point. The theoretical curvature plot along the track length is shown in Fig. 4.8.

By randomly correcting the calculated Cartesian coordinates of the model geometric layout, two variants of the "deformed" railway track were obtained for further analysis. Variant I is characterized by a maximum error of ±10 mm, and Variant II is characterized by an error of ±25 mm (i.e., worse quality of measurement data). Determination of the curvature for both variants was carried out at different

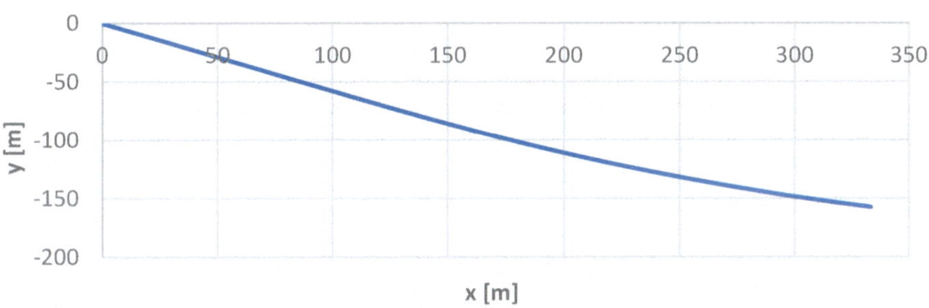

Fig. 4.7 Considered fragment of the model geometric layout

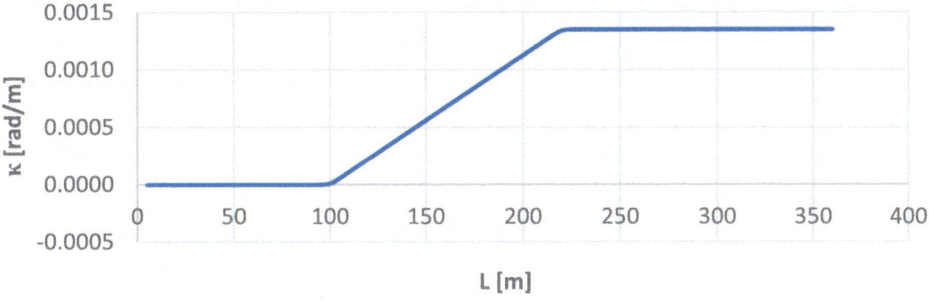

Fig. 4.8 Theoretical curvature graph along the length of the geometric layout

lengths of the moving chord. Calculations were made for $l_c = 5$ m, 10 m, 20 m, 30 m, 40 m and 50 m.

4.6 Comparison of Obtained Curvature Values

A summary of the results obtained regarding the curvature of the track axis is presented in Table 4.2. For each chord length l_c, the values of the curvature κ_{CA} and the radius R_{CA} of the circular arc, as well as the standard deviation of the curvature σ_κ and the percentage indicator $\sigma_\kappa/\kappa_{CA}$ (adopted as a numerical comparative parameter) are given therein.

Table 4.2 shows that the values of the radius of the circular arc, determined on the basis of the determined curvature for the chord length $l_c = 5$ m, clearly deviate from the model radius, i.e., $R_{CA} = 740$ m. For the chord $l_c = 10$ m it is much better, but only for $l_c \geq 20$ m the correct solution is obtained, i.e., slight deviations from the initial R_{CA} value. Further analysis will explain the reasons for this situation.

Figure 4.9 shows the curvature diagram obtained along the length of the test layout in variant I, determined with the length of the moving chord $l_c = 5$ m, and Fig. 4.10—for variant II with the same length of the chord. Both drawings were made in the same scale, against the background of hypothetical curvature distributions, determined according to the rules presented in works (Koc 2022b, 2023b); these distributions are marked in red.

Table 4.2 Identification of the circular arc in both variants of the test geometrical layout at different chord lengths

Chord length l_c [m]	Variant of the test layout	Curvature of the arc κ_{CA} [rad/m]	Radius of the arc R_{CA} [m]	Standard deviation σ_κ [rad/m]	Percentage indicator $\sigma_\kappa/\kappa_{CA}$ [%]
5	I	0.001377848	725.769503	0.0003621867	26.286406
5	II	0.001421427	703.518514	0.0010302871	72.482604
10	I	0.001359864	735.367689	0.0000686577	5.048864
10	II	0.001353549	738.798676	0.0002099314	15.509703
20	I	0.001352841	739.185175	0.0000220327	1.628622
20	II	0.001352548	739.345587	0.0000599687	4.433756
30	I	0.001351177	740.095452	0.0000077916	0.576649
30	II	0.001351298	740.029114	0.0000238386	1.764124
40	I	0.001351399	739.974055	0.0000044687	0.330676
40	II	0.001351005	740.189997	0.0000153885	1.139040
50	I	0.001351428	739.958285	0.0000030411	0.225025
50	II	0.001351800	739.754201	0.0000098876	0.731439

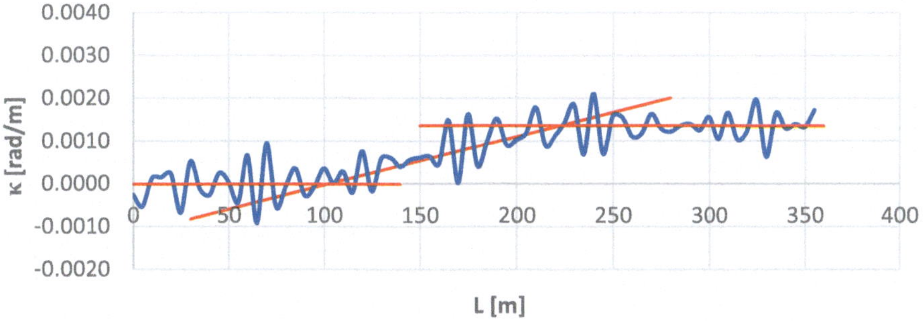

Fig. 4.9 Curvature diagram along the length of the test layout in variant I determined at $l_c = 5$ m (hypothetical curvature distribution is marked in red)

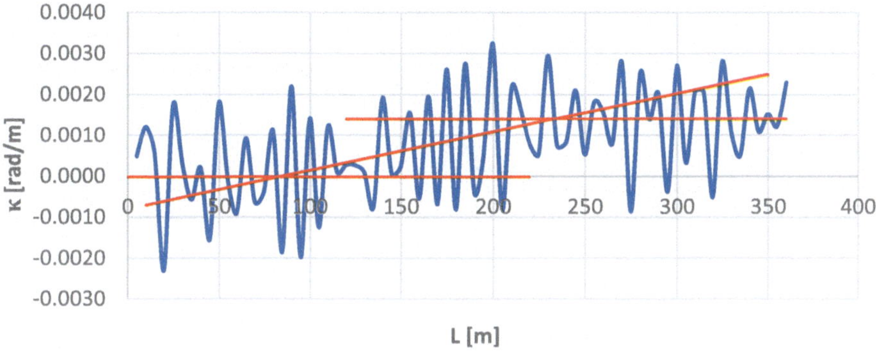

Fig. 4.10 Curvature diagram along the length of the test layout in variant II determined at $l_c = 5$ m (hypothetical curvature distribution is marked in red)

As can be seen, the oscillation of the curvature in relations to the hypothetical values in variant II, with lower quality of measurement data, is characterized by much greater amplitudes than in variant I. This is confirmed by the percentage ratio $\sigma_\kappa/\kappa_{CA}$ in Table 4.2. On the other hand, full identification of the geometric layout (i.e., determination of the length of the transition curve and the location of its beginning and end) on the basis of the curvature diagram would be very difficult (and burdened with a large error) both in variant I and variant II. Therefore, the findings contained in the work (Koc 2022a) are confirmed, which show that in operated railway track, the length of the moving chord should meet the condition $l_c \geq 20$ m. This is also related to the situation related to the R_{CA} radius values in Table 4.2.

Figures 4.11 and 4.12 show the curvature diagram obtained with the length of the moving chord $l_c = 20$ m—respectively—for the test layout in variants I and II. Figures 4.13 and 4.14 show analogous graphs obtained at $l_c = 40$ m. All drawings were made to the same scale, against the background of hypothetical curvature distributions.

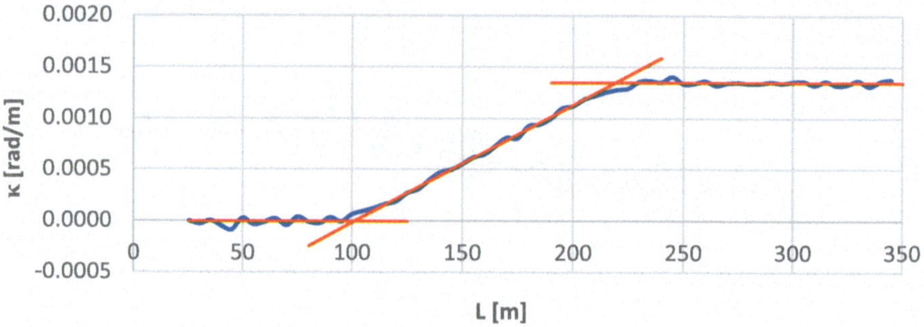

Fig. 4.11 Curvature diagram along the length of the test layout in variant I determined at $l_c = 20$ m (hypothetical curvature distribution is marked in red)

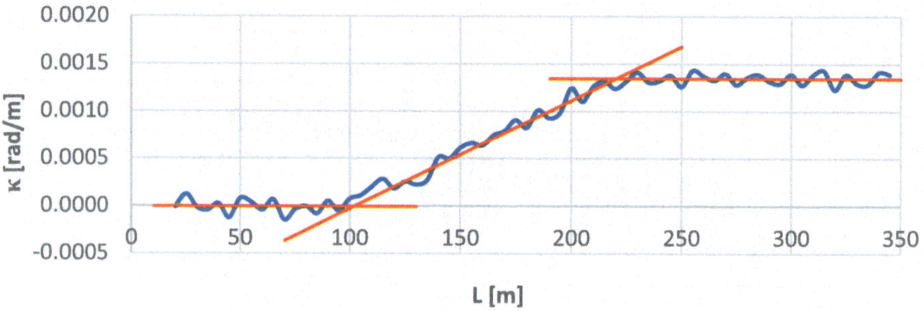

Fig. 4.12 Curvature diagram along the length of the test layout in variant II determined at $l_c = 20$ m (hypothetical curvature distribution is marked in red)

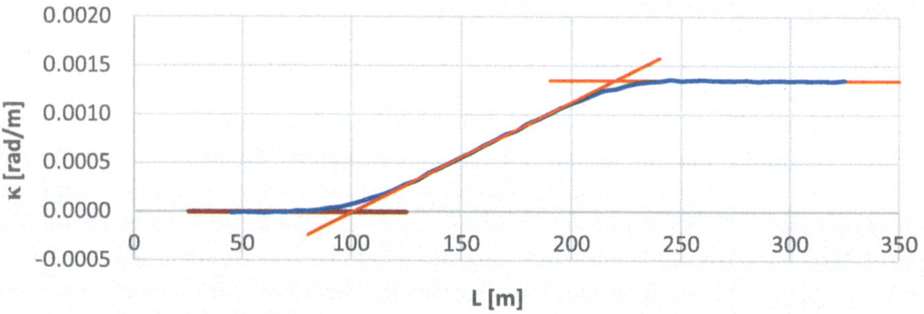

Fig. 4.13 Curvature diagram along the length of the test layout in variant I determined at $l_c = 40$ m (hypothetical curvature distribution is marked in red)

There is no doubt that the curvature diagrams in Figs. 4.11, 4.12, 4.13 and 4.14 can be the basis for identifying the geometrical layout. Extending the moving chord alleviates the perturbations in the curvature diagram. However, the absolute values of the curvature of the test layout in variant II are still greater than in variant I. This is confirmed by the graphs of the percentage ratio $\sigma_\kappa / \kappa_{CA}$ in Fig. 4.15.

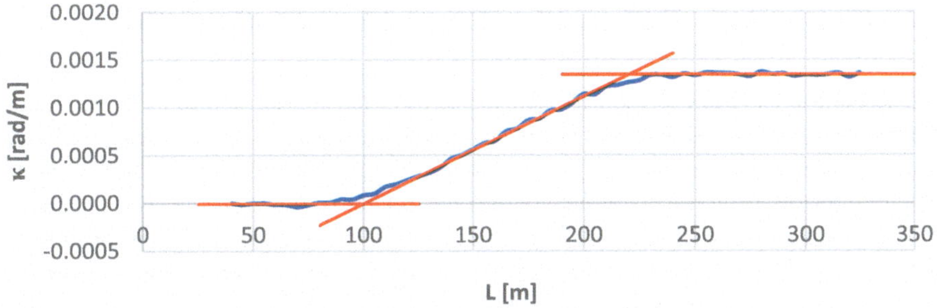

Fig. 4.14 Curvature diagram along the length of the test layout in variant II determined at $l_c = 40$ m (hypothetical curvature distribution is marked in red)

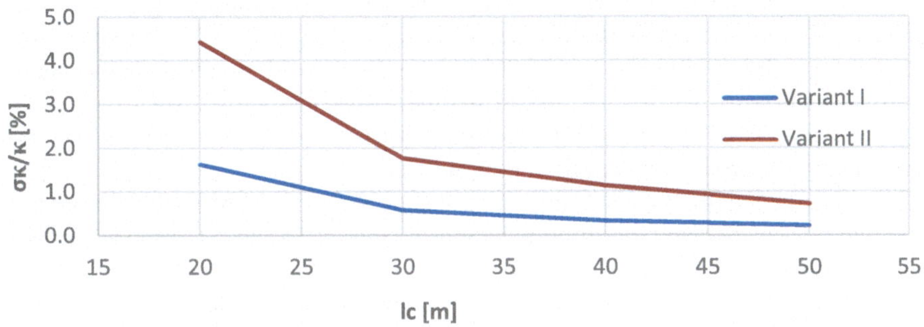

Fig. 4.15 Graphs of the percentage ratio $\sigma_\kappa/\kappa_{CA}$ for different lengths of the moving chord

The analysis shows that the quality of the measurement data plays an important role in determining the curvature value using the moving chord method. The poorer quality of these data, characterized by a greater value of the maximum error, causes greater disturbances in the curvature plot. However, these disturbances can be significantly reduced by appropriately extending the chord length used.

Taking into account the most favorable l_c given in Sect. 4.3, depending on the range of R_{CA} values, in the cases considered in this paper, the chord $l_c = 30$ m should be used. When deciding to use a longer chord than would result from the radius of the arc, one should take into account the limitations resulting from the theoretical basis of the method of curvature determination used. As shown in Sect. 4.3, the greater the length of the chord used, the greater the difference between the length of the chord (used in the curvature formula) and the length of the corresponding circular arc.

4.7 Comparison of Transition Curve Identification Effects

The transition curve is also identified on the basis of the curvature diagrams obtained. It consists in determining the position of the beginning (L_{BTC} coordinate) and end (L_{ETC} coordinate) of the transition curve in a linear coordinate system; the length of the curve results from the difference in the coordinates of these points. The summary of the obtained results regarding the identification of the transition curve is presented in Table 4.3. For each appropriate chord length l_c, the values of curvature κ_{CA} of the circular arc, linear coordinates L_{BTC} and L_{ETC} and the length of the curve l_{TC} are given therein.

The following design values were in force in the model geometric layout: $L_{BTC} = 100$ m, $L_{ETC} = 220$ m and $l_{TC} = 120$ m. As a result of creating variants, I and II, these values undoubtedly changed, but it could not be a radical change. As can be seen from Table 4.3, the largest deviation occurred in variant II of the test layout, for the chord length $l_c = 5$ m ($l_{TC} = 150.902944$ m). For $l_c = 10$ m it was clearly better in this variant ($l_{TC} = 131.420182$ m). In all other cases, including variant I with $l_c =$ 5 m and $l_c = 10$ m, the situation was completely different—it was very similar and only slightly differed from the initial design values.

Limiting the length of the moving chord to $l_c \geq 20$ m, graphs of linear coordinates L_{BTC} and L_{ETC} were made for different lengths l_c (Figs. 4.16 and 4.17) and analogous graphs l_{TC} (Fig. 4.18).

As shown in Fig. 4.16, with the length of the moving chord $l_c \geq 20$ m, the determined values of the linear coordinate of the beginning of the transition curve tend to reach $L_{BTC} = 100$ m. For variant I, they are in the range of 1 m and regularly

Table 4.3 Identification of the transition curve in both variants of the test geometric layout at different chord lengths

Chord length l_c [m]	Variant of the test layout	Curvature of the arc κ_{CA} [rad/m]	Linear coordinate L_{BTC} [m]	Linear coordinate L_{ETC} [m]	Curve length l_{TC} [m]
5	I	0.001377848	101.622975	223.043003	121.420028
5	II	0.001421427	83.242574	234.145518	150.902944
10	I	0.001359864	102.094964	220.529044	118.434080
10	II	0.001353549	96.013029	227.433211	131.420182
20	I	0.001352841	101.012621	218.985241	117.972620
20	II	0.001352548	101.845477	220.435085	118.589608
30	I	0.001351177	100.845246	218.965940	118.120695
30	II	0.001351298	100.896480	219.541052	118.644572
40	I	0.001351399	100.490984	219.670554	119.179570
40	II	0.001351005	99.899260	220.264771	120.365511
50	I	0.001351428	100.012080	220.118360	120.106285
50	II	0.001351800	99.337387	219.873610	120.536222

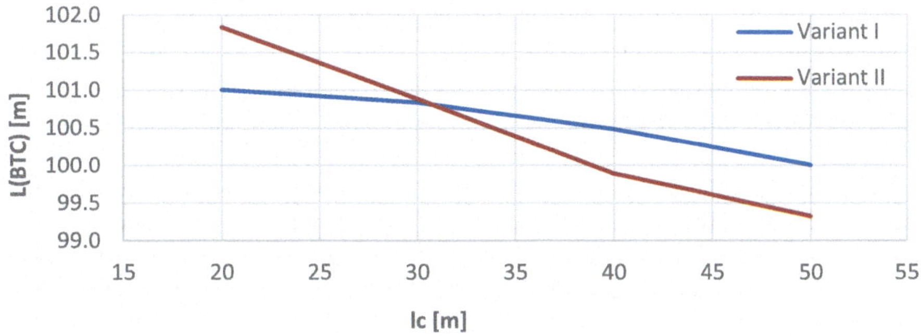

Fig. 4.16 Diagrams of the L_{BTC} linear coordinate for different lengths of the moving chord

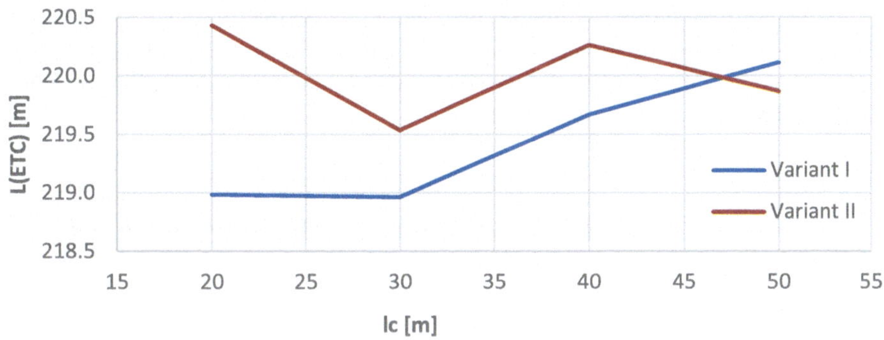

Fig. 4.17 Diagrams of the L_{ETC} linear coordinate for different lengths of the moving chord

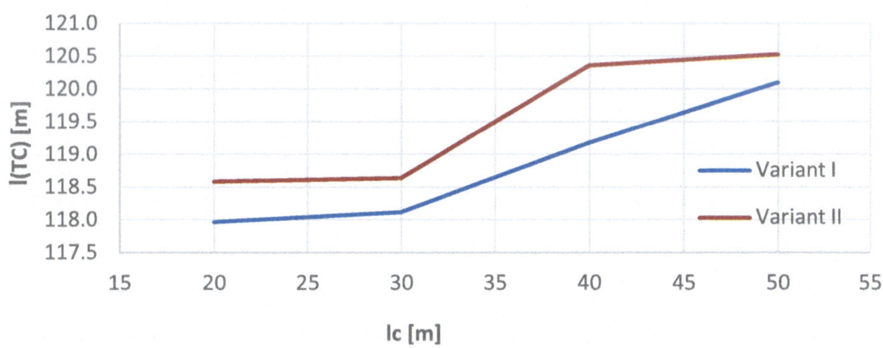

Fig. 4.18 Graphs of the determined length l_{TC} of transition curve for different lengths of the moving chord

decrease with the increase of l_c. For variant II, the L_{BTC} coordinates also regularly decrease with the increase of l_c, but they occupy a larger range, i.e., 2.5 m.

Figure 4.17 shows that the determined values of the linear coordinates of the end of the transition curve tend to reach $L_{ETC} = 220$ m. For variant I, they regularly increase with the increase of l_c, while for variant II their change is less regular. For the length of the moving chord $l_c = 50$ m, the obtained L_{ETC} values are not much different.

Determined as the difference of L_{ETC} and L_{BTC} coordinates, the length of the transition curve increases with the increase of l_c (Fig. 4.18). All of them are in the range 2.5 m. For the chord length $l_c = 50$ m, they reach values close to $l_{TC} = 120$ m (this applies in particular to variant I, in variant II the l_{TC} is higher by 43 cm than in variant I).

The presented analysis shows that with the length of the moving chord $l_c \geq 20$ m, the quality of the measurement data—while maintaining the maximum error within reasonable limits—does not play a significant role in identifying the transition curve, i.e., determining the linear coordinates L_{BTC} and L_{ETC} and the length l_{TC} of the curve. This means that a given geometric layout can be fully identified. Similarly, to the identification of a circular arc, better results are obtained by using a longer chord than would result from the radius of the arc.

4.8 Direction of Further Development of the Moving Chord Method

The possibility of using the moving chord method to determine the curvature of the track axis on the basis of mobile satellite measurements remains to be clarified. This will be the direction of further development of this method in the near future. Determining the coordinates of the track axis using mobile GNSS measurements undoubtedly has a future ahead of it, despite the problems associated with the decrease in signal quality due to the presence of terrain barriers. However, the identified inconveniences when using this measurement technique cannot hide the fact that they provided—in a very short time—a huge number of coordinates of measurement points. This in turn should increase the accuracy of the moving chord method. In this method, the key problem is to determine the coordinates of the end of the chord—this is done by interpolation carried out in the appropriate interval, and the length of such an interval decreases with the increase in the number of points.

The first attempts related to this direction of research have already been undertaken. Data from the campaign of mobile satellite measurements described in Specht and Koc (2016) were used. Figure 4.19 shows the effect of identifying a selected track section.

The measurements underlying the curvature graphs in Fig. 4.19 were made using satellite receivers with a frequency of 5 Hz. Coordinate values were therefore obtained at intervals of 0.2 s. Today receivers with a frequency of 100 Hz are already commonly used. Therefore, new measurements, using current technical possibilities, should provide rich research material. It would made it possible to draw valuable conclusions

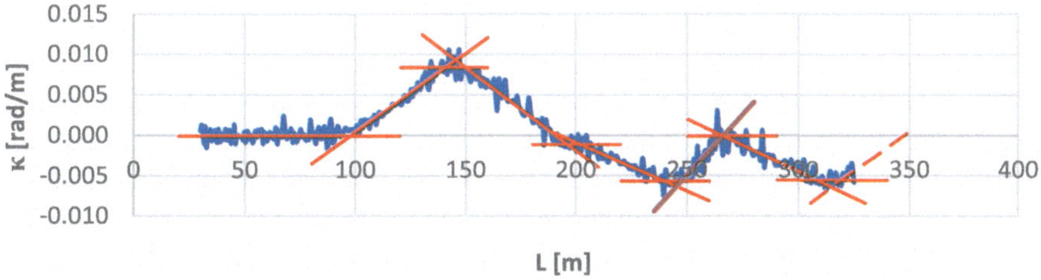

Fig. 4.19 Identification of a selected track section using the moving chord method based on mobile satellite measurements

regarding both the effectiveness of the moving chord method and the accuracy of mobile satellite measurements.

4.9 Conclusion

The solution to the problem of identifying the measured railway route currently consists of generating an optimized geometrical layout in the considered area by minimizing the deviations of this layout from the measurement points, while meeting the appropriate maintenance and operational requirements. The traditional solution to the problem is largely based on human experience and requires overcoming many difficulties, which is why numerous attempts are made to automate this process. There are two variants of the procedure: geometric identification and alignment optimization. These variants may be implemented separately or simultaneously.

An effective solution to the problem of identifying the geometric layout of the track is possible thanks to the use of the obtained measurement data (i.e., Cartesian coordinates) to determine the curvature of the track axis in the horizontal plane. A number of works have demonstrated the high efficiency of the proposed new method for determining the curvature of the railway track axis—the so-called moving chord method. This paper presents the theoretical basis of this method and the defined scope of its applicability. As it turns out, the moving chord method corresponds to the conditions on railway roads.

Appropriate analyses concerning the application of the moving chord method for model geometrical layouts (described with mathematical equations) showed complete compliance of the obtained curvature diagrams with the expected values, which are the basis of the given geometric solution. This clearly confirmed the correctness of the method from the substantive point of view. The main advantage of the discussed method is its unambiquity, resulting from a direct reference to the definition of curvature. Used in the process of geometric identification of the railway route, it allows for direct determination of the location of the so-called segmentation points, i.e., places where individual geometric elements connect with each other (e.g., straight sections with transition curves, transition curves with circular arcs, etc.). In other identification methods, this is the biggest problem with it.

This paper also addresses the issue of explaining the influence of deviations from the model (theoretical) shape of the track on the obtained curvature values, occurring in the real (i.e., in operation) railway track. This can be treated as an element of the analysis of the sensitivity of the discussed method to these deviations which are the result not only the existing deformations of the track, but—mainly—of the measurement error. The work compares the curvature values for two generated geometrical layouts with different quality of measurement data, expressed in different values of the maximum error. Different lengths of the moving chord were assumed in the calculations. It turned out that in all cases, the oscillation of the curvature relative to the hypothetical values in the variants with poorer quality of measurement data is characterized by higher amplitude values. For short chords ($l_c = 5$ m and $l_c = 10$ m) full identification of the geometric layout based on the curvature diagram would be very difficult (and burdened with the large error) regardless of the quality of the data used. They were confirmed by the finding contained in the work (Koc 2022a) that in the operated railway track the length of the moving chord should meet the condition $l_c \geq 20$ m.

The analysis shows that the poorer quality of the measurement data, characterized by a higher value of the maximum error, causes greater disturbances in the curvature diagram determined by the moving chord method. These disturbances can be significantly reduced by appropriately increasing the chord length used. However, when deciding to use a longer chord than would result from the radius of the arc, one should take into account the limitations resulting from the theoretical basis of the method of curvature determination used. As shown, the greater the length of the chord used, the greater the differences between the length of the chord (used in the curvature formula) and the length of the corresponding circular arc.

As stated, with the length of the moving chord $l_c \geq 20$ m, the quality of the measurement data—while maintaining the maximum error within reasonable limits—basically does not play a significant role in determining the linear coordinates of the segmentation points. This means that a given geometric layout can be fully identified.

The possibility of using the moving chord method to determine the curvature of the track axis on the basis of mobile satellite measurements remains to be clarified. This will be the direction of further development of this method in the near future. The first attempts related to this direction of research have already been undertaken.

References

Camacho-Torregrosa FJ, Pérez-Zuriaga AM, Campoy-Ungría JM, García A, Tarko AP (2015) Use of heading direction for recreating the horizontal alignment of an existing road. Comp Aided Civil Infrastruct Eng 30(4):282–299

Cellmer S, Rapinski J, Skala M, Palikowska K (2016) New approach to arc fitting for railway track realignment. J Surv Eng 142(2):1–24
European Committee for Standardization (CEN): Railway applications-Track—Track alignment design parameters—Track gauges 1435 mm and wider. Part 1: Plain line. EN 13803-1. Brussels (2010)
Deutsche Bahn: 883.2000 DB_REF-Festpunktfeld [in German]. Deutsche Bahn Netz AG, Berlin (2016)
Dong H, Easa SM, Li J (2007) Approximate extraction of spiraled horizontal curves from satellite imagery. J Surv Eng 133(1):36–40
Easa SM, Wang F (2011) Fitting composite horizontal curves using the total least-squares method. Surv Rev 43(319):67–79
Federal Railroad Administration: Code of federal regulations title 49 transportation. US Government Printing Office, Washington (2008)
Fu BT (2016) A chord curvature model for silhouette curvature calculations. In: Electronics, communications and networks V. Springer, Singapore, pp 11–17
Gikas V, Stratakos J (2012) A novel geodetic engineering method for accurate and automated road/railway centerline geometry extraction based on the bearing diagram and fractal behawior. IEEE Trans Intell Transp Syst 13(1):115–126
Guimarães-Steinicke C, Weigelt A, Ebeling A, Eisenhauer N, Duque-Lazo J, Reu B, Roscher C, Schumacher J, Wang C (2019) Chapter Four—Terrestrial laser scanning reveals temporal changes in biodiversity mechanisms driving grassland productivity. Adv Ecol Res 61:133–161
Guo J, Zhong B (2014) U-chord curvature: a computational method of discrete curvature. Int J Pattern Recognit Artif Intell 27(8):683–691
Koc W (2020) The method of determining horizontal curvature in geometrical layouts of railway track with the use of moving chord. Arch Civ Eng 66(4):579–591
Koc W (2021a) Analysis of the effectiveness of determining the horizontal curvature of a track axis using a moving chord. Problemy Kolejnictwa – Railway Rep 65(190):77–86
Koc W (2021b) Analysis of moving chord inclination angles when determining curvature of track axis. Curr J Appl Sci Technol 40(10):92–103
Koc W (2022a) Determination of moving chord length for determining the curvature of in-service railway track. Problemy Kolejnictwa – Railway Rep 66(197):147–160
Koc W (2022b) Estimation of the horizontal curvature of the railway track axis with the use of a moving chord based on geodetic measurements. J Surv Eng 148(4):04022007
Koc W (2022c) Identification of geometrical parameters of an operational railway route determined by the curvature of the track axis. Eur J Appl Sci 10(5):129–148
Koc W (2022d) Designing the railway line direction changing area in linear and Cartesian coordinate systems. Eur J Appl Sci 10(4):277–296
Koc W (2023a) Verification of the theoretical basis for determining the curvature of the track axis using the moving chord method. Eur J Appl Sci 11(3):577–597
Koc W (2023b) The procedure of identifying the geometrical layout of an exploited railway route based on the determined curvature of the track axis. Sensors 23(1):274
Lai X, Schonfeld P (2016) Concurrent optimization of rail transit alignments and station locations. Urban Rail Transit 2(1):1–15
Lenda G (2014) Determining the geometrical parameters of exploited rail track using approximating spline functions. Arch Civ Eng 60(3):295–305
Li W, Pu H, Schonfeld P, Song Z, Zhang H, Wang L, Wang J, Peng X, Peng L (2019) A method for automatically recreating the horizontal alignment geometry of existing railways. Comp Aided Civil Infrastruct Eng 34(1):71–94
Li W, Zhen S, Schonfeld P, Pu H, Zhang Z, Zhao L, Qiu X, Wei F, Yan W (2022) Recreating existing railway horizontal alignments automatically using overall swing iteration. J Transp Eng Part A Syst 148(8):04022046
Luo W, Li L, Wang K (2018) Automatic horizontal curve identification and measurement using mobile mapping system. J Surv Eng 144(4):04018007

Moritz H (2000) Geodetic reference system 1980. J Geodesy 74:128–133
Network Rail: NR/L3/TRK/0030 NR reinstatement of absolute track geometry (WCRL Routes), issue 1. London (2008)
Othman S, Thomson R, Lannér G (2012) Using naturalistic field operational test data to identify horizontal curves. J Transp Eng 138(9):1151–1160
Quan Y, Lau L (2019) Development of a trajectory constrained rotating arm rig for testing GNSS kinematic positioning. Measurement 140:479–485
Regulation of the Council of Ministers of 15 October 2012 on the national spatial reference system [in Polish], Warszawa. J Laws 2012:1247 (2012)
Skala-Szymanska M, Cellmer S, Rapinski J (2014) Use of Nelder-Mead simplex method to arc fitting for railway track realignment. In: Selected papers of the 9th international conference environmental engineering 2014, Vilnius, Vilnius Gediminas Technical University Press Technika
Song Z, Yang F, Schonfeld P, Li J, Pu H (2020) Heuristic strategies of modified Levenberg-Marquardt algorithm for fitting transition curves. J Surv Eng 146(2):04020001
Song Z, Fang T, Schonfeld P, Li J (2021) Effect of point configurations on parameter estimation analysis of circles. J Surv Eng 147(3):04021010
Specht C, Koc W (2016) Mobile satellite measurements in designing and exploitation of rail roads. Transp Res Proc 14:625–634
Starlino Electronics Homepage: A guide to using IMU (accelerometer and gyroscope devices) in embedded applications. http://www.starlino.com/imu_guide.html. Accessed July 18, 2023
Polish State Railways: Technical Standards – detailed technical conditions for the modernization or construction of railway lines for speed $V_{max} \leq 200$ km/h (for conventional rolling stock)/ 250 km/h (for rolling stock with a tilting box). Volume I—Attachment ST-T1-A6: Geometrical layouts of tracks [in Polish]. Polish Railway Lines, Warszawa (2018)
Szwilski AB, Dailey P, Sheng Z, Begley RD (2005) Employing HADGPS to survey track and monitor movement at curves. In: Forde MC (ed) The eight international conference railway engineering–2005, London
Turiño CE (2008) Gauss Krüger projection for areas of wide longitudinal extent. Int J Geogr Inf Sci 22(6):703–719
Wu S, Zhao X, Pang C, Zhang L, Xu Z, Zou K (2020) Improving ambiguity resolution success rate in the joint solution of GNSS-based attitude determination and relative positioning with multivariate constraints. GPS Solutions 24(1):31
Zhang H, Pu H, Schonfeld P, Song T, Li W, Wang J, Peng X, Hu J (2020) Multi-objective railway alignment optimization considering costs and environmental impacts. Appl Soft Comput 89:106105
Zhong B (2010) On the stability of refined L-curvature under rotation transformations. Appl Mech Mater 20–23:401–406

Chapter 5
Deterministic and Probabilistic Forced Response Analysis of a Railway Coach

Srihari Palli, Ramji Koona, Azad Duppala, Chintada Vinod Babu, Rakesh Chandmal Sharma, and Ramajogi Naidu Doddi

5.1 Introduction

Study of dynamic behavior of railway vehicle or any other ground vehicle includes free as well as forced vibration analysis. Free vibration analysis of railway coach determines the natural frequencies and mode shapes of the railway coach. Forced vibration analysis consists of harmonic, transient dynamic and random response analyses. The innovative approaches to transportation, from the feasibility of the Mumbai-Pune Hyperloop corridor and waste-to-biofuel production to advanced vibration control in marine and rail systems, highlight significant strides in sustainable and intelligent transportation technologies (Badgujar et al. 2024; Kokkinos et al. 2024; Raju Rallabandi et al. 2024; Rallabandi et al. 2024; Sharma et al. 2024a, b, c, d, e, f, g; Upadhyay et al. 2024a, b). Study of harmonic response of railway coach is a characteristic of a vibratory system, which is the result of a dynamic load applied with a known frequency. Harmonic response analysis gives the ability to predict the sustained dynamic behavior of structures, thus enabling to verify if the structure designs can successfully overcome resonance, fatigue and other harmful effects of forced vibrations. Vibration response of the railway vehicle subjected to cyclic force

S. Palli · A. Duppala (✉) · R. N. Doddi
Aditya Institute of Technology and Management, Tekkali, Andhra Pradesh 532201, India
e-mail: duppalaazadin@gmail.com

R. Koona
Andhra University College of Engineering, Visakhapatnam, Andhra Pradesh 530003, India

C. V. Babu
GMR Institute of Technology, Rajam, Andhra Pradesh 532127, India

R. C. Sharma
Mechanical Engineering Department, Graphic Era (Deemed to Be) University, Dehradun 248002, India

Mechanical Engineering Department, Graphic Era Hill University, Dehradun 248002, India

S. K. Sharma et al. (eds.), *Dynamics of Transportation Ecosystem, Modeling, and Control*, Energy, Environment, and Sustainability,
https://doi.org/10.1007/978-981-97-0437-8_5

or sinusoidal inputs is carried in some research studies (Cooperrider 1975; Durali and Mahdi 2008; Li and Young 2003; Lu and Gill 1993; Pombo and Ambrósio 2005; Popp et al. 2003, 2005; Ramji et al. 2007; Sewall 1972). Equations of motion developed in formulating the bogie model by researchers (Popp et al. 2003, 2005) were reduced analytically to algebraic equations under vertical harmonic excitation to calculate the frequency response function. Recent studies have significantly contributed to the understanding of rail vehicle and the improvement of railway systems (Palli et al. 2018; Sharma et al. 2015, 2017, 2018; Sharma and Sharma 2018a, b, c; Sharma and Kumar 2017a, 2018a, b).

Due to the settlement of subgrade, aging, rail wear, wheel flats, sharp and deep wheel flanges, and ballast loosening variations in geometry of the track get developed. In the vertical and lateral profiles, track irregularity or abnormality is a variation from a long-term average level. When random rail irregularities are considered, the random response of the railway vehicle can be readily computed using frequency response functions. The random response studies (Catherines and Clevenson 1969; Cooperrider 1975; Gangadharan et al. 2004; Goel et al. 2005; Hemantha Kumar et al. 2005; Iyengar and Jaiswal 1995; Karuppaiah et al. 1999, 2003, 2011b, 1968; Law and Cooperrider 1974; Lee and Pradko 1968; Lu and Gill 1993; MacMillan 1962; Mixson and Steiner 1969; Pearce and May 1969; Pervozvanskii 1965; Sharma 2011a; Stassen 1967; Van Deusen 1967) conducted on railway vehicle consider random irregularities in track, ballast loosening and aging. Recent advancements in the fields of vehicle dynamics and railway infrastructure have been documented in various studies (Bhardawaj et al. 2023; Hossain et al. 2023; Lee and Sharma 2023; Sharma et al. 2023c, d, e, f, g, h; Upadhyay et al. 2023a, b; Wu et al. 2023). The author (Sharma 2011a, b) studied the coupled vertical-lateral dynamics of rail vehicle using Lagrangian method for a 37 DOF Indian railway full coach model subjected to standard PSD (Wickens 1971) excitation.

In the present study, harmonic response of an Indian railway ICF (Integral Coach Factory) sleeper coach is found by giving sinusoidal excitation to the identified wheel rail contact nodes in the form of frequency and amplitude. Rail vehicle dynamics generally takes into account low frequency in the range of 0–100 Hz (Cooperrider 1975; Kouroussis et al. 2015; Li and Young 2003). However, dominance of amplitude for locomotives lies in the frequency range of 0–50 Hz (Indian Railways Institute of Civil Engineering 2016; Ministry of Indian Railways Website 2023; Pradko et al. 1966) due to which this range is preferable for analysis.

5.2 Harmonic Analysis of Railway Coach

The CAD model of the ICF railway sleeper coach is translated into ANSYS for FE analysis. Two nodes on each of the four wheels of the railway bogie are given a maximum vertical amplitude of 25 mm with reference to Indian railway standard literature (Gangadharan 2001; Ramji et al. 2007) in a frequency range of 0–50 Hz as shown in Fig. 5.1.

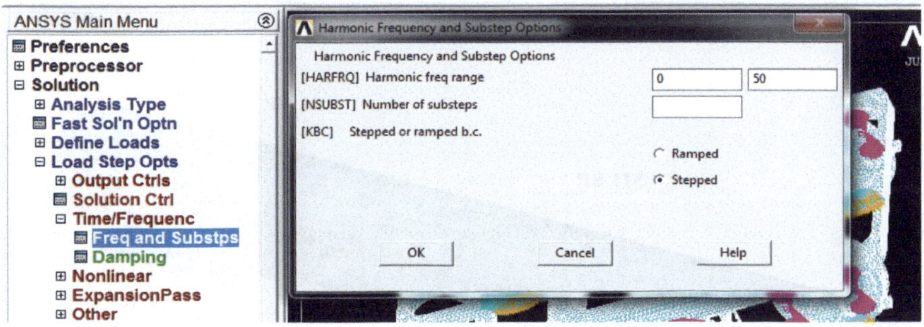

Fig. 5.1 Harmonic frequency range of the railway coach

The response to the harmonic excitation at salient locations on the car body is studied. As the effect of stiffness and damping provided on the bogie is of interest against the harmonic loads, accordingly these key locations are identified. Those salient points are:

- Three points on the left side panel of coach shell, namely FLSP (Front left side panel), MLSP (Middle left side panel) and RLSP (Rear left side panel) approximately on the front portion, center portion and rear portion of the left side panel of the coach shell. The comprehensive studies on wear mitigation in locomotive components, lateral stability, ride comfort, and advanced suspension systems have significantly contributed to enhancing the performance and safety of rail and road vehicles (Kumar et al. 2022; Sharma et al. 2022e, f, g; Sharma and Sharma 2022; Vishwakarma et al. 2022a, b). Due to symmetric geometry, right-side panel portions of the coach shell are not considered (Adeoye et al. 2023a; Gopala Rao et al. 2022; Kohli et al. 2022; Mohapatra et al. 2016a, b; Sharma and Kumar 2017b; Sharma et al. 2020c).
- Three points on the front, middle and rear portions of the underframe of the coach, i.e., CFUF (Coach front underframe), CMUF (Coach middle underframe) and CRUF (Coach rear underframe).

Passenger berths are fixed to the underframe and hence locations on underframe are selected for analysis (Sharma and Palli 2016). Side panel locations of the coach are selected to determine the dynamic effect on the sheet metal body of the coach shell (Bandekar et al. 2017; Bhardawaj et al. 2020a; Hossain et al. 2017; Mohapatra et al. 2017a; Sharma and Kumar 2018c; Sharma and Sharma, 2019; Sharma et al. 2021a, 2022a). These locations are selected to analyze the response at different positions of the car body so as to determine which of those locations are relatively comfortable. Figure 5.2 indicates the identified salient locations of the coach.

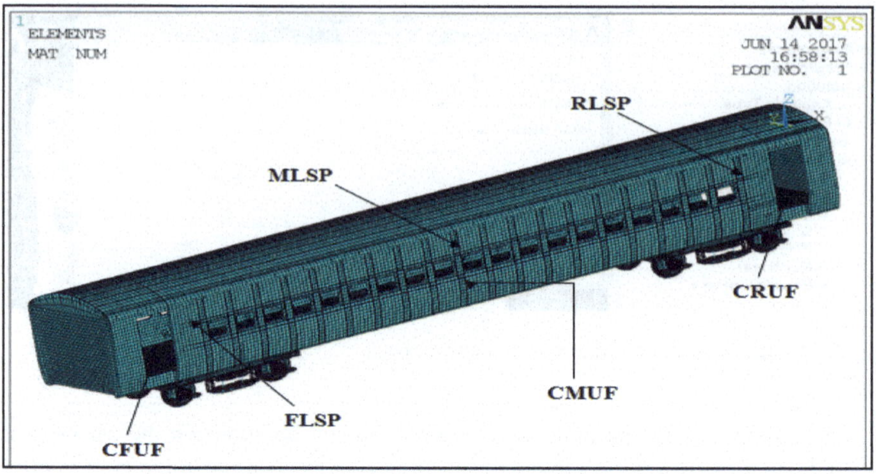

Fig. 5.2 Salient points on the railway coach FE model for harmonic response

5.2.1 Harmonic Response of Coach in Unladen Condition

Variations in harmonic response of the coach in unladen condition, i.e., without considering the 72 passengers' weight are plotted for the identified salient locations of the coach. The recent advancements in rail vehicle system modeling, structural analysis, and performance evaluation have been extensively studied in various contexts, including crashworthiness, lateral stability, critical speed analysis, and multi-objective optimization (Goyal et al. 2019; Sharma 2019; Sharma et al. 2019; Bhardawaj et al. 2020b, 2021; Lee and Sharma 2020; Choi et al. 2021; Lee et al. 2021; Sharma and Sharma 2021). Under the influence of sinusoidal excitation fed to the wheel bottom points, the vertical displacement response at the key locations is obtained (Goswami et al. 2019; Mohanty et al. 2016; Nibedita et al. 2021; Pattnaik et al. 2022; Samantaray et al. 2014; Sharma et al. 2022b; Sinha et al. 2019; Swapna Rekha et al. 2020). The frequency response is obtained through the time history post-processor module in ANSYS in CSV (comma separated value) format. This time history data are used in plotting the harmonic response plots of the underframe and side panel of the coach model.

Variations in the vertical displacement of the front, middle and rear portions of the underframe, i.e., the floor of the coach are plotted in Fig. 5.3. The response peaks are observed to be attained at frequencies matching with the natural frequencies. The first harmonic peak is observed at around 0.6 Hz, which matches with the bounce mode frequency (0.616 Hz) in unladen condition (Agrahari et al. 2021a; Biswas et al. 2021; Das et al. 2023; Pundir et al. 2023a; Sharma et al. 2020b, 2021b, 2022c). The second response peak of CFUF, CMUF and CRUF is observed at around 3.25 Hz, which is close to roll mode frequency (3.33 Hz) of the unladen coach. The amplitude difference of middle portion of underframe with that of front and rear portions of underframe is there because at middle portion of the underframe the excitation forces are balanced (Devadarshini et al. 2022; Kohli et al. 2020; Maity et al. 2020; Misra et al. 2022; Mohapatra et al. 2017b; Sharma and Lee 2020d; Sharma et al. 2021d, 2022d) whereas the front and rear portions of the underframe are vulnerable to excitations. Subsequent response peaks of CFUF, CMUF and CRUF also are observed to be matching with natural frequencies of the coach.

Figure 5.4 illustrates the response on the front, rear and middle portions of the side panel to the left of the coach shell. The points FLSP, RLSP on the front and rear portions of the left side panel along with the middle portion of side panel MLSP display the first response peak at around 0.6 Hz, which agrees with the bounce mode frequency (0.616 Hz) of the coach in unladen condition. The second harmonic peak is observed at 3.25 Hz which is close to roll mode frequency (3.33 Hz). The substantial difference in amplitude of MLSP located at the middle portion of side panel compared with that of FLSP and RLSP indicates that front and rear ends of the side panel of the coach shell are more vulnerable to excitations as per literature (Acharya et al. 2020; Bhardawaj et al. 2019a, 2020b, 2021; Sarkar et al. 2018; Palli et al. 2022; Pundir et al. 2023b; Sharma and Lee 2020b). The ensuing response peaks of the side panel key points also agree with the natural frequencies obtained for unladen coach in modal analysis. The amplitude pattern for front and rear portions of coach is similar and the amplitude pattern of middle portion of coach is at lesser height over the entire frequency range. The middle location of coach is in vicinity to the CG of the coach and the force distribution is balanced (Mohanpatra et al. 2020a, b; Pattnaik et al. 2020; Sharma et al. 2020a) due to which the middle portion's displacement response is less.

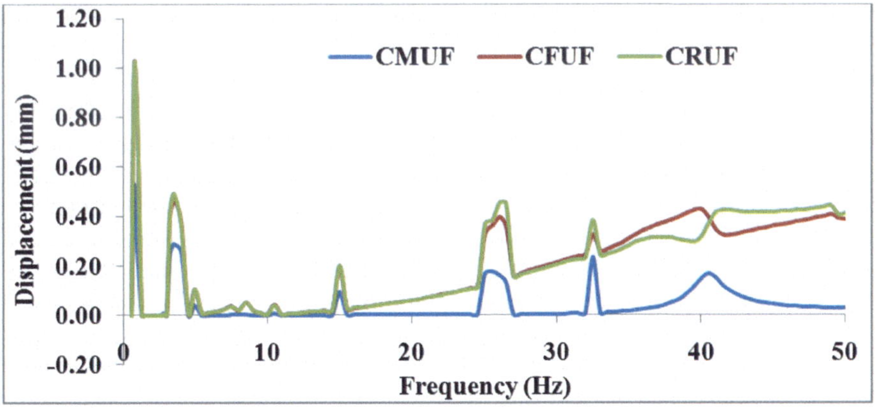

Fig. 5.3 Amplitude versus frequency plot for coach underframe (unladen condition)

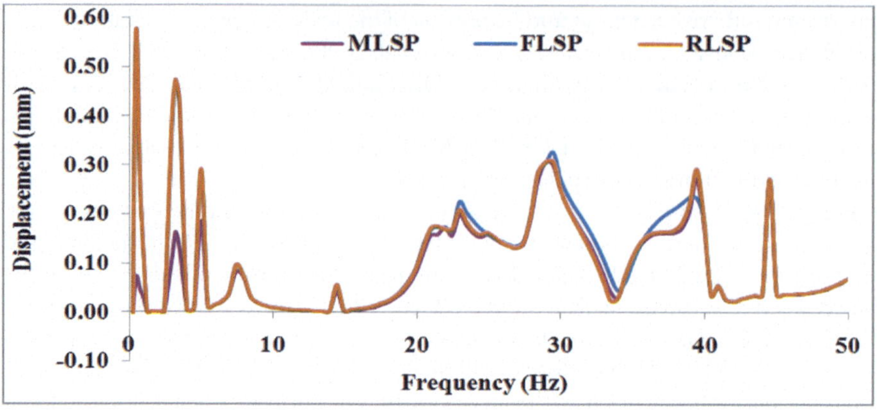

Fig. 5.4 Amplitude versus frequency plot for coach side panel (unladen condition)

5.2.2 Harmonic Response of Coach in Laden Condition

The harmonic response studies of the coach are carried out by considering the weight of all the 72 passengers of the sleeper coach. The average weight of a passenger is assumed as 65 kg, thus lumping 4.68 tons at the CG of the coach. In ANSYS APDL, mass coupling is done to apply the total passenger load using CEqn (Constraint Equation) element, so that the laden condition is simulated.

The plot in Fig. 5.5 indicates amplitude variations with respect to frequency of the car body underframe under laden condition of the coach. The first resonant peak is attained at 0.58 Hz, which is in agreement with bounce mode frequency (0.583 Hz) for the front (CFUF), rear (CRUF) and center or middle (CMUF) locations of the underframe. The next resonant peak is observed at around 2.9 Hz, which is close to the roll mode frequency (2.872 Hz). Subsequent response peaks are also in agreement with the natural frequencies of the railway coach model subjected to laden condition.

The variation in amplitude with respect to frequency for the side panel of the coach shell under the influence of passenger load in laden condition is plotted in Fig. 5.6. The side panel's front (FLSP), rear (RLSP) and middle (MLSP) portions attain first harmonic peak at 0.58 Hz, which is in agreement with the bounce mode frequency (0.583 Hz) of the coach. The overall response of coach in laden condition shows low disturbances than the unladen coach due to the contribution of passenger mass (Gangadharan 2001; Poop et al. 2005; Ramji et al. 2007). The lower amplitude of MLSP compared to that of FLSP and RLSP is because at the middle of coach shell's side panel the excitation forces are balanced whereas the front and rear portions of the side panel are more susceptible to disturbances as these locations are farther from center of gravity of the coach.

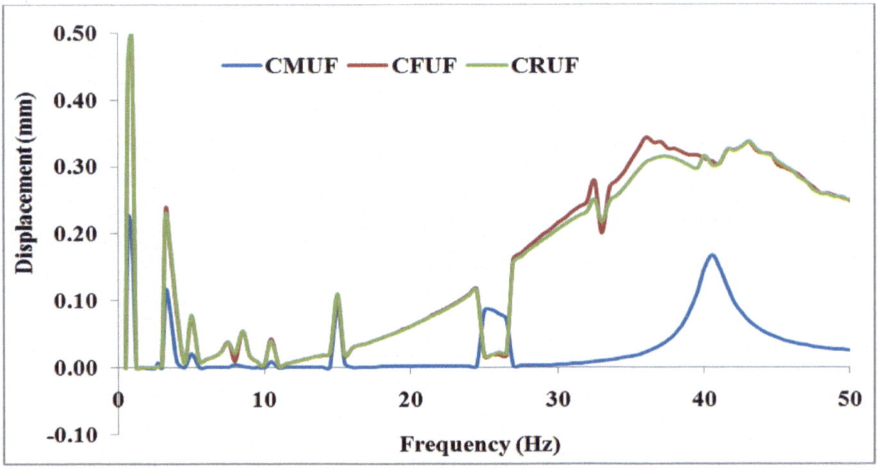

Fig. 5.5 Amplitude versus frequency plot for coach underframe (laden condition)

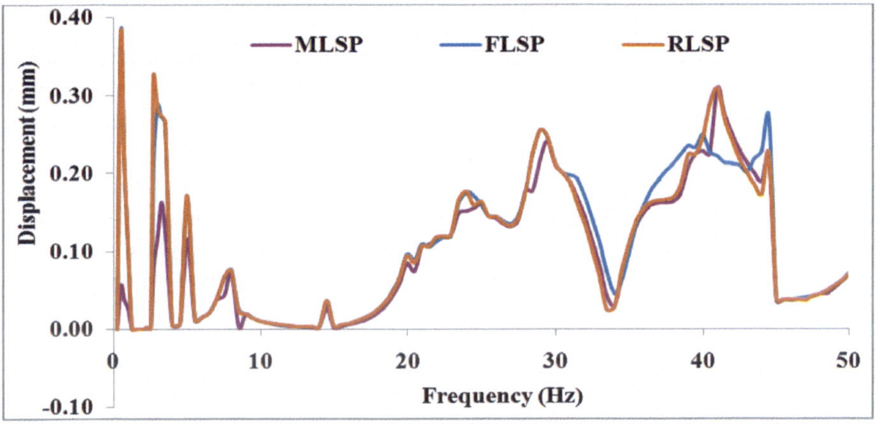

Fig. 5.6 Amplitude versus frequency plot for coach side panel (laden condition)

5.2.3 Results and Discussion

Harmonic response of the bogie and coach models subjected to sinusoidal excitation at wheel contact points is obtained considering both unladen and laden conditions. From the results observed of harmonic analysis carried out using FEA in ANSYS, it can be concluded that the first resonant peak attained at the bounce mode frequencies of bogie and coach under unladen and laden conditions, respectively. Further, response peaks attained also match with the respective natural frequencies of the bogie and coach in both unladen and laden conditions.

Table 5.1 Coach harmonic response (first peak at bounce mode frequency and displacement)

Location on coach		First harmonic response peak frequency (Hz)	Matched natural frequency from modal analysis in literature (Hz)	Vertical displacement (mm)
Unladen coach underframe	CFUF	0.6	0.616	1.007
	CRUF	0.6	0.616	1.007
	CMUF	0.6	0.616	0.523
Unladen coach side panel	FLSP	0.6	0.616	0.571
	RLSP	0.6	0.616	0.572
	MLSP	0.6	0.616	0.072
Laden coach underframe	CFUF	0.58	0.583	0.442
	CRUF	0.58	0.583	0.456
	CMUF	0.58	0.583	0.223
Laden coach side panel	FLSP	0.58	0.583	0.382
	RLSP	0.58	0.583	0.379
	MLSP	0.58	0.583	0.056

Table 5.1 shows the first resonant peak at bounce mode frequency values and vertical displacements of the coach underframe and side panel. From the railway coach harmonic analysis, it is stated that the first resonant peak is attained at frequency which is in good agreement with bounce mode frequency at various locations throughout the coach. The results obtained from the harmonic analysis of the coach indicate that the side panel and underframe of the coach under both unladen and laden conditions are subjected to disturbances with higher displacement in the low frequency range of 0–10 Hz. The front and rear portions of both the underframe and side panel of coach are subjected to the highest displacement at the first resonant peak frequency due to their vulnerability to excitations supported by research studies (Adeoye et al. 2022, 2023b, c; Chatterjee et al. 2022; Mohapatra et al. 2017c, d, 2022; Nayak et al. 2022; Pati et al. 2020, 2022).

5.2.4 Summary of Harmonic Analysis

Harmonic response of the ICF railway bogie and coach models is extracted in both the load conditions as further step to the modal analysis. Sinusoidal input as per research studies of Indian railways context (Bhardawaj et al. 2020c; Kanth Choppara et al. 2019; Palli and Koona 2015; Palli et al. 2021; Sharma et al. 2020d, e, 2021c) is fed as input to two nodes on each wheel of the bogie and the harmonic response amplitude is obtained for the coach. The displacement at prominent locations on the coach side panel and underframe are plotted against frequency. In both unladen and laden conditions, the middle or center portion of the coach is subjected to lower displacements indicating the uniform and balanced distribution of dynamic forces at the middle of the coach due to its proximity to center of gravity of the coach.

5.3 Random Response Analysis

Track longitudinal profile is also subjected to creep forces in the direction of the train movement due to the wheel-rail contact. Track irregularity is one of the major causes that influence the train-track dynamics. Profile irregularities make the vehicles vibrate, which in turn induce stochastic forces on the track leading to further deterioration of the track and reduction of comfort. Track irregularities include four random parameters, namely, vertical profile, alignment, cross-level and gauge (Goel et al. 2005) as shown in Fig. 5.7. These four random fields are correlated. However, in practice, the vertical irregularity is used extensively.

5.3.1 Power Spectral Density

In vibration analysis, PSD is signal Power Spectral Density. Each phrase represents a crucial PSD component. The PSD magnitude is the signal's mean-square value, which represents power. It doesn't mean power (watts or horsepower). Since power is proportional to the mean-square value of some quantity (such as the square of current or voltage in an electrical circuit), any quantity's mean-square value equals its power. PSD is spectral because frequency affects it. PSD indicates a signal's frequency distribution, like a rainbow's wavelength dispersion. PSD magnitude is standardized to a single hertz bandwidth, defining density. A signal measuring acceleration in g's has a PSD of g^2/Hz (Bhardawaj et al. 2019b; Mohapatra et al. 2015, 2021).

PSD does not mention the amount being measured; hence power is sometimes replaced by its name. The PSD of an acceleration signal is often called its Acceleration

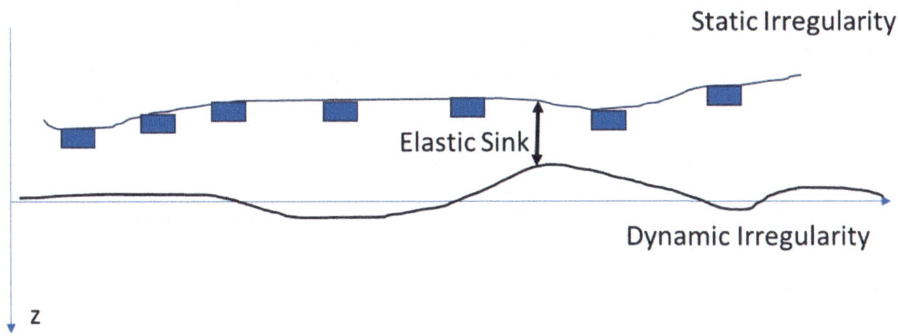

Fig. 5.7 Track irregularities

Spectral Density. The mean-square value (power) is a handy signal strength metric. Since the mean value is approaching 0, it cannot determine the signal's average amplitude. Instead, the signal is squared (resulting in a positive number) and the mean is calculated. To get a linear number (g's in this example), take the square root to get the RMS (Agarwal et al. 2020; Agrahari et al. 2021b; Bhattarai et al. 2022; PanduRanga Vital et al. 2019; Sharma et al. 2023a).

Combine signals of different frequencies using the mean-square value. In resonance-based systems, signal frequency distribution is useful. From the signal's temporal history, beam resonance frequencies are hard to calculate. However, the tip vibration frequency spectrum peaks clearly demonstrate resonance frequencies. Signal frequency dispersion magnitude relies on frequency band count. Squared spectra magnitudes are proportional to frequency bandwidth. The PSD divides the squared magnitude by the frequency bandwidth to provide a constant value regardless of bandwidth (Lokanadham et al. 2022; Satyanarayana et al. 2022; Sharma et al. 2023b).

Railway track profiles are random and characterized statistically. Track variation may be described statistically using power spectral density. Power spectral density functions show railway track abnormalities. PSD curves are usually specified or measured and computed using vibration analysis equipment.

5.3.2 Random Inputs to the Railway Coach

Linear Variable Differential Transducers (LVDT) in Track Recording Cars (TRCs) by the research wing of Indian Railways, Research Designs and Standards Organization (RDSO), Lucknow (India), measure random track geometry variations at 0.404 m. Rail engineers are interested in track profiles with wave lengths from 0.1 to 100 m. TRC's software analyses transducer signals processed by hardware signal conditioners.

The regular track maintenance is carried out by tamping machines with nominal periodicity (Indian Railways Institute of Civil Engineering 2016). Tamping machines are heavy on track machines which are deployed by zonal railways as per the predefined schedule which is circulated to railway divisions. The track maintenance periodicity is dependent on the following relevant factors:

a. Traffic density, axle load and maximum permissible speed of trains.
b. Track geometry (gradients and curvatures).
c. Track structure and condition of track components.
d. Depth of ballast cushion, availability of clean cushion and arrears of deep screening, shoulder cleaning.
e. Type of formation, i.e., embankment or cutting, type of soil, condition and stability of formation.
f. Climate condition such as rainfall, snowfall, etc.

g. Local conditions such as adjoining built-up area, trespassing and drainage problems.
h. Mode of tamping, i.e., smoothening mode or design mode and dynamic track stabilization after tamping.

Track geometry variations due to rail deterioration, rail corrugation, shelling and surface squats are identified and accordingly maintained in the pre-monsoon, during monsoon and post-monsoon seasons. Such identification of track geometry variation is convenient for construction of new tracks or maintenance of old tracks, but is not sufficient for the evaluation of vehicle ride behavior and passenger comfort. Vertical irregularities are represented by the equation of the type $S(\Omega) = C_{sp}\Omega^{-N}$ (Goel et al. 2005). $S(\Omega)$ is the spectral density at frequency Ω, and C_{sp} is the empirical constant. These PSD estimates are specified in mm^2/cycle and are converted into root mean square acceleration estimates specified in g^2/Hz by the in-built software in track recording car of RDSO (Goel et al. 2005; Indian Railways Institute and of Civil Engineering 2016). The flowchart of the PSD conversion in software is explained in Fig. 7.2. Signals of interest for track geometry parameters are in the frequency range from 0.01 to 10 cycles/m. Since the signals are sampled at a distance of 0.404 m, the maximum frequency that could be analyzed is $\frac{1}{2\times0.404}$ cycles/m, i.e., 1.2376 cycles/m. In other words, the wavelength less than 0.808 m cannot be measured and analyzed (Goel et al. 2005).

Response spectrum analysis is performed by giving base excitation in the form of power spectral densities (PSDs) at identified wheel track contact points, two each on one wheel of a bogie, thus making eight points for bogie analysis and sixteen points for a coach. PSD of acceleration also means square acceleration response and expressed in terms of g^2/Hz given as base excitation (Goel et al. 2005; Sharma 2011a, b) to the wheel track contact points which are identified in Eigen frequency modal analysis in vertical (Z) direction. These vertical PSD of acceleration values are specified within the range of 0–60 Hz as per RDSO standards and specified in Indian standard literature (Gangadharan et al. 2004; Goel et al. 2005; Ramji et al. 2007; Sharma 2011a, b). The PSD (Z) values fed at the wheel contact points are presented in Table 5.2.

The response to the spectral inputs given to the wheel contact points of the ICF railway bogie model is extracted in terms of PSD. In ANSYS APDL, spectrum analysis is performed on both the bogie as well as coach models to find the response. For the spectrum PSD analysis, the number of modes to be considered is specified as shown in Fig. 5.8. Spectrum type is chosen as "P.S.D". The first hundred modes are considered for spectrum analysis using option "NMODE". The track input values in terms of g^2/Hz with respect to frequency are set in the PSD input settings as well as PSD versus Frequency in Figs. 5.9 and 5.10. Since PSD of acceleration is fed as wheel inputs, the units are considered "g^2/Hz" in the option "Type of Response Spectra". Since the entire coach model is built in millimeters, the acceleration due to gravity value 9810 mm/sec^2, is input in the option "GVALUE" in the PSD Settings.

Table 5.2 PSD track to wheel inputs in vertical (Z) direction

Serial number	Frequency (Hz)	PSD of acceleration (g^2/Hz)	Serial number	Frequency (Hz)	PSD of acceleration (g^2/Hz)
1	1.95E−01	1.06E−03	26	2.03E+01	1.11E−02
2	3.16E−01	3.77E−03	27	2.15E+01	1.46E−02
3	3.85E−01	7.20E−03	28	2.30E+01	1.22E−02
4	4.58E−01	1.56E−02	29	2.35E+01	8.40E−03
5	5.21E−01	4.05E−02	30	2.52E+01	1.01E−02
6	5.92E−01	1.02E−01	31	2.61E+01	1.29E−02
7	6.42E−01	1.53E−01	32	2.70E+01	1.42E−02
8	7.03E−01	1.19E−01	33	2.83E+01	1.14E−02
9	7.35E−01	6.05E−02	34	2.86E+01	8.40E−03
10	9.66E−01	1.66E−02	35	2.95E+01	6.17E−03
11	1.23E+00	9.22E−03	36	3.06E+01	7.66E−03
12	1.62E+00	5.80E−03	37	3.20E+01	8.66E−03
13	2.08E+00	4.68E−03	38	3.35E+01	6.56E−03
14	2.44E+00	3.89E−03	39	3.43E+01	4.53E−03
15	3.04E+00	3.89E−03	40	4.07E+01	2.94E−03
16	4.45E+00	1.14E−02	41	4.36E+01	2.45E−03
17	5.35E+00	1.66E−02	42	4.62E+01	2.37E−03
18	6.28E+00	1.66E−02	43	4.78E+01	2.77E−03
19	7.90E+00	1.38E−02	44	5.00E+01	1.91E−03
20	9.72E+00	2.06E−02	45	5.30E+01	2.52E−03
21	1.07E+01	2.06E−02	46	5.48E+01	1.80E−03
22	1.24E+01	1.56E−02	47	5.48E+01	1.54E−03
23	1.39E+01	1.56E−02	48	5.60E+01	1.36E−03
24	1.65E+01	1.38E−02	49	5.94E+01	1.97E−03
25	1.83E+01	1.14E−02	50	6.00E+01	1.06E−03

5.3.3 Random Response Analysis of Railway Coach

In the PSD response analysis of the railway coach it is assumed that the coach under unladen condition does not consider the passenger load whereas under laden condition considers the full 72 passengers load of 4.68 tons which is coupled through mass element at the C.G. of the general sleeper coach model. The random vertical track inputs specified in Table 5.1 are fed as input to eight-wheel contact points in terms of g^2/Hz for both the bogie wheels of the coach. Vertical acceleration response for a point on the coach underframe is extracted in terms of PSD in g^2/Hz. This

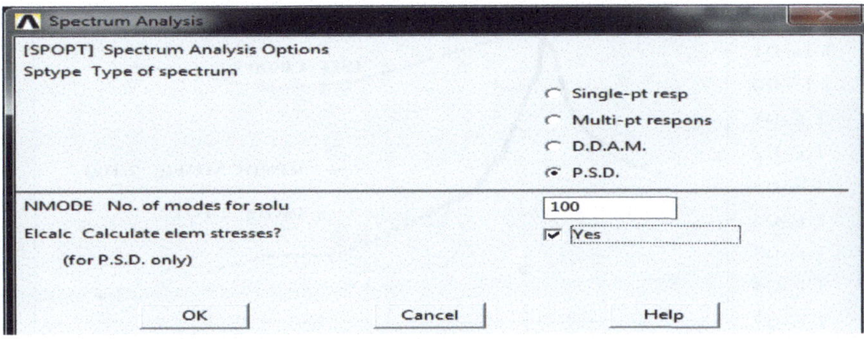

Fig. 5.8 PSD spectrum analysis in ANSYS

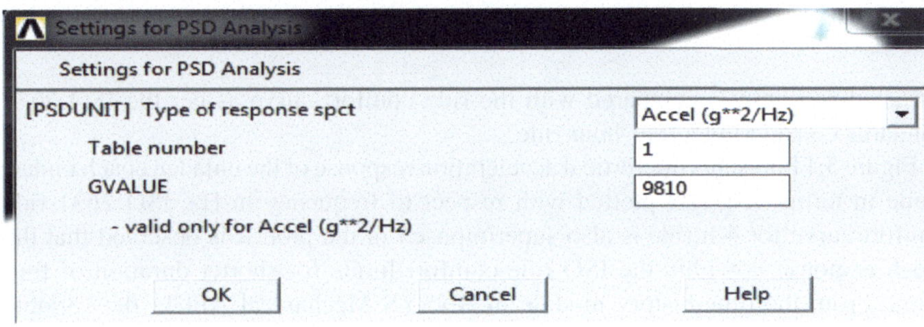

Fig. 5.9 PSD settings in ANSYS

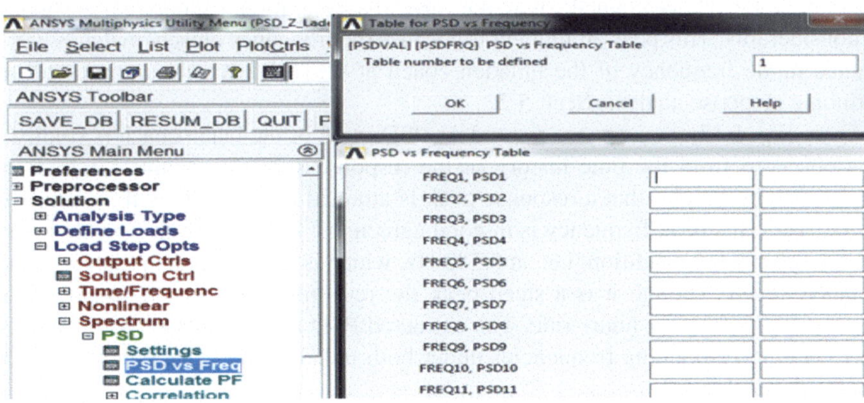

Fig. 5.10 PSD (g^2/Hz) versus frequency (Hz) in ANSYS

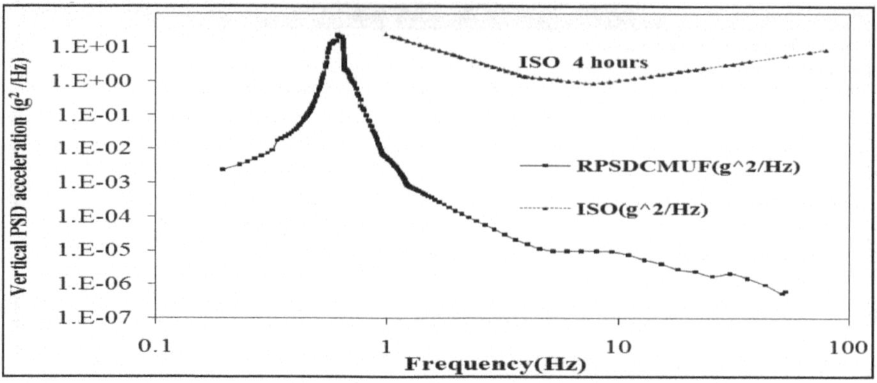

Fig. 5.11 PSD acceleration versus frequency of the coach (unladen)

extracted response is compared with the ride comfort curve as per the ISO 2631 standards considering a four-hour ride.

Figure 5.11 presents the vertical acceleration response of the unladen coach underframe in terms of g^2/Hz plotted with respect to frequency in Hz. ISO 2631 ride comfort curve for 4 h ride is also superimposed in the plot. It is observed that the coach response is within the ISO ride comfort limits for shorter duration of four hours. From the time-history module in ANSYS Mechanical APDL, the variable viewer presents the output PSD of acceleration response of the coach underframe at different frequencies. However, at frequency of 0.611 Hz, a PSD response peak value of 22.4 g^2/Hz is attained where the vertical acceleration response curve shoots up considerably. This peak attainment is because the frequency closely matches with bounce mode frequency of the unladen coach at 0.616 Hz which is presented in harmonic response analysis Sect. 5.2.

From Fig. 5.12, the response to random PSD input of the laden coach is studied. It is observed from the time history of the response spectrum analysis, which is also shown in the plot that a response peak is attained at 0.584 Hz with a value of 19.6 g^2/Hz. This peak frequency is in good agreement with the coach bounce mode frequency in laden condition, i.e., at 0.583 Hz, which is presented in modal analysis of laden coach. Though it is a steep peak the response still lies within ISO ride comfort limits for four hours ride. The extracted PSD acceleration response of the coach and corresponding frequencies under both unladen and laden conditions are tabulated in Table 5.3.

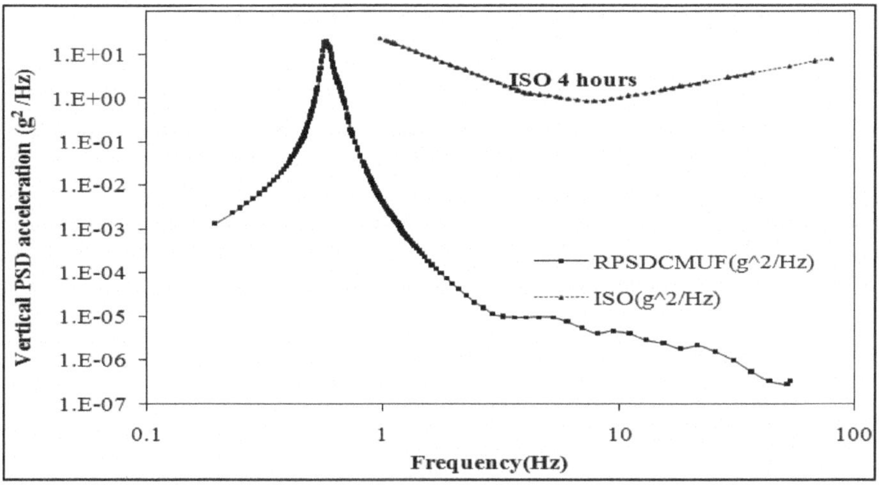

Fig. 5.12 PSD acceleration versus frequency of the coach (laden)

Table 5.3 PSD of acceleration response of railway coach model

Coach load condition	PSD of acceleration at response peak (g^2/Hz)	PSD peak attained frequency (Hz)	Matched natural frequency (Hz)
Unladen	22.4	0.611	0.616 (bounce mode)
Laden	19.6	0.584	0.583 (bounce mode)

5.4 Results and Discussion

The results from the vertical random response analysis of the coach indicate that coach response pattern to random excitation in both unladen and laden conditions is similar. However, with the passenger mass, the PSD of acceleration output response decreases. It is concluded that bounce mode shows dominant effect on the vertical random dynamic behavior of the unladen and laden coach. Though the acceleration peak response is less than the ISO ride comfort limits this can be avoided by riding at either higher or lower frequencies of the coach bounce mode frequency (Sharma 2011a, b). The peak attainment is only at lower frequency, i.e., at 0.611 Hz for unladen coach and 0.584 Hz for laden coach; whereas at higher frequencies the PSD output response significantly comes down supported by literature (Sharma 2011a, b). Hence it is concluded that at lower frequencies (less than 1 Hz) vertical profile irregularities have more impact on the ride comfort of the sleeper coach.

5.5 Summary of Random Response Analysis

PSD response spectrum analysis is carried out on the ICF railway bogie and coach models in ANSYS using FEA. It is concluded that under the random excitations in the form of vertical PSD of acceleration inputs indicating the vertical profile irregularities of the track the response of the bogie and coach lies within ISO ride comfort limits for a short duration of four hours. The bounce mode significantly influences the passenger ride with steep rise in the PSD of acceleration response in g^2/Hz at the bounce mode frequencies for both bogie and coach in unladen and laden conditions.

5.6 Conclusions

The harmonic analysis of coach indicates that the underframe and side panel of the coach attain first peak around frequency 0.6 Hz in both unladen and laden conditions matching with the bounce and pitch mode frequencies of the coach. It is also concluded that the highest amplitude is attained for the front and rear locations of the underframe at the first resonant peak with a displacement of 1.007 mm in unladen condition and 0.45 mm in the laden condition. Similarly, the side panel of the coach attains the highest amplitude at the first resonant peak with a displacement of 0.571 mm in unladen condition and 0.38 mm in laden condition. The reason for this response is due to the dominance of bounce mode (Banarjee 2006; Ramji et al. 2007), contributing to the vertical amplitude of the side panel and underframe of the coach. For both the underframe and side panel, the second dominant peak is attained at 3.25 Hz in unladen condition and 3 Hz in laden condition matching with the roll mode frequencies 3.33 Hz and 3.02 Hz, respectively.

From the PSD response analysis of the bogie, it is stated that the initial peak response is matching with bounce mode frequency 7.68 Hz in unladen condition and 7.67 Hz in the laden condition of the bogie, which means that bounce mode is predominant in vertical dynamic behavior (Sharma 2011a, b). Though there are peak responses beyond 10 Hz frequency, those lie well within the ISO 2631 ride comfort limits and hence passengers will feel comfortable with current bogie arrangement as specified in studies (Gangadharan et al. 2004; Hemantha Kumar et al. 2005; Sharma 2011a, b). Further, it is concluded from PSD acceleration response of the coach that the coach underframe response on which the berths are located lies within ISO 2631 ride comfort limits as per standard Indian Railways-specific studies (Sharma 2011a, b). However, the underframe is still susceptible due to a steep rise in response at its bounce mode frequency, i.e., at 0.61 Hz in unladen and 0.583 Hz in laden conditions, drawing attention towards modifications in the car body of the coach.

References

Acharya A, Gahlaut U, Sharma K, Sharma SK, Vishwakarma PN, Phanden RK (2020) Crashworthiness analysis of a thin-walled structure in the frontal part of automotive chassis. Int J Veh Struct Syst 12. https://doi.org/10.4273/ijvss.12.5.06

Adeoye AO, Quadri RO, Lawal OS (2022) Assessment of biofuel potential of tenera palm kernel shell via fixed bed pyrolysis and thermal characterization. Results Surf Interfaces 9:100091. https://doi.org/10.1016/j.rsurfi.2022.100091

Adeoye AO, Quadri RO, Lawal OS (2023a) Wet synthesis, characterization of goethite nanoparticles and its application in catalytic pyrolysis of palm kernel shell in TGA. Results Surf Interfaces 11:100118. https://doi.org/10.1016/j.rsurfi.2023.100118

Adeoye AO, Quadri RO, Lawal OS (2023b) Biomass waste pyrolysis in the context of Nigeria's oil and gas industry: an overview. Int J Ambient Energy 44:2525–2541. https://doi.org/10.1080/01430750.2023.2256325

Adeoye AO, Lawal OS, Quadri RO, Malomo D, Aliyu MT, Dang GE, Emojevu EO, Maikato MJ, Yahaya MG, Omonije OO, Edidem VG, Abubakar YK, Offor OF, Sochima EH, Peter BE, Hikon BN (2023c) Sustainable energy via thermochemical and biochemical conversion of biomass wastes for biofuel production. In: Transportation energy and dynamics. Energy, environment, and sustainability. Springer, Singapore, pp 245–306. https://doi.org/10.1007/978-981-99-2150-8_11

Agarwal KM, Mohapatra S, Sharma P, Sharma S, Bhatia D, Mishra A (2020) Study and overview of the novel corona virus disease (COVID-19). Sens Int 1:100037. https://doi.org/10.1016/j.sintl.2020.100037

Agrahari R, Sarraf G, Joshi NC, Mohapatra S, Varma A (2021a) Insight of biopolymers and applications of polyhydroxyalkanoates. In: Microbial polymers. Springer Singapore, Singapore, pp 177–191. https://doi.org/10.1007/978-981-16-0045-6_8

Agrahari R, Mohanty S, Vishwakarma K, Nayak SK, Samantaray D, Mohapatra S (2021b) Update vision on COVID-19: structure, immune pathogenesis, treatment and safety assessment. Sens Int 2:100073. https://doi.org/10.1016/j.sintl.2020.100073

Badgujar P, Sharma S, Rahul D, Sharma SK, Upadhyay RK (2024) Feasibility study of Mumbai-Pune hyperloop corridor: challenges and implementation. Res Transp Bus Manage 52:101089

Banarjee N (2006) Modeling and dynamics of railway vehicles: a bond graph approach. PhD thesis, Mechanical Engineering, IIT Kharagpur

Bandekar D, Chouhan OP, Mohapatra S, Hazra M, Hazra S, Biswas S (2017) Putative protein VC0395_0300 from Vibrio cholerae is a diguanylate cyclase with a role in biofilm formation. Microbiol Res 202:61–70. https://doi.org/10.1016/j.micres.2017.05.003

Bhardawaj S, Chandmal Sharma R, Kumar Sharma S (2019a) Development and advancement in the wheel-rail rolling contact mechanics. IOP Conf Ser Mater Sci Eng 691:012034. https://doi.org/10.1088/1757-899X/691/1/012034

Bhardawaj S, Chandmal Sharma R, Kumar Sharma S (2019b) A survey of railway track modelling. Int J Veh Struct Syst 11. https://doi.org/10.4273/ijvss.11.5.08

Bhardawaj S, Sharma RC, Sharma SK (2020a) Ride analysis of track-vehicle-human body interaction subjected to random excitation. J Chin Soc Mech Eng Trans Chin Inst Eng Ser C/Chung-Kuo Chi Hsueh K. Ch'eng Hsuebo Pao 41:229–238

Bhardawaj S, Sharma RC, Sharma SK (2020b) Development in the modeling of rail vehicle system for the analysis of lateral stability. Mater Today Proc 25:610–619. https://doi.org/10.1016/j.matpr.2019.07.376

Bhardawaj S, Sharma RC, Sharma SK (2020c) Analysis of frontal car crash characteristics using ANSYS. Mater Today Proc 25:898–902. https://doi.org/10.1016/j.matpr.2019.12.358

Bhardawaj S, Sharma RC, Sharma SK, Sharma N (2021) On the planning and construction of railway curved track. Int J Veh Struct Syst 13(2):151–159. https://doi.org/10.4273/ijvss.13.2.04

Bhardawaj S, Sharma RC, Sharma SK, Gopala Rao L, Vashist A (2023) Modeling of biomechanical human body model for seat to head transmissibility analysis. Noise Vib Worldwide 54(2–3):62–74. https://doi.org/10.1177/09574565231154243

Bhattarai S, Mhamdi A, Hossain I, Raoui Y, Pandey R, Madan J, Bouazizi A, Maiti M, Gogoi D, Sharma A (2022) A detailed review of perovskite solar cells: Introduction, working principle, modelling, fabrication techniques, future challenges. Micro Nanostruct 172:207450. https://doi.org/10.1016/j.micrna.2022.207450

Biswas N, Saha S, Biswas BK, Chowdhury M, Rahaman A, Junghare V, Mohapatra S, Hazra S, Zangrando E, Roy Choudhury R, Roy Choudhury C (2021) The DNA- and protein-binding properties and cytotoxicity of a new copper(II) hydrazone Schiff base complex. J Coord Chem 74:1482–1504. https://doi.org/10.1080/00958972.2021.1913128

Catherines JJ, Clevenson SA (1969) Vibration measurements of the Penn central metroliner. In: Langley working paper LWP-770, Langley Research Center, Hampton

Chatterjee A, Nirwan S, Mohapatra S, Sharma P, Agnihotri A, Shrivastava N (2022) Biochemical aspects of pathogenic variability in white rust infected Indian mustard. Mycologia 114:757–768. https://doi.org/10.1080/00275514.2022.2060007

Choi S, Lee J, Sharma SK (2021) A study on the performance evaluation of hydraulic tank injectors. In: Joshi P, Gupta SS, Shukla AK, Gautam SS (eds) Advances in engineering design. Lecture notes in mechanical engineering, pp 183–190. https://doi.org/10.1007/978-981-33-4684-0_19

Cooperrider NK (1975) Railway truck response to random rail irregularities. J Eng Ind ASME 97(3):957–964. https://doi.org/10.1115/1.3438710

Das S, Swapnarekha H, Vimal S (2023) Integration of blockchain technology with renewable energy for sustainable development: issues, challenges and future direction. https://doi.org/10.1007/978-981-99-3734-9_48

Devadarshini D, Mohapatra S, Pati S, Maity S, Rath CC, Jena PK, Samantaray D (2022) Evaluation of PHAs production by mixed bacterial culture under submerged fermentation. Biologia (bratisl) 78:895–902. https://doi.org/10.1007/s11756-022-01302-5

Durali M, Mahdi JM (2008) Investigation of wagon derailment moving on random rail irregularities using nonlinear 3-dimentional model (research note). Int J Eng Trans B Appl 21(4):385–400

Gangadharan KV (2001) Analytical and experimental studies on dynamics of railroad vehicles. PhD thesis, Machine Dynamics Laboratory, Department of Applied Mechanics, IIT Madras

Gangadharan KV, Sujatha C, Ramamurti V (2004) Experimental and analytical ride comfort evaluation of a railway coach. In: Paper # 249, Proceedings of IMAC-XXII A conference & exposition on structural dynamics, Dearborn, Michigan, Jan 26–29

Goel VK, Thakur M, Deep K, Awasthi BP (2005) Mathematical model to represent the track geometry variation using PSD. Indian Railway Tech Bull LXI(312–313):1–10

Goyal S, Anand CS, Sharma SK, Sharma RC (2019) Crashworthiness analysis of foam filled star shape polygon of thin-walled structure. Thin-Walled Struct 144:106312. https://doi.org/10.1016/j.tws.2019.106312

Gopala Rao L, Sharma RC, Satyanarayana V, Palli S (2022) Stochastic optimal preview control response of a quarter-car nonlinear suspension model using spectral decomposition method. Noise Vib Worldw 53:225–232. https://doi.org/10.1177/09574565221093234

Goswami B, Rathi A, Sayeed S, Das P, Sharma RC, Sharma SK (2019) Optimization design for aerodynamic elements of Indian locomotive of passenger train. In: Advances in engineering design. Lecture notes in mechanical engineering. Springer, Singapore, pp 663–673. https://doi.org/10.1007/978-981-13-6469-3_61

Hemantha Kumar B, Kumar S, Sujatha C (2005) Lateral dynamic analysis of a typical Indian rail-road vehicle. In: Proceedings of 12th national conference on machines and mechanisms (NaCoMM-2005), IIT Guwahati, Dec 16–17, pp 208–214

Hossain I, Ahmed YE, Sharma SK (2023) The use of both metallic-insert and TiO_2/water nanofluid to improve the performance of photovoltaic modules; experimental study. Therm Sci Eng Prog 42:101944. https://doi.org/10.1016/j.tsep.2023.101944

Hossain I, Velkin VI, Shcheklein SE (2017) The study of passive vibration dampers in pipelines using Piv-methodology for single phase flow. In: WIT transactions on ecology and the environment, pp 565–570. https://doi.org/10.2495/ESUS170521

Indian Railways Institute of Civil Engineering: Handbook for track maintenance, First Edition (2016)

Iyengar RN, Jaiswal OR (1995) Random field modelling of railway track irregularities. J Transp Eng 121(4):304–308

Kanth Choppara R, Chandmal Sharma R, Kumar Sharma S, Gupta T (2019) Aero dynamic cross wind analysis of locomotive. IOP Conf Ser Mater Sci Eng 691:012035. https://doi.org/10.1088/1757-899X/691/1/012035

Karuppaiah N, Sujatha C, Ramamurthi V (1999) Modal and vibration/stress analysis of a passenger vehicle by FEM. In: Proceedings of symposium on international automotive technology. SAE Tech., pp 129–133

Karuppaiah N, Deshpande PS, Sujatha C, Ramamurthi V (2003) Vibration analysis in a light passenger vehicle by rigid body/finite element modelling. Adv Vib Eng 2(2):106–120

Kohli I, Joshi NC, Mohapatra S, Varma A (2020) Extremophile—an adaptive strategy for extreme conditions and applications. Curr Genomics 21:96–110. https://doi.org/10.2174/1389202921666200401105908

Kohli, I., Mohapatra, S., Kumar, P., Goel, A., Varma, A., Joshi, N.C.: Role of Bacterial Endophytes in the Promotion of Plant Growth. In: Bacterial Endophytes for Sustainable Agriculture and Environmental Management. pp. 235–253. Springer Singapore, Singapore (2022). https://doi.org/10.1007/978-981-16-4497-9_12.

Kokkinos NC, Emmanouilidou E, Sharma SK (2024) Waste-to-biofuel production for the transportation sector. In: Intelligent transportation system and advanced technology. Springer, pp 99–122

Kouroussis G, Connolly DP, Alexandrou G, Vogiatzis K (2015) The effect of railway local irregularities on ground vibration. Transp Res Part D Transp Environ 39:17–30

Kumar U, Kasvekar R, Sharma SK, Upadhyay RK (2022) Wear of wheels and axle in locomotive and measures taken by Indian railway. In: Advances in engine tribology. Springer, pp 77–96. https://doi.org/10.1007/978-981-16-8337-4_5

Law EH, Cooperrider NK (1974) A survey of railway vehicle dynamics research. J Dyn Syst Measur Control Trans ASME 96(2):132, 134–146

Lee RA, Pradko F (1968) Analytical analysis of human vibration. SAE paper issue 680091

Lee J, Sharma SK (2020) Numerical investigation of critical speed analysis of high-speed rail vehicle. 한국정밀공학회학술발표대회 논문집 (Korean Society of Precision Engineering), p 696

Lee J, Sharma SK (2023) Intelligent systems for railway infrastructure. In: Applied sciences, vol 13, issue 20. MDPI, p 11274

Lee J, Han J, Sharma SK (2021) Structural analysis on the separated and integrated differential gear case for the weight reduction. In: Advances in engineering design: select proceedings of FLAME 2020, p 175

Li CY, Young TH (2003) Vertical vibration analysis of vehicle/imperfect track systems. Veh Syst Dyn 40(5):329–349

Lokanadham D, Sharma RC, Palli S, Bhardawaj S (2022) Wear rate modelling and analysis of limestone slurry particulate composites using the fuzzy method. Int J Recent Innov Trends Comput Commun 10(1):133–143. https://doi.org/10.17762/ijritcc.v10i1s.581

Lu G, Gill KF (1993) Track-transmission system dynamic analysis. Proc Inst Mech Eng Part F J Rail Rapid Transit 207(2):99–113

MacMillan RH (1962) Non-linear control systems analysis. Pergamon Press, London, pp 76–79

Maity S, Das S, Mohapatra S, Tripathi AD, Akthar J, Pati S, Pattnaik S, Samantaray DP (2020) Growth associated polyhydroxybutyrate production by the novel Zobellellae tiwanensis strain DD5 from banana peels under submerged fermentation. Int J Biol Macromol 153:461–469. https://doi.org/10.1016/j.ijbiomac.2020.03.004

Ministry of Indian Railways Website, Government of India, track recording and analysis. http://www.indianrailways.gov.in/railwayboard/uploads/codesmaual/IRPWM/PermanentWayManualCh6_data.htm. Accessed Aug 14, 2023

Misra S, Mohanty D, Mohapatra S (2022) Food-based probiotics: functional dietary ingredients. In: Probiotics. Elsevier, pp 257–275. https://doi.org/10.1016/B978-0-323-85170-1.00006-3

Mixson JS, Steiner R (1969) Optimization of a simple dynamic model of a railroad car under random and sinusoidal inputs. FRA-RT-72-04, Office of High SpeedGround Transportation, Washington, DC (PB-201 620)

Mohanty DP, Mohapatra S, Misra S, Sahu PS (2016) Milk derived bioactive peptides and their impact on human health—a review. Saudi J Biol Sci 23:577–583. https://doi.org/10.1016/j.sjbs.2015.06.005

Mohapatra S, Samantaray DP, Samantaray SM (2015) Study on polyhydroxyalkanoates production using rhizospheric soil bacterial isolates of sweet potato. Indian J Sci Technol 8:57. https://doi.org/10.17485/ijst/2015/v8iS7/64027

Mohapatra S, Rath SN, Pradhan SK, Samantaray DP, Rath CC (2016a) Secondary structural models (16S rRNA) of polyhydroxyalkanoates producing bacillus species isolated from different rhizospheric soil: phylogenetics and chemical analysis. Int J Bioautom 20:329–338

Mohapatra S, Samantaray DP, Samantaray SM, Mishra BB, Das S, Majumdar S, Pradhan SK, Rath SN, Rath CC, Akthar J, Achary KG (2016b) Structural and thermal characterization of PHAs produced by Lysinibacillus sp. through submerged fermentation process. Int J Biol Macromol 93:1161–1167. https://doi.org/10.1016/j.ijbiomac.2016.09.077

Mohapatra S, Mohanta PR, Sarkar B, Daware A, Kumar C, Samantaray DP (2017a) Production of polyhydroxyalkanoates (PHAs) by bacillus strain isolated from waste water and its biochemical characterization. Proc Natl Acad Sci India Sect B Biol Sci 87:459–466. https://doi.org/10.1007/s40011-015-0626-6

Mohapatra S, Mishra R, Roy P, Yadav KL, Satapathi S (2017b) Systematic investigation and in vitro biocompatibility studies on implantable magnetic nanocomposites for hyperthermia treatment of osteoarthritic knee joints. J Mater Sci 52:9262–9268. https://doi.org/10.1007/s10853-017-1136-0

Mohapatra S, Sarkar B, Samantaray DP, Daware A, Maity S, Pattnaik S, Bhattacharjee S (2017c) Bioconversion of fish solid waste into PHB using Bacillus subtilis based submerged fermentation process. Environ Technol 38:3201–3208. https://doi.org/10.1080/09593330.2017.1291759

Mohapatra S, Maity S, Dash HR, Das S, Pattnaik S, Rath CC, Samantaray D (2017d) Bacillus and biopolymer: prospects and challenges. Biochem Biophys Rep 12:206–213. https://doi.org/10.1016/j.bbrep.2017.10.001

Mohapatra S, Mishra Y, Dash L (2020a) COVID-19: analysing the legal nuances of the lockdown order. Indian J Forensic Med Toxicol (2020). https://doi.org/10.37506/ijfmt.v14i4.11553

Mohapatra S, Pattnaik S, Maity S, Mohapatra S, Sharma S, Akhtar J, Pati S, Samantaray DP, Varma A (2020b) Comparative analysis of PHAs production by Bacillus megaterium OUAT 016 under submerged and solid-state fermentation. Saudi J Biol Sci 27:1242–1250. https://doi.org/10.1016/j.sjbs.2020.02.001

Mohapatra S, Vishwakarma K, Joshi NC, Maity S, Kumar R, Ramchander M, Pattnaik S, Samantaray DP (2021) A review on PHAs: the future biopolymer. In: Environmental and agricultural microbiology. Wiley, pp 83–100. https://doi.org/10.1002/9781119525899.ch5

Mohapatra S, Mohapatra S, Han H, Ariza-Montes A, López-Martín M (2022) Climate change and vulnerability of agribusiness: assessment of climate change impact on agricultural productivity. Front Psychol 13. https://doi.org/10.3389/fpsyg.2022.955622

Nayak J, Swapnarekha H, Routray AR, Nayak SR, Behera HS (2022) Big data in education: present and future. In: Computational intelligence in data mining. Smart innovation, systems and technologies, vol 281. Springer, Singapore, pp 721–739. https://doi.org/10.1007/978-981-16-9447-9_54

Nibedita S, Swati P, Pattnaik M, Mohapatra S (2021) Methane emission and strategies for mitigation in livestock. In: Environmental and agricultural microbiology. Wiley, pp 257–274. https://doi.org/10.1002/9781119525899.ch12

Palli S, Koona R (2015) Analyses of dynamic response of a railway bogie. Int J Veh Noise Vib 11:103. https://doi.org/10.1504/IJVNV.2015.070015

Palli S, Dontikurti R, Sharma RC, Sharma N (2021) Analysis of dynamic response of a railway locomotive using ANSYS. Int J Veh Struct Syst 13. https://doi.org/10.4273/ijvss.13.2.21

Palli S, Duppala A, Sharma RC, Gopala Rao LVV (2022) Dynamic simulation of automotive vehicle suspension using MATLAB simulink. Int J Veh Struct Syst 14. https://doi.org/10.4273/ijvss.14.3.04

Palli S, Koona R, Sharma SK, Sharma RC (2018) A review on dynamic analysis of rail vehicle coach. Int J Veh Struct Syst 10(3):204–211. https://doi.org/10.4273/ijvss.10.3.10

PanduRanga Vital T, Lakshmi BG, Swapna Rekha H, DhanaLakshmi M (2019) Student performance analysis with using statistical and cluster studies. In: Soft computing in data analytics. Advances in intelligent systems and computing, vol 758. Springer, Singapore, pp 743–757. https://doi.org/10.1007/978-981-13-0514-6_71

Pati S, Maity S, Dash A, Jema S, Mohapatra S, Das S, Samantaray DP (2020) Biocompatible PHB production from bacillus species under submerged and solid-state fermentation and extraction through different downstream processing. Curr Microbiol 77:1203–1209. https://doi.org/10.1007/s00284-020-01922-7

Pati S, Mohanty MK, Mohapatra S, Samantaray D (2022) Bioethanol production by Enterobacter hormaechei through carbon monoxide-rich syngas fermentation. Int J Energy Res 46:20096–20106. https://doi.org/10.1002/er.8678

Pattnaik S, Dash D, Mohapatra S, Pattnaik M, Marandi AK, Das S, Samantaray DP (2020) Corrigendum to "Improvement of rice plant productivity by native Cr(VI) reducing and plant growth promoting soil bacteria Enterobacter cloacae" [Chemosphere 240:124895 (2020)]. Chemosphere 238:125011. https://doi.org/10.1016/j.chemosphere.2019.125011

Pattnaik S, Mohapatra S, Pati S, Dash D, Devadarshini D, Tanaya K, Mishra BB, Samantaray D (2022) Microbial bioremediation of Cr(VI)-contaminated soil for sustainable agriculture. In: Microbial biodegradation and bioremediation. Elsevier, pp 395–407. https://doi.org/10.1016/B978-0-323-85455-9.00008-4

Pearce TG, May BJ (1969) A study of the stability and dynamic response of the linear induction motor test vehicle. FRA-RT-70-25, Office of High SpeedGround Transportation, Washington, DC (PB-192718)

Pervozvanskii AA (1965) Random processes in nonlinear control systems. Academic Press, New York, pp 77–106. ISBN 9780080955216

Pombo J, Ambrósio J (2005) Dynamic analysis of a railway vehicle in real operation conditions using a new wheel-rail contact detection model. Int J Veh Syst Model Test 1(1/2/3):79–105

Popp K, Kaiser I, Kruse H (2003) System dynamics of railway vehicles and track. Arch Appl Mech 72(11):949–961. Springer

Popp K, Knothe K, Pöpper C (2005) System dynamics and long term behaviour of railway vehicles, track and subgrade: report on the DFG priority programme in Germany and subsequent research. Int J Veh Mech Mob Veh Syst Dyn 43(6):485–521

Pradko F, Lee R, Kaluza V (1966) Theory of human vibration response. ASME paper 66-WA/BHF-15

Pundir S, Upadhyay S, Kumar N, Chandra Joshi N, Priya R, Ahmad Mir R, Hossain I, Pandey OP (2023a) Interconnected CuO nanoplates as a highly efficient electrocatalyst for oxygen evolution reaction. Mater Lett 336:133921. https://doi.org/10.1016/j.matlet.2023.133921

Pundir S, Upadhyay S, Priya R, Kumar N, Chetana S, Hossain I, Joshi NC, Pandey OP (2023b) Synthesis of 1D β-MnO_2 for high-performance supercapacitor application. J Solid State Electrochem 27:531–538. https://doi.org/10.1007/s10008-022-05347-z

Ramji K, Goel VK, Rao S, Naidu MK (2007) Dynamic behaviour of railway coach and bogie frame using finite element analysis. J Inst Eng (India), Part MC 87:7–17

Raju Rallabandi S, Pilla DP, Dowluru S, Palli S, Sharma N, Sharma SK, Sharma RC (2024) Critical evaluation of epoxy-hemp-pineapple-palm fiber composites using hybrid AHM-TOPSIS technique for sustainable structural applications. J Chin Inst Eng 1–12

Rallabandi SR, Srinivas L, Palli S, Sharma RC, Sharma N, Sharma A, Sharma SK (2024) A contrastive characterization of pure Mg and AZ91D alloy based on the testing of mechanical, corrosion, wear, and erosion properties. Eng Res Express 6(1):015017

Samantaray D, Mohapatra S, Mishra BB (2014) Microbial bioremediation of industrial effluents. In: Microbial biodegradation and bioremediation. Elsevier, pp 325–339. https://doi.org/10.1016/B978-0-12-800021-2.00014-5

Sarkar MS, Segu H, Bhaskar JV, Jakher R, Mohapatra S, Shalini K, Shivaji S, Reddy PA (2018) Ecological preferences of large carnivores in remote, high-altitude protected areas: insights from Buxa Tiger Reserve, India. Oryx 52:66–77. https://doi.org/10.1017/S0030605317000060

Satyanarayana VSV, Sharma RC, Sateesh B, Gopala Rao LVV, Mohan Rao N, Palli S (2022) Optimum response of a nonlinear passive vehicle suspension system under random road excitations. SAE Int J Commer Veh 16:02-16-01–0004. https://doi.org/10.4271/02-16-01-0004

Sewall JL (1972) On the application of parameter identification to high-speed ground transportation systems. In: System identification of vibrating structures, ASME winter annual meeting, pp 133–148

Sharma RC (2011a) Ride analysis of an Indian railway coach using Lagrangian dynamics. Int J Veh Struct Syst 3(4):219–224

Sharma RC (2011b) Parametric analysis of rail vehicle parameters influencing ride behavior. Int J Eng Sci Technol 3(8):54–65

Sharma SK (2019) Multibody analysis of longitudinal train dynamics on the passenger ride performance due to brake application. Proc Inst Mech Eng Part K J Multi-Body Dyn 233(2):266–279. https://doi.org/10.1177/1464419318788775

Sharma SK, Kumar A (2017a) Impact of electric locomotive traction of the passenger vehicle ride quality in longitudinal train dynamics in the context of Indian railways. Mech Ind 18(2):222. https://doi.org/10.1051/meca/2016047

Sharma SK, Kumar A (2017b) Ride performance of a high speed rail vehicle using controlled semi active suspension system. Smart Mater Struct 26:055026. https://doi.org/10.1088/1361-665X/aa68f7

Sharma SK, Kumar A (2018a) Disturbance rejection and force-tracking controller of nonlinear lateral vibrations in passenger rail vehicle using magnetorheological fluid damper. J Intell Mater Syst Struct 29(2):279–297. https://doi.org/10.1177/1045389X17721051

Sharma SK, Kumar A (2018b) Impact of longitudinal train dynamics on train operations: a simulation-based study. J Vib Eng Technol 6(3):197–203. https://doi.org/10.1007/s42417-018-0033-4

Sharma SK, Kumar A (2018c) Ride comfort of a higher speed rail vehicle using a magnetorheological suspension system. Proc Inst Mech Eng Part K J Multi-body Dyn 232:32–48. https://doi.org/10.1177/1464419317706873

Sharma SK, Lee J (2020a) Finite element analysis of a fishplate rail joint in extreme environment condition. Int J Veh Struct Syst 12. https://doi.org/10.4273/ijvss.12.5.03

Sharma SK, Lee J (2020b) Design and development of smart semi active suspension for nonlinear rail vehicle vibration reduction. Int J Struct Stab Dyn 20:2050120. https://doi.org/10.1142/S0219455420501205

Sharma SK, Lee J (2021) Crashworthiness analysis for structural stability and dynamics. Int J Struct Stab Dyn 21:2150039. https://doi.org/10.1142/S0219455421500395

Sharma RC, Palli S (2016) Analysis of creep force and its sensitivity on stability and vertical-lateral ride for railway vehicle. Int J Veh Noise Vib 12:60. https://doi.org/10.1504/IJVNV.2016.077474

Sharma SK, Sharma RC (2018a) An investigation of a locomotive structural crashworthiness using finite element simulation. SAE Int J Commercial Veh 11(4):235–244. https://doi.org/10.4271/02-11-04-0019

Sharma SK, Sharma RC (2018b) Simulation of quarter-car model with magnetorheological dampers for ride quality improvement. Int J Veh Struct Syst 10(3):169–173. https://doi.org/10.4273/ijvss.10.3.03

Sharma RC, Sharma SK (2018c) Sensitivity analysis of three-wheel vehicle's suspension parameters influencing ride behavior. Noise Vib Worldwide 49(7–8), 272–280. https://doi.org/10.1177/0957456518796846

Sharma SK, Sharma RC (2019) Pothole detection and warning system for Indian roads. In: Advances in interdisciplinary engineering. Lecture notes in mechanical engineering. Springer, Singapore, pp 511–519. https://doi.org/10.1007/978-981-13-6577-5_48

Sharma SK, Sharma RC (2021) Multi-objective design optimization of locomotive nose. SAE Technical Paper, p 10. https://doi.org/10.4271/2021-01-5053

Sharma RC, Sharma SK (2022) Ride analysis of road surface-three-wheeled vehicle-human subject interactions subjected to random excitation. SAE Int J Commercial Veh 15(3):299–3120, 2-15-03-0017. https://doi.org/10.4271/02-15-03-0017

Sharma SK, Sharma RC, Kumar A, Palli S (2015) Challenges in rail vehicle-track modeling and simulation. Int J Veh Struct Syst 7(1):1–9. https://doi.org/10.4273/ijvss.7.1.01

Sharma RC, Palli S, Sharma SK, Roy M (2017) Modernization of railway track with composite sleepers. Int J Veh Struct Syst 9(5):321–329. https://doi.org/10.4273/ijvss.9.5.10

Sharma RC, Sharma SK, Palli S (2018) Rail vehicle modelling and simulation using Lagrangian method. Int J Veh Struct Syst 10(3):188–194. https://doi.org/10.4273/ijvss.10.3.07

Sharma SK, Saini U, Kumar A (2019) Semi-active control to reduce lateral vibration of passenger rail vehicle using disturbance rejection and continuous state damper controllers. J Vib Eng Technol 7(2):117–129. https://doi.org/10.1007/s42417-019-00088-2

Sharma SK, Sharma RC, Sharma N (2020a) Combined multi-body-system and finite element analysis of a rail locomotive crashworthiness. Int J Veh Struct Syst 12. https://doi.org/10.4273/ijvss.12.4.15

Sharma RC, Sharma SK, Palli S (2020b) Linear and non-linear stability analysis of a constrained railway wheelaxle. Int J Veh Struct Syst 12. https://doi.org/10.4273/ijvss.12.2.04

Sharma S, Sharma RC, Sharma SK, Sharma N, Palli S, Bhardawaj S (2020c) Vibration isolation of the quarter car model of road vehicle system using dynamic vibration absorber. Int J Veh Struct Syst 12. https://doi.org/10.4273/ijvss.12.5.05

Sharma RC, Sharma SK, Sharma N, Sharma S (2020d) Analysis of ride and stability of an ICF railway coach. Int J Veh Noise Vib 16:127. https://doi.org/10.1504/IJVNV.2020.117820

Sharma RC, Sharma S, Sharma SK, Sharma N (2020e) Analysis of generalized force and its influence on ride and stability of railway vehicle. Noise Vib Worldw 51:95–109. https://doi.org/10.1177/0957456520923125

Sharma RC, Palli S, Sharma N, Sharma SK (2021a) Ride behaviour of a four-wheel vehicle using H infinity semi-active suspension control under deterministic and random inputs. Int J Veh Struct Syst 13. https://doi.org/10.4273/ijvss.13.2.18

Sharma RC, Sharma S, Sharma N, Sharma SK (2021b) Linear and nonlinear analysis of ride and stability of a three-wheeled vehicle subjected to random and bump inputs using bond graph and Simulink methodology. SAE Int J Commer Veh 14:02-15-01–0001. https://doi.org/10.4271/02-15-01-0001

Sharma RC, Sharma S, Sharma SK, Sharma N, Singh G (2021c) Analysis of bio-dynamic model of seated human subject and optimization of the passenger ride comfort for three-wheel vehicle using random search technique. Proc Inst Mech Eng Part K J Multi-body Dyn 235:106–121. https://doi.org/10.1177/1464419320983711

Sharma SK, Sharma RC, Lee J (2021d) Effect of rail vehicle-track coupled dynamics on fatigue failure of coil spring in a suspension system. Appl Sci 11:2650. https://doi.org/10.3390/app11062650

Sharma SK, Mohapatra S, Sharma RC, Alturjman S, Altrjman C, Mostarda L, Stephan T (2022a) Retrofitting existing buildings to improve energy performance. Sustainability 14:666. https://doi.org/10.3390/su14020666

Sharma SK, Sharma RC, Lee J, Jang H-L (2022b) Numerical and experimental analysis of DVA on the flexible-rigid rail vehicle carbody resonant vibration. Sensors 22:1922. https://doi.org/10.3390/s22051922

Sharma SK, Lee J, Jang H-L (2022c) Mathematical modeling and simulation of suspended equipment impact on car body modes. Machines 10:192. https://doi.org/10.3390/machines10030192

Sharma SK, Sharma RC, Lee J (2022d) In situ and experimental analysis of longitudinal load on carbody fatigue life using nonlinear damage accumulation. Int J Damage Mech 31:605–622. https://doi.org/10.1177/10567895211046043

Sharma RC, Gopala Rao LVV, Sharma SK, Palli S, Satyanarayana VSV (2022e) Analysis of lateral stability and ride of an Indian railway constrained dual-axle bogie frame. SAE Int J Commercial Veh 16(2):213–228, 02-16-02-0014. https://doi.org/10.4271/02-16-02-0014

Sharma RC, Palli S, Gopala Rao LVV, Duppala A, Sharma SK (2022f) Four-wheel vehicle response under bump, pothole, harmonic, and random excitations using bond graph/simulink technique. SAE Int J Commercial Veh 16(2):129–139, 02-16-02-0008. https://doi.org/10.4271/02-16-02-0008

Sharma RC, Palli S, Jha AK, Bhardawaj S, Sharma SK (2022g) Vibration and ride comfort analysis of railway vehicle system subjected to deterministic inputs. Resmilitaris 12(5):1345–1355

Sharma RC, Palli S, Bhardawaj S (2023a) A review of wheel-rail dynamic behavior on curved tracks. Int J Veh Struct Syst 15. https://doi.org/10.4273/ijvss.15.1.04

Sharma RC, Palli S, Sharma SK (2023b) Ride analysis of railway vehicle considering rigidity and flexibility of the carbody. J Chin Inst Eng 46:355–366. https://doi.org/10.1080/02533839.2023.2194918

Sharma RC, Bhardawaj S, Sharma SK, Bahl S (2023c) Ride behaviour of three-wheel vehicle using semi-active suspension using skyhook control strategy. Int J Veh Struct Syst 15(1). https://doi.org/10.4273/ijvss.15.1.05

Sharma SK, Sharma RC, Choi Y, Lee J (2023d) Experimental and mathematical study of flexible–rigid rail vehicle riding comfort and safety. Appl Sci 13(9):5252

Sharma SK, Sharma RC, Choi Y, Lee J (2023e) Modelling and dynamic analysis of adaptive neuro-fuzzy inference system-based intelligent control suspension system for passenger rail vehicles using magnetorheological damper for improving ride index. Sustainability 15(16):12529

Sharma SK, Sharma RC, Lavania S, Palli S, Avesh M (2023f) Industry-driven approach for ANFIS-based intelligent control suspension system with MR damper for enhanced ride quality in passenger rail vehicles for technological investigations. Natl Acad Sci Lett 1–4

Sharma SK, Sharma RC, Upadhyay RK, Lee J (2023g) Coupled torsional–transverse vibration reduction in marine propulsion dynamics with novel approach using magnetorheological damping and adaptive control system. J Vib Eng Technol 1–11

Sharma SK, Upadhyay RK, Kumar V, Valera H (2023h) Introduction to sustainable transportation system. In: Transportation energy and dynamics. Springer, pp 3–6

Sharma RC, Palli S, Dharmana L, Sharma SK, Sharma A, Kharola A, Sharma N (2024a) Simulation of transient dynamic response of a locomotive considering its flexibility and rigidity. Noise Vib Worldwide 55(1–2):33–47

Sharma RC, Palli S, Duppala A, Bhardawaj S, Sharma SK, Sharma N, Sharma A (2024b) Rigid body analysis of ride behaviour of roadway vehicle using Lagrangian dynamics. Int J Veh Struct Syst (IJVSS) 16(1)

Sharma RC, Sharma N, Palli S, Sharma SK, Joshi S, Rallabandi SR, Sharma A (2024c) Dynamic behavior of three-wheel vehicle with inerter-based dynamic absorber suspension system. Noise Vib Worldwide 55(4–5):253–263. https://doi.org/10.1177/09574565241243398

Sharma SK, Kumar N, Avesh M, Sharma RC, Siddiqui MIH, Lee J (2024d) Navigating tranquillity with $H\infty$ controller to mitigate ship propeller shaft vibration. J Vib Eng Technol 1–11

Sharma SK, Sharma RC, Lee J (2024e) Modelling and simulation of MR damper characterization and uncertainty assessment in marine engine vibration isolation with FLPID control. Int J Model Simul 1–11

Sharma SK, Sharma RC, Lee J (2024f) Propelling precision of longitudinal vibration mitigation in ship propeller shafts through advanced nonlinear intelligent semi-active control leveraging adaptive neuro-fuzzy inference system with linear quadratic regulator. J Vib Control. https://doi.org/10.1177/10775463241244836

Sharma SK, Sharma RC, Upadhyay RK, Lee J (2024g) Enhanced ship engine vibration reduction using magnetorheological dampers and adaptive neuro-fuzzy control system. Ships Offshore Struct 1–17

Sinha AK, Sengupta A, Gandhi H, Bansal P, Agarwal KM, Sharma SK, Sharma RC, Sharma SK (2019) Performance enhancement of an all-terrain vehicle by optimizing steering, powertrain and brakes. In: Advances in engineering design. Lecture notes in mechanical engineering. Springer, Singapore, pp 207–215. https://doi.org/10.1007/978-981-13-6469-3_19

Stassen HG (1967) Random lateral motions of railway vehicles. Doctoral dissertation. Technische Hogeschool, Delft, Netherlands

Swapna Rekha H, Nayak J, Behera HS (2020) Pi-Sigma neural network: survey of a decade progress. In: Computational intelligence in pattern recognition. Advances in intelligent systems and computing, vol 1120. Springer, Singapore, pp 429–441. https://doi.org/10.1007/978-981-15-2449-3_37

Upadhyay RK, Sharma SK, Kumar V, Valera H (2023a) Introduction to transportation systems technology and integrated management. In: Transportation systems technology and integrated management. Springer, pp 1–4. https://doi.org/10.1007/978-981-99-1517-0_1

Upadhyay RK, Sharma SK, Kumar V, Valera H (2023b) Transportation systems technology and integrated management. Edited by RK Upadhyay, SK Sharma, V Kumar, H Valera. Springer Nature, Singapore. https://doi.org/10.1007/978-981-99-1517-0

Upadhyay RK, Sharma SK, Kumar V (Eds) (2024a) Intelligent transportation system and advanced technology. Springer Nature Singapore. https://doi.org/10.1007/978-981-97-0515-3

Upadhyay RK, Sharma SK, Kumar V (2024b) Introduction to intelligent transportation system and advanced technology. In: Intelligent transportation system and advanced technology. Energy, Environment, and Sustainability. pp 3–6. Springer, Singapore. https://doi.org/10.1007/978-981-97-0515-3_1

Van Deusen BD (1967) Analytical techniques for designing ride quality into automotive vehicles. SAE paper issue 670021

Van Deusen BD (1968) Human response to vehicle vibration. SAE paper issue 680090

Vishwakarma PN, Mishra P, Sharma SK (2022a) Characterization of a magnetorheological fluid damper a review. Mater Today: Proc 56:2988–2994. https://doi.org/10.1016/j.matpr.2021.11.143

Vishwakarma PN, Mishra P, Sharma SK (2022b) Formulation of semi-active suspension system and controls in rail vehicle. SSRN Electron J. https://doi.org/10.2139/ssrn.4159616

Wickens AH (1971) Advanced passenger train. In: Preprint issue 1461, Paper presented at ASCE-ASME transportation engineering meeting, July 26–30, Seattle, Washington

Wu Q, Cole C, Spiryagin M, Chang C, Wei W, Ursulyak L, Shvets A, Murtaza MA, Mirza IM, Zhelieznov K, Mohammadi S, Serajian H, Schick B, Berg M, Sharma RC, Aboubakr A, Sharma SK, Melzi S, Di Gialleonardo E, Cantone L et al (2023) Freight train air brake models. Int J Rail Transp 11(1):1–49. https://doi.org/10.1080/23248378.2021.2006808

Chapter 6
Reliability Model of Railroad Traffic Control System Equipped with Peltier Cell

Jerzy Kisilowski, Rafał Kowalik, and Emil Sadowski

6.1 Introduction

Publications (ASONG 2013; Bałuch 2011) address the application of cells for cooling purposes in rail vehicles and station cabinets, but lack algorithms as well as solutions for determining temperature and humidity. Equivalent models for engineering simulations for cell modules are presented in papers (Alaoui 2011; Aoki et al. 2018; Astrain and Martínez 2012; Bar-Cohe 2013; Bazu and Bajenescu 2009; Belov et al. 1999; Esarte et al. 2001; Kandasamy et al. 2004; Kobari et al. 2015; Kocur and Šuriansky 1847; Królicka et al. 2012; Kuper 2015; Lee 2010; Lineykin and Ben-Yaakov 2005; Min and Rowe 2000; Mitrani et al. 2007; Nuaire 2019; Press et al. 2002; Skrzypczak 2012; Whitaker 2001; Xin and Liang-Jun 2007; Xu et al. 2007; Yager and Filev 1995). The effect of humidity on the reliability of electronic systems is discussed in publications (Bariciński 2015; Bayle and Mettas 2010; Bazu and Bajenescu 2011; Bennet 2002; Będkowski and Dąbrowski 2006a; BMP280 2015; Buist 1995; Caillat et al. 1997; Çengel and Boles 2008; Ćwirko and Ćwirko 2008a; Dyduch and Moczarski 2012; Yang et al. 2008; Zybała and Pietrzak 2013). The effect of temperature on the reliability of electronic systems is presented in papers (Escobar and Meeker 2006; Filin 2002; Filin and Owsicki 2000; Filin et al. 2010; Flakiewicz

J. Kisilowski (✉)
Kazimierz Pulaski University of Technology and Humanities in Radom, Malczewskiego, Radom, Poland
e-mail: jerzykisilow@kisilowscy.waw.pl

R. Kowalik
Polish Air Force University, Dywizjonu 303 Street No. 35, Deblin, Poland
e-mail: r.kowalik@law.mil.pl

E. Sadowski
P.U.H. HESTA, Podchorążych Street No. 7, 26-600 Radom, Poland
e-mail: emil@hesta.com.pl

S. K. Sharma et al. (eds.), *Dynamics of Transportation Ecosystem, Modeling, and Control*, Energy, Environment, and Sustainability,
https://doi.org/10.1007/978-981-97-0437-8_6

2009; Fleurial et al. 1997; Gabert 2008; Mannella 2014; Goldsmid 2010; Gould and Shammas 2012; Gromov 1979; Haidar et al. 2016; Harman 2003; Karimi et al. 2011; Kijak 1975). In papers (Çengel and Boles 2008; Ćwirko and Ćwirko 2008b; Fairbanks 2008; Sharma 2019; Haruyama 2001; Hnatek 2003; Huebener and Tsuei 1998; Hur et al. 2008; Hwang et al. 2009; Jacob 2016; Jowett 1972; Kijak 1975; Zalewski 1998; Zheng et al. 2011), the influence of environmental parameters on changes in the physical properties of electronic components is presented. The publications (Będkowski and Dąbrowski 2006b; Dąbrowska-Bajon et al. 1981; Kisiel and Bajera 1988; Lexicon of Railway Terms 2011; Markowski 2008; Matkowski 2009; Siergijczyk and Rosiński 2014; Singh 2003; Smith et al. 2010; Sotskov 1973; Ważyńska-Fiok and Jaźwiński 1990) define system reliability. The Technical Guidelines for Construction of Railway Traffic Control Devices of PKP S.A. indicate to "ensure resistance to harmful environmental influences (temperature, humidity, pollution)." Instruction Ie-100a "Conditions for safe installation and operation of railroad traffic control devices on railroad lines managed by PKP Polskie Linie Kolejowe S.A." clarifies the range of ambient temperature and humidity (−40 to 70 °C).

Manufacturers of trackside traffic control systems in their documentation (Documentation of EON-3 2023; Driankow et al. 1996; MAXIM 2008; Mogielnicki 1983; ZWUA 2008; Węgierski 1971; Wojtas 2010; Zajączkowski et al. 1982; Zakrzewski et al. 2015) specify acceptable operating conditions, but there is no information about the impact of selected environmental factors on the reliability of equipment. There is also a lack of information and research on the effect of temperatures in a wider range (e.g., −45 °C, and/or thermal shock), allowing the installation of equipment in more difficult climatic zones (e.g., for the "Trans-Siberian" railroad or railroads in Finland), and improving the reliability of this system component. Excessively high temperatures have been the most common cause of TCS equipment unreliability in Australia. Based on the analysis of the state of the art, it was found that there were no studies simultaneously of the effects of temperature and humidity on electronic devices used in railroad traffic control. Also, no such solution was found to improve selected environmental conditions in TCS devices with the help of Peltier cells.

6.2 Introduction

6.2.1 Exponential Distribution Sample

One of the simplest probabilistic models of the uptime of a non-renewable object is a random variable T, whose damage intensity is constant, i.e., independent of time (Macha 2001):

$$\lambda(t) = \lambda = \text{const} \tag{6.1}$$

Then, based on the relationships given in the tables (Macha 2001) is obtained (for $t \geq 0$) (Macha 2001):

$$R(t) = \exp(-\lambda t), R(0) = 1 \tag{6.2}$$

$$Q(t) = F(t) = 1 - \exp(-\lambda t) \tag{6.3}$$

Such unreliability (distribution) has a random variable with an exponential distribution. Other functional and numerical characteristics are as follows (Fig. 6.1) (Macha 2001):

$$f(t) = \lambda \exp(-\lambda t)\text{---probability density,} \tag{6.4}$$

$$\Lambda(t) = \lambda t\text{---cumulative damage intensity,} \tag{6.5}$$

$$\hat{t} = \lambda^{-1}\text{---expected time of suitability,} \tag{6.6}$$

$$\sigma_t^2 = \lambda^{-2}\text{---variance,} \tag{6.7}$$

$$M_e = \lambda^{-1} \ln 2\lambda\text{---mediana,} \tag{6.8}$$

$$M_o = 0\text{---mode.} \tag{6.9}$$

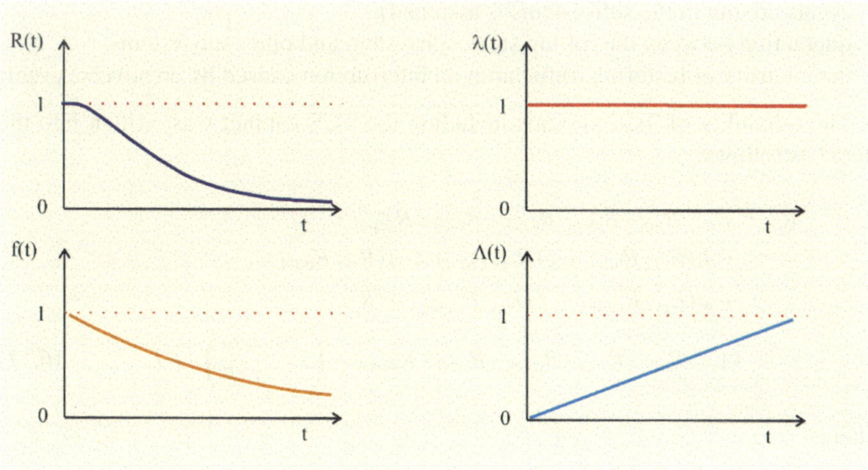

Fig. 6.1 Reliability of TCS systems considering TCS cabinet equipment

The exponential distribution is characterized by a property called non-memory, which is that the probability of an object's fitness is independent of how long it has been in use. In other words, an undamaged used object is as reliable as a new object. This model does not take into account changes in the object due to adaptation and aging processes, such as fatigue and material loss. Objects for which an exponential distribution can be applied are those with purely random damage.

The paper (Węgierski 1971) relates the issues of modeling mass service systems to aspects of time reserves needed in the proper operation of TCS systems. The author of the paper (Węgierski 1971) starts from the concept of relative time reserve $b = z/ z\acute{s}r$, which is a random variable with a distribution:

$$h(b) = e^{-b} \tag{6.10}$$

The distribution of the sum of B_n stocks is expressed as follows:

$$H(B_n) = \int_0^{B_n} \frac{|(B_n)^{n-1}}{(n-1)!} e^{-Bn} dB_n \tag{6.11}$$

Knowing the relative value of primary delay p1/zśr, it is possible to determine the number of time reserves needed to eliminate the traffic effects of primary delay.

A similar issue is the capacity of track junctions, which considers the occupancy of critical infrastructure points in the context of conflict runs.

More detailed is the analysis of the effects of events as primary damage and associated secondary damage (ignoring the traffic impact).

The model of rail service reliability consists of three components:

- reliability of the technical objects that make up the system (the independence of events arising in the subsystems is assumed),
- interaction between the rolling stock subsystem and other subsystems,
- the intensity of restoring traffic after an interruption caused by an adverse event.

The reliability of TCS systems including the TCS cabinet was written into the model as follows:

$$\begin{aligned} R_{tr} = {} & R_{rs} \cdot R_{nrs} \cdot R_{\text{stat}} \cdot R_{\text{track}} \cdot R_{\text{civil}} \cdot R_{\text{depot}} \\ & + b(R_{rs}, R_{nrs}) + c(R_{rs}, R_{\text{stat}}) + d(R_{rs}, R_{\text{track}}) \\ & + e(R_{rs}, R_{\text{civil}}) + f\left(R_{rs}, R_{\text{depot}}\right) + R_t^n \cdot \\ & (1 - R_{nrs} \cdot R_{stat} \cdot R_{\text{track}} \cdot R_{\text{civil}} \cdot R_{\text{depot}}) \cdot \left(1 - \frac{rt}{OT}\right) \cdot PT \end{aligned} \tag{6.12}$$

where

- R_{rs}—reliability of rail vehicles,

- R_{nrs}—reliability of TCS cabinet infrastructure (signaling, communication, electrical power), also depends on environmental conditions, which will be further studied,
- R_{stat}—reliability of station functions,
- R_{track}—reliability of track superstructure,
- R_{civil}—reliability of engineering facilities,
- R_{depot}—reliability of maintenance facilities,
- R_t—reliability of a single train that has completed its daily tasks,
- n—daily number of trains,
- rt—the sum of time during which the system was unfit to run services,
- OT—sum of the daily time of running train traffic,
- PT—frequency of restoration of traffic as a result of repairs carried out using standard repair crews,
- $b()$, $c()$, $d()$, $e()$, $f()$—relations between rolling stock and other subsystems.

The method of determining the value of dependency coefficients was not explained. Next, a model of timeliness was presented:

$$
\begin{aligned}
TCS = {} & \left[\prod_{i=1}^{TP} R_i^{N_{ij}} + \sum_{k=1}^{FT} g(k,j)\right] \cdot R_{\text{nrs}} \cdot R_{\text{stat}} \cdot R_{\text{track}} \cdot R_{\text{civil}} \cdot R_{\text{depot}} \\
& + h(R_{rs}, R_{\text{nrs}}) + l(R_{rs}, R_{\text{stat}}) + p(R_{rs}, R_{\text{track}}) \\
& + s(R_{rs}, R_{\text{civil}}) + t\big(R_{rs}, R_{\text{depot}}\big) \\
& + \left[\prod_{i=1}^{TP} R_i^{N_{ij}} \cdot \big(1 - R_{\text{nrs}} \cdot R_{\text{stat}} \cdot R_{\text{track}} \cdot R_{\text{civil}} \cdot R_{\text{depot}}\big) \cdot \left(1 - \frac{rt}{OT}\right) \cdot PQ\right]
\end{aligned}
\tag{6.13}
$$

where

- R_i—reliability of the TCS system performing the task of traffic control at the station,
- TP—number of model courses in one day,
- N_{ij}—number of courses on the j-th day for the i-th relation,
- FT—the daily number of errors in the TCS cabinet,
- $g(k,j)$—the number of delays that occurred for different relations (it is assumed that for k greater than or equal to three relations $g(k,j)$ tends to zero),
- PQ—the frequency of suppression of delays due to time reserves,
- $h()$, $l()$, $p()$, $s()$, $t()$—coefficients representing on-time arrival of trains despite the occurrence of damage of a given type.

6.2.2 Weibull Distribution

The Weibull distribution is characterized by greater generality than the exponential. It is used when the damage intensity is a monotonic variable. Among other things, this distribution is used to describe the fatigue life of materials and mechanical structures. The damage intensity is determined by the formula (Macha 2001):

$$\lambda(t) = \alpha\beta t^{(\alpha-1)} \tag{6.14}$$

where

- $(\alpha, \beta > 0)$—fixed numbers,
- α—shape parameter,
- β—scale parameter.

when

- $\alpha > 1$, $\lambda(t)$ increases
- $\alpha = 1$, $\lambda(t) = \text{const} = \beta$
- $\alpha < 1$, $\lambda(t)$ decreases.

Other functional and numerical characteristics are as follows (Macha 2001):

$$R(t) = exp(-\beta t^{\alpha}), t \geq 0, \tag{6.15}$$

$$Q(t) = F(t) = 1 - \exp(-\beta t^{\alpha}), \tag{6.16}$$

$$f(t) = \alpha\beta t^{(\alpha-1)} \exp(-\beta t^{\alpha}), \tag{6.17}$$

$$\Lambda(t) = \beta t^{\alpha}, \tag{6.18}$$

$$\hat{t} = \Gamma\left(1 + \frac{1}{\alpha}\right)\beta^{-\frac{1}{\alpha}} \tag{6.19}$$

where $\Gamma(x)$—the values of the Euler gamma function (read from tables) (Macha 2001):

e.g., "$\Gamma(n) = (n - 1)$"!,

$$\begin{aligned} &\Gamma\left(\tfrac{1}{2}\right) = \sqrt{\pi} \\ &\Gamma\left(n + \tfrac{1}{2}\right) = \tfrac{1\cdot 3\cdot 5\ldots(2n-1)\sqrt{\pi}}{2^n} \\ &\Gamma(n+1) = n! \end{aligned} \tag{6.20}$$

When $\alpha = 1$ an exponential distribution is obtained, while when $\alpha = 2$ a Rayleigh distribution is obtained. For exponential distribution (Macha 2001):

$$\hat{t} = \Gamma(2)\beta^{-1} = \frac{1}{\beta} = \frac{1}{\lambda}(when\ \lambda = \beta) \tag{6.21}$$

For Rayleigh distribution (Macha 2001):

$$\hat{t} = \Gamma\left(\frac{3}{2}\right)\beta^{-\frac{1}{2}} = \frac{\sqrt{\pi}}{2\sqrt{\beta}} = \frac{0.8862}{\sqrt{\beta}} \tag{6.22}$$

Variance (Macha 2001):

$$\sigma_t^2 = \left[\Gamma\left(1 + \frac{2}{\alpha}\right) - \Gamma^2\left(1 + \frac{1}{\alpha}\right)\right]\beta^{-\frac{2}{\alpha}} \tag{6.23}$$

For exponential distribution (Macha 2001):

$$\sigma_t^2 = \frac{1}{\beta^2} = \frac{1}{\lambda^2} \tag{6.24}$$

because (Macha 2001)

$$\Gamma(3) - \Gamma^2(2) = 2! - (1!)^2 = 1 \tag{6.25}$$

For Rayleigh distribution (Macha 2001):

$$\sigma_t^2 = \frac{4 - \pi}{4\beta} \tag{6.26}$$

because (Macha 2001)

$$\Gamma(2) - \Gamma^2\left(\frac{3}{2}\right) = 1 - \left(\frac{1}{2}\sqrt{\pi}\right)^2 = \frac{4 - \pi}{4} = 0.2146 \tag{6.27}$$

Median (Macha 2001):

$$M_e = \sqrt[\alpha]{\frac{\ln 2}{\beta}} \tag{6.28}$$

Quantile (Macha 2001):

$$t_p = \sqrt[\alpha]{\frac{-\ln(1 - p)}{\beta}} \tag{6.29}$$

A three-parameter Weibull distribution is also used (Macha 2001), for which the

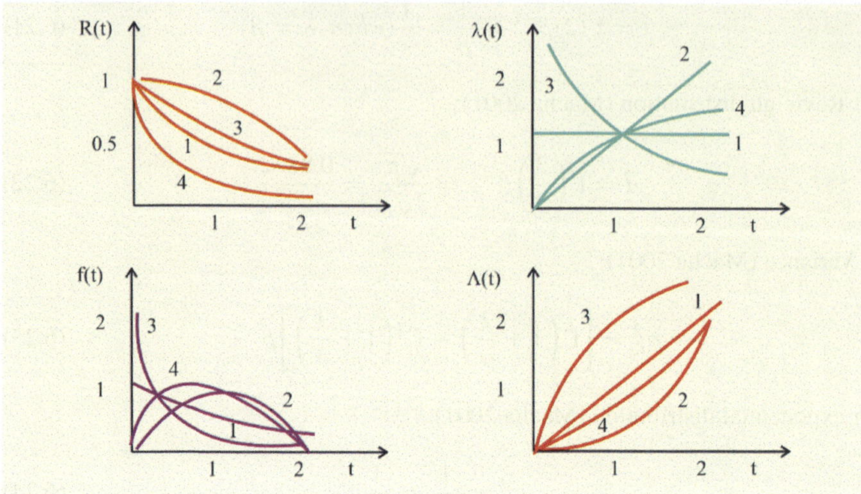

Fig. 6.2 Weibull family of distributions: 1—exponential ($\alpha = \beta = 1$); 2—Rayleigh's ($\alpha = 2, \beta = 0.5$); 3—of decreasing damage intensity ($\alpha = 0.5, \beta = 2$); 4—of increasing damage intensity ($\alpha = 3/2, \beta = 2/3$). The function flows relate to $\alpha\beta = 1$

$$\lambda(t) = \begin{cases} \alpha\beta(t - t_0)^{\alpha-1} \\ 0 \text{ for } t < t0 \end{cases} \tag{6.30}$$

where t_0—offset parameter.

In the case of a three-dimensional Weibull distribution, failure of the object can occur only after the initial failure-free operation in the range of $[0, t_0)$ (Fig. 6.2).

The calculations were performed for the parameters shown in Figs. 6.3 and 6.4 and for 1000 TCS cabinet engagements.

6.3 Expected Remaining Facility Uptime

The reliability of an object that has already been in use for a certain period of time is generally different from the reliability of the same object at the start of its use. The reliability of an object in use is determined by a function whose value at any time is equal to the conditional expected value of the object's remaining airworthiness time. This function is called the expected remaining serviceability time and is defined by the formula (Macha 2001):

$$r(t) = E[T - t|T \geq t], t \geq 0 \tag{6.31}$$

	Statistic	P-Value
Anderson-Darling	2.17521	0.0738137
Cramér-von Mises	0.440295	0.056674
Kolmogorov-Smirnov	0.104262	0.0713016

data mean =3.58686, st.dev=0.979515
Weibull mean=3.43857, st.dev=1.02783

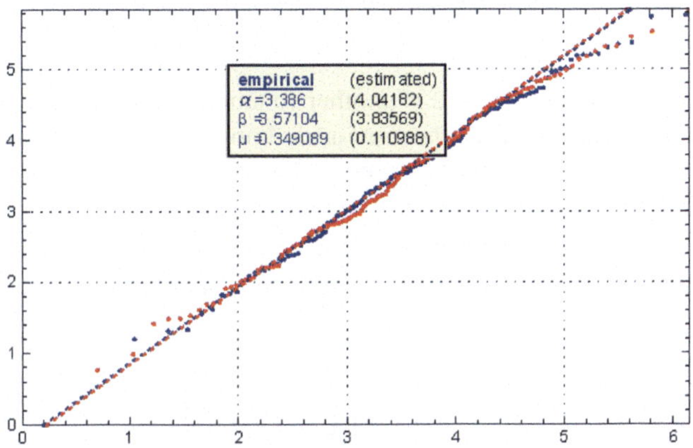

Fig. 6.3 Occurrence of failure in the system from the Weibull distribution for 1000 TCS cabinet switch-ups [Own elaboration]

No.	$L(h)$	$x = \log L$	$R(\%)$	$F = R/100$	$y = \log\log[1/(1-F)]$
1	157.99	2.1986	9.46	0.0946	−1.3649
2	163.88	2.2145	22.97	0.2297	−0.9456
3	234.36	2.3699	36.49	0.3649	−0.7052
4	574.28	2.7591	50.00	0.5000	−0.5214
5	698.41	2.8441	63.51	0.6351	−0.3587
6	1255.20	3.0987	77.03	0.7703	−0.1946
7	1826.20	3.2615	90.54	0.9054	0.0104

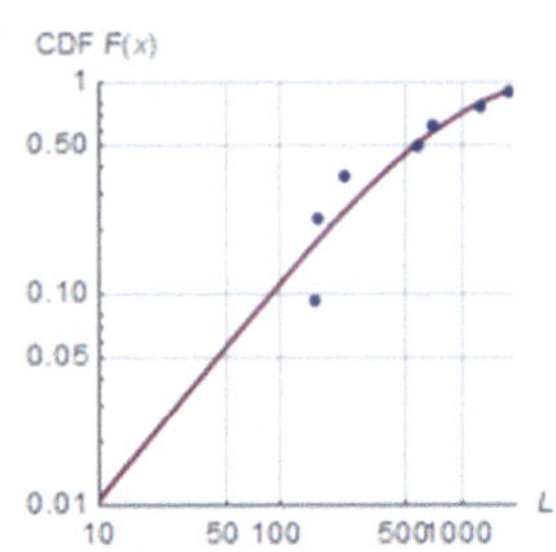

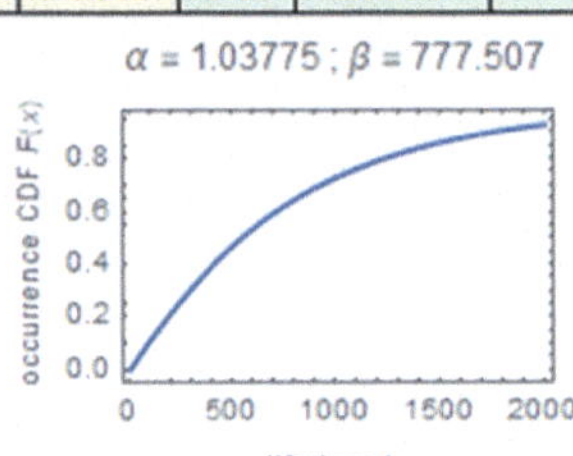

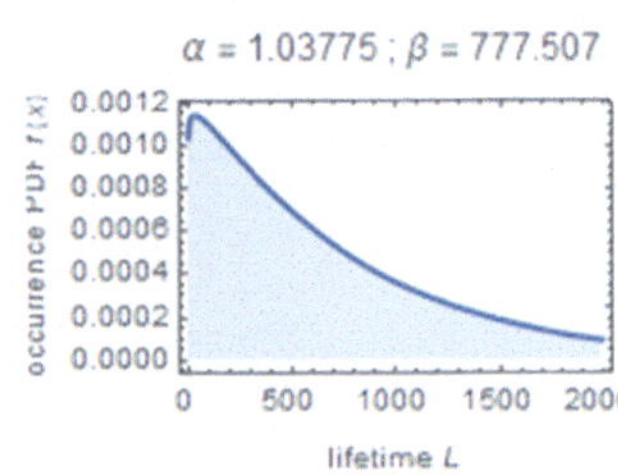

Fig. 6.4 Incidence of failure in the system from the Weibull distribution [Own elaboration]

$$r(t) = \int_t^\infty \frac{R(x)}{R(t)} dx = \frac{1}{R(t)} \int_t^\infty R(x)dx \tag{6.32}$$

It can be noted that for $t = 0$ one obtains (Macha 2001)

$$r(0) = \int_t^\infty R(x)dx = E[T] = \hat{t} \text{ if only } R(0) = 1 \tag{6.33}$$

The expected remaining uptime is another functional characteristic of the reliability of a non-renewable object. The previously known functional characteristics of reliability (Macha 2001) can also be expressed by it:

$$Q(t) = F(t) = 1 - \frac{r(0)}{r(t)} \exp\left(-\int_t^1 \frac{dx}{r(x)}\right) \tag{6.34}$$

$$R(t) = \frac{r(0)}{r(t)} \exp\left(-\int_t^1 \frac{dx}{r(x)}\right) \tag{6.35}$$

$$f(t) = \frac{r(0)}{r^2(t)}\left[1 + \frac{dr(t)}{dt}\right] \exp\left(-\int_0^t \frac{dx}{r(x)}\right) \tag{6.36}$$

$$\lambda(t) = \frac{1}{r(t)}\left[1 + \frac{dr(t)}{dt}\right] \tag{6.37}$$

$$\Lambda(t) = \int_0^t \frac{dx}{r(x)} - \ln\frac{r(0)}{r(t)} \tag{6.38}$$

For exponential distribution (Macha 2001):

$$\begin{aligned} r(t) &= \int_t^\infty \frac{\exp(-\lambda x)}{\exp(-\lambda t)} dx \\ &= \frac{1}{e^{-\lambda t}}\left(-\frac{1}{\lambda^{\lambda t}} e^{-\lambda x}\right|_t^\infty = \frac{1}{e^{-\lambda t}}\left(-\frac{1}{\lambda e^{\lambda t}}\right|_t^\infty = \\ &= \frac{1}{e^{-\lambda t}}\left[-\frac{1}{\lambda e^{\lambda \infty}} - \left(-\frac{1}{\lambda e^{\lambda t}}\right)\right] = \frac{1}{\lambda} = \hat{t} \end{aligned} \tag{6.39}$$

For the Weibull distribution (Macha 2001):

$$r(t) = \int_t^{\infty} \frac{\exp(-\beta x^{\alpha})}{\exp(-\beta t^{\alpha})} dx = \int_t^{\infty} \exp\left[\beta\left(t^{\alpha} - x^{\alpha}\right)\right] dx \tag{6.40}$$

When modeling the survival function, an auxiliary concept is introduced—gambling, that is, a quantity that tells what is the probability of observing an event at the next moment Δt (if it has not been observed until now (until t)). It is assumed that the value of the hazard is related to the characteristics of the object under study (depends on the value of explanatory variables—independent variables). The most common assumption is the form of the influence of explanatory variables on the value of hazards.

The graphs of the Hazard function shown in Fig. 6.5 are determined for the same parameter values, differing only in the type of probability distribution of the random variable T denoting time to failure. Analysis of the graphs shows that both criterion functions reach a minimum value for similar time moments. However, there are some differences in the correct values of the criterion function. The minimum value of the Hazard function for the Gamma distribution is smaller than for the Weibull distribution. This shows that identifying the type of time-to-failure probability distribution is important in operation tasks.

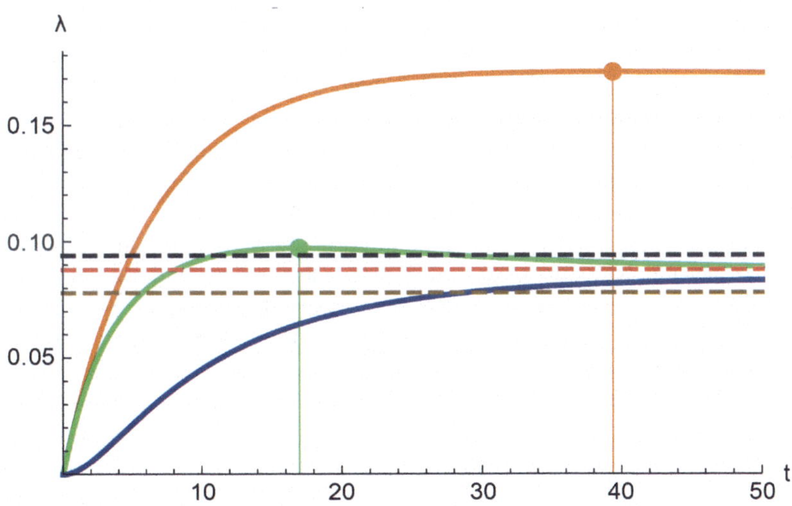

Fig. 6.5 Hazard function [Own elaboration]

6.4 Expected Remaining Facility Uptime

Operation, and therefore maintenance and operation of control systems with a higher level of reliability including TCS cabinets, requires additional qualifications of personnel related to the guarantee of maintaining the level of safety integrity (Documentation of EON-3 devices 2023; Driankow et al. 1996; ZWUS 2008). Access to the equipment of the control system should be carefully defined in the documentation of the system, where there should be provisions regarding both the authority and responsibility of each group of personnel. The division into groups can be made, in a very general way, based on the range of activities that individuals will perform while working with the control system, into

- operators,
- operating personnel,
- supplier personnel.

Operators are the group working the longest with the control system, but at the same time have very limited authority. Their role is limited to issuing commands and monitoring the status of the system.

Service personnel have less frequent contact with the system, and their role is limited to periodic inspection of the correctness of the operation of the equipment (inspections) and making ongoing repairs. Persons belonging to this group have greater authority, e.g., they can delete errors, and are authorized to run the system's self-test function. They can also change the configuration of the system, with the reconfiguration procedure being in accordance with the provisions of the control system documentation. This group of personnel should know the principles of interaction of environmental conditions (temperature, humidity, ...) on the correct functioning of TCS systems located in track cabinets.

The supplier's personnel have the most authority, most often unlimited, with their access to the system being the most limited and reduced to the commissioning of the control system and commissioning. During this period, procedures must be implemented to determine the environmental conditions affecting the correct operation of TCS systems located in trackside cabinets. During operation, people from this group of users can gain access to elements of the control system only with the knowledge and consent of people from other groups and the consent of the owner of the infrastructure. This is dictated by the possibility of influence on the course of the controlled process by persons who do not have sufficient knowledge of the implementation of the control process and/or do not bear direct responsibility in the event of leading to a hazardous event.

The described safety guaranteed by the separation of authority and responsibility of individual persons working with the control system is only one element of the safety policy that the control system owner should have.

An equally important factor affecting the safety of the control system and playing a key role in the process of determining the safety level (safety integrity) of the control system is the reliability parameter of the hardware used in the control system and the organization of safety procedures in the software. In this regard, procedures should be implemented to determine the impact of environmental conditions on the reliability of TCS system operation.

The value of hardware defectiveness, determined using the *Mean Time Between Failure* (MTBF) or *Mean Time To Failure* (MTTF) indicator, together with software safety procedures, makes up the functional safety of the system. Functional safety is defined as the part of total safety relating to the controlled device and control system that depends on the proper operation of safety-related control systems.

For rail traffic control systems, which are control systems with a higher level of reliability, the level of safety integrity is estimated based on formula (6.41):

$$THR \approx \frac{FR_A}{SDR_A} \cdot \frac{FR_B}{SDR_B} \cdot (SDR_A \cdot SDR_B) \tag{6.41}$$

where

- *FR*—damage index, respectively for channels A and B, for computer systems the damage intensity value is adopted as λ,
- *SDR*—safe shutdown factor, for channels A and B, respectively.

The SDR safe shutdown factor is defined according to Eq. (6.42), with average periodic testing times assumed to be equal to fault detection times:

$$\frac{T}{2} + NT = SDT = \frac{1}{SDR} \tag{6.42}$$

where

- *T*—testing time,
- *NT*—blocking time,
- *SDT*—average system response time (from detection to lockout).

The value of MTBF operating in 2oo2 mode (two channels functioning in the system and in each channel two blocks connected in series), assuming that both channels have the same damage intensity, is 3/2λ. With different damage intensities in each channel, the MTBF value of the entire system is (6.43)

$$MTBF = \int_0^\infty R(t)dt = \frac{1}{\lambda_a} + \frac{1}{\lambda_b} - \frac{1}{\lambda_a - \lambda_b} \tag{6.43}$$

where

- $R(t)$—damage probability distribution,
- λ_a—intensity of damage to the component (system) *a*,
- λ_b—intensity of damage to the component (system) *b*.

At the same time, it should be noted that certain assumptions have been made in making the MTBF = 1 comparison, since these values cannot be compared directly. The damage intensity value is not constant over time and does not take into account service or restoration time, unlike the MTBF value. An assumption comparing the two values can be made by, among other things, assuming maintenance-free equipment up to the first fault, i.e., after the first fault occurs, a system component is replaced, and assuming that the fault rate over the life of the system is constant. In addition, when assuming the MTBF value, no account is taken of the service time, i.e., system recovery, which enters into the time reported as MTBF. Therefore, it can be said that the MTBF value is equal to the MTTF for a given system component.

In terms of reliability, the TCS cabinet is a complex object. It consists of subsystems and elements with a serial reliability structure, i.e., failure of any of them causes failure of all elements of the TCS cabinet.

Figure 6.6 shows an example model of an n-element object with finite renewal times. In this figure, t_1, t_2, t_3 .... denote the moments of damage to the object, and u_1, u_2, u_3, ... denote moments of renewals. If the moments of damage and renewal of all elements are marked on a common axis, we get the damage flux and renewal flux, which is the sum of n damage (renewal) processes, because there are so many elements in the object (Figs. 6.7, 6.8, 6.9, 6.10, 6.11 and 6.12).

Analyzing the TCS cabinet as a complex object consisting of n renewable elements and assuming that:

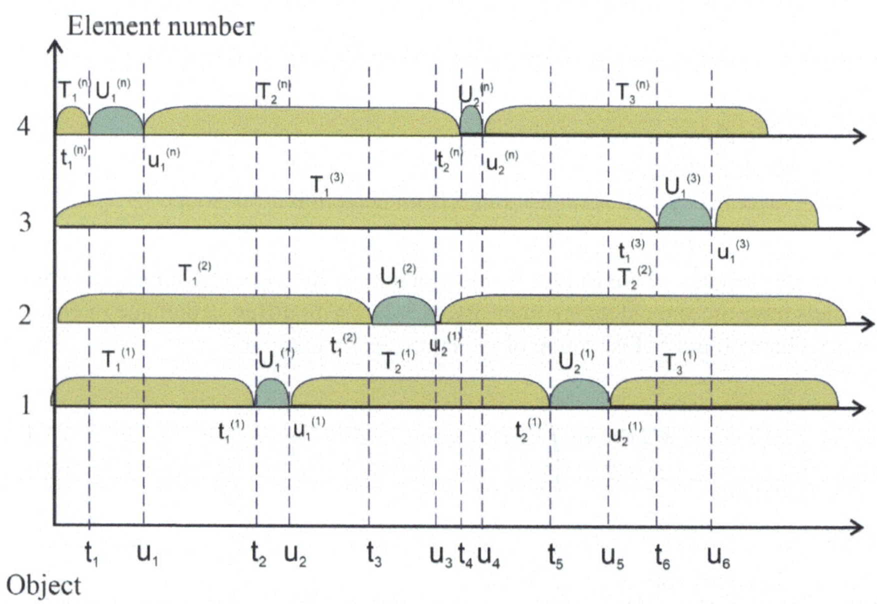

Fig. 6.6 Damage and renewal model of an object composed of n renewable elements connected in series

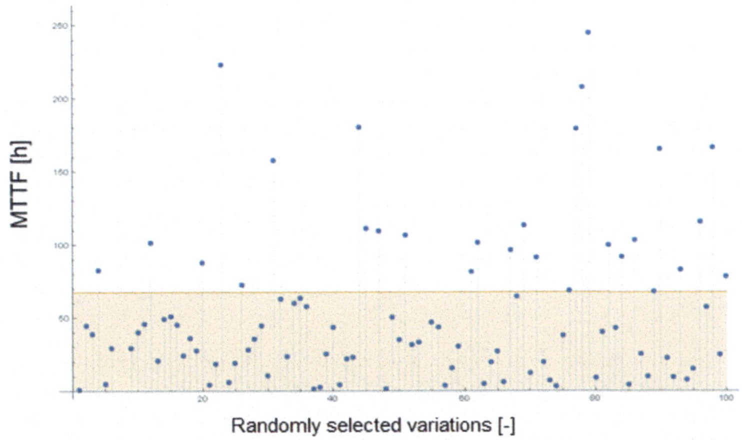

Fig. 6.7 Mean time to failure (MTTF) function determination waveforms for a Peltier cell cabinet for 30,000 cooling operations [Own elaboration]

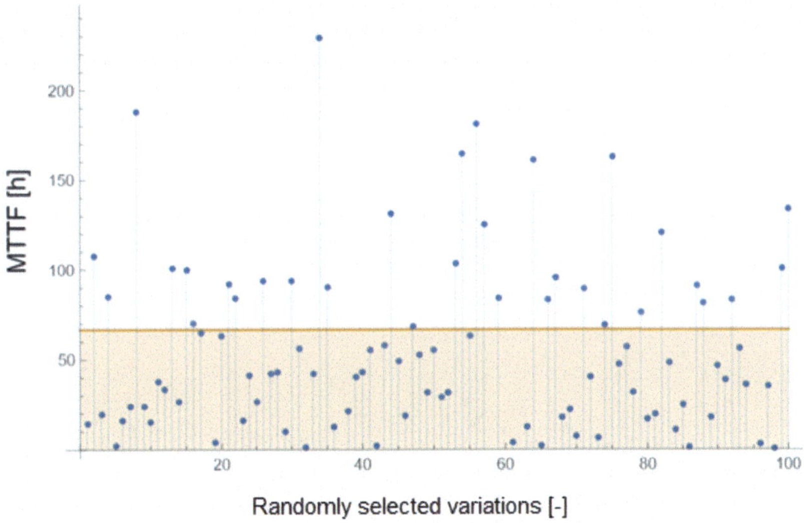

Fig. 6.8 Mean time to failure (MTTF) function determination waveforms for a cabinet without a Peltier cell for 30,000 cooling operations [Own elaboration]

- operation, damage and renewals of one type of element do not affect the reliability of the others,
- the moments of damage (renewals) of each element constitute the damage (renewal) process,
- all times of correct operation of the i-th element have an equal distribution,
- all renewal times of the i-th element also have an equal distribution.

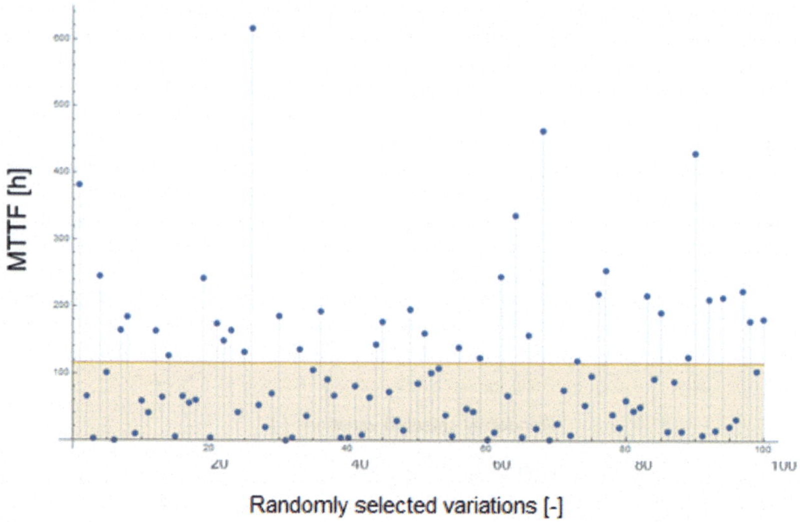

Fig. 6.9 Mean time to failure (MTTF) function determination waveforms for a Peltier cell cabinet for 40,000 cooling operations [Own elaboration]

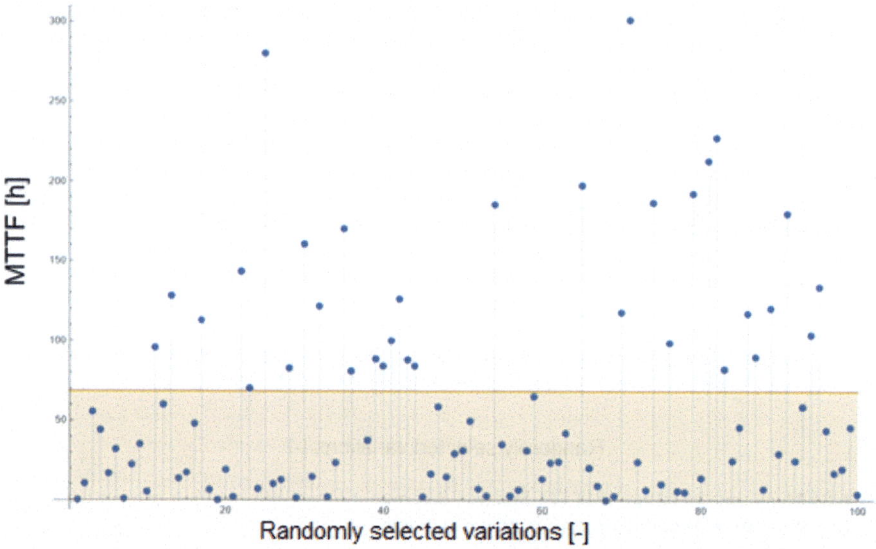

Fig. 6.10 Mean time to failure (MTTF) function determination waveforms for a cabinet without a Peltier cell for 40,000 cooling operations [Own elaboration]

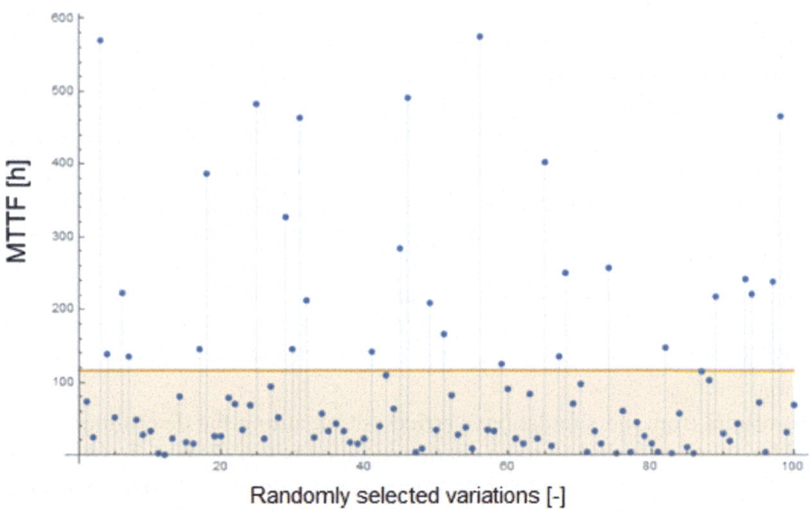

Fig. 6.11 Mean time to failure (MTTF) function determination waveforms for a Peltier cell cabinet for 60,000 cooling operations [Own elaboration]

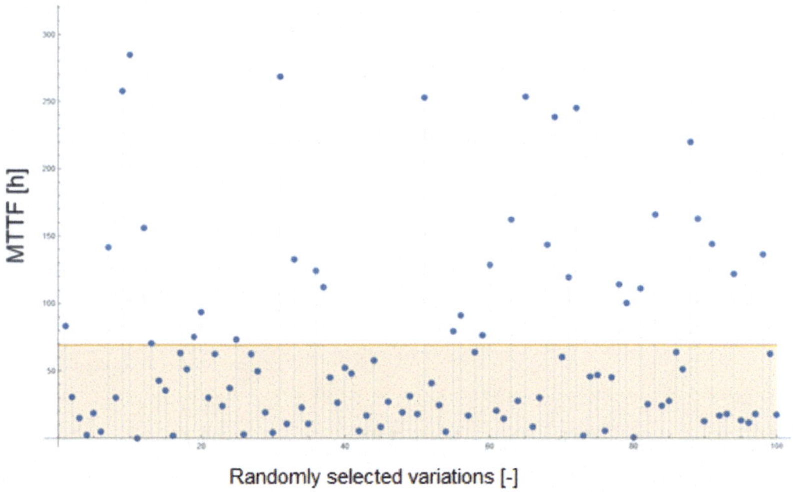

Fig. 6.12 Mean time to failure (MTTF) function determination waveforms for a cabinet without a Peltier cell for 60,000 cooling operations [Own elaboration]

The determination of the MTTF function was developed according to the procedure in earlier chapters of the dissertation.

First, the identification of risks occurring during all activities related to the operation of the TCS cabinet was carried out. Undesirable events (ZN) were selected, for which the probability of their occurrence per one switch on the track cabinet and per one hour of track cabinet operation was calculated. The data for the calculations were defined based on the analysis of foreign manufacturers of railroad automation equipment, taking into account a 2% higher number of TCS cabinet failure occurrences. During the entire life of the TCS cabinet, 1010 adverse events occurred. The equipment comprising the TCS cabinet operated for about 96,169 h during its lifetime. During this time, it was assumed that the Peltier cell carried out the cooling process: for the first case, it was switched on 30,000 times, for the second case 40,000 and in the third case 60,000 times.

Based on the above data, the probability of ZN is

Q(1 cabinet switching) $= 12.6 \cdot 10^{-3}$ and Q(1 h) $= 26.2 \cdot 10^{-3}$.

The probability of non-occurrence of ZN after 10, 20, 50 and 100 switchings of the TCS cabinet was calculated, as well as after 10, 20, 50 and 100 h of TCS cabinet operation.

The results from the simulations carried out regarding the determination of the MTTF function are included in the charts presented below (Figs. 6.7, 6.8, 6.9, 6.10, 6.11 and 6.12). The average time to an adverse event was MTTF = 69.45 h in the first case. An adverse event was recorded on average every 79 TCS cabinet activations. In the second case, the average time to occurrence of a TCS cabinet failure was MTTF = 110.45 h. An adverse event was recorded, on average, every 20 TCS cabinet switch-ups. For the third case, the average time of failure occurrence in the TCS cabinet was MTTF = 112.42 h. An adverse event was recorded on average every 11 TCS cabinet switch-ups.

In general, adverse events do not cause any consequences, or the consequences of events are classified as incidents. The total number of undesirable events is 950, that is, the probability of an undesirable event occurring during one TCS cabinet switch-on is $1.01 \bullet 10^{-3}$. This is a significant number compared to the operation of land-based equipment.

In order to increase the level of safety, attention should be paid to adverse events with the highest risk. That's why proper operation measures are so important. Inspections should be carried out at times in accordance with the instructions, and repairs should be done properly.

6.5 Verification of the Developed Program and Comparison of Calculation Results

Calculations of the indicators needed to assess the level of safety integrity were also performed using Mathematica software. The use of the simulation method for the analysis of the time of correct operation in the Mathematica environment requires input data, which include, first of all, the reliability structure of the system, the reliability characteristics of the objects of this system, the duration of the simulation, expressed in the assumed units of work performed, and the number of repetitions of the simulation. As a result of performing the calculations, the damage probability

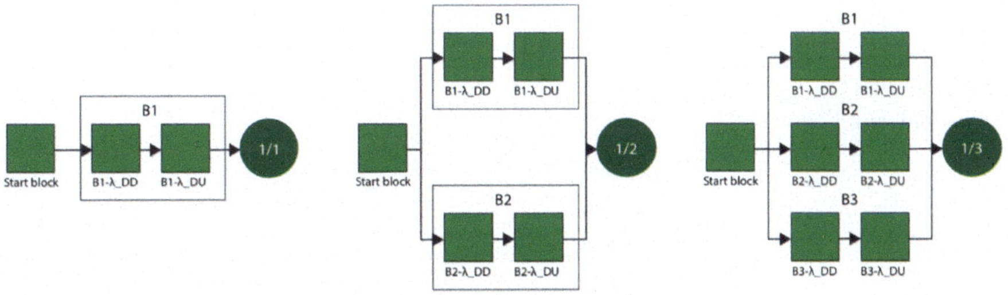

Fig. 6.13 Selected alternate block diagrams of the analyzed kzn structures

values of the structure's objects are obtained. Subsequent simulation activities of the program lead to the calculation of the reliability of the system on the basis of its reliability structure declared earlier by the analyst.

For the purpose of this study, a periodic test interval of $T_1 = 17.520$ [h] was assumed, and then Monte Carlo simulation parameters were applied. An end-of-life time of 20,000 [h] was assumed. This time was set larger than T1 due to the property of the calculation process, the results of which at the boundaries of the intervals show a significant scatter of values compared to those inside the interval. The computational step expressed by time increment = 10 [h] and the repetition rate of 100,000 cycles of transitions were selected as a result of optimizing the accuracy and duration of the process of generating simulation values. Due to the need to take into account the specified fractions of damage λ_{DD} and λ_{DU}, which make up λ_D, calculations were carried out for the substitute structures of kzn systems prepared for this purpose. Selected from the structures are shown in Fig. 6.13.

Models of classical reliability theory can be used to describe the reliability of the TCS cabinet or the functional assemblies mentioned above. In such a case, various structures are used, most often serial. However, comparing these models with reliability models such as state-transition graphs, it can be seen that the latter are more useful for the reliability description of the TCS cabinet. This is because they allow for a more in-depth study of the reliability of the TCS cabinet than the models of classical reliability theory, since they reproduce the changes resulting from the specifics of their operation and functional assemblies. For this reason, transition graph models are called reliability-functional models of equipment, including the TCS cabinet. In addition, the usefulness of these models is also due to the possibility of relatively easy determination of reliability indicators, both of the TCS cabinet and its assemblies. These indicators can be obtained using analytical relationships, or computer simulation methods. Under appropriate assumptions regarding the properties of probability distributions of random quantities (variables) and other properties that characterize TCS cabinets as objects of reliability studies, the theory of semi-Markov processes can be used.

In practice, failed functional assemblies and their structural components, for any TCS cabinet, are generally refurbished (if cost-effective, of course) by performing

Table 6.1 Assumed values of input data for reliability calculations of TCS system without and with Peltier cell

Data set 1 (without Peltier cell)	
λ_D [1/h]	1.80E−06
DC [%]	93
λ_{DD} [1/h]	1.67E−06
λ_{DU} [1/h]	1.26E−07
MTTR [h]	2.40E+01
μ_{DD} [1/h]	$MTTR^{-1}$
μ_{DD} [1/h]	$((T_1/2) + MTTR)^{-1}$
Data set 2 (with Peltier cell)	
λ_D [1/h]	1.25E−04
DC [%]	85
λ_{DD} [1/h]	1.06E−04
λ_{DU} [1/h]	1.88E−05
MTTR [h]	2.40E+01
μ_{DD} [1/h]	$MTTR^{-1}$
μ_{DD} [1/h]	$((T_1/2) + MTTR)^{-1}$

appropriate maintenance. An additional limitation of using a serial reliability structure to describe the reliability of a TCS cabinet is the need to adopt, as already mentioned, only a bivariate division of the reliability state (i.e., fit-unfit) of the TCS cabinet and its individual functional assemblies.

The presented reliability description of any TCS cabinet can, of course, be expanded, specifying as many states of intermediate (partial) fitness as will be needed by the user of a given TCS cabinet to ensure its reasonable operation.

Calculations were carried out for two values of T1 times 8760 [h] and 17,520 [h] and for two values of damage intensity λ_D, expressed by fractions λ_{DD} and λ_{DU}. It was assumed that all objects in the kzn structure are identical and have the same reliability parameters. The values of the parameters used are shown in Table 6.1.

The results of the performed calculations of the probability of being in an unfit state of systems with certain structures are shown in Figs. 6.14, 6.16 and 6.17 with a Peltier cell, while Figs. 6.15 and 6.18 without a Peltier cell.

The results of the conducted tests are presented in the form of characteristics of the probability of the system remaining in an unfit state, which are plotted in Figs. 6.19, 6.20 and 6.21 with a semi-logarithmic grid with Peltier elements. Figure 6.22 includes a graph without the Peltier element.

The results obtained from analytical calculations performed with the developed program contain comparable values. Differences that occur can only be due to rounding in calculations performed using digital technology.

The probability function of unserviceable states, according to the assumed intensity of damage in the initial stage decreases, then stabilizes at a value of 0.1. In the

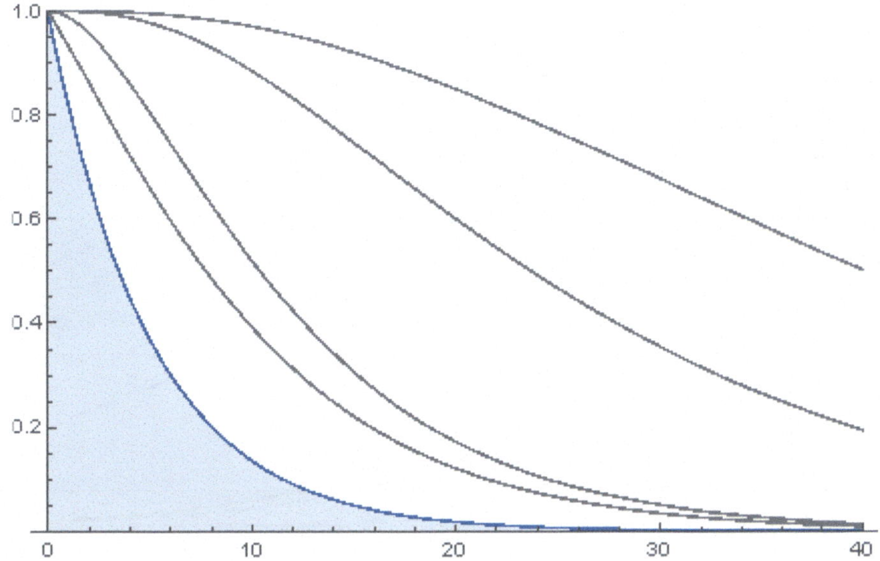

Fig. 6.14 Probability waveforms of the unfitness state of parallel and series element 1 (2oo2) with Peltier cell [Own elaboration]

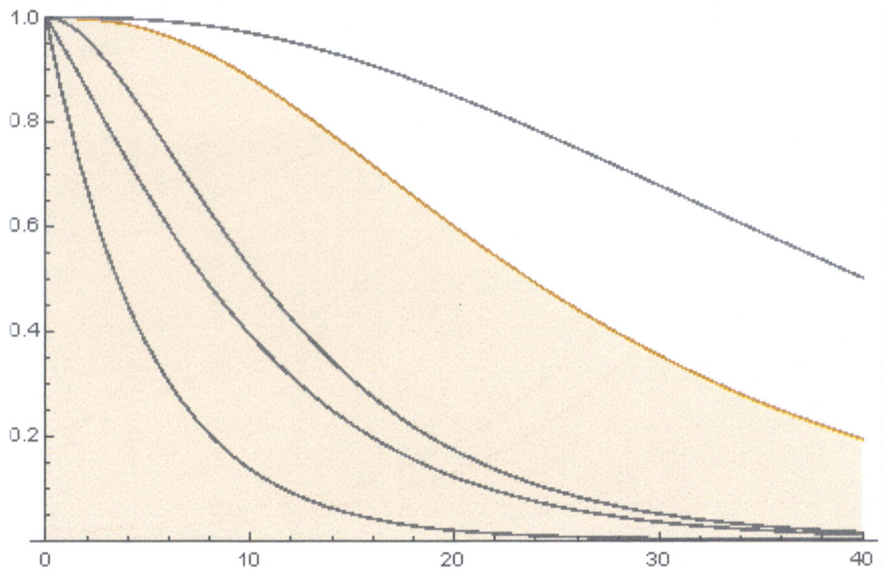

Fig. 6.15 Probability waveforms of being in an unfit state of parallel and series element 5 (2oo2) without Peltier cell [Own elaboration]

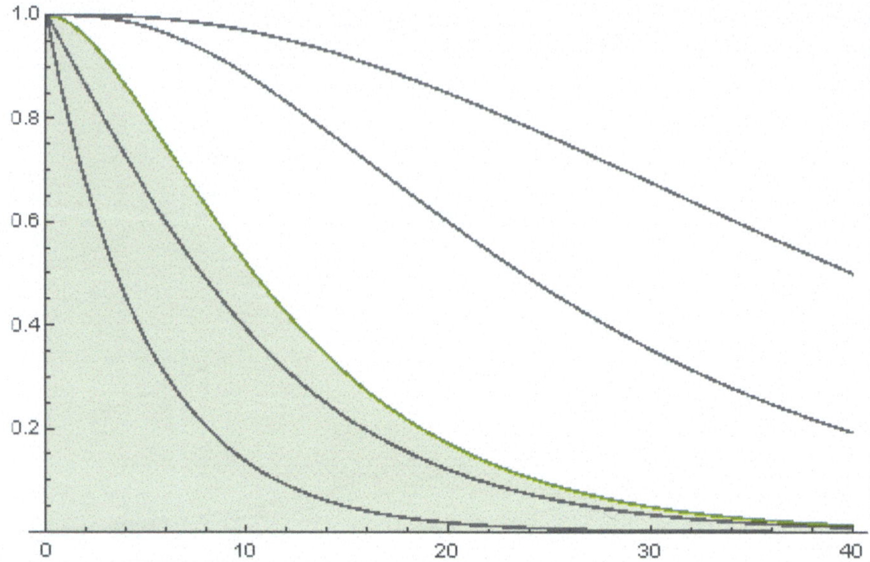

Fig. 6.16 Probability waveforms of the unfitness state of parallel and series element 8 (3oo4) with a Peltier cell [Own elaboration]

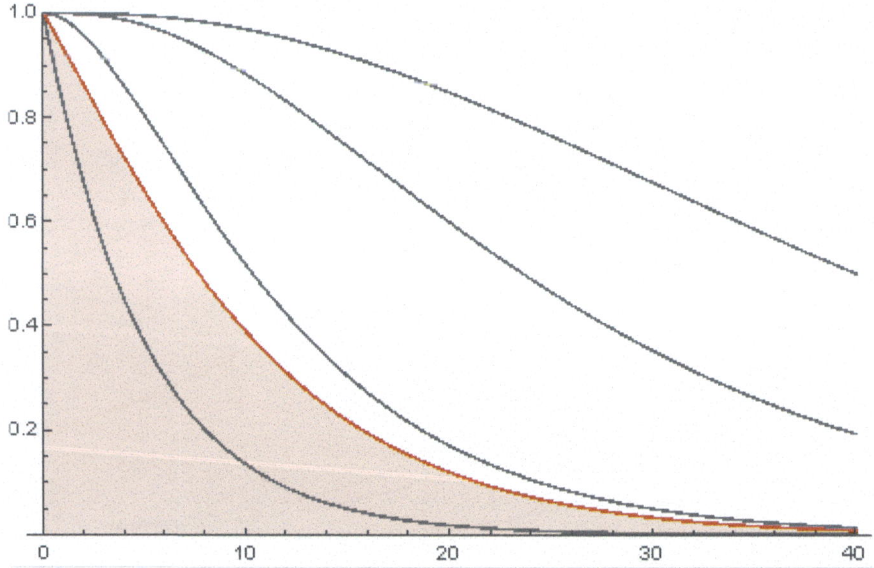

Fig. 6.17 Probability waveforms of being in an unfit state of parallel and series element 5 (2oo2) with a Peltier cell [Own elaboration]

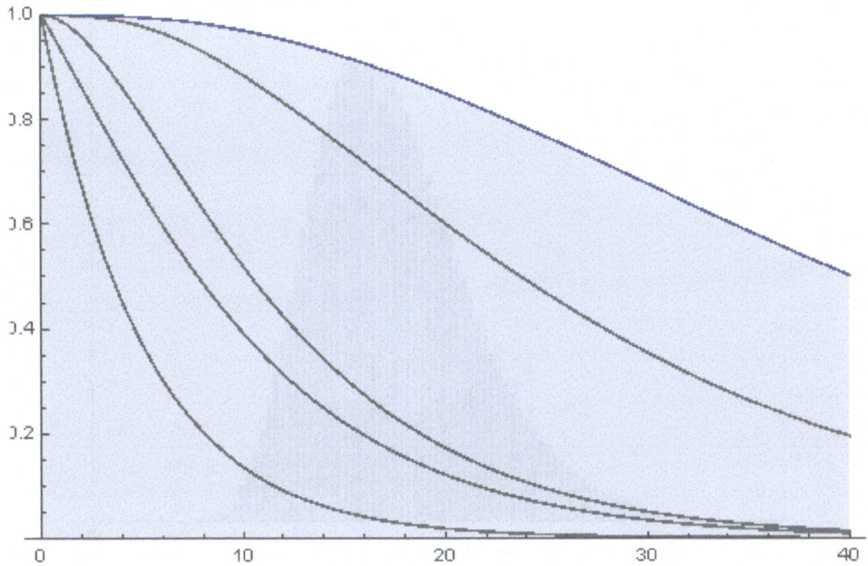

Fig. 6.18 The probability waveforms of staying in the unfit state of parallel and series element 4 (3oo4) without a Peltier cell [Own elaboration]

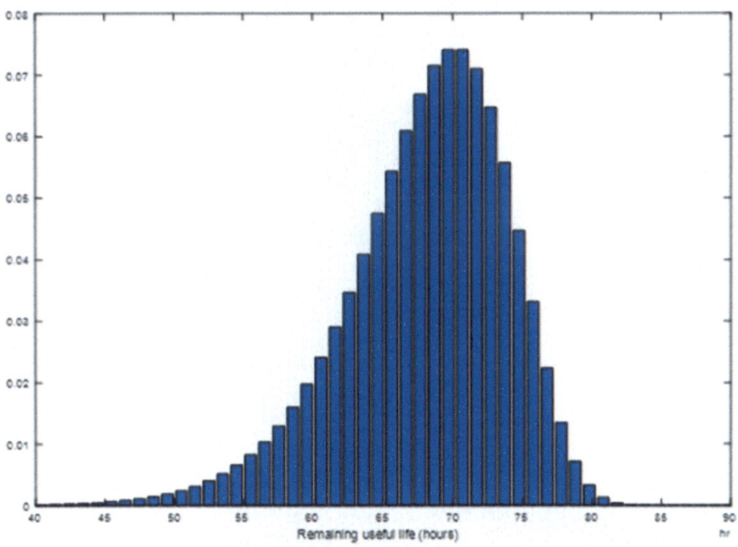

Fig. 6.19 Results of calculations of probabilities of unfitness states of TCS systems according to the developed program in cooling mode [Own elaboration]

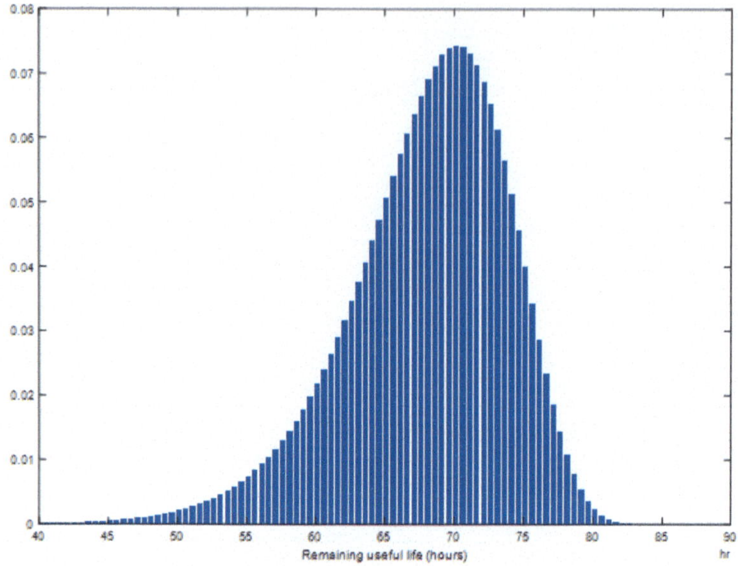

Fig. 6.20 Results of calculations of probabilities of unfitness states of TCS systems according to the developed program in heating mode [Own elaboration]

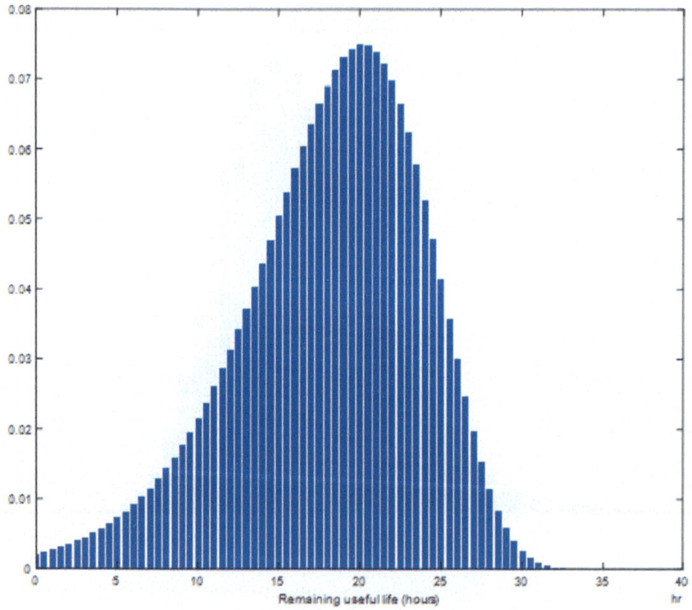

Fig. 6.21 Results of calculations of probabilities of inoperability states of TCS systems according to the developed program in walking and heating mode [Own elaboration]

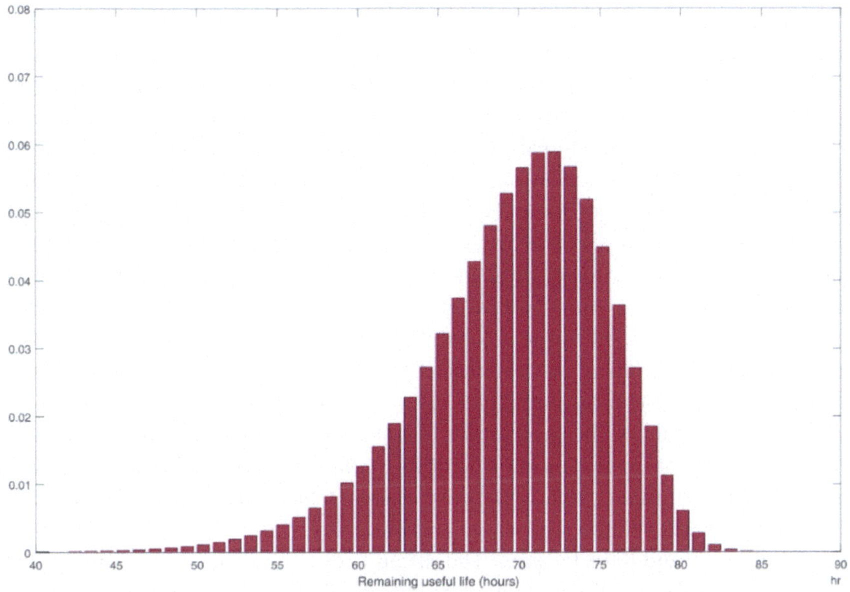

Fig. 6.22 Results of calculations of probabilities of unfitness states of TCS systems without Peltier cell [Own elaboration]

final stage of operation, with the increase in the intensity of the transition to an unserviceable state, an increase in the probability function of staying in an unserviceable state can be seen. The rate of growth of this function in the final stage depends on the duration of operation, when the probabilities are obtained without the Peltier cell it is difficult to obtain the constancy of operation at a value of 0.2 between the states of SP constant functions, and on the course of the function beyond these periods. In the case without a Peltier cell, the probability function of unserviceable states, according to the assumed damage intensity, in the initial stage is decreasing, then converges to 0.9. In the initial period, it is apparent, during the marking of the different values of the intensity of change of states, that nonlinear functions converge more quickly to the limit than linear functions (linear graphs of the function of the intensity of change of states lie above the assumed second-degree function). In this range, one can clearly see the lower value of state probability for nonlinear functions, which is due to faster convergence. The convergence of the two functions is then marked. In the further course, it is observed that the intensity of state changes in the case of the nonlinear function will converge faster to the limit of 1 than it does in the linear function. Once both intensities reach the time at which the transition from any state to the unfit state occurs with probability 1, a clear convergence of the course of the function is noted.

The calculations show that the use of Markov and semi-Markov processes gives sufficient information in the considered facility. In order to be able to better determine the efficiency of the TCS cabinet elements, it is necessary to make a thorough analysis of the relationships between the processes and to find appropriate statistical tools that are able to adequately represent the course of the process. The paper presents graphs comparing the waveforms of the probability functions of the state of unfitness determined for Markov and semi-Markov processes and semi-Markov processes with the assumption of variable intensity of state changes.

Thus, the numerical solution does not provide an answer to a very important, for the operational practice of the TCS cabinet, question: how do the probabilities of the states of a semi-Markov process change when t is large? It follows from the theory of semi-Markov processes that these probabilities, in the case of ergodic semi-Markov processes, tend, with the passage of time, to well-defined, fixed numbers. These numbers are called limiting probabilities of states, and the sequence of these numbers forms the limiting distribution of the process. This distribution makes it possible to define the technical readiness factor of the TCS cabinet.

The estimated data of the probability of unfitness states were analyzed only for adverse events for rail transportation with the inclusion of the TCS cabinet and the Peltier cell installed in it or not. In the case of the TCS cabinet system with a Peltier cell, these are events related to damage and hazards resulting from the operation of electronic devices and their failures. The difference in the structure of damages and hazards is due to the conditions and contribution of individual events to the correct operation of the TCS cabinet. For the estimated values of damage intensity and repair intensity, the task of correct operation of the TCS cabinet lasting 40 h was analyzed. The algorithm performed the analysis for 21 h of operation in the cabinet and 19 h for the operation of the cabinet equipment interacting with infrastructure elements. These results are shown in Figs. 6.14, 6.15, 6.16, 6.17 and 6.18. The data obtained at this stage was used to determine the probability of the inoperable states of the cabinet with and without the Peltier cell.

As can be seen, the algorithm correctly captures the change in TCS cabinet operation over 21 h of correct TCS cabinet operation, and takes into account the increase in readiness of the entire system resulting from improved reliability of the railroad infrastructure. Of course, the reliability of the obtained estimate of the readiness of the track cabinet depends on the reliability of the input data. Performing a statistical evaluation of each state, the characteristics of the system's suitability and unsuitability times were obtained under the condition of changing the operating state of the TCS cabinet with or without the Peltier cell. From the presented graphs of the probabilities of unfitness states of TCS systems, it follows that the technical state of each TCS cabinet with a Peltier cell changes continuously, and for this reason it is possible to consider recalculable sets of their states, that is, sets consisting of an infinite number of elementary technical states. The value of the probabilities of unfitness states of TCS systems with the Peltier cell applied exceeds the range of 0.7, while that without it is below this value. The histograms (Figs. 6.19, 6.20, 6.21 and 6.22) confirm that the dwell times of unfit states can be described by a Hazard function. The presented histograms of the probability distribution function of the residence time in the fit and unfit states of the considered TCS cabinet system confirm the validity of the preliminary initial assumptions made about the distribution of the system's operation time in the given reliability states and the assumed failures of individual electronic components of the TCS cabinet system. The presented calculations show that the time of correct operation of the system is about 2.8 times longer than the time of system unfitness. To better assess the state of unfitness, there is a need to divide this set, and their states into a small number of classes (subsets) of technical states. The peculiarity of the method discussed is the replacement of the random residence time of an object in each state with its average value and the random determination of the transition to the next state of the object. The apparatus of Markov processes is thus used to determine the expected residence times of the system in different states and the stationary probabilities between states.

6.6 Conclusions

An effective process of supervising eclectic mechanical equipment in the TCS cabinet requires the selection of the right strategy for managing the technical infrastructure and indicators for evaluating its performance. The compilation of data in the form of MTTF function graphs shown in Figs. 6.7, 6.8, 6.9, 6.10, 6.11 and 6.12 enables failure rate analysis for specific time intervals. Survival analysis tools also make it possible to carry out a process of comparing survival times in two or more samples. Also noteworthy is the parameter Hazard rate. It is defined as the probability per unit time that a case that has survived from the beginning of a given interval will fail in that interval. In the algorithm, this value was randomly taken from the interval from 1 to 100. This indicator clearly increases its value in intervals where the number of failures increases, and therefore the observation of its variation can be a good element of prediction of future events. The process involves fitting a theoretical distribution of failure probabilities over time for empirical data. In this way, graphs of the mean failure time function and probability density distribution can be obtained. In this way, graphs of the mean failure time function and probability density distribution can be obtained. Since the Weibull distribution is a distribution that is used quite often in analyses related to failure rates, it was this distribution that was chosen when generating the graphs. The simulations carried out showed that the devices equipped with the Peltier cell TCS cabinet had a longer time to failure in the system under study.

In the case of the failure rate waveform for the TCS cabinet for different hours of continuous operation (30,000, 40,000 and 60,000), we also observe that, for the data with the Peltier cell used, significantly lower numbers of failures were obtained, which translates into an extension of the service life of the electronic components on the cabinet's equipment. In the case of data from the TCS cabinet without the Peltier cell, the obtained function graphs assume a similar shape, but the failures are much more numerous and occur after about 50 h of continuous operation. In both cases, it should be concluded that the obtained graphs have similar waveforms, but they are distinguished by the number of failures that occurred. Analyzing the graphs of the tested samples, it should be noted that the fact of the occurrence of the second and subsequent failures significantly affects the probability of survival.

Its value drops below 0.5 as early as 100 days, while for machines experiencing only 1 failure it remains at 0.7 even up to day 220. This is valuable information for identifying systems that potentially threaten the smooth operation of the TCS system. The results of comparing samples of the MTTF function allow us to conclude that both the occurrence of a second failure and the number of days that elapse from inspection to the time of failure have a significant impact on the course of the survival function. Systems that have experienced only one failure have a higher probability of survival than those that have failed a second time. Similarly, systems whose time of failure occurrence was greater than 90 days since the last inspection are characterized by higher values of the survival function than cases in which the failure occurred within 90 days of inspection.

References

Alaoui C (2011) Peltier thermoelectric modules modeling and evaluation. Int J Eng (IJE) 5(1)
AM2302 Temperature and humidity module, ASONG (2013)
Aoki Y, Tsujie I, Nagai T (2018) The effect of ramp rate on temperature cycle fatigue in solder joints, Espec Technology Report, no 25
Astrain D, Martínez Á (2012) Heat exchangers for thermoelectric devices. InTech
Bałuch H (2011) Leksykon terminów kolejowych, Kolejowa Oficyna Wydawnicza, Warszawa
Bar-Cohe A (2013) Encyclopedia of thermal packaging: set 1. In: Thermal packaging techniques. World Scientific Publishing Co. Pte. Ltd.
Bariciński A (2015) Control cabinet cooling systems. Automatyka 5:44–51
Bayle F, Mettas A (2010) Temperature acceleration models in reliability predictions: justification & improvements. In: Reliability and maintainability symposium, San Jose, CA, USA, Jan 25–28
Bazu M, Bajenescu T (2009) Reliability of electronic components: a practical guide to electronic systems manufacturing. Springer, Berlin
Bazu M, Bajenescu T (2011) Failure analysis: a practical guide for manufacturers of electronic components and systems. Wiley & Sons Ltd.
Będkowski L, Dąbrowski T (2006a) Basics of operation. In: Part II. Basics of operational reliability, Warsaw. ISBN: 978-83-89399-31-1
Będkowski L, Dąbrowski T (2006b) Basics of operation. In: Part II: Basics of operational reliability, Warszawa. ISBN 978-83-89399-31-1
Belov IM, Volkov MP, Manyakin SM (1999) Optimization of Peltier thermocouple using distributed Peltier effect. In: 18th international conference on thermoelectrics
Bennet G (2002) In: Encyclopedia of physical science and technology, 3rd edn, vol 15
BMP280 (2015) Digital pressure sensor, BOSCH Sensortec
Buist RJ (1995) The extrinsic thomson effect. In: Proceedings of 14th international conference on thermoelectrics, St. Petersburg
Caillat T, Fleurial J-P, Borschevsky A (1997) Preparation and thermoelectric properties of semiconducting Zn_4Sb_3. J Phys Chem Solids 58:1119
Çengel YA, Boles MA (2008) Thermodynamics: an engineering approach. McGraw-Hill Companies
Ćwirko J, Ćwirko R (2008a) Temperature tests of electronic modules. Bull Mil Univ Technol LVII(2)
Ćwirko J, Ćwirko R (2008b) Badania temperaturowe modułów elektronicznych, Biuletyn WAT, vol LVII, Nr 2
Dąbrowska-Bajon M, Karbowiak H, Grochowski K (1981) Principles of designing railway traffic control systems and devices, WKŁ, Warsaw. ISBN 83-206-0188-6

Documentation of EON-3 devices (Accessed Jan 18, 2018). http://automatyka.ndl.pl/TCS/przekaznikowe/dteon3/index.htm. Accessed Sept 21, 2023
Dokumentacja Techniczno-rozruchowa, Czujnik Koła ELS-95, Bombardier Transportation (ZWUS) Polska, DTR-2008/ELS-95, wersja 01
Driankow D, Helendoorn H, Reinfrank M (1996) Introduction to fuzzy control, WNT Warszawa
DS18B20 programmable resolution 1-wire digital thermometer, MAXIM (2008)
Dyduch J, Moczarski J (2012) Basics of operation of railway traffic control systems. Radom University of Technology. ISBN 978-83
Laboratory Equipment. Nuaire. Process 9—1037P Rev. 5 4/02. Accessed 27 Jan 2019
Esarte J, Min G, Rowe DM (2001) Modelling heat exchangers for thermoelectric generators. J Pow Sour 93
Escobar LA, Meeker WQ (2006) A review of accelerated test models. Stat Sci 21(4):552–577
Fairbanks J (2008) Thermoelectric applications in vehicles status. U.S. Department of Energy
Filin S (2002) Thermoelectric refrigeration devices, IPPU Masta, Gdańsk, 1st edn. ISBN 83-913895-6-1
Filin S, Owsicki A (2000) Principles of design and operation of thermoelectric refrigerators
Filin S, Owsicki A, Zakrzewski B (2010) Experimental research on stationary thermoelectric refrigerators
Flakiewicz K (2009) Air and gas humidity. Measurements and calibration. Metrol Bull Central Office Meas 4(15)
Fleurial J-P, Borschevsky A, Caillat T, Morelli DT, Meisner GP (1997) In: Proceedings of 16th international conference on thermoelectrics. IEEE, Piscataway, NJ
Gabert A (2008) Temperature stabilization of electronics module, Luleå University of Technology. ISSN 1402–1617
Goldsmid HJ (2010) Introduction to thermoelectricity. Springer
Gould C, Shammas N (2012) Three dimensional TCAD simulation of a thermoelectric module suitable for use in a thermoelectric energy harvesting system, InTech
Gromov G (1979) Thermoelectric cooling modules. In: Business briefing: global photonics applications & technology. RMT Ltd., Moskwa
Haidar S, Isaac I, Singleton T (2016) Thermoelectric cooling using Peltier cells in cascade. Undergraduate Laboratories, Department of Physics, University of Alberta, Edmonton
Harman T (2003) Quantum dot superlattice thermoelectric unicouples for conversion of waste heat to electrical power, presented at the 2003 MRS Fall Meeting, Boston, MA, Dec 1–5
Haruyama T (2001) Performance of Peltier elements as a cryogenic heat¯ux sensor at temperatures down to 60 K. Cryogenics 41:335–339
Hnatek ER (2003) Practical reliability of electronic equipment and products. Marcel Dekker Inc., NY
Huebener RP, Tsuei CC (1998) Prospects for Peltier cooling of superconducting electronics. Cryogenics 38(3)
Hur J-H, Lee T-G, Moon S-A, Lee S-J, Yoo H (2008) Thermal reliability analysis of a BLDC motor in a high-speed axial fan by the accelerated-life test and numerical methods. Heat Mass Trans 44:1355–1369
Hwang GS, Gross AJ, Kim H, Lee SW, Ghafouri N, Huang BL, Lawrence C, Uher C, Najafi K, Kaviany M (2009) Micro thermoelectric cooler: planar multistage. Int J Heat Mass Transf 52:1843–1852
Jacob P (2016) Failure mechanisms and precautions in plug connectors and relays. Microelectron Reliab 64:693–698
Jowett CE (1972) Compability and testing of electronic components. Butterworths, London
Kandasamy S, Kalantarzadeh K, Rosengarten G, Wlodarski W (2004) Modelling of a thin film thermoelectric micro-Peltier module. In: 2004 IEEE region 10 conference TENCON 2004, Chiang Mai, Thailand, vol 4, pp 310–313
Karimi G, Culham JR, Kazerouni V (2011) Performance analysis of multi-stage thermoelectric-tric coolers. Int J Refrig 34:2129–2135

Kijak J, [author. Borkowski A et al] (1975) Construction of electronic devices: collective work. Wydawnictwa Naukowo-Techniczne, Warsaw
Kisiel R, Bajera A (1988) Basics of designing electronic devices. Warsaw University of Technology, Warsaw
Kobari T, Okajima J, Komiya A, Maruyama Sh (2015) Development of guarded hot plate apparatus utilizing Peltier module for precise thermal conductivity measurement of insulation materials. Int J Heat Mass Transf 91:1157–1166
Kocur V, Šuriansky J (2011) Simulations and modelling of absorption cooling unit by means of Peltier module. In: Transactions of the VŠB—Technical University of Ostrava, Mechanical Series, no 1, vol LVII, article no 1847
Królicka A, Hruban A, Mirowska A (2012) Modern thermoelectric materials—literature review. Institute of Electronic Materials Technology, Materialy Electronics, Warsaw, vol 40, no 4/2012
Kuper CM (2015) Thermal testing and quality assurance of BGA, LCC, & QFN electronic packages, The University of New Mexico, Department of Mechanical Engineering
Lee HS (2010) Thermal design heat sinks, thermoelectrics, heat pipes, compact heat exchangers, and solar cells. Wiley & Sons Inc., Hoboken, New Jersey
Lexicon of Railway Terms, KOW Sp. z o. o., Warsaw (2011).
Lineykin S, Ben-Yaakov S (2005) PSPICE-compatible equivalent circuit of thermoelectric coolers
Macha E (2001) Reliability of machines. Opole University of Technology
Mannella GA, La Carrubba V, Brucato V (2014) Peltier cells as temperature control elements: experimental characterization and modeling. Appl Thermal Eng 63:234–245
Markowski P (2008) Doctoral thesis, Thermoelectric properties of thick-film composites, Wrocław University of Science and Technology, Faculty of Microsystem Electronics and Photonics, Wrocław
Matkowski PK (2009) Thermomechanical analysis of microelectronic connectors made using lead-free solders, Wrocław
Min G, Rowe DM (2000) Improved model for calculating the coefficient of performance of a Peltier module. Energy Convers Manage 41:163–171
Mitrani D, Salazar J, Turó A, García MJ, Chávez JA (2007) Lumped and distributed parameter SPICE models of TE devices considering temperature dependent material properties. ISBN: 978-2-35500-002-7, ©EDA Publishing/Therminic
Mogielnicki H (1983) Materiały do chłodzenia termoelektrycznego, Instytut Technologii Materiałów Elektronicznych, Materiały Elektroniczne, Zeszyt 3, Nr 43/1983, Warszawa
Press WH, Teukolsky SA, Vetterling WT, Flannery BP (2002) Numerical recipes in C. The art of scientific computing. Cambridge University Press
Sharma, SK (2019) Multibody analysis of longitudinal train dynamics on the passenger ride performance due to brake application. Proceedings of the Institution of Mechanical Engineers, Part K: Journal of Multi-Body Dynamics, 233(2):266–279. https://doi.org/10.1177/1464419318788775
Siergijczyk M, Rosiński A (2014) The concept of supporting the maintenance of transport telematics devices. In: Proceedings of the winter school of reliability conference, Szczyrk
Singh G (2003) Steady state and transient analysis of thermoelectric devices using finite element method. Rochester Institute of Technology, NY
Skrzypczak P (2012) Analysis of cooling systems for high-power light-emitting diodes using Peltier cells. Doctoral thesis, Poznań University of Technology
Smith DJ, Kenneth G, Simpson KGL (2010) Safety critical systems handbook. A straightforward guide to functional safety, IEC 61508 and related standards. Elsevier, Oxford
Sotskov BS (1973) Translation Roman Maniewski, Reliability of automation elements and devices, Wydawnictwa Naukowo-Techniczne, Warsaw
Ważyńska-Fiok K, Jaźwiński J (1990) Reliability of technical systems. National Scientific Publishing House, Warsaw
Węgierski J (1971) Probabilistic methods in the design of rail transport, Wydawnictwo Komunikacji i Connectivity, Warsaw

Whitaker JC (2001) The resource handbook of electronics. CRC Press
Wojtas P (2010) The impact of the integration of dispatch systems and telecommunications network configuration on their reliability and functionality. Doctoral thesis, Silesian University of Technology, Gliwice
Xin W, Liang-Jun X (2007) Finite element model analysis of thermal failure in connector. J Zhejiang Univ Science A 8(3):397–402. ISSN 1009-3095
Xu X, Van Dessel S, Messac A (2007) Study of the performance of thermoelectric modules for use in active building envelopes. Build Environ 42:1489–1502
Yager RR, Filev DP (1995) Basics of fuzzy modeling and control, WNT, Warszawa
Yang B, Ahuja H, Tran TN (2008) Thermoelectric technology assessment: application to air conditioning and refrigeration, vol 14, no 5. HVAC&R RESEARCH
Zajączkowski A, Kalicińska K, Olendrzyński W (1982) Electrical railway traffic protection devices: station devices, 2nd edn, amended, WKiŁ, Warszawa
Zakrzewski B, Filin S, Zeńczak W, Łokietek T, Wasiutinskij S, Chmielowski M, Tuchowski W, Złoczowska E (2015) Refrigeration and air conditioning—Volume I, Faculty of maritime technology and transport, West Pomeranian Voivodeship University of Technology in Szczecin
Zalewski W (1998) Theoretical foundations and examples of applications. Heat pumps, Script, Krakow University of Technology Publishing House, ISBN 83-903878-3-2
Zheng Y et al (2011) Carrying capacity reliability of railway networks. J Transp Syst Eng Inform Technol 11. Elsevier
Zybała R, Pietrzak K (2013) Electrical connectors in thermoelectric modules. Electr Mater 41(2)

Part III
Innovative Technologies in Transportation

Chapter 7
Blockchain Solutions for Cost-Efficiency and Traceability in Indian Food Distribution

Mohd Avesh, Srihari Palli, Ismail Hossain, and Rakesh Chandmal Sharma

7.1 Introduction

The Indian food supply chain, as a critical component of the nation's agricultural landscape, has long grappled with intricate challenges. India, the world's second-largest producer of fruits and vegetables, possesses immense potential to alleviate issues related to hunger and food scarcity, both domestically and internationally. However, this potential remains largely untapped, hampered by a multitude of persistent challenges that continue to afflict the food supply chains. Recent advancements in the fields of vehicle dynamics and railway infrastructure for food supply chain have been documented in various studies (Bhardawaj et al. 2023; Hossain et al. 2023; Lee and Sharma 2023; Sharma et al. 2023c, d, e, f, g, h; Upadhyay et al. 2023a, b; Wu et al. 2023). Central to these challenges is the presence of intermediaries, commonly known as middlemen, who operate as intermediaries between farmers and consumers. While intermediaries are a common feature in supply chains

M. Avesh
Mazedan International Research Academy, Okhla, Delhi, New Delhi 110025, India

S. Palli
Aditya Institute of Technology and Management, Tekkali, AP 532201, India

I. Hossain
Department of Nuclear and Renewable Energy, Ural Federal University, Yekaterinburg 620002, Russia

R. C. Sharma (✉)
Mechanical Engineering Department, Graphic Era (Deemed to Be University), Dehradun 248002, India
e-mail: rcsharmaiitr@gmail.com

Mechanical Engineering Department, Graphic Era Hill University, Dehradun 248002, India

S. K. Sharma et al. (eds.), *Dynamics of Transportation Ecosystem, Modeling, and Control*, Energy, Environment, and Sustainability,
https://doi.org/10.1007/978-981-97-0437-8_7

worldwide, their prevalence within India's food distribution network carries significant repercussions (Adeoye et al. 2023a). These intermediaries often result in a scenario where a substantial portion of profits dissipates before reaching the diligent farmers responsible for cultivating the crops. Compounding this issue are the substantial transaction costs incurred throughout the supply chain. Transaction costs encompass a broad spectrum of expenses, spanning transportation, storage, negotiations, and information management. Within the context of the food supply chain, these costs translate into diminished profits for farmers and augmented prices for consumers, creating a situation detrimental to all stakeholders. The need for a transformative solution to address these challenges has never been more pressing. Enter blockchain technology—a revolutionary innovation that has disrupted industries on a global scale. Blockchain has gained prominence not only as the foundational technology behind cryptocurrencies such as Bitcoin but also as a promising instrument for enhancing transparency, diminishing costs, and augmenting traceability within supply chains (Gopala Rao et al. 2022; Mohapatra et al. 2016a, 2016b; Sharma et al. 2020a).

This article embarks on a comprehensive exploration of the role of blockchain technology within the Indian food supply chain, all from a third-party perspective. It delves deeply into the intricacies of reducing transaction costs and enhancing traceability, two central challenges that have long impeded the progress and prosperity of this pivotal sector. The journey commences with an examination of transaction costs, shedding light on their significance and understanding their consequences on the livelihoods of farmers and the broader food industry. Following that, it transitions into the realm of traceability, unraveling the complexities of this issue within the Indian context (Kohli et al. 2022; Sharma and Kumar 2017a). Subsequently, blockchain technology is introduced as a potential game-changer, demonstrating how it possesses the capability to eliminate middlemen, consequently fostering more equitable profits for farmers. Furthermore, the innate security features of blockchain technology are highlighted, making a compelling case for its suitability in addressing traceability concerns and assuring consumers of the safety and authenticity of the food supply. To substantiate the claims made, findings from a comprehensive questionnaire-based study involving industry professionals from various Indian cities are presented (Bhardawaj et al. 2020a; Sharma and Kumar 2018a; Sharma et al. 2021a). Through descriptive and quantitative comparison analyses, insights into the potential impact of blockchain on transaction costs and traceability within the Indian food supply chain are provided. As the article navigates through the research findings, it paints a compelling picture of the transformative potential of blockchain within the Indian food supply chain. Real-world examples and case studies of successful blockchain implementations are showcased, offering a glimpse into a future where the industry operates more efficiently, profitably, and securely. Hence, this article serves as a roadmap for understanding how blockchain technology can effectively address the persistent challenges of transaction costs and traceability within the Indian food supply chain. It is the aspiration that this exploration will not only inform but also inspire further adoption and collaboration within India's food industry, ultimately leading to a more equitable and sustainable food supply chain for all parties involved.

Subsequent sections will delve deeper into the intricacies of transaction costs, traceability issues, research methodologies, and research findings, all within the context of blockchain's transformative potential.

7.2 Transaction Costs

Within the context of the Indian food supply chain, it becomes evident that transaction costs are a pervasive challenge that warrants thorough examination. Transaction costs, in this intricate network, encompass a myriad of expenses incurred throughout the various stages of production, distribution, and consumption (Bandekar et al. 2017; Hossain et al. 2017; Mohapatra et al. 2017a; Sharma and Sharma 2019; Sharma et al. 2022a). It is essential to emphasize that these costs extend beyond monetary outlays and often encompass intangible factors such as time, effort, and the exchange of information. A fundamental facet of transaction costs in the food supply chain is transportation costs. These encompass the expenses associated with the physical movement of goods from farms to processing facilities, distribution centers, and ultimately to retailers or consumers. They entail costs related to fuel, vehicle maintenance, labor, and logistical coordination. Storage costs also feature prominently in the equation. Recent studies have significantly contributed to the understanding of rail vehicle dynamics and the improvement of railway systems (Palli et al. 2018; Sharma et al. 2015, 2017, 2018; Sharma and Sharma 2018a, b, c; Sharma and Kumar 2017b, 2018b, c). Ensuring the safe and quality-preserving storage of food products is a critical component of the supply chain. This necessitates investments in storage facilities, refrigeration infrastructure, and monitoring systems to prevent spoilage or wastage (Goswami et al. 2019; Mohanty et al. 2016; Pattnaik et al. 2022; Samantaray et al. 2014; Sharma et al. 2022b; Sinha et al. 2019; Swapna Rekha et al. 2020). Negotiation costs emerge as a consequential aspect, particularly in transactions involving multiple parties. Whether it be farmers, wholesalers, or retailers, negotiations form a pivotal aspect of the supply chain process. These negotiations require valuable time and resources for all parties involved in reaching mutually beneficial agreements. Figure 7.1 shows the Blockchain Implementation Impact on Traceability Problem (Biswas et al. 2021; Nibedita et al. 2021; Sharma et al. 2020b, 2021b, 2022c).

Information costs, too, cannot be understated. Timely and accurate information plays a pivotal role in decision-making within the supply chain. Information costs encompass the expenses associated with the gathering, processing, and dissemination of crucial information regarding product quality, market demand, pricing dynamics, and logistical considerations. Additionally, quality assurance costs are a vital element within the transaction cost framework. Ensuring the quality and safety of food products necessitates rigorous testing, stringent quality control measures, and compliance with regulatory standards (Agrahari et al. 2021a; Das et al. 2023; Pundir et al. 2023a; Sharma et al. 2022d). These quality assurance costs are indispensable in maintaining consumer confidence and trust in the food supply chain. Moving beyond the conceptual framework, it is imperative to recognize the tangible impact of transaction costs

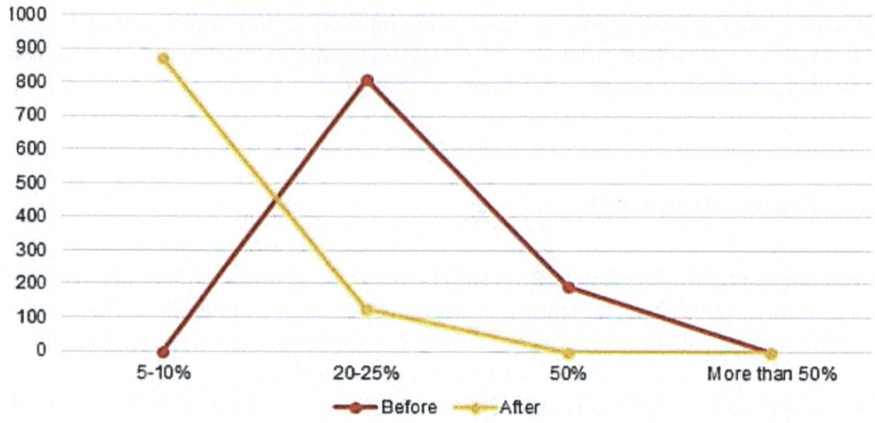

Fig. 7.1 The blockchain implementation impact on traceability problem

on farmers, particularly those operating on a smaller scale or marginalized producers. These effects are multifaceted, encompassing several dimensions. Firstly, the burden of high transaction costs leads to reduced profit margins for farmers. The disproportionate share of expenses within the supply chain means that farmers often receive a meager portion of the final retail price of their products. This, in turn, affects their economic viability and sustainability. The comprehensive studies on wear mitigation in locomotive components, lateral stability, ride comfort, and advanced suspension systems have significantly contributed to enhancing the performance and safety of rail and road vehicles (Kumar et al. 2022; Sharma et al. 2022e, f, g; Sharma and Sharma 2022; Vishwakarma et al. 2022a, b).

Secondly, transaction costs can curtail farmers' market access. The substantial expenses associated with transporting goods to distant markets can be financially prohibitive for small-scale farmers. Consequently, they may find themselves confined to local or regional markets, limiting their potential for growth and profitability. Furthermore, the volatility in transaction costs can subject farmers to unpredictable price fluctuations (Devadarshini et al. 2022; Kohli et al. 2020; Maity et al. 2020; Misra et al. 2022; Mohapatra et al. 2017b; Sharma and Lee 2020a). Variations in transportation and storage costs, coupled with broader market dynamics, can result in erratic and potentially unfavorable pricing conditions for farmers, adding to their financial instability. Lastly, the allocation of resources by farmers is heavily influenced by the management of transaction costs. Significant portions of financial resources, as well as time and labor, are directed toward mitigating these costs. This allocation, while necessary, can divert valuable resources from critical aspects of farming, such as improving crop quality, expanding production, and investing in sustainable practices. In light of these challenges and the substantial impact of transaction costs on the food supply chain, there emerges a compelling need for innovative and cost-efficient solutions. In the subsequent sections of this article, we will explore these solutions in greater depth, particularly within the context of how blockchain technology can play a pivotal role in addressing transaction costs within the Indian food supply chain.

7.3 Traceability Issues in the Indian Food Supply Chain

In the intricate landscape of the Indian food supply chain, traceability issues loom large, demanding meticulous scrutiny. Traceability, in this context, denotes the capacity to meticulously monitor and authenticate the origin, voyage, and handling of food products as they traverse the supply chain. The recent studies on adsorption techniques for water remediation, biodiesel effects on carbon steel, dye absorption using biochar, green hydrogen production, and nanoparticle applications in combating multidrug-resistant bacteria underscore the advancements in environmental and material sciences (Ukhurebor et al. 2024; Adama et al. 2024; Eleryan et al. 2024; Rathinavelu et al. 2023, 2024; Iheme et al. 2024; Elmaghraby et al. 2023). It holds the key to ensuring food safety, upholding quality standards, and nurturing consumer trust. Currently, there are two new initiatives for traceability in the Indian food Industry which are two standard autonomous bodies called the GS1 and APEDA: GS1 is basically for the packaged food and packaged beverages available in grocery stores; this comes under the Ministry of Commerce and Industry of Government of India and it has founded many big organizations like FICCI, CII, ASSOCHAM, BIS, etc. but the drawback with this is that the buyer cannot trace back the product in the supply chain (Palli et al. 2022; Sarkar et al. 2018; Sharma et al. 2021c). Companies do put the GS1 standard numbers on the packaging of generic products, but they are of no use to the consumer. Whereas the APEDA has emerged in recent years in the agricultural food supply chain which has increased the emphasis of technology in farming and traceability particularly the role of information technology is being increased by the APEDA in farming of grapes, banana, and ladyfinger all around the country (Acharya et al. 2020; Bhardawaj et al. 2019a, 2020b; Mohapatra et al. 2020a; Pundir et al. 2023b; Sharma and Lee 2020b, 2021). Tracenet is a system that works under APEDA, and it stores all the data related to traceability and information about the product in one place which is accessible to everyone. Tracenet is put to service for organic products in India. Figure 7.2 shows a traceability barcode released by GS1 and APEDA, but this is only applicable to horticulture proceeds (Table 7.1).

7.3.1 Lack of Standardized Traceability Systems

One of the fundamental challenges within the Indian food supply chain is the absence of standardized traceability systems. The lack of a universal framework complicates the establishment of consistent tracking mechanisms. In a diverse country like India, where different regions employ varying practices and technologies, standardization

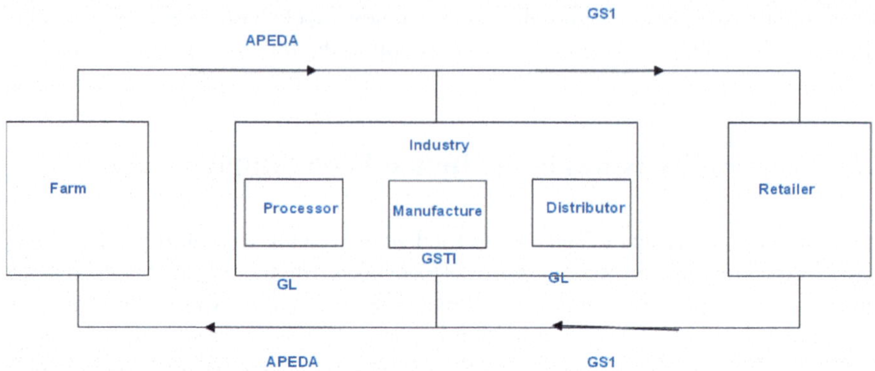

Fig. 7.2 Traceability barcode released by GS1 and APEDA

Table 7.1 Key challenges in traceability within the Indian food supply chain

Challenge	Impact
Lack of standardized traceability systems	Hinders the ability to establish consistent tracking
Incomplete or inconsistent record-keeping	Complicates the verification of product provenance
Multiplicity of intermediaries and handoffs	Introduces potential for errors, gaps, and discrepancies
Limited transparency in sourcing and procurement	Conceals risks and inefficiencies
Vulnerability to fraud and adulteration	Undermines consumer trust and product integrity

becomes crucial to effective traceability. The innovative approaches to transportation, from the feasibility of the Mumbai-Pune Hyperloop corridor and waste-to-biofuel production to advanced vibration control in marine and rail systems, highlight significant strides in sustainable and intelligent transportation technologies (Badgujar et al. 2024; Kokkinos et al. 2024; Raju Rallabandi et al. 2024; Rallabandi et al. 2024; Sharma et al. 2024a, b, c, d, e, f, g; Upadhyay et al. 2024a, b). As shown in Fig. 7.2, what blockchain will do is it will integrate both of these standards (GS1 and the APEDA) seamlessly as all the information will be on a single platform regarding that particular food item whether packaged or unpackaged (Fig. 7.3).

7.3.2 Incomplete or Inconsistent Record-Keeping

Another significant hurdle contributing to traceability challenges is the issue of incomplete or inconsistent record-keeping (Adeoye et al. 2023b; Mohapatra et al. 2020b, 2022; Pattnaik et al. 2020; Sharma et al. 2020c). Throughout the supply chain,

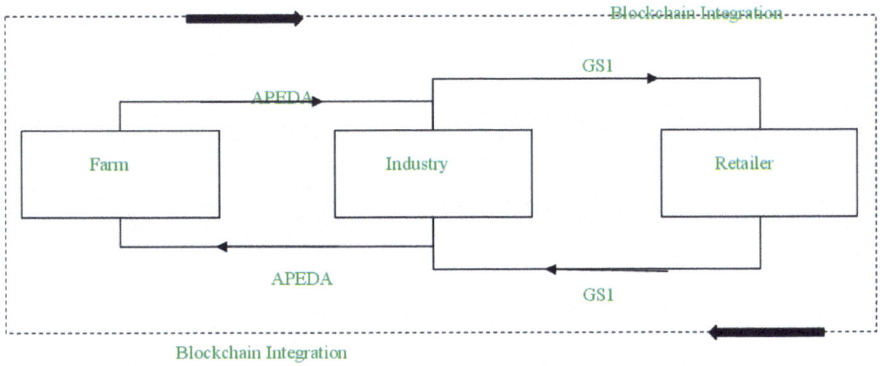

Fig. 7.3 Blockchain integration of APEDA and GS1

Table 7.2 Traceability gaps within the Indian food supply chain

Supply chain stage	Traceability challenges
Production and farming	Lack of farm-level tracking systems
Processing and packaging	Inadequate recording of processing details
Transportation and logistics	Inconsistent data on transportation routes
Wholesale and distribution	Limited visibility into distribution networks
Retail and consumer	Challenges in verifying product origins

there are multiple points where information regarding the product's journey should be documented. However, the reliability and comprehensiveness of these records can vary widely, making it arduous to establish a comprehensive traceability trail (Table 7.2).

7.3.3 Multiplicity of Intermediaries and Handoffs

The Indian food supply chain is a complex web, involving a multitude of intermediaries and handoffs at various stages of production, distribution, and consumption. This complexity introduces a significant challenge when it comes to traceability. At each transition point, where goods change hands or move from one facility to another, there is an inherent potential for errors, omissions, or miscommunication to occur. Imagine a product's journey from a small-scale farmer's field to the shelves of a grocery store. Along the way, it may pass through the hands of multiple stakeholders, such as wholesalers, distributors, processors, and retailers. Each of these interactions presents an opportunity for information to be inaccurately recorded, details to be overlooked, or data to be lost (Chatterjee et al. 2022; Mohapatra et al. 2017c; Pati et al. 2020, 2022). For instance, when a farmer sells their produce to a

Table 7.3 Vulnerabilities associated with traceability gaps

Vulnerability	Associated risks
Food safety	Difficulty in tracing contamination sources
Quality assurance	Inability to verify product quality and freshness
Regulatory compliance	Challenges in meeting traceability requirements
Consumer trust	Erosion of trust due to lack of transparency
Economic losses	Losses due to inefficiencies, fraud, and recalls

wholesaler, the accurate recording of the product's origin, quantity, and quality is crucial. Any mistakes made at this stage can have cascading effects down the supply chain. If the wholesaler's records do not align with what the retailer expects, it can lead to confusion and potential disputes.

Furthermore, as products change hands, they may be repackaged, combined with other goods, or subjected to various storage conditions. Each of these actions introduces additional layers of complexity to the traceability process. Without a robust system in place, tracking a product's origin and ensuring its quality and safety become increasingly challenging (Adeoye et al. 2022, 2023c; Mohapatra et al. 2017d; Nayak et al. 2022; Sharma et al. 2020d). In essence, the Indian food supply chain's complexity not only amplifies traceability issues but also underscores the urgent need for innovative solutions, such as blockchain, to address these challenges comprehensively. Blockchain's ability to provide a transparent, immutable, and decentralized ledger can significantly alleviate these complexities and enhance traceability throughout the supply chain (Sharma et al. 2020e).

7.3.4 Limited Transparency in Sourcing and Procurement

Transparency in sourcing and procurement practices remains a concern. Limited transparency conceals potential risks and inefficiencies within the supply chain (Bhardawaj et al. 2020c; Kanth Choppara et al. 2019; Mohapatra et al. 2015; Palli and Koona 2015; Palli et al. 2021; Sharma et al. 2021d). For instance, it becomes challenging to identify the origin of a product or assess the conditions under which it was produced, which is vital for ensuring quality and safety (Table 7.3).

7.3.5 Vulnerability to Fraud and Adulteration

The Indian food supply chain is susceptible to fraudulent activities and adulteration. Without robust traceability mechanisms, unscrupulous actors may introduce counterfeit or substandard products into the market. This not only undermines consumer trust but also poses significant health and safety risks (Agarwal et al. 2020; Agrahari et al. 2021b; Bhardawaj et al. 2019b; Bhattarai et al. 2022; Mohapatra et al. 2021; PanduRanga Vital et al. 2019). Addressing these traceability challenges within the Indian food supply chain is paramount to ensure the integrity of the system, bolster consumer confidence, and guarantee the safety and quality of food products. In the following sections, we will explore how blockchain technology, with its innate transparency and immutability, can serve as a transformative solution to comprehensively address these traceability concerns.

7.4 Blockchain Technology: A Transformative Solution for Traceability

Within the intricate landscape of the Indian food supply chain, the integration of blockchain technology emerges as a transformative solution to the persistent traceability challenges (Avesh and Srivastava 2020; Satyanarayana et al. 2022; Sharma and Palli 2016; Sharma et al. 2023a, b). Blockchain, in this context, represents a decentralized and transparent digital ledger that records transactions across a network of computers. Each transaction, once recorded, becomes immutable and linked to the preceding one, creating an unbroken chain of information (Table 7.4).

a. Immutable Record-Keeping: Blockchain technology excels in ensuring the permanence and accuracy of traceability data. Once information is recorded on the blockchain, it becomes immutable, meaning it cannot be altered or deleted without consensus from the network participants (Avesh and Srivastava 2019; Avesh et al. 2019a). This characteristic assures the integrity of traceability records, providing a reliable source of information.
b. Decentralization: A notable advantage of blockchain is its decentralization. Unlike traditional systems reliant on a central authority for verification, blockchain operates on a peer-to-peer network. Each participant in the network maintains a copy of the ledger, reducing the reliance on a single entity. This decentralization enhances trust and security within the supply chain.

Table 7.4 Key advantages of blockchain for traceability

Advantages	Impact
Immutable record-keeping	Ensures the permanence and accuracy of traceability data
Decentralization	Reduces the dependence on a central authority for verification
Transparency	Provides real-time visibility into the supply chain
Enhanced security	Guards against tampering, fraud, and unauthorized changes
Smart contracts	Automates and enforces contract terms and agreements

c. Transparency: Blockchain technology offers real-time visibility into the supply chain. Each transaction is recorded chronologically, enabling stakeholders to trace the journey of a product from its origin to its current location. This transparency not only enhances traceability but also enables quick identification of any discrepancies or issues (Avesh and Srivastava 2012).
d. Enhanced Security: Blockchain is renowned for its robust security features. Transactions on the blockchain are secured through cryptography, making them resistant to tampering, fraud, and unauthorized changes (Mohd and Srivastava 2019). This level of security ensures the authenticity and reliability of traceability data.
e. Smart Contracts: Another compelling aspect of blockchain is the utilization of smart contracts. Smart contracts are self-executing agreements with the terms of the contract directly written into code (Avesh et al. 2019b). In the context of traceability, smart contracts can automate and enforce contract terms and agreements, streamlining processes and ensuring adherence to predefined rules.

By implementing blockchain-powered traceability within the Indian food supply chain, several benefits can be realized. These include improved food safety through rapid identification and containment of contaminated products, enhanced product quality with real-time monitoring of conditions, streamlined compliance with traceability regulations, increased consumer confidence through transparency, and heightened operational efficiency by reducing errors and automating processes (Tables 7.5 and 7.6).

Table 7.5 Key components of blockchain-powered traceability

Component	Description
Digital identities	Assigning unique digital identities to products, enabling tracking
Timestamped transactions	Recording the time and date of each transaction for chronological tracing
Supply chain visibility	Real-time visibility into product locations and status
Authentication mechanisms	Verification of product authenticity through cryptographic methods
Data sharing permissions	Controlled sharing of information with authorized parties

Table 7.6 Benefits of implementing blockchain-powered traceability

Benefits	Impact
Improved food safety	Swift identification and containment of contaminated products
Enhanced product quality	Real-time monitoring of product conditions and freshness
Regulatory compliance	Facilitates adherence to traceability regulations
Consumer confidence	Reinforces trust through transparency and authenticity
Operational efficiency	Streamlines supply chain processes and reduces errors

7.5 Practical Applications of Blockchain-Powered Traceability in the Indian Food Supply Chain

In the intricate landscape of the Indian food supply chain, the integration of blockchain technology is poised to usher in transformative changes, specifically in the realm of traceability. The recent advancements in rail vehicle system modeling, structural analysis, and performance evaluation have been extensively studied in various contexts, including crashworthiness, lateral stability, critical speed analysis, and multi-objective optimization (Goyal et al. 2019; Sharma 2019; Sharma et al. 2019; Bhardawaj et al. 2020b, 2021; Lee and Sharma 2020; Choi et al. 2021; Lee et al. 2021; Sharma and Sharma 2021). This section elucidates practical applications of blockchain within this context, showcasing its potential to comprehensively address the challenges outlined earlier (Tables 7.7 and 7.8).

1. Product Tracking: Blockchain empowers real-time tracking of products throughout their journey from farm to table. Each product is assigned a unique digital identity that is recorded on the blockchain. This identity accompanies the product at every stage, ensuring transparency and enabling consumers to trace the origin and journey of the food they consume.
2. Tamper-Proof Records: One of the hallmark features of blockchain is its creation of immutable records. Once data is recorded on the blockchain, it cannot be altered or deleted without consensus from the network participants. This feature

Table 7.7 Practical applications of blockchain for traceability

Application	Description
Product tracking	Real-time tracking of products from farm to table, providing transparency
Tamper-proof records	Immutable records that safeguard against tampering, fraud, and unauthorized changes
Supply chain visibility	Enhanced visibility into product locations and conditions
Automated compliance	Streamlined adherence to traceability regulations
Quality assurance	Monitoring product quality and freshness in real time

Table 7.8 Benefits of blockchain-powered traceability for stakeholders

Stakeholder	Benefits
Farmers	Fairer compensation and market access through reduced transaction costs
Distributors and retailers	Enhanced transparency and efficient inventory management
Regulators	Improved oversight and streamlined compliance monitoring
Consumers	Assurance of food safety, quality, and authenticity
Food industry	Enhanced operational efficiency and reduced risk of recalls

Table 7.9 Outcomes of implementing blockchain-powered traceability

Outcomes	Impact
Reduced foodborne illnesses	Swift identification and recall of contaminated products
Increased consumer confidence	Reinforced trust in product safety and authenticity
Efficient recall management	Expedited removal of unsafe products from the market
Streamlined regulatory audits	Simplified compliance monitoring and reporting
Supply chain optimization	Enhanced efficiency, reduced waste, and cost savings

provides a robust defense against tampering, fraud, and unauthorized changes, ensuring the reliability and authenticity of traceability records.

3. Supply Chain Visibility: Blockchain offers enhanced visibility into product locations and conditions. This real-time transparency enables stakeholders to monitor products as they move through the supply chain, making it possible to identify any discrepancies or issues promptly.
4. Automated Compliance: Blockchain simplifies the task of adhering to traceability regulations. By automating data recording and verification processes, it streamlines compliance efforts, ensuring that all necessary information is accurately documented and readily accessible for audits.
5. Quality Assurance: Blockchain technology facilitates the monitoring of product quality and freshness in real time. Sensors and IoT devices integrated with the blockchain can provide continuous updates on factors such as temperature, humidity, and shelf life, thereby ensuring that products meet established quality standards.

By implementing blockchain-powered traceability solutions within the Indian food supply chain, a range of benefits can be realized. These include reduced instances of foodborne illnesses through swift identification and recall of contaminated products, increased consumer confidence stemming from trust in product safety and authenticity, efficient recall management by expediting the removal of unsafe products from the market, streamlined regulatory audits through simplified compliance monitoring and reporting, and supply chain optimization through enhanced efficiency, reduced waste, and cost savings (Table 7.9).

7.6 Real-World Examples of Blockchain Implementation in the Indian Food Supply Chain

This section highlights real-world examples and case studies of successful blockchain implementations within the Indian food supply chain. These instances serve as tangible illustrations of how blockchain technology has been harnessed to enhance traceability, reduce transaction costs, and bring about transformative changes (Table 7.10).

Table 7.10 Real-world examples of blockchain implementation

Example	Description
Farmer connect	Collaborative platform connecting coffee farmers and consumers through blockchain technology, providing transparency on product origin and supporting fair compensation for farmers
The mahindra samriddhi initiative	Initiative by Mahindra Group leveraging blockchain to connect farmers with urban consumers, ensuring traceability, and fair pricing for agricultural produce
Walmart's blockchain-powered supply chain	Walmart India's adoption of blockchain to monitor and trace the sourcing and distribution of fruits and vegetables, enhancing product quality and safety
IBM food trust	Collaboration between IBM and retailers like Walmart, enabling end-to-end traceability in the food supply chain, bolstering consumer trust and safety
CropIn technology solutions	Company offering blockchain-enabled solutions for agribusinesses to monitor crop cultivation, ensuring transparency and quality control
Ninjacart	Indian agritech startup employing blockchain for tracking the sourcing and delivery of fresh produce to retailers, streamlining the supply chain
Food safety and standards authority of India (FSSAI)	FSSAI exploring blockchain to bolster food safety, track adulteration, and ensure compliance with food regulations

These examples showcase the versatility of blockchain technology in addressing the specific challenges of the Indian food supply chain. Through these implementations, transparency, traceability, and efficiency have been significantly enhanced, ultimately benefitting farmers, consumers, and the food industry as a whole.

7.7 Conclusion and Future Prospects

The integration of blockchain technology into the Indian food supply chain has yielded substantial benefits and transformative potential. Through enhanced traceability, blockchain instills consumer trust in food safety and quality, while simultaneously reducing transaction costs, ensuring fairer compensation for farmers, and streamlining regulatory compliance. Improved supply chain efficiency further expedites recall management in the face of safety concerns. Looking ahead, the future prospects for blockchain in the Indian food supply chain are promising. As more stakeholders recognize its advantages, broader adoption is likely. This, in turn, may lead to increased investment in blockchain infrastructure, bolstering the capabilities and reach of traceability and transaction cost reduction initiatives. The trajectory is clear: blockchain is poised to revolutionize the Indian food supply chain, benefiting farmers, consumers, and the entire food industry ecosystem in the process.

Acknowledgements The research funding from the Ministry of Science and Higher Education of the Russian Federation (Ural Federal University Program of Development within the Priority-2030 Program) is gratefully acknowledged.

References

Acharya A, Gahlaut U, Sharma K, Sharma SK, Vishwakarma PN, Phanden RK (2020) Crashworthiness analysis of a thin-walled structure in the frontal part of automotive chassis. Int J Veh Struct Syst 12. https://doi.org/10.4273/ijvss.12.5.06

Adama KK, Ukhurebor KE, Pal K, Hossain I (2024) Effect of neem oil biodiesel on the surface and structural integrity of carbon steel alloy: chromatographic, spectroscopic, and morphological investigations. Int J Biol Macromol 269(Part 2):132199. https://doi.org/10.1016/j.ijbiomac.2024.132199

Adeoye AO, Quadri RO, Lawal OS (2022) Assessment of biofuel potential of tenera palm kernel shell via fixed bed pyrolysis and thermal characterization. Results Surf Interf 9:100091. https://doi.org/10.1016/j.rsurfi.2022.100091

Adeoye AO, Quadri RO, Lawal OS (2023a) Wet synthesis, characterization of goethite nanoparticles and its application in catalytic pyrolysis of palm kernel shell in TGA. Results Surf Interfaces 11:100118. https://doi.org/10.1016/j.rsurfi.2023.100118

Adeoye AO, Quadri RO, Lawal OS (2023b) Biomass waste pyrolysis in the context of Nigeria's oil and gas industry: an overview. Int J Ambient Energy 44:2525–2541. https://doi.org/10.1080/01430750.2023.2256325

Adeoye AO, Lawal OS, Quadri RO, Malomo D, Aliyu MT, Dang GE, Emojevu EO, Maikato MJ, Yahaya MG, Omonije OO, Edidem VG, Abubakar YK, Offor OF, Sochima EH, Peter BE, Hikon BN (2023c) Sustainable energy via thermochemical and biochemical conversion of biomass wastes for biofuel production. In: Transportation energy and dynamics. Energy, environment, and sustainability. Springer, Singapore, pp245–306. https://doi.org/10.1007/978-981-99-2150-8_11

Agarwal KM, Mohapatra S, Sharma P, Sharma S, Bhatia D, Mishra A (2020) Study and overview of the novel corona virus disease (COVID-19). Sensors Int 1:100037. https://doi.org/10.1016/j.sintl.2020.100037

Agrahari R, Sarraf G, Joshi NC, Mohapatra S, Varma A (2021a) Insight of biopolymers and applications of polyhydroxyalkanoates. In: Microbial polymers. Springer Singapore, Singapore, pp 177–191. https://doi.org/10.1007/978-981-16-0045-6_8

Agrahari R, Mohanty S, Vishwakarma K, Nayak SK, Samantaray D, Mohapatra S (2021b) Update vision on COVID-19: structure, immune pathogenesis, treatment and safety assessment. Sensors Int 2:100073. https://doi.org/10.1016/j.sintl.2020.100073

Avesh M, Srivastava R (2012) Modeling simulation and control of active suspension system in Matlab Simulink environment. In: 2012 students conference on engineering and systems. IEEE, pp 1–6. https://doi.org/10.1109/SCES.2012.6199124

Avesh M, Srivastava R (2019) Parametric optimization to design a passenger car suspension system for better dynamic performance. Eur Transp - Trasp Eur

Avesh M, Srivastava R (2020) Passenger car active suspension system model for better dynamic characteristics. Natl Acad Sci Lett 43:37–41. https://doi.org/10.1007/s40009-019-00807-z

Avesh M, Srivastava R, Chandmal Sharma R, Sharma N (2019a) Effective design of active suspension system using fuzzy logic control approach. Int J Veh Struct Syst 11. https://doi.org/10.4273/ijvss.11.5.13

Avesh M, Srivastava R, Sharma RC (2019b) Parametric optimization for the design of passenger vehicle suspension system with the application of genetic algorithm. Int J Veh Struct Syst 11. https://doi.org/10.4273/ijvss.11.2.08

Badgujar P, Sharma S, Rahul D, Sharma SK, Upadhyay RK (2024) Feasibility study of Mumbai-Pune hyperloop corridor: challenges and implementation. Res Transp Bus Manage 52:101089

Bandekar D, Chouhan OP, Mohapatra S, Hazra M, Hazra S, Biswas S (2017) Putative protein VC0395_0300 from Vibrio cholerae is a diguanylate cyclase with a role in biofilm formation. Microbiol Res 202:61–70. https://doi.org/10.1016/j.micres.2017.05.003

Bhardawaj S, Chandmal Sharma R, Kumar Sharma S (2019a) Development and advancement in the wheel-rail rolling contact mechanics. IOP Conf Ser Mater Sci Eng 691:012034. https://doi.org/10.1088/1757-899X/691/1/012034

Bhardawaj S, Chandmal Sharma R, Kumar Sharma S (2019b) A survey of railway track modelling. Int J Veh Struct Syst 11. https://doi.org/10.4273/ijvss.11.5.08

Bhardawaj S, Sharma RC, Sharma SK (2020a) Ride analysis of track-vehicle-human body interaction subjected to random excitation. J Chin Soc Mech Eng Trans Chinese Inst Eng Ser C/ Chung-Kuo Chi Hsueh K. Ch'eng Hsuebo Pao 41:229–238

Bhardawaj S, Sharma RC, Sharma SK (2020b) Development in the modeling of rail vehicle system for the analysis of lateral stability. Mater Today Proc 25:610–619. https://doi.org/10.1016/j.matpr.2019.07.376

Bhardawaj S, Sharma RC, Sharma SK (2020c) Analysis of frontal car crash characteristics using ANSYS. Mater Today Proc 25:898–902. https://doi.org/10.1016/j.matpr.2019.12.358

Bhardawaj S, Sharma RC, Sharma SK, Sharma N (2021) On the planning and construction of railway curved track. Int J Veh Struct Syst 13(2):151–159. https://doi.org/10.4273/ijvss.13.2.04

Bhardawaj S, Sharma RC, Sharma SK, Gopala Rao L, Vashist A (2023) Modeling of biomechanical human body model for seat to head transmissibility analysis. Noise Vib Worldwide 54(2–3):62–74. https://doi.org/10.1177/09574565231154243

Bhattarai S, Mhamdi A, Hossain I, Raoui Y, Pandey R, Madan J, Bouazizi A, Maiti M, Gogoi D, Sharma A (2022) A detailed review of perovskite solar cells: Introduction, working principle, modelling, fabrication techniques, future challenges. Micro Nanostruct 172:207450. https://doi.org/10.1016/j.micrna.2022.207450

Biswas N, Saha S, Biswas BK, Chowdhury M, Rahaman A, Junghare V, Mohapatra S, Hazra S, Zangrando E, Roy Choudhury R, Roy Choudhury C (2021) The DNA- and protein-binding properties and cytotoxicity of a new copper(II) hydrazone Schiff base complex. J Coord Chem 74:1482–1504. https://doi.org/10.1080/00958972.2021.1913128

Chatterjee A, Nirwan S, Mohapatra S, Sharma P, Agnihotri A, Shrivastava N (2022) Biochemical aspects of pathogenic variability in white rust infected Indian mustard. Mycologia 114:757–768. https://doi.org/10.1080/00275514.2022.2060007

Choi S, Lee J, Sharma SK (2021) A study on the performance evaluation of hydraulic tank injectors. In: Joshi P, Gupta SS, Shukla AK, Gautam SS (eds) Advances in engineering design. Lecture notes in mechanical engineering, pp 183–190. https://doi.org/10.1007/978-981-33-4684-0_19

Das S, Swapnarekha H, Vimal S (2023) Integration of blockchain technology with renewable energy for sustainable development: issues, challenges and future dire. https://doi.org/10.1007/978-981-99-3734-9_48

Devadarshini D, Mohapatra S, Pati S, Maity S, Rath CC, Jena PK, Samantaray D (2022) Evaluation of PHAs production by mixed bacterial culture under submerged fermentation. Biologia (bratisl) 78:895–902. https://doi.org/10.1007/s11756-022-01302-5

Eleryan A, Hassaan M, Aigbe UO, Ukhurebor K, Onyancha RB, El-Nemr MA, Ragab S, Hossain I, El Nemr A (2024) Kinetic and isotherm studies of Acid Orange 7 dye absorption using sulphonated mandarin biochar treated with TETA. Biomass Convers Biorefinery 14(9):10599–10610. https://doi.org/10.1007/s13399-023-04089-w

Elmaghraby NA, Omer AM, Kenawy ER, Gaber M, Hassaan MA, Ragab S, Hossain I, Nemr AE (2023) Electrospun cellulose acetate/activated carbon composite modified by EDTA (rC/AC-EDTA) for efficient methylene blue dye removal. Sci Rep 13(1):9919. https://doi.org/10.1038/s41598-023-36994-5

Gopala Rao L, Sharma RC, Satyanarayana V, Palli S (2022) Stochastic optimal preview control response of a quarter-car nonlinear suspension model using spectral decomposition method. Noise Vib Worldw 53:225–232. https://doi.org/10.1177/09574565221093234

Goswami B, Rathi A, Sayeed S, Das P, Sharma RC, Sharma SK (2019) Optimization design for aerodynamic elements of indian locomotive of passenger train. In: Advances in engineering design. Lecture notes in mechanical engineering. Springer, Singapore, pp 663–673. https://doi.org/10.1007/978-981-13-6469-3_61

Goyal S, Anand CS, Sharma SK, Sharma RC (2019) Crashworthiness analysis of foam filled star shape polygon of thin-walled structure. Thin-Walled Struct 144:106312. https://doi.org/10.1016/j.tws.2019.106312

Hossain I, Ahmed YE, Sharma SK (2023) The use of both metallic-insert and TiO_2/water nanofluid to improve the performance of photovoltaic modules; experimental study. Therm Sci Eng Prog 42:101944. https://doi.org/10.1016/j.tsep.2023.101944

Hossain I, Velkin VI, Shcheklein SE (2017) The study of passive vibration dampers in pipelines using Piv-methodology for single phase flow. In: WIT transactions on ecology and the environment, pp 565–570. https://doi.org/10.2495/ESUS170521

Iheme CI, Elemike EE, Igwe CU, Ujowundu FN, Ogbonna CU, Uche ZC, Okoro OC, Hossain I, Onwudiwe DC (2024) Synthesis, characterization, radical scavenging properties of zinc oxide nanoparticles and inhibitory effect of ZnONPs-ciprofloxacin nanoconjugates on multidrug-resistant Staphylococcus aureus (MRSA) enzyme. Inorg Chem Commun 160:111864. https://doi.org/10.1016/j.inoche.2023.111864

Kanth Choppara R, Chandmal Sharma R, Kumar Sharma S, Gupta T (2019) Aero dynamic cross wind analysis of locomotive. IOP Conf Ser Mater Sci Eng 691:012035. https://doi.org/10.1088/1757-899X/691/1/012035

Kohli I, Joshi NC, Mohapatra S, Varma A (2020) Extremophile—An adaptive strategy for extreme conditions and applications. Curr Genomics 21:96–110. https://doi.org/10.2174/1389202921666200401105908

Kohli I, Mohapatra S, Kumar P, Goel A, Varma A, Joshi NC (2022) Role of bacterial endophytes in the promotion of plant growth. In: Bacterial endophytes for sustainable agriculture and environmental management. Springer Singapore, Singapore, pp 235–253. https://doi.org/10.1007/978-981-16-4497-9_12

Kokkinos NC, Emmanouilidou E, Sharma SK (2024) Waste-to-biofuel production for the transportation sector. In: Intelligent transportation system and advanced technology. Springer, pp 99–122

Kumar U, Kasvekar R, Sharma SK, Upadhyay RK (2022) Wear of wheels and axle in locomotive and measures taken by Indian railway. In: Advances in engine tribology. Springer, pp 77–96. https://doi.org/10.1007/978-981-16-8337-4_5

Lee J, Sharma SK (2020) Numerical investigation of critical speed analysis of high-speed rail vehicle. 한국정밀공학회학술발표대회 논문집 (Korean Society of Precision Engineering), p 696

Lee J, Sharma SK (2023) Intelligent systems for railway infrastructure. In: Applied sciences, vol 13, issue 20. MDPI, p 11274

Lee J, Han J, Sharma SK (2021) Structural analysis on the separated and integrated differential gear case for the weight reduction. In: Advances in engineering design: select proceedings of FLAME 2020, p 175

Maity S, Das S, Mohapatra S, Tripathi AD, Akthar J, Pati S, Pattnaik S, Samantaray DP (2020) Growth associated polyhydroxybutyrate production by the novel Zobellellae tiwanensis strain

DD5 from banana peels under submerged fermentation. Int J Biol Macromol 153:461–469. https://doi.org/10.1016/j.ijbiomac.2020.03.004

Misra S, Mohanty D, Mohapatra S (2022) Food-based probiotics: functional dietary ingredients. In: Probiotics. Elsevier, pp 257–275. https://doi.org/10.1016/B978-0-323-85170-1.00006-3

Mohanty DP, Mohapatra S, Misra S, Sahu PS (2016) Milk derived bioactive peptides and their impact on human health—a review. Saudi J Biol Sci 23:577–583. https://doi.org/10.1016/j.sjbs.2015.06.005

Mohapatra S, Samantaray DP, Samantaray SM (2015) Study on polyhydroxyalkanoates production using rhizospheric soil bacterial isolates of sweet potato. Indian J Sci Technol 8:57. https://doi.org/10.17485/ijst/2015/v8iS7/64027

Mohapatra S, Samantaray DP, Samantaray SM, Mishra BB, Das S, Majumdar S, Pradhan SK, Rath SN, Rath CC, Akthar J, Achary KG (2016a) Structural and thermal characterization of PHAs produced by Lysinibacillus sp. through submerged fermentation process. Int J Biol Macromol 93:1161–1167. https://doi.org/10.1016/j.ijbiomac.2016.09.077

Mohapatra S, Rath SN, Pradhan SK, Samantaray DP, Rath CC (2016b) Secondary structural models (16S rRNA) of polyhydroxyalkanoates producing bacillus species isolated from different rhizospheric soil: Phylogenetics and chemical analysis. Int J Bioautom 20:329–338

Mohapatra S, Mohanta PR, Sarkar B, Daware A, Kumar C, Samantaray DP (2017a) Production of Polyhydroxyalkanoates (PHAs) by bacillus strain isolated from waste water and its biochemical characterization. Proc Natl Acad Sci India Sect B Biol Sci 87:459–466. https://doi.org/10.1007/s40011-015-0626-6

Mohapatra S, Mishra R, Roy P, Yadav KL, Satapathi S (2017b) Systematic investigation and in vitro biocompatibility studies on implantable magnetic nanocomposites for hyperthermia treatment of osteoarthritic knee joints. J Mater Sci 52:9262–9268. https://doi.org/10.1007/s10853-017-1136-0

Mohapatra S, Sarkar B, Samantaray DP, Daware A, Maity S, Pattnaik S, Bhattacharjee S (2017c) Bioconversion of fish solid waste into PHB using bacillus subtilis based submerged fermentation process. Environ Technol 38:3201–3208. https://doi.org/10.1080/09593330.2017.1291759

Mohapatra S, Maity S, Dash HR, Das S, Pattnaik S, Rath CC, Samantaray D (2017d) Bacillus and biopolymer: prospects and challenges. Biochem Biophys Rep 12:206–213. https://doi.org/10.1016/j.bbrep.2017.10.001

Mohapatra S, Mishra Y, Dash L (2020a) COVID-19: analysing the legal nuances of the lockdown order. Indian J Forensic Med Toxicol. https://doi.org/10.37506/ijfmt.v14i4.11553

Mohapatra S, Pattnaik S, Maity S, Mohapatra S, Sharma S, Akhtar J, Pati S, Samantaray DP, Varma A (2020b) Comparative analysis of PHAs production by Bacillus megaterium OUAT 016 under submerged and solid-state fermentation. Saudi J Biol Sci 27:1242–1250. https://doi.org/10.1016/j.sjbs.2020.02.001

Mohapatra S, Vishwakarma K, Joshi NC, Maity S, Kumar R, Ramchander M, Pattnaik S, Samantaray DP (2021) A review on PHAs: the future biopolymer. In: Environmental and agricultural microbiology. Wiley, pp 83–100. https://doi.org/10.1002/9781119525899.ch5

Mohapatra S, Mohapatra S, Han H, Ariza-Montes A, López-Martín M (2022) Climate change and vulnerability of agribusiness: Assessment of climate change impact on agricultural productivity. Front Psychol 13. https://doi.org/10.3389/fpsyg.2022.955622

Mohd A, Srivastava R (2019) Parametric optimization and experimental validation for nonlinear characteristics of passenger car suspension system. Period Polytech Transp Eng 49:103–113. https://doi.org/10.3311/PPtr.12999

Nayak J, Swapnarekha H, Routray AR, Nayak SR, Behera HS (2022) Big data in education: present and future. In: Computational intelligence in data mining. Smart innovation, systems and technologies, vol 281. Springer, Singapore, pp 721–739. https://doi.org/10.1007/978-981-16-9447-9_54

Nibedita S, Swati P, Pattnaik M, Mohapatra S (2021) Methane emission and strategies for mitigation in livestock. In: Environmental and agricultural microbiology. Wiley, pp 257–274. https://doi.org/10.1002/9781119525899.ch12

Palli S, Dontikurti R, Sharma RC, Sharma N (2021) Analysis of dynamic response of a railway locomotive using ANSYS. Int J Veh Struct Syst 13. https://doi.org/10.4273/ijvss.13.2.21

Palli S, Duppala A, Sharma RC, Gopala Rao LVV (2022) Dynamic simulation of automotive vehicle suspension using MATLAB simulink. Int J Veh Struct Syst 14. https://doi.org/10.4273/ijvss.14.3.04

Palli S, Koona R (2015) Analyses of dynamic response of a railway bogie. Int J Veh Noise Vib 11:103. https://doi.org/10.1504/IJVNV.2015.070015

Palli S, Koona R, Sharma SK, Sharma RC (2018) A review on dynamic analysis of rail vehicle coach. Int J Veh Struct Syst 10(3):204–211. https://doi.org/10.4273/ijvss.10.3.10

PanduRanga Vital T, Lakshmi BG, Swapna Rekha H, DhanaLakshmi M (2019) Student performance analysis with using statistical and cluster studies. In: Soft computing in data analytics. Advances in intelligent systems and computing, vol 758. Springer, Singapore, pp 743–757. https://doi.org/10.1007/978-981-13-0514-6_71

Pati S, Maity S, Dash A, Jema S, Mohapatra S, Das S, Samantaray DP (2020) Biocompatible PHB production from bacillus species under submerged and solid-state fermentation and extraction through different downstream processing. Curr Microbiol 77:1203–1209. https://doi.org/10.1007/s00284-020-01922-7

Pati S, Mohanty MK, Mohapatra S, Samantaray D (2022) Bioethanol production by Enterobacter hormaechei through carbon monoxide-rich syngas fermentation. Int J Energy Res 46:20096–20106. https://doi.org/10.1002/er.8678

Pattnaik S, Dash D, Mohapatra S, Pattnaik M, Marandi AK, Das S, Samantaray DP (2020) Corrigendum to "Improvement of rice plant productivity by native Cr(VI) reducing and plant growth promoting soil bacteria Enterobacter cloacae". [Chemosphere 240:124895]. Chemosphere 238:125011. https://doi.org/10.1016/j.chemosphere.2019.125011

Pattnaik S, Mohapatra S, Pati S, Dash D, Devadarshini D, Tanaya K, Mishra BB, Samantaray D (2022) Microbial bioremediation of Cr(VI)-contaminated soil for sustainable agriculture. In: Microbial biodegradation and bioremediation. Elsevier, pp 395–407. https://doi.org/10.1016/B978-0-323-85455-9.00008-4

Pundir S, Upadhyay S, Kumar N, Chandra Joshi N, Priya R, Ahmad Mir R, Hossain I, Pandey OP (2023a) Interconnected CuO nanoplates as a highly efficient electrocatalyst for oxygen evolution reaction. Mater Lett 336:133921. https://doi.org/10.1016/j.matlet.2023.133921

Pundir S, Upadhyay S, Priya R, Kumar N, Chetana S, Hossain I, Joshi NC, Pandey OP (2023b) Synthesis of 1D β-MnO_2 for high-performance supercapacitor application. J Solid State Electrochem 27:531–538. https://doi.org/10.1007/s10008-022-05347-z

Raju Rallabandi S, Pilla DP, Dowluru S, Palli S, Sharma N, Sharma SK, Sharma RC (2024) Critical evaluation of epoxy-hemp-pineapple-palm fiber composites using hybrid AHM-TOPSIS technique for sustainable structural applications. J Chin Inst Eng 1–12

Rallabandi SR, Srinivas L, Palli S, Sharma RC, Sharma N, Sharma A, Sharma SK (2024) A contrastive characterization of pure Mg and AZ91D alloy based on the testing of mechanical, corrosion, wear, and erosion properties. Eng Res Express 6(1):015017

Rathinavelu V, Hariharan R, Marimuthu K, Prabagaran S, Chathapuram Balasubramanian P, De Poures MV, Hossain I, Seikh AH, Kalam MA (2023) Effect of steam gasification parameters on hydrogen extraction from indirect land use change crops. Energy Sources Part A: Recovery Util Environ Eff 45(4):12263–12276. https://doi.org/10.1080/15567036.2023.2272661

Rathinavelu V, Kulandaivel A, Pandey AK, Bhatt R, De Poures MV, Hossain I, Seikh AH, Iqbal A, Murugan P (2024) Production of green hydrogen from sewage sludge/algae in agriculture diesel engine: performance evaluation. Heliyon 10(1):e23988. https://doi.org/10.1016/j.heliyon.2024.e23988

Samantaray D, Mohapatra S, Mishra BB (2014) Microbial bioremediation of industrial effluents. In: Microbial biodegradation and bioremediation. Elsevier, pp 325–339. https://doi.org/10.1016/B978-0-12-800021-2.00014-5

Sarkar MS, Segu H, Bhaskar JV, Jakher R, Mohapatra S, Shalini K, Shivaji S, Reddy PA (2018) Ecological preferences of large carnivores in remote, high-altitude protected areas: insights from Buxa Tiger Reserve, India. Oryx 52:66–77. https://doi.org/10.1017/S0030605317000060

Satyanarayana VSV, Sharma RC, Sateesh B, Gopala Rao LVV, Mohan Rao N, Palli S (2022) Optimum response of a nonlinear passive vehicle suspension system under random road excitations. SAE Int J Commer Veh 16:02-16-01–0004. https://doi.org/10.4271/02-16-01-0004

Sharma SK (2019) Multibody analysis of longitudinal train dynamics on the passenger ride performance due to brake application. Proc Inst Mech Eng Part K J Multi-Body Dyn 233(2):266–279. https://doi.org/10.1177/1464419318788775

Sharma SK, Kumar A (2017a) Ride performance of a high speed rail vehicle using controlled semi active suspension system. Smart Mater Struct 26:055026. https://doi.org/10.1088/1361-665X/aa68f7

Sharma SK, Kumar A (2017b) Impact of electric locomotive traction of the passenger vehicle ride quality in longitudinal train dynamics in the context of Indian railways. Mech Ind 18(2):222. https://doi.org/10.1051/meca/2016047

Sharma SK, Kumar A (2018a) Ride comfort of a higher speed rail vehicle using a magnetorheological suspension system. Proc Inst Mech Eng Part K J Multi-Body Dyn 232:32–48. https://doi.org/10.1177/1464419317706873

Sharma SK, Kumar A (2018b) Disturbance rejection and force-tracking controller of nonlinear lateral vibrations in passenger rail vehicle using magnetorheological fluid damper. J Intell Mater Syst Struct 29(2):279–297. https://doi.org/10.1177/1045389X17721051

Sharma SK, Kumar A (2018c) Impact of longitudinal train dynamics on train operations: a simulation-based study. J Vib Eng Technol 6(3):197–203. https://doi.org/10.1007/s42417-018-0033-4

Sharma SK, Lee J (2020a) Finite element analysis of a fishplate rail joint in extreme environment condition. Int J Veh Struct Syst 12. https://doi.org/10.4273/ijvss.12.5.03

Sharma SK, Lee J (2020b) Design and development of smart semi active suspension for nonlinear rail vehicle vibration reduction. Int J Struct Stab Dyn 20:2050120. https://doi.org/10.1142/S0219455420501205

Sharma SK, Lee J (2021) Crashworthiness analysis for structural stability and dynamics. Int J Struct Stab Dyn 21:2150039. https://doi.org/10.1142/S0219455421500395

Sharma RC, Palli S (2016) Analysis of creep force and its sensitivity on stability and vertical-lateral ride for railway vehicle. Int J Veh Noise Vib 12:60. https://doi.org/10.1504/IJVNV.2016.077474

Sharma SK, Sharma RC (2018a) An investigation of a locomotive structural crashworthiness using finite element simulation. SAE Int J Commercial Veh 11(4):235–244. https://doi.org/10.4271/02-11-04-0019

Sharma SK, Sharma RC (2018b) Simulation of quarter-car model with magnetorheological dampers for ride quality improvement. Int J Veh Struct Syst 10(3):169–173. https://doi.org/10.4273/ijvss.10.3.03

Sharma RC, Sharma SK (2018c) Sensitivity analysis of three-wheel vehicle's suspension parameters influencing ride behavior. Noise Vib Worldwide 49(7–8), 272–280. https://doi.org/10.1177/0957456518796846

Sharma SK, Sharma RC (2019) Pothole detection and warning system for Indian roads. In: Advances in interdisciplinary engineering. Lecture notes in mechanical engineering. Springer, Singapore, pp 511–519. https://doi.org/10.1007/978-981-13-6577-5_48

Sharma SK, Sharma RC (2021) Multi-objective design optimization of locomotive nose. SAE Technical Paper, p 10. https://doi.org/10.4271/2021-01-5053

Sharma RC, Sharma SK (2022) Ride analysis of road surface-three-wheeled vehicle-human subject interactions subjected to random excitation. SAE Int J Commercial Veh 15(3):299–312, 02-15-03-0017. https://doi.org/10.4271/02-15-03-0017

Sharma SK, Sharma RC, Kumar A, Palli S (2015) Challenges in rail vehicle-track modeling and simulation. Int J Veh Struct Syst 7(1):1–9. https://doi.org/10.4273/ijvss.7.1.01

Sharma RC, Palli S, Sharma SK, Roy M (2017) Modernization of railway track with composite sleepers. Int J Veh Struct Syst 9(5):321–329. https://doi.org/10.4273/ijvss.9.5.10

Sharma RC, Sharma SK, Palli S (2018) Rail vehicle modelling and simulation using Lagrangian method. Int J Veh Struct Syst 10(3):188–194. https://doi.org/10.4273/ijvss.10.3.07

Sharma SK, Saini U, Kumar A (2019) Semi-active control to reduce lateral vibration of passenger rail vehicle using disturbance rejection and continuous state damper controllers. J Vib Eng Technol 7(2):117–129. https://doi.org/10.1007/s42417-019-00088-2

Sharma S, Sharma RC, Sharma SK, Sharma N, Palli S, Bhardawaj S (2020a) Vibration isolation of the quarter car model of road vehicle system using dynamic vibration absorber. Int J Veh Struct Syst. 12. https://doi.org/10.4273/ijvss.12.5.05

Sharma RC, Sharma SK, Palli S (2020b) Linear and non-linear stability analysis of a constrained railway wheelaxle. Int J Veh Struct Syst 12. https://doi.org/10.4273/ijvss.12.2.04

Sharma SK, Sharma RC, Sharma N (2020c) Combined multi-body-system and finite element analysis of a rail locomotive crashworthiness. Int J Veh Struct Syst 12. https://doi.org/10.4273/ijvss.12.4.15

Sharma RC, Sharma SK, Sharma N, Sharma S (2020d) Analysis of ride and stability of an ICF railway coach. Int J Veh Noise Vib 16:127. https://doi.org/10.1504/IJVNV.2020.117820

Sharma RC, Sharma S, Sharma SK, Sharma N (2020e) Analysis of generalized force and its influence on ride and stability of railway vehicle. Noise Vib Worldw 51:95–109. https://doi.org/10.1177/0957456520923125

Sharma RC, Palli S, Sharma N, Sharma SK (2021a) Ride behaviour of a four-wheel vehicle using H infinity semi-active suspension control under deterministic and random inputs. Int J Veh Struct Syst 13.https://doi.org/10.4273/ijvss.13.2.18

Sharma RC, Sharma S, Sharma N, Sharma SK (2021b) Linear and nonlinear analysis of ride and stability of a three-wheeled vehicle subjected to random and bump inputs using bond graph and simulink methodology. SAE Int J Commer Veh 14:02-15-01–0001. https://doi.org/10.4271/02-15-01-0001

Sharma SK, Sharma RC, Lee J (2021c) Effect of rail vehicle-track coupled dynamics on fatigue failure of coil spring in a suspension system. Appl Sci 11:2650. https://doi.org/10.3390/app11062650

Sharma RC, Sharma S, Sharma SK, Sharma N, Singh G (2021d) Analysis of bio-dynamic model of seated human subject and optimization of the passenger ride comfort for three-wheel vehicle using random search technique. Proc Inst Mech Eng Part K J Multi-Body Dyn 235:106–121. https://doi.org/10.1177/1464419320983711

Sharma SK, Mohapatra S, Sharma RC, Alturjman S, Altrjman C, Mostarda L, Stephan T (2022a) Retrofitting existing buildings to improve energy performance. Sustainability 14:666. https://doi.org/10.3390/su14020666

Sharma SK, Sharma RC, Lee J, Jang HL (2022b) Numerical and experimental analysis of DVA on the flexible-rigid rail vehicle carbody resonant vibration. Sensors 22:1922. https://doi.org/10.3390/s22051922

Sharma SK, Lee J, Jang HL (2022c) Mathematical modeling and simulation of suspended equipment impact on car body modes. Machines 10:192. https://doi.org/10.3390/machines10030192

Sharma SK, Sharma RC, Lee J (2022d) In situ and experimental analysis of longitudinal load on carbody fatigue life using nonlinear damage accumulation. Int J Damage Mech 31:605–622. https://doi.org/10.1177/10567895211046043

Sharma RC, Gopala Rao LVV, Sharma SK, Palli S, Satyanarayana VSV (2022e) Analysis of lateral stability and ride of an Indian railway constrained dual-axle bogie frame. SAE Int J Commercial Veh 16(2):213–228, 02-16-02-0014. https://doi.org/10.4271/02-16-02-0014

Sharma RC, Palli S, Gopala Rao LVV, Duppala A, Sharma SK (2022f) Four-wheel vehicle response under bump, pothole, harmonic, and random excitations using bond graph/simulink technique. SAE Int J Commercial Veh 16(2):129–139, 02-16-02-0008. https://doi.org/10.4271/02-16-02-0008

Sharma RC, Palli S, Jha AK, Bhardawaj S, Sharma SK (2022g) Vibration and ride comfort analysis of railway vehicle system subjected to deterministic inputs. Resmilitaris 12(5):1345–1355

Sharma RC, Palli S, Bhardawaj S (2023a) A review of wheel-rail dynamic behavior on curved tracks. Int J Veh Struct Syst 15. https://doi.org/10.4273/ijvss.15.1.04

Sharma RC, Palli S, Sharma SK (2023b) Ride analysis of railway vehicle considering rigidity and flexibility of the carbody. J Chin Inst Eng 46:355–366. https://doi.org/10.1080/02533839.2023.2194918

Sharma RC, Bhardawaj S, Sharma SK, Bahl S (2023c) Ride behaviour of three-wheel vehicle using semi-active suspension using skyhook control strategy. Int J Veh Struct Syst 15(1). https://doi.org/10.4273/ijvss.15.1.05

Sharma SK, Sharma RC, Choi Y, Lee J (2023d) Experimental and mathematical study of flexible–rigid rail vehicle riding comfort and safety. Appl Sci 13(9):5252

Sharma SK, Sharma RC, Choi Y, Lee J (2023e) Modelling and dynamic analysis of adaptive neuro-fuzzy inference system-based intelligent control suspension system for passenger rail vehicles using magnetorheological damper for improving ride index. Sustainability 15(16):12529

Sharma SK, Sharma RC, Lavania S, Palli S, Avesh M (2023f) Industry-driven approach for ANFIS-based intelligent control suspension system with MR damper for enhanced ride quality in passenger rail vehicles for technological investigations. Natl Acad Sci Lett 1–4

Sharma SK, Sharma RC, Upadhyay RK, Lee J (2023g) Coupled torsional–transverse vibration reduction in marine propulsion dynamics with novel approach using magnetorheological damping and adaptive control system. J Vib Eng Technol 1–11

Sharma SK, Upadhyay RK, Kumar V, Valera H (2023h) Introduction to sustainable transportation system. In: Transportation energy and dynamics. Springer, pp 3–6

Sharma RC, Palli S, Dharmana L, Sharma SK, Sharma A, Kharola A, Sharma N (2024a) Simulation of transient dynamic response of a locomotive considering its flexibility and rigidity. Noise Vib Worldwide 55(1–2):33–47

Sharma RC, Palli S, Duppala A, Bhardawaj S, Sharma SK, Sharma N, Sharma A (2024b) Rigid body analysis of ride behaviour of roadway vehicle using Lagrangian dynamics. Int J Veh Struct Syst (IJVSS) 16(1)

Sharma RC, Sharma N, Palli S, Sharma SK, Joshi S, Rallabandi SR, Sharma A (2024c) Dynamic behavior of three-wheel vehicle with inerter-based dynamic absorber suspension system. Noise Vib Worldwide 55(4–5):253–263. https://doi.org/10.1177/09574565241243398

Sharma SK, Kumar N, Avesh M, Sharma RC, Siddiqui MIH, Lee J (2024d) Navigating tranquillity with $H\infty$ controller to mitigate ship propeller shaft vibration. J Vib Eng Technol 1–11

Sharma SK, Sharma RC, Lee J (2024e) Modelling and simulation of MR damper characterization and uncertainty assessment in marine engine vibration isolation with FLPID control. Int J Model Simul 1–11

Sharma SK, Sharma RC, Lee J (2024f) Propelling precision of longitudinal vibration mitigation in ship propeller shafts through advanced nonlinear intelligent semi-active control leveraging adaptive neuro-fuzzy inference system with linear quadratic regulator. J Vib Control. https://doi.org/10.1177/10775463241244836

Sharma SK, Sharma RC, Upadhyay RK, Lee J (2024g) Enhanced ship engine vibration reduction using magnetorheological dampers and adaptive neuro-fuzzy control system. Ships Offshore Struct 1–17

Sinha AK, Sengupta A, Gandhi H, Bansal P, Agarwal KM, Sharma SK, Sharma RC, Sharma SK (2019) Performance enhancement of an all-terrain vehicle by optimizing steering, powertrain and brakes. In: Advances in engineering design. Lecture notes in mechanical engineering. Springer, Singapore, pp 207–215. https://doi.org/10.1007/978-981-13-6469-3_19

Swapna Rekha H, Nayak J, Behera HS (2020) Pi-Sigma neural network: survey of a decade progress. In: computational intelligence in pattern recognition. Advances in intelligent systems and computing, vol 1120. Springer, Singapore, pp 429–441. https://doi.org/10.1007/978-981-15-2449-3_37

Ukhurebor KE, Hossain I, Pal K, Jokthan G, Osang F, Ebrima F, Katal D (2024) Applications and contemporary issues with adsorption for water monitoring and remediation: a facile review. Top Catal 67(1–4):140–155. https://doi.org/10.1007/s11244-023-01817-4

Upadhyay RK, Sharma SK, Kumar V (Eds) (2024a) Intelligent transportation system and advanced technology. Springer Nature Singapore. https://doi.org/10.1007/978-981-97-0515-3

Upadhyay RK, Sharma SK, Kumar V (2024b) Introduction to intelligent transportation system and advanced technology. In: Intelligent transportation system and advanced technology. Energy, Environment, and Sustainability. pp 3–6. Springer, Singapore. https://doi.org/10.1007/978-981-97-0515-3_1

Upadhyay RK, Sharma SK, Kumar V, Valera H (2023a) Introduction to transportation systems technology and integrated management. In: Transportation systems technology and integrated management. Springer, pp 1–4. https://doi.org/10.1007/978-981-99-1517-0_1

Upadhyay RK, Sharma SK, Kumar V, Valera H (2023b) Transportation systems technology and integrated management. Edited by RK Upadhyay, SK Sharma, V Kumar, H Valera. Springer Nature, Singapore. https://doi.org/10.1007/978-981-99-1517-0

Vishwakarma PN, Mishra P, Sharma SK (2022a) Characterization of a magnetorheological fluid damper a review. Mater Today: Proc 56:2988–2994. https://doi.org/10.1016/j.matpr.2021.11.143

Vishwakarma PN, Mishra P, Sharma SK (2022b) Formulation of semi-active suspension system and controls in rail vehicle. SSRN Electron J. https://doi.org/10.2139/ssrn.4159616

Wu Q, Cole C, Spiryagin M, Chang C, Wei W, Ursulyak L, Shvets A, Murtaza MA, Mirza IM, Zhelieznov K, Mohammadi S, Serajian H, Schick B, Berg M, Sharma RC, Aboubakr A, Sharma SK, Melzi S, Di Gialleonardo E, Cantone L et al (2023) Freight train air brake models. Int J Rail Transp 11(1):1–49. https://doi.org/10.1080/23248378.2021.2006808

Chapter 8
Streamlining Aerodynamic Optimization with Generalized ANN Support

Sarfraz Ahmad Awan and Mohd Avesh

8.1 Introduction

Over the past few years, the transportation industry has heavily adopted the use of aerodynamic shape optimization in the process of designing ever more efficient vehicles. In general, of the total propulsive power required by an airship, about 2/3rd is used up in the total drag of the body (Adeoye et al. 2023a). In another study, it was found that as much as 50% of the energy in the passenger cars was used to overcome drag. While the application of ASO is seen extensively in the aircraft designing, roadways, railways, and waterways have realized the importance of ASO as well. Recent studies have significantly contributed to the understanding of Aerodynamic Optimization of rail vehicle and the improvement of railway systems (Palli et al. 2018; Sharma et al. 2018; Sharma and Sharma 2018; Sharma and Kumar 2018b, c; Sharma and Sharma 2018a, b). The ASO is important not just to the vehicle design, but crucial to the design of bridges and high-rise buildings. On a unitary level, the loss of the energy might not amount to much for an average car user and possibly not even noticeable, but the combined wastage would still be extremely high (Adeoye et al. 2023a; Gopala Rao et al. 2022; Mohapatra et al. 2016a; Sharma et al. 2020a). Despite the clear advantages of applying some form of ASO, the degree of benefits achieved are highly versatile (Mohapatra et al. 2016b). While using traditional optimization techniques is simple, the use case is limited to the simple 2D air foil shapes. Not to mention the high computation costs and the tendency of getting stuck in a local optimum. An adjoint solver will provide increased accuracy for a fraction of time,

S. A. Awan
Department of Computer Science & Information Technology, Virtual University of Pakistan, Lahore 54000, Pakistan

M. Avesh (✉)
Mazedan International Research Academy, Okhla, Delhi, New Delhi 110025, India
e-mail: mail2avesh@gmail.com

S. K. Sharma et al. (eds.), *Dynamics of Transportation Ecosystem, Modeling, and Control*, Energy, Environment, and Sustainability,
https://doi.org/10.1007/978-981-97-0437-8_8

but the cost is still considerably high (Kohli et al. 2022). A simple aircraft could take days to be optimized and even then, one would not know if it was close to the global optimal solution (Sharma and Kumar 2017a, 2018a; Sharma et al. 2021a).

These are serious issues in successfully implementing an ASO framework in the industry which only get more complicated with the increase in the design spaces. Another difficulty in using an ASO algorithm is the suitability of the algorithm (Bandekar et al. 2017; Bhardawaj et al. 2020a; Hossain et al. 2017; Mohapatra et al. 2017a; Sharma and Sharma 2019; Sharma et al. 2022a; Swapna Rekha et al. 2020). This research attempts to look at the latest techniques and some older popular techniques of the industry, and discuss the challenges faced and the solutions and aims to produce a framework suitable for use in the industry without being bound by the domain knowledge of aerodynamics and instead be able to perform high-quality optimization, leading to more efficiency in the sector over all (Goswami et al. 2019; Mohanty et al. 2016; Nibedita et al. 2021; Pattnaik et al. 2022; Samantaray et al. 2014; Sharma et al. 2022b; Sinha et al. 2019). Over the years, there have been developments both in the optimization algorithms and the computing technology. In the preliminary research, a recent development utilizing different combinations of the algorithms, and how the advancement in computing technology is being exploited to make the optimization more efficient. The recent advancements in rail vehicle system modeling, structural analysis, and performance evaluation have been extensively studied in various contexts, including crashworthiness, lateral stability, critical speed analysis, and multi-objective optimization (Goyal et al. 2019; Sharma 2019; Sharma et al. 2019; Bhardawaj et al. 2020c; Lee and Sharma 2020; Bhardawaj et al. 2021; Choi et al. 2021; Lee et al. 2021; Sharma and Sharma 2021).

8.1.1 The Concept of Aerodynamic Shape Optimization

Aerodynamic shape optimization is a method referring to optimization process used to maximize the performance of a given aerodynamic component/body by changing its shape. A typical ASO process can be broken down in the following subprocesses (Biswas et al. 2021; Das et al. 2023; Pundir et al. 2023a; Sharma et al. 2020b, 2021b, 2022c).

- Problem Formulation
- Geometric Representation
- Constraint Handling
- Optimization of the objective function
- Problem Formulation.

The solution to a typical optimization problem, particularly in the context of aerodynamic design optimization, is contingent upon several factors, including the nature of the problem, available resources, and the desired outcome. The comprehensive studies on wear mitigation in locomotive components, lateral stability, ride comfort, and advanced suspension systems have significantly contributed to enhancing the performance and safety of rail and road vehicles (Kumar et al. 2022; Sharma et al. 2022e, f, g; Sharma and Sharma 2022; Vishwakarma et al. 2022a, b). To achieve optimal results, it is crucial to make informed decisions regarding the choice of solving strategy and the algorithms to be employed. However, the very first and foundational step in this process is the formulation of the objective function, which serves as the guiding principle for the entire optimization endeavor (Agrahari et al. 2021a; Maity et al. 2020; Misra et al. 2022; Sharma and Lee 2020a; Sharma et al. 2022d).

In the optimization, a solution algorithm typically iteratively strives to attain either a global minimum or maximum over a convex objective function (Kohli et al. 2020; Mohapatra et al. 2017b). The outcome of this endeavor hinges on the specific characteristics of the objective function and the chosen solver. Broadly, the usual outcomes can be categorized as follows:

a Solver Finds a Local Optimum Close to the Global Optima: In some instances, the solver successfully converges to a local optimum that is in close proximity to the global optimum. This outcome is indicative of a relatively successful optimization process, albeit within the vicinity of the global solution (Bhardawaj et al. 2020b; Devadarshini et al. 2022; Palli et al. 2022; Sarkar et al. 2018; Sharma et al. 2021c).
b Solver Finds a Local Optima Far from the Global Optima: In other cases, the solver may converge to a local optimum that is quite distant from the global optimum (Acharya et al. 2020; Bhardawaj et al. 2019a; Mohapatra et al. 2020; Pundir et al. 2023b; Sharma and Lee 2020b, 2021). While this signifies that the optimization process has found a solution, it may not be the best possible outcome in terms of the global objective.
c Solver Fails to Converge on an Optima After Multiple Iterations: Unfortunately, there are scenarios where the solver struggles to converge to any meaningful optima, despite multiple iterations (Pattnaik et al. 2020). This situation can be particularly challenging and may necessitate further action.

In light of these possible outcomes, the designer faces three distinct options:

- Re-start the Process/Perform a Full Evaluation of the Solution: This involves going back to the initial stages of the optimization process, either starting afresh or rigorously evaluating the existing solution to validate its reproducibility (Mohapatra et al. 2020). It is a prudent step to ensure the reliability of the solution.
- Opt for a Different Optimization Strategy: An alternative approach is to consider a change in the chosen optimization strategy. This option acknowledges that the current strategy may not be yielding the desired results and advocates exploring different avenues for optimization (Sharma et al. 2020c). It demonstrates adaptability and a willingness to experiment with new methodologies.

It is crucial to emphasize that option C should be approached with caution, as it is often characterized by a high demand for computational resources. Implementing this option can be resource-intensive and should be carefully evaluated in terms of cost and potential benefits (Adeoye et al. 2023b; Mohapatra et al. 2017c, 2022; Pati et al. 2022).

Furthermore, it is imperative to acknowledge that when accepting a solution, one must recognize the inherent challenge: the likelihood of the solution being a global optimum is exceedingly low in many complex optimization problems (Adeoye et al. 2023c; Chatterjee et al. 2022; Nayak et al. 2022; Pati et al. 2020). Therefore, when assessing and selecting solutions, it is essential to do so with the awareness that achieving a global optimum may be elusive, and settling for a satisfactory local optimum is often the practical approach.

8.1.2 Geometric Representation

Variables and parameters are vital for any optimization solution. Number of variables and parameters decide the complexity of the problem. Too few parameters lead to loss of important potential improvements and too many lead to redundant variables (Adeoye et al. 2022; Mohapatra et al. 2017d; Sharma et al. 2020d). Often times, the definition of the geometry and the variables considered lead to sub-optimality in the solution. The usual methods used for performing geometric parameterization are based on the recent advancements in the fields of aerodynamics and its infrastructure have been documented in various studies (Bhardawaj et al. 2023; Hossain et al. 2023; Lee and Sharma 2023; Sharma et al. 2023c, d, e, f, g, h; Upadhyay et al. 2023a, b; Wu et al. 2023).

a Constructive Models:
- Spline Methods
- Partial Differential Equations

b Deformative Models:
- Free-form definition
- Volumetric Models.

A proper geometric representation model choice can prove to be resource-efficient. It is because of this that there have been multiple efforts to develop some tool to assist in making a choice of which technique to use (Sharma et al. 2020e).

8.1.3 Constraint Handling

After deciding upon a suitable parameter, the next stage in the ASO/ADO process is constraint handling (Bhardawaj et al. 2020c). In this stage, depending on the type of geometric model chosen, the parameters and the strictness of the physical requirements make up for the decisions regarding the constraint handling (Palli et al. 2021).

8.1.4 Optimization of the Objective Function

This step is arguably the most important step of the ASO process, and the most resource-extensive. Under this step, a suitable algorithm is used (depending on the objective function, design constraints, and expected output), which returns the optimized design variables, hopefully as per the requirement (Sharma et al. 2021d). The popular useful types of algorithms are

a. Gradient-based Method
b Gradient-Free Method
c Hybrid Algorithms
d Surrogate Modeling.

Other than the standard categories mentioned above, a framework that is later discussed in the later section does not strictly fit in these categories, although the framework could loosely be said to be a hybrid algorithm. The different methods at using ANN with any of the existing multi-objective design optimization techniques that have been previously done (Bhardawaj et al. 2019b; Kanth Choppara et al. 2019; Mohapatra et al. 2015; Palli and Koona 2015). It has been highlighted that while several attempts were made to use ANN and DNN for determining the objective functions, one particularly interesting approach was to use the ANN to guide the MOPSO algorithm by predicting the region of global optima. They were successfully able to design a framework that would accurately predict up to 94% of times the precise location of global optimization (Agarwal et al. 2020; Agrahari et al. 2021b; Bhattarai et al. 2022; Mohapatra et al. 2021; PanduRanga Vital et al. 2019; Sharma et al. 2023a, b). The objective of this research is to identify a technology, or a combination of a few technologies to allow for creating an industrial tool that requires minimum setup and be able to achieve a global minimum solution efficiently and feasibly. As mentioned in the following sections, only with the recent development of Statistical Learning models, advanced cheap GPUs and the affinity for data-driven engineering have enabled the possibility for such a tool to be developed.

8.2 Literature Review

Given the importance of the process, there exist plenty of research and academic material on the topic of aerodynamic shape optimization Skinner, Behtash 2018. As is with any scientific development, most popular optimization technologies are not suitable for general case. Instead, all of the technologies are well suited for a very particular situation, and may not be able to work efficiently, if not worse out of that niche situation (Avesh and Srivastava 2012, 2019a, 2020; Avesh et al. 2019a; Satyanarayana et al. 2022; Sharma and Palli 2016).

In the preliminary research for major frameworks and algorithms, the resulting observations can be categorized in the following section. Of the mentioned algorithms, the most promising algorithms for achieving the objectives are

- Surrogate model
- Hybrid framework
- ANN-assisted system.

On first glance, the ANN system could be mistaken for the standard form of Hybrid system; however, the nature of application of the assistive algorithm aiding the optimizer algorithm works in a fundamentally different manner. The innovative approaches to transportation, from the feasibility of the Mumbai-Pune Hyperloop corridor and waste-to-biofuel production to advanced vibration control in marine and rail systems, highlight significant strides in sustainable and intelligent transportation technologies (Badgujar et al. 2024; Kokkinos et al. 2024; Rallabandi et al. 2024a, b; Sharma et al. 2024a, b, c, d, e f, g; Upadhyay et al. 2024a, b).

8.2.1 Gradient-based Methods (GBM)

Fluid mechanics are described by the Navier–Stokes equations, which are typically written as

$$\partial u/\partial t + (u \cdot \nabla)u = -\nabla P + \nu \nabla^2 u \tag{8.1}$$

This equation represents the conservation of momentum for a fluid. Here, u is the velocity vector, t is time, ∇ represents the gradient operator, P is pressure, and ν is the kinematic viscosity. It's a set of three equations (one for each spatial dimension) that relate the velocity and pressure fields in a fluid.

8.2.2 Computational Fluid Dynamics (CFD)

CFD is a numerical technique used to solve the Navier–Stokes equations and simulate fluid flow. It involves discretizing the equations into a finite number of discrete points or cells in a computational grid or mesh. The equations are then solved iteratively for each of these points to approximate the behavior of the fluid over time (Avesh and Srivastava 2019b; Avesh et al. 2019b).

8.2.3 Gradient-based Optimization

Gradient-based optimization algorithms aim to find the minimum or maximum of a given function. In this context, let's consider a function $f(V)$, where V is a vector of design variables. Gradient descent is a common gradient-based method, and it is mathematically represented as Eq. 8.2:

$$V_{\text{new}} = V_{\text{old}} - \alpha \nabla f(V_{\text{old}}) \tag{8.2}$$

Here, $\nabla f(V_{\text{old}})$ is the gradient of the function f at the current point V_{old}, and α is a small positive number known as the learning rate, which determines the step size in the direction of the negative gradient. This process is repeated iteratively to find the optimal values for V that minimize or maximize the objective function f.

8.2.4 Challenges of Gradient-based Optimization

Gradient-based optimization techniques are highly effective when dealing with design spaces that exhibit continuity and mathematical predictability. In such scenarios, where the possible values for the design variables smoothly transition from one point to another, gradient-based methods can excel. These methods rely on the mathematical gradient of the objective function and work best when the design space follows a well-behaved mathematical pattern (Sharma et al. 2020a). However, when applied to complex and high-fidelity design spaces, gradient-based optimization methods encounter challenges. These spaces often involve intricate interactions between variables and exhibit a multitude of local optima, making it difficult for the algorithm to swiftly converge to the global optimum. The slow convergence in these cases can hinder the efficiency of gradient-based methods, necessitating careful consideration of alternative approaches or strategies. Furthermore, the success of gradient-based optimization critically depends on the user's familiarity with the design space. Users must possess a sound understanding of the space's characteristics and nuances. Additionally, they are required to provide an initial reference point for the optimization algorithm to commence its search. This initial point serves as a starting position from which the algorithm explores and iteratively refines its solution. Thus, while gradient-based optimization offers a powerful tool for finding optimal solutions, it also places a degree of responsibility on the user's domain knowledge and input in guiding the optimization process effectively.

8.2.5 Evaluation of Gradients

In gradient-based optimization, calculating the gradient $\nabla f(V)$ is crucial. One traditional approach for evaluating gradients is to use numerical methods like Finite Difference Method (FDM). This involves perturbing the design variables slightly and observing how the objective function changes, then approximating the gradient based on these differences. FDM is commonly used in industries like aviation. These mathematical concepts describe the fundamentals of fluid mechanics, gradient-based optimization, and the challenges associated with these methods in various engineering applications (Swapna Rekha et al. 2020). Fluid mechanics are governed by the Navier–Stokes formula. However, these equations are a set of coupled, nonlinear partial differential equations. This means that it is difficult to calculate the analytical solution to a point that, if possible, it is preferable to perform an experiment in the laboratory conditions. With the development of CFD, solving fluid mechanics becomes fairly easy. However, the basics of the CFD depend on being able to discretize the equations. To do so, CFD depends on numerical methods such as FDM, FEM, and PDE. Geometrically, discretization can be represented as a single unit in a mesh.

Gradient-based algorithms make use of calculus to choose the direction of the optimum solution based on the derivative (slope or gradient) of the objective function. It is a traditionally used method, and has been popularly used in the aerospace ecosystem. GBM poses low computational requirements, especially for deformative geometric parameterizations. This is probably the most common type of optimization algorithm with similarities to linear convex optimization. The major challenge faced by this method however is that the design space is required to be continuous and mathematically predictable. Another issue with this method is that for high-fidelity design spaces, the gradient-based method is usually slow-moving and is unlikely to achieve global optima (Misra et al. 2022). Finally, the algorithm depends on the user's knowledge of the design space, i.e., the user must provide a reference point for the algorithm to start from. While the optimization algorithm is fairly straightforward, a helper algorithm is required to calculate the gradient for the solution. The traditional approach to evaluating gradients was to use numerical methods, such as the FDM, and is still the most popular approach, especially in the aircraft industry. However, the major drawback is in their geometric inflexibility which complicates their applications to general complex domains. These can be alleviated by the use of either mapping techniques and/or masking to fit the computational mesh to the computational domain. That however does not properly match our needs.

8.2.6 Complex-Step Derivative Approximation

The complex-step derivative approximation is a numerical method for approximating the derivative of a real-valued function of a real variable. It is based on the Cauchy-Riemann equations, which relate the real and imaginary parts of a complex function. The complex-step derivative approximation works as follows:

1. Evaluate the function at a point x and a complex number $x + ih$, where h is a small step size.
2. Calculate the difference between the two function values as in Eq. 8.3:

$$f'(x) \approx \frac{(f(x + ih) - f(x))}{h} \tag{8.3}$$

The imaginary part of the difference is ignored, since it is a second-order error term. The complex-step derivative approximation is a second-order accurate method, meaning that the error term is proportional to h. This is in contrast to the forward and backward difference approximations, which are first-order accurate and have errors that are proportional to h. The complex-step derivative approximation is a simple and robust method for approximating derivatives (Sarkar et al. 2018). It is particularly useful for functions that are difficult to differentiate analytically, such as functions that are defined implicitly or numerically. In many situations, we need to evaluate the derivative of a function but the explicit formula for the derivative is not available. The complex-step approximation approximates the derivative (and the function value itself) from a single function evaluation. Compared to FDM, the complex-step derivative performs much better in the case when the function is not explicitly available.

8.2.7 Automatic Differentiation

Automatic differentiation refers to a general method of taking a function which computes a value, and automatically constructing a procedure for computing derivatives of that value. Finite differences are expensive, since you need to do a forward pass for each derivative. It also induces huge numerical error (Mohapatra et al. 2020). Autodiff is both efficient (linear in the cost of computing the value) and numerically stable.

8.3 Analytic Methods in Optimization

Analytic Methods in Optimization refer to a diverse set of mathematical techniques and algorithms employed to find optimal solutions to complex problems across various domains. These methods offer systematic approaches for navigating intricate search spaces, often involving nonlinear constraints and high-dimensional design spaces. From Sequential Quadratic Programming (SQP) and the Adjoint Method to Gradient-Free Algorithms and Surrogate Modeling, these analytic methods empower engineers, scientists, and practitioners to tackle challenging optimization tasks efficiently and effectively as shown in Fig. 8.1.

8.3.1 Sparse Nonlinear Optimization (SNOPT) Sequential Quadratic Programming

The method was first suggested for the special case of convex optimization. The approach was popularized for general nonlinear constraints. The concept behind the sequential quadratic programming (SQP) methods is to find the solution to the nonlinearly constrained problem using a sequence of quadratic programming (QP)

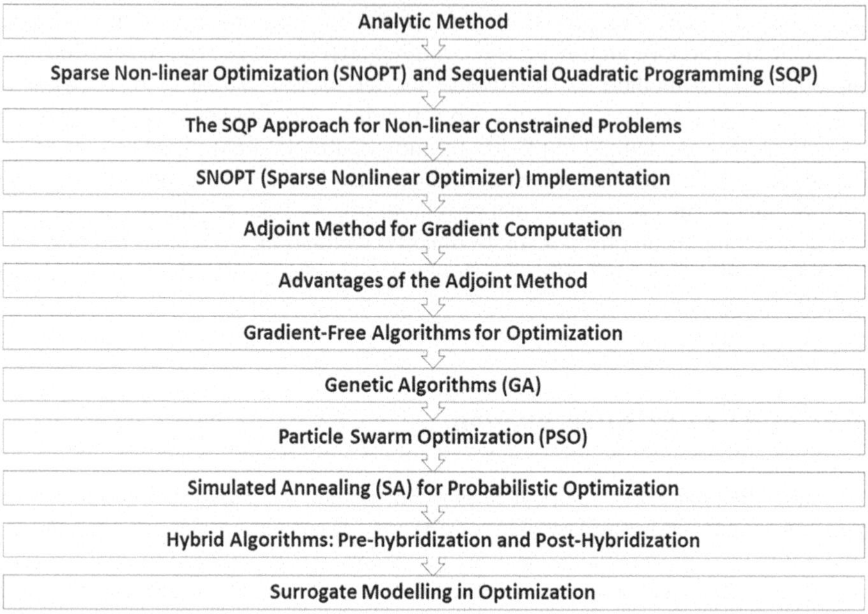

Fig. 8.1 The different methods of analytic methods in optimization

sub-problems (Mohapatra et al. 2020). Each QP sub-problem constraint is a linearized version of the constraints of the original problem, and the objective function of the sub-problem is a quadratic approximation to the Lagrangian function. Sparse Nonlinear Optimizer (SNOPT) is the implementation of a particular SQP algorithm that exploits sparsity in the constraint Jacobian and maintains a limited memory. The treatment of infeasible nonlinear constraints using elastic programming, use of a well-conditioned nonorthogonal basis for the null space of the QP working set (assisted by sparse rank-revealing LU factors), early termination of the QP sub-problems, and finite-difference estimates of missing gradients.

8.3.2 Adjoint Method

The adjoint method is one of the most popular methods in the ASO in use today, and is considered one of the newer technologies that gird application in the industrial setting. The idea behind the adjoint method is to find the gradient using a partial derivative and find the relation between the parameters and the resulting value. Unlike FDM, the adjoint method requires a very low number of calculations to reach the solution. For a very large number of parameters, the adjoint method easily outperforms FDM and FDM-like methods.

8.3.3 Gradient-Free Algorithms

This method utilizes meta-heuristics for achieving the optimum solution without depending on the gradient or the derivative of the objective function for finding the location of the global optima. Independence from the gradient means that the algorithm can be used to optimize discontinuous design spaces, without a need for a baseline design (Sharma et al. 2020c). This also results in some innovative designs. Some popular algorithms of gradient-free methods are.

8.3.4 Genetic Algorithms

Among the meta-heuristic algorithms, the genetic algorithm (GA) is a well-known algorithm, which is inspired by the biological evolution process. GA mimics the Darwinian theory of survival of the fittest in nature. GA was proposed by J. H. Holland in 1992. The basic elements of GA are chromosome representation, fitness selection, and biological-inspired operators. Holland also introduced a novel element, namely Inversion that is generally used in implementations of GA. The chromosomes take the binary string format. In chromosomes, each locus (specific position on chromosome) has two possible alleles (variant forms of genes). Chromosomes are considered as points in the solution space. These are processed using genetic operators by iteratively replacing their population. The fitness function is used to assign a value for all the chromosomes in the population (Bhardawaj et al. 2019b). The biological-inspired operators are selection, mutation, and crossover. In selection, the chromosomes are selected on the basis of their fitness value for further processing. In the crossover operator, a random locus is chosen and it changes the subsequences between chromosomes to create offspring. In mutation, some bits of the chromosomes will be randomly flipped on the basis of probability. The further development of GA based on operators, representation, and fitness has diminished. Therefore, these elements of GA are focused in this paper.

8.3.5 Particle Swarm Optimization

Particle Swarm Optimization (PSO) is a computational method that optimizes a problem by iteratively trying to improve a candidate solution with regard to a given measure of quality. It solves a problem by having a population of candidate solutions, here dubbed particles, and moving these particles around in the search space according to a simple mathematical formula over the particle's position and velocity. Each particle's movement is influenced by its local best-known position, but is also guided toward the best-known positions in the search space, which are updated as better positions found by other particles. This is expected to move the swarm toward the best solutions.

8.3.6 Simulated Annealing

Simulated annealing (SA) is a probabilistic technique for approximating the global optimum of a given function. Specifically, it is a meta-heuristic to approximate global optimization in a large search space for an optimization problem. It is often used when the search space is discrete (e.g., the traveling salesman problem). For problems where finding an approximate global optimum is more important than finding a precise local optimum in a fixed amount of time, simulated annealing may be preferable to exact algorithms such as gradient descent or branch and bound. The algorithm is named after the annealing process in metallurgy, a technique to increase the size of its crystals and reduce their defects. Simulated annealing can be used for very hard computational optimization problems where exact algorithms like FDM and PDE fail. It usually achieves an approximate solution to the global minimum and has better results than other techniques for many practical problems, even in accuracy. The problems solved by SA are currently formulated by an objective function of many variables, subject to several constraints.

8.3.7 Hybrid Algorithms

In application, traditional techniques used some gradient-based algorithms, to optimize low-fidelity panel design spaces. However, this technique was limited by the computational resources available at that time. This meant that often multiple potential optimal solutions could not be considered because the algorithm would consume too much time to have any practical application. On the other hand, metaheuristics allow the gradient-free algorithms to compute high-resolution solutions from scratch with higher success at finding global optima. There were then attempts to improve these techniques by complimenting the weaknesses of one algorithm with the strengths of other algorithms. Hybridization methods can be categorized into one of three groups.

8.3.8 Pre-hybridization

Pre-hybridization refers to using an algorithm on the function that has been pre-optimized by another algorithm. A pre-optimization strategy uses the GA's stochastic search capability along with the GBM optimizer SNOPT to efficiently identify local optima and enforce constraints. They implement the developed HM into a series of different aerodynamic optimization problems including airfoil optimization, transsonic wing section optimization, subsonic wing-section optimization, and blended wing-body optimization. The HM's initial population is optimized by SNOPT for a limited number of design iterations on each candidate solution in order to balance biased and insufficient solution development. The improved candidate solutions are then stochastically perturbed by the GA's crossover operations to define a new population.

8.3.9 Post-hybridization

Similar to previously mentioned HMs, post-hybridization refers to the combination of algorithms that are used in a post-optimization setting, i.e., the final generation selected by the GAs provides an input, a baseline for the GBM operator. This type of HM is one of the first types of hybridization to be used in the aerodynamic shape optimization industry.

8.3.10 Surrogate Modeling

Surrogate modeling is a technique employed in optimization processes, particularly in domains like aerodynamic design optimization, where the primary aim is to replace the use of resource-intensive algorithms with statistical models that provide similar results. Essentially, it involves creating a statistical approximation of the complex simulations performed by the original optimization algorithms. In the context of optimizing the shape of an aircraft wing, the optimization process entails running simulations of airflow (in this case, air) around the wing with various shape variables. However, real-world simulations of this nature often demand a significant amount of time, ranging from minutes to hours or even days for a single run. This time-consuming nature of simulations renders routine tasks, such as design optimization, design space exploration, sensitivity analysis, and what-if analysis, extremely challenging, as they require thousands or even millions of simulation evaluations. The approach surrogacy adopts to alleviate this computational burden involves the construction of approximation models, commonly known as surrogate models, response surface models, meta-models, or emulators. These surrogate models mimic the behavior of the original simulation model but are considerably less resource-intensive to evaluate. Surrogate models are typically constructed using a data-driven, bottom-up approach, where the focus is on modeling the input–output relationship of the simulator. This approach is often referred to as behavioral modeling or black-box modeling because it does not assume knowledge of the inner workings of the simulation code, only the input–output behavior.

It's worth noting that while the categories mentioned are prevalent, they are by no means exhaustive. The field of optimization is diverse, and various concepts and implementations are used extensively in industry applications, such as Control Theory and the ALU method. These techniques, while popular and robust, often have specific use cases and constraints. This framework was developed to address the computationally intensive nature of optimization processes, particularly in industrial contexts, where highly accurate results are essential but traditionally take hours or even days to obtain. The primary goal of this framework was to expedite the optimization process by enabling the Optimization Model to make approximate predictions of feasible solutions without necessitating a full evaluation. This novel hybrid optimization approach combines an Artificial Neural Network (ANN) with Multi-Objective Particle Swarm Optimization (MOPSO). The ANN serves to enhance the optimization's performance when a known objective function exists, without sacrificing correctness. The motivation behind this innovation was the realization that the lengthy evaluation cycles required to verify all solutions in the optimization process were a major bottleneck. The proposed solution was to attach an ANN to guide the process and reduce full evaluations only for predicted global optimum solutions.

The AI-assisted framework comprises three main components:

Artificial Neural Network (ANN): The framework employs a neural network implementation to aid in the optimization process. This neural network is a multilayer perceptron with adjustable parameters. Each training of the neural network starts with a randomly generated initial seed.

Multi-Objective Particle Swarm Optimization (MOPSO): MOPSO is an extension of Particle Swarm Optimization (PSO) tailored for multi-objective optimization. PSO is an optimization technique based on a simplified version of bird flocking. In PSO, a swarm of particles moves through parameter space, guided by the global best result found by the swarm, individual memories of their best results, and inertia. This results in multiple sets of solutions, known as Pareto-optimal solutions, representing trade-offs between objectives.

The ANN-MOPSO framework effectively reduces computational run-time. Sharma et al. demonstrated a 33% reduction in computational time compared to the standard MOPSO algorithm when applying the ANN-MOPSO hybrid to optimize the Aegis UAV as a proof of concept. However, the successful implementation of the ANN-MOPSO framework also presents certain challenges. One significant consideration is managing the confidence and error associated with the ANN's output. In practical applications of ANNs, a skepticism level of approximately 15% has been found to be appropriate. This skepticism parameter reflects the level of doubt toward solutions generated by the ANN that may be invalid.

Additionally, the training method for the ANN is a crucial aspect of its performance. ANNs require sufficient data for learning, and, in many cases, large datasets are not readily available. To overcome this challenge, a hybrid training system was considered. In this approach, the ANN is initially provided with standard datasets, but it also undergoes live training, learning from data generated in real time. The training process concludes once an acceptable dataset size has been reached, preventing overfitting. Continuous live training, as used in this work, involves classifying trial solutions suggested by the optimizer as valid or invalid in real time. This approach is deemed suitable for aerospace shape optimization (ASO) in an industrial setting.

However, the effectiveness of continuous training may vary based on the availability of existing datasets, as training on historical data can provide the model with valuable insights. Despite the advancements made in AI-assisted optimization frameworks like ANN-MOPSO, it's essential to acknowledge that no single algorithm or surrogate model can universally address all optimization challenges. Moreover, the choice of optimization technique often depends on the specific problem domain, design space, and objectives.

In the aerospace industry, for instance, optimization algorithms and frameworks are often tailored to specific scenarios and objectives. These solutions are typically not generalized but rather designed to address particular challenges within aerospace design. The integration of AI, neural networks, and optimization algorithms represents a promising avenue for addressing the computational challenges associated with aerodynamic design optimization. While the ANN-MOPSO framework has shown significant potential, it is important to continue exploring and adapting these techniques to specific industrial settings to achieve high accuracy in aerodynamic shape optimization while minimizing computational costs. The field of optimization remains dynamic, with ongoing research and development aimed at automating and streamlining complex optimization processes in various domains, including aerospace. Surrogate modeling is a technique employed in optimization processes, particularly in domains like aerodynamic design optimization, where the primary aim is to replace the use of resource-intensive algorithms with statistical models that provide similar results. Essentially, it involves creating a statistical approximation of the complex simulations performed by the original optimization algorithms.

8.4 Methodology and Material

In pursuit of the overarching objective of incorporating cutting-edge optimization technology into the field of vehicle shape optimization, as well as identifying the inherent challenges, an exhaustive review of optimization processes pertaining to aerodynamics was deemed imperative. While the literature on this subject is abundant, the task required a structured approach for the thorough evaluation and comparison of various optimization technologies. Achieving this necessitated an intensive literature review, through which both quantitative and qualitative data were meticulously gathered as a result of secondary research. Particular emphasis was placed on ensuring that the selected articles were not only relevant but also made meaningful contributions to the central aim of this research.

As a result of this rigorous review process, numerous technologies and algorithms were sifted through and excluded. Among the assortment of sorted algorithms and technologies, the most suitable candidate was singled out based on a comprehensive assessment, taking into consideration the following critical criteria.

Availability: The selected technology or framework had to be readily accessible and obtainable, ensuring that it could be readily applied to the task at hand.

Adaptability: The framework needed to exhibit flexibility and adaptability to cater to the specific requirements and nuances of vehicle shape optimization.

Usability: Usability played a pivotal role, as the chosen technology had to be user-friendly and conducive to efficient implementation within the context of vehicle shape optimization.

Computation Cost: While assessing suitability, computational cost was a vital factor. However, it's worth noting that, at the time of writing, only two frameworks were publicly available from their respective authors, making it challenging to calculate computation costs for benchmarking purposes. Nonetheless, insights from a review article authored by Skinner and Behtash provided valuable guidance on which algorithms were best suited to various situations. In conducting a qualitative evaluation of the available technologies based on these parameters, it was determined that the model emerged as the most suitable choice for the envisioned objectives. As established in the existing literature, it is evident that some modifications to the selected framework are essential to transform it into a generalized framework capable of accommodating the specific needs of this research. For instance, while integrating a solver of their choice with the ANN-MOPSO model, the requirements of this research necessitate the development of a suitable wrapper to enhance the framework's functionality.

Additionally, a noteworthy enhancement to the framework involves the incorporation of a simple classifier. This classifier would endow the framework with the ability to make real-time decisions regarding the most appropriate solver, thereby facilitating seamless transitions between different solvers as dictated by the specific optimization scenario at hand. This refinement is expected to further elevate the framework's adaptability and overall performance in the complex of vehicle shape optimization.

8.5 Results and Discussion

The proposed framework is a complex and multifaceted system designed to address the intricate challenges of Aerodynamic Shape Optimization (ASO). This framework comprises four distinct modules, each playing a crucial role in the optimization process as a classifier, fast approximator, optimizer, and flow solver placeholder.

A notable feature of this framework is its adaptability, particularly in the choice of the Flow Solver module. The classifier, fast approximator, and optimizer modules maintain their core functionalities but remain open to potential improvements or adjustments. The Flow Solver Placeholder is intentionally designed to accommodate various flow solver options, allowing for flexibility in addressing diverse optimization problems. Let's delve deeper into each of these modules.

8.5.1 Classifier

The classifier module serves as a pivotal component of the framework, responsible for categorizing and making preliminary assessments of potential solutions. In the current proposed setup, a k-nearest means classifier is the chosen implementation. To function effectively, this classifier requires training on an existing dataset that includes information from flow solvers, the objective function, and any pertinent constraints. The classifier's role is to categorize solutions and guide subsequent steps within the framework.

8.5.2 Fast Approximator

The fast approximator plays a critical role in the framework. It operates as an Artificial Neural Network (ANN) and is pre-trained using existing panel data. However, the uniqueness of this implementation is its capability for live training during run-time. This dynamic training approach allows the fast approximator to provide real-time approximations of whether a given solution is a global optimum. An impressive accuracy rate of up to 94% in their Aegis UAV case study was reported, resulting in a substantial reduction in computational time (Samantaray et al. 2014).

8.5.3 Optimizer

The optimizer module is a fundamental component of the framework responsible for driving the search for optimal solutions. In the original framework, Multi-Objective Particle Swarm Optimization (MOPSO), a type of Particle Swarm Optimization (PSO) algorithm developed by Coello and Coello, was used. The PSO algorithm is well-suited for this framework due to its inherent optimization approach. As of the current proposal, no changes are suggested for the optimizer. However, future refinements may be considered if an alternative optimizer emerges that synergizes effectively with the approximation techniques. Simulated annealing, for instance, is an intriguing candidate that warrants exploration (Sharma et al. 2020b).

8.5.4 Flow Solver Placeholder

The Flow Solver Placeholder module remains open-ended, allowing for the incorporation of various flow solver options. This flexibility is vital as different optimization problems may necessitate different flow solvers to simulate and analyze fluid dynamics around the given shape. The placeholder is designed with a wrapper or compatibility layer, enabling the seamless interchangeability of flow solver modules as dictated by the classifier's categorization. Despite the promise and potential of this framework, it's essential to acknowledge certain challenges and limitations.

8.5.5 *Geometric Parameterization*

One of the most significant challenges that remains unresolved in the framework is the issue of geometric parameterization. This aspect of ASO is highly problem-specific and often requires a deep understanding of domain knowledge. To optimize a specific objective, it's essential to know what parameters influence the optimization process. Unfortunately, as of now, there is no universally applicable technology that can easily and accurately parameterize the shape of 2D or 3D objects. Therefore, this framework does not address the geometric parameterization challenge, which remains a significant hurdle in fully automating the ASO process. In summary, the proposed framework for Aerodynamic Shape Optimization is a sophisticated and adaptable system composed of four essential modules (Chatterjee et al. 2022). It offers the potential to streamline and expedite the optimization process, with flexibility in choosing flow solvers and optimization techniques. While there is confidence in the framework's potential, particularly based on the technologies it incorporates, it must be acknowledged that it has not yet been rigorously tested in real-world scenarios. The modular nature of this framework allows for future enhancements and refinements to address specific challenges or take advantage of emerging technologies. However, the most pressing challenge, the parameterization of geometric shapes, remains unaddressed, highlighting the need for further advancements in this area to achieve complete automation in the field of Aerodynamic Shape Optimization.

8.6 Limitations of the Research

While every effort and meticulous care have been dedicated to assimilating the articles, literature, and code, it is crucial to acknowledge the possibility that a more efficient technology or approach may exist, capable of surpassing the capabilities of the current model. It is also worth recognizing that specific techniques might excel in particular cases where the proposed model performs less optimally. This variability is accepted as inherent in the framework. However, it's important to emphasize that the framework's primary purpose is to generalize the optimization process, reducing the need for extensive domain knowledge and enabling the utilization of data-driven techniques. The trade-off for this generality is that it may not achieve the same level of performance as specialized techniques in specific scenarios. Nevertheless, the framework allows for the possibility of fine-tuning to better suit particular cases, albeit at the expense of some degree of generality. One of the most significant limitations of this research is the absence of first-hand experimental data. As of the time of writing, the implementation of the proposed framework is still in progress. Consequently, claims regarding the framework's efficiency may not yet be fully substantiated. However, there is a degree of confidence in the novelty and potential of the framework for general applications in the field of optimization. It is expected that ongoing implementation and experimentation will further validate the framework's utility and efficacy in diverse scenarios.

8.7 Conclusion

In a more detailed examination, it becomes evident that despite the wealth of research materials and the existence of various optimization techniques, many of these technologies face challenges when applied in real-life scenarios. These challenges can range from issues related to scalability and adaptability to difficulties in achieving the desired level of optimization performance. Furthermore, the technologies that do manage to excel often do so within highly specialized contexts, and their effectiveness tends to be confined to those specific scenarios. When attempts are made to apply them to more general or diverse situations, their performance may degrade, and they may fail to meet the expectations of the optimization task at hand. The development of the ANN-MOPSO model sought to address some of these limitations by harnessing the power of Artificial Intelligence, particularly through the utilization of Artificial Neural Networks (ANNs). The primary objective of this model was to enhance optimization efficiency by incorporating intelligent decision-making capabilities into the optimization process. This approach aimed to strike a balance between computational efficiency and optimization accuracy.

The inclusion of the "AVL" component in this model signifies the potential for broader adaptability and utility. This adaptation leverages the ANN-MOPSO framework's core principles while enhancing its capabilities to address a wider range of optimization challenges. Expanding on this concept, the research presented here proposes a modified framework designed to facilitate the general application of Aerospace Shape Optimization (ASO). This framework aims to overcome the limitations of traditional optimization methods by incorporating elements from the ANN-MOPSO model. By doing so, it offers the potential to enhance performance and efficiency in the transportation sector at large. In essence, this modified framework seeks to bridge the gap between theoretical advancements in optimization techniques and their practical application in real-world settings. It strives to provide a versatile and adaptable solution that can cater to the diverse optimization needs within the transportation industry, ultimately contributing to more efficient and effective transportation solutions.

References

Acharya A, Gahlaut U, Sharma K, Sharma SK, Vishwakarma PN, Phanden RK (2020) Crashworthiness analysis of a thin-walled structure in the frontal part of automotive chassis. Int J Veh Struct Syst 12 (2020). https://doi.org/10.4273/ijvss.12.5.06

Adeoye AO, Quadri RO, Lawal OS (2022) Assessment of biofuel potential of tenera palm kernel shell via fixed bed pyrolysis and thermal characterization. Res Surf Interf 9:100091. https://doi.org/10.1016/j.rsurfi.2022.100091

Adeoye AO, Quadri RO, Lawal OS (2023a) Wet synthesis, characterization of goethite nanoparticles and its application in catalytic pyrolysis of palm kernel shell in TGA. Results Surf Interface 11:100118. https://doi.org/10.1016/j.rsurfi.2023.100118

Adeoye AO, Quadri RO, Lawal OS (2023b) Biomass waste pyrolysis in the context of Nigeria's oil and gas industry: an overview. Int J Ambient Energy 44:2525–2541. https://doi.org/10.1080/01430750.2023.2256325

Adeoye AO, Lawal OS, Quadri RO, Malomo D, Aliyu MT, Dang GE, Emojevu EO, Maikato MJ, Yahaya MG, Omonije OO, Edidem VG, Abubakar YK, Offor OF, Sochima EH, Peter BE, Hikon BN (2023c) Sustainable energy via thermochemical and biochemical conversion of biomass wastes for biofuel production. In: Transportation energy and dynamics. Energy, environment, and sustainability. Springer, Singapore, pp 245–306. https://doi.org/10.1007/978-981-99-2150-8_11

Agarwal KM, Mohapatra S, Sharma P, Sharma S, Bhatia D, Mishra A (2020) Study and overview of the novel corona virus disease (COVID-19). Sens Int 1:100037. https://doi.org/10.1016/j.sintl.2020.100037

Agrahari R, Sarraf G, Joshi NC, Mohapatra S, Varma A (2021a) Insight of biopolymers and applications of polyhydroxyalkanoates. In: Microbial polymers. Springer Singapore, Singapore, pp 177–191. https://doi.org/10.1007/978-981-16-0045-6_8

Agrahari R, Mohanty S, Vishwakarma K, Nayak SK, Samantaray D, Mohapatra S (2021b) Update vision on COVID-19: structure, immune pathogenesis, treatment and safety assessment. Sens Int 2:100073. https://doi.org/10.1016/j.sintl.2020.100073

Avesh M, Srivastava R (2012) Modeling simulation and control of active suspension system in Matlab Simulink environment. In: 2012 students conference on engineering and systems. IEEE, pp 1–6. https://doi.org/10.1109/SCES.2012.6199124

Avesh M, Srivastava R (2019a) Parametric optimization to design a passenger car suspension system for better dynamic performance. Eur Transp Trasp Eur

Avesh M, Srivastava R (2019b) Parametric optimization and experimental validation for nonlinear characteristics of passenger car suspension system. Period Polytech Transp Eng 49:103–113. https://doi.org/10.3311/PPtr.12999

Avesh M, Srivastava R (2020) Passenger car active suspension system model for better dynamic characteristics. Natl Acad Sci Lett 43:37–41. https://doi.org/10.1007/s40009-019-00807-z

Avesh M, Srivastava R, Sharma RC, Sharma N (2019a) Effective design of active suspension system using fuzzy logic control approach. Int J Veh Struct Syst 11. https://doi.org/10.4273/ijvss.11.5.13

Avesh M, Srivastava R, Sharma RC (2019b) Parametric optimization for the design of passenger vehicle suspension system with the application of genetic algorithm. Int J Veh Struct Syst 11. https://doi.org/10.4273/ijvss.11.2.08

Badgujar P, Sharma S, Rahul D, Sharma SK, Upadhyay RK (2024) Feasibility study of Mumbai-Pune Hyperloop corridor: challenges and implementation. Res Transp Bus Manage 52:101089

Bandekar D, Chouhan OP, Mohapatra S, Hazra M, Hazra S, Biswas S (2017) Putative protein VC0395_0300 from Vibrio cholerae is a diguanylate cyclase with a role in biofilm formation. Microbiol Res 202:61–70. https://doi.org/10.1016/j.micres.2017.05.003

Bhardawaj S, Chandmal Sharma R, Kumar Sharma S (2019a) Development and advancement in the wheel-rail rolling contact mechanics. IOP Conf Ser Mater Sci Eng 691:012034. https://doi.org/10.1088/1757-899X/691/1/012034

Bhardawaj S, Chandmal Sharma R, Kumar Sharma S (2019b) A survey of railway track modelling. Int J Veh Struct Syst 11. https://doi.org/10.4273/ijvss.11.5.08

Bhardawaj S, Sharma RC, Sharma SK (2020a) Ride analysis of track-vehicle-human body interaction subjected to random excitation. J Chin Soc Mech Eng Trans Chin Inst Eng Ser C/Chung-Kuo Chi Hsueh K. Ch'eng Hsuebo Pao. 41:229–238

Bhardawaj S, Sharma RC, Sharma SK (2020b) Development in the modeling of rail vehicle system for the analysis of lateral stability. Mater Today Proc 25:610–619. https://doi.org/10.1016/j.matpr.2019.07.376

Bhardawaj S, Sharma RC, Sharma SK (2020c) Analysis of frontal car crash characteristics using ANSYS. Mater Today Proc 25:898–902. https://doi.org/10.1016/j.matpr.2019.12.358

Bhardawaj S, Sharma RC, Sharma SK (2020d) Development in the modeling of rail vehicle system for the analysis of lateral stability. Mater Today: Proc 25:610–619. https://doi.org/10.1016/j.matpr.2019.07.376

Bhardawaj S, Sharma RC, Sharma SK, Sharma N (2021) On the planning and construction of railway curved track. Int J Veh Struct Syst 13(2):151–159. https://doi.org/10.4273/ijvss.13.2.04

Bhardawaj S, Sharma RC, Sharma SK, Gopala Rao L, Vashist A (2023) Modeling of biomechanical human body model for seat to head transmissibility analysis. Noise Vib Worldwide 54(2–3):62–74. https://doi.org/10.1177/09574565231154243

Bhattarai S, Mhamdi A, Hossain I, Raoui Y, Pandey R, Madan J, Bouazizi A, Maiti M, Gogoi D, Sharma A (2022) A detailed review of perovskite solar cells: introduction, working principle, modelling, fabrication techniques, future challenges. Micro Nanostruct 172:207450. https://doi.org/10.1016/j.micrna.2022.207450

Biswas N, Saha S, Biswas BK, Chowdhury M, Rahaman A, Junghare V, Mohapatra S, Hazra S, Zangrando E, Roy Choudhury R, Roy Choudhury C (2021) The DNA- and protein-binding properties and cytotoxicity of a new copper(II) hydrazone Schiff base complex. J Coord Chem 74:1482–1504. https://doi.org/10.1080/00958972.2021.1913128

Chatterjee A, Nirwan S, Mohapatra S, Sharma P, Agnihotri A, Shrivastava N (2022) Biochemical aspects of pathogenic variability in white rust infected Indian mustard. Mycologia 114:757–768. https://doi.org/10.1080/00275514.2022.2060007

Choi S, Lee J, Sharma SK (2021) A study on the performance evaluation of hydraulic tank injectors. In Joshi P, Gupta SS, Shukla AK, Gautam SS (eds) Advances in engineering design. Lecture Notes in Mechanical Engineering, pp 183–190. https://doi.org/10.1007/978-981-33-4684-0_19

Das S, Swapnarekha H, Vimal S (2023) Integration of blockchain technology with renewable energy for sustainable development: issues, challenges and future direction.https://doi.org/10.1007/978-981-99-3734-9_48

Devadarshini D, Mohapatra S, Pati S, Maity S, Rath CC, Jena PK, Samantaray D (2022) Evaluation of PHAs production by mixed bacterial culture under submerged fermentation. Biologia (bratisl) 78:895–902. https://doi.org/10.1007/s11756-022-01302-5

Gopala Rao L, Sharma RC, Satyanarayana V, Palli S (2022) Stochastic optimal preview control response of a quarter-car nonlinear suspension model using spectral decomposition method. Noise Vib Worldw 53:225–232. https://doi.org/10.1177/09574565221093234

Goswami B, Rathi A, Sayeed S, Das P, Sharma RC, Sharma SK (2019) Optimization design for aerodynamic elements of Indian locomotive of passenger train. In: Advances in engineering design. Lecture notes in mechanical engineering. Springer, Singapore, pp 663–673. https://doi.org/10.1007/978-981-13-6469-3_61

Goyal S, Anand CS, Sharma SK, Sharma RC (2019) Crashworthiness analysis of foam filled star shape polygon of thin-walled structure. Thin-Walled Struct 144:106312. https://doi.org/10.1016/j.tws.2019.106312

Hossain I, Velkin VI, Shcheklein SE (2017) The study of passive vibration dampers in pipelines using Piv-methodology for single phase flow. In: WIT transactions on ecology and the environment, pp 565–570. https://doi.org/10.2495/ESUS170521

Hossain I, Ahmed YE, Sharma SK (2023) The use of both metallic-insert and TiO2/water nanofluid to improve the performance of photovoltaic modules; experimental study. Thermal Sci Eng Progress 42:101944. https://doi.org/10.1016/j.tsep.2023.101944

Kanth Choppara R, Chandmal Sharma R, Kumar Sharma S, Gupta T (2019) Aero dynamic cross wind analysis of locomotive. IOP Conf Ser Mater Sci Eng 691:012035. https://doi.org/10.1088/1757-899X/691/1/012035

Kohli I, Joshi NC, Mohapatra S, Varma A (2020) Extremophile—an adaptive strategy for extreme conditions and applications. Curr Genomics 21:96–110. https://doi.org/10.2174/1389202921666200401105908

Kohli I, Mohapatra S, Kumar P, Goel A, Varma A, Joshi NC (2022) Role of bacterial endophytes in the promotion of plant growth. In: Bacterial endophytes for sustainable agriculture and environmental management. Springer Singapore, Singapore, pp 235–253. https://doi.org/10.1007/978-981-16-4497-9_12

Kokkinos NC, Emmanouilidou E, Sharma SK (2024) Waste-to-biofuel production for the transportation sector. In: Intelligent transportation system and advanced technology. Springer, pp 99–122

Kumar U, Kasvekar R, Sharma SK, Upadhyay RK (2022) Wear of wheels and axle in locomotive and measures taken by Indian railway. In: Advances in engine tribology Springer, pp 77–96. https://doi.org/10.1007/978-981-16-8337-4_5

Lee J, Sharma SK (2020) Numerical investigation of critical speed analysis of high-speed rail vehicle. 한국정밀공학회학술발표대회 논문집 (Korean Society of Precision Engineering), 696

Lee J, Sharma SK (2023) Intelligent systems for railway infrastructure. Appl Sci 13(20):11274. MDPI

Lee J, Han J, Sharma SK (2021) Structural analysis on the separated and integrated differential gear case for the weight reduction. Advances in engineering design: select proceedings of FLAME 2020, 175

Maity S, Das S, Mohapatra S, Tripathi AD, Akthar J, Pati S, Pattnaik S, Samantaray DP (2020) Growth associated polyhydroxybutyrate production by the novel Zobellellae tiwanensis strain DD5 from banana peels under submerged fermentation. Int J Biol Macromol 153:461–469. https://doi.org/10.1016/j.ijbiomac.2020.03.004

Misra S, Mohanty D, Mohapatra S (2022) Food-based probiotics: functional dietary ingredients. In: Probiotics. Elsevier, pp 257–275. https://doi.org/10.1016/B978-0-323-85170-1.00006-3

Mohanty DP, Mohapatra S, Misra S, Sahu PS (2016) Milk derived bioactive peptides and their impact on human health—a review. Saudi J Biol Sci 23:577–583. https://doi.org/10.1016/j.sjbs.2015.06.005

Mohapatra S, Samantaray DP, Samantaray SM (2015) Study on polyhydroxyalkanoates production using rhizospheric soil bacterial isolates of sweet potato. Indian J Sci Technol 8:57. https://doi.org/10.17485/ijst/2015/v8iS7/64027

Mohapatra S, Samantaray DP, Samantaray SM, Mishra BB, Das S, Majumdar S, Pradhan SK, Rath SN, Rath CC, Akthar J, Achary KG (2016a) Structural and thermal characterization of PHAs produced by Lysinibacillus sp. through submerged fermentation process. Int J Biol Macromol 93:1161–1167. https://doi.org/10.1016/j.ijbiomac.2016.09.077

Mohapatra S, Rath SN, Pradhan SK, Samantaray DP, Rath CC (2016b) Secondary structural models (16S rRNA) of polyhydroxyalkanoates producing bacillus species isolated from different rhizospheric soil: phylogenetics and chemical analysis. Int J Bioautomation 20:329–338

Mohapatra S, Mohanta PR, Sarkar B, Daware A, Kumar C, Samantaray DP (2017a) Production of polyhydroxyalkanoates (PHAs) by bacillus strain isolated from waste water and its biochemical characterization. Proc Natl Acad Sci India Sect B Biol Sci 87:459–466. https://doi.org/10.1007/s40011-015-0626-6

Mohapatra S, Mishra R, Roy P, Yadav KL, Satapathi S (2017b) Systematic investigation and in vitro biocompatibility studies on implantable magnetic nanocomposites for hyperthermia treatment of osteoarthritic knee joints. J Mater Sci 52:9262–9268. https://doi.org/10.1007/s10853-017-1136-0

Mohapatra S, Sarkar B, Samantaray DP, Daware A, Maity S, Pattnaik S, Bhattacharjee S (2017c) Bioconversion of fish solid waste into PHB using Bacillus subtilis based submerged fermentation process. Environ Technol 38:3201–3208. https://doi.org/10.1080/09593330.2017.1291759

Mohapatra S, Maity S, Dash HR, Das S, Pattnaik S, Rath CC, Samantaray D (2017d) Bacillus and biopolymer: prospects and challenges. Biochem Biophys Rep 12:206–213. https://doi.org/10.1016/j.bbrep.2017.10.001

Mohapatra S, Mishra Y, Dash L (2020a) COVID-19: Analysing the legal nuances of the lockdown order. Ind J Forensic Med Toxicol. https://doi.org/10.37506/ijfmt.v14i4.11553

Mohapatra S, Pattnaik S, Maity S, Mohapatra S, Sharma S, Akhtar J, Pati S, Samantaray DP, Varma A (2020b) Comparative analysis of PHAs production by Bacillus megaterium OUAT 016 under submerged and solid-state fermentation. Saudi J Biol Sci 27:1242–1250. https://doi.org/10.1016/j.sjbs.2020.02.001

Mohapatra S, Vishwakarma K, Joshi NC, Maity S, Kumar R, Ramchander M, Pattnaik S, Samantaray DP (2021) A review on PHAs: the future biopolymer. In: Environmental and agricultural microbiology. Wiley, pp 83–100. https://doi.org/10.1002/9781119525899.ch5

Mohapatra S, Mohapatra S, Han H, Ariza-Montes A, López-Martín M (2022) Climate change and vulnerability of agribusiness: assessment of climate change impact on agricultural productivity. Front Psychol 13. https://doi.org/10.3389/fpsyg.2022.955622

Nayak J, Swapnarekha H, Routray AR, Nayak SR, Behera HS (2022) Big data in education: present and future. In: Computational intelligence in data mining. Smart innovation, systems and technologies, vol 281. Springer, Singapore, pp 721–739. https://doi.org/10.1007/978-981-16-9447-9_54

Nibedita S, Swati P, Pattnaik M, Mohapatra S (2021) Methane emission and strategies for mitigation in livestock. In: Environmental and agricultural microbiology. Wiley, pp 257–274. https://doi.org/10.1002/9781119525899.ch12

Palli S, Koona R (2015) Analyses of dynamic response of a railway bogie. Int J Veh Noise Vib 11:103. https://doi.org/10.1504/IJVNV.2015.070015

Palli S, Koona R, Sharma SK, Sharma RC (2018) A review on dynamic analysis of rail vehicle coach. Int J Veh Struct Syst 10(3):204–211. https://doi.org/10.4273/ijvss.10.3.10

Palli S, Dontikurti R, Sharma RC, Sharma N (2021) Analysis of dynamic response of a railway locomotive using ANSYS. Int J Veh Struct Syst 13. https://doi.org/10.4273/ijvss.13.2.21

Palli S, Duppala A, Sharma RC, Gopala Rao LVV (2022) Dynamic simulation of automotive vehicle suspension using MATLAB simulink. Int J Veh Struct Syst 14. https://doi.org/10.4273/ijvss.14.3.04

PanduRanga Vital T, Lakshmi BG, Swapna Rekha H, DhanaLakshmi M (2019) Student performance analysis with using statistical and cluster studies. In: Soft computing in data analytics. Advances in intelligent systems and computing, vol 758. Springer, Singapore, pp 743–757. https://doi.org/10.1007/978-981-13-0514-6_71

Pati S, Maity S, Dash A, Jema S, Mohapatra S, Das S, Samantaray DP (2020) Biocompatible PHB production from bacillus species under submerged and solid-state fermentation and extraction through different downstream processing. Curr Microbiol 77:1203–1209. https://doi.org/10.1007/s00284-020-01922-7

Pati S, Mohanty MK, Mohapatra S, Samantaray D (2022) Bioethanol production by Enterobacter hormaechei through carbon monoxide-rich syngas fermentation. Int J Energy Res 46:20096–20106. https://doi.org/10.1002/er.8678

Pattnaik S, Dash D, Mohapatra S, Pattnaik M, Marandi AK, Das S, Samantaray DP (2020) Corrigendum to "Improvement of rice plant productivity by native Cr(VI) reducing and plant growth promoting soil bacteria Enterobacter cloacae" [Chemosphere 240 (2020) 124895]. Chemosphere 238:125011. https://doi.org/10.1016/j.chemosphere.2019.125011

Pattnaik S, Mohapatra S, Pati S, Dash D, Devadarshini D, Tanaya K, Mishra BB, Samantaray D (2022) Microbial bioremediation of Cr(VI)-contaminated soil for sustainable agriculture. In: Microbial biodegradation and bioremediation. Elsevier, pp 395–407. https://doi.org/10.1016/B978-0-323-85455-9.00008-4

Pundir S, Upadhyay S, Kumar N, Chandra Joshi N, Priya R, Ahmad Mir R, Hossain I, Pandey OP (2023a) Interconnected CuO nanoplates as a highly efficient electrocatalyst for oxygen evolution reaction. Mater Lett 336:133921. https://doi.org/10.1016/j.matlet.2023.133921

Pundir S, Upadhyay S, Priya R, Kumar N, Chetana S, Hossain I, Joshi NC, Pandey OP (2023b) Synthesis of 1D β-MnO2 for high-performance supercapacitor application. J Solid State Electrochem 27:531–538. https://doi.org/10.1007/s10008-022-05347-z

Rallabandi SR, Pilla DP, Dowluru S, Palli S, Sharma N, Sharma SK, Sharma RC (2024a) Critical evaluation of epoxy-hemp-pineapple-palm fiber composites using hybrid AHM-TOPSIS technique for sustainable structural applications. J Chinese Inst Eng 1–12

Rallabandi SR, Srinivas L, Palli S, Sharma RC, Sharma N, Sharma A, Sharma SK (2024b) A contrastive characterization of pure Mg and AZ91D alloy based on the testing of mechanical, corrosion, wear, and erosion properties. Eng Res Express 6(1):015017

Samantaray D, Mohapatra S, Mishra BB (2014) Microbial bioremediation of industrial effluents. In: Microbial biodegradation and bioremediation. Elsevier, pp 325–339. https://doi.org/10.1016/B978-0-12-800021-2.00014-5

Sarkar MS, Segu H, Bhaskar JV, Jakher R, Mohapatra S, Shalini K, Shivaji S, Reddy PA (2018) Ecological preferences of large carnivores in remote, high-altitude protected areas: insights from Buxa Tiger Reserve, India. Oryx 52:66–77. https://doi.org/10.1017/S0030605317000060

Satyanarayana VSV, Sharma RC, Sateesh B, Gopala Rao LVV, Mohan Rao N, Palli S (2022) Optimum response of a nonlinear passive vehicle suspension system under random road excitations. SAE Int J Commer Veh 16:02-16-01–0004. https://doi.org/10.4271/02-16-01-0004

Sharma SK (2019) Multibody analysis of longitudinal train dynamics on the passenger ride performance due to brake application. Proc Inst Mech Eng, Part K: J Multi-Body Dynam 233(2):266–279. https://doi.org/10.1177/1464419318788775

Sharma SK, Kumar A (2017a) Ride performance of a high speed rail vehicle using controlled semi active suspension system. Smart Mater Struct 26:055026. https://doi.org/10.1088/1361-665X/aa68f7

Sharma SK, Kumar A (2017b) Impact of electric locomotive traction of the passenger vehicle Ride quality in longitudinal train dynamics in the context of Indian railways. Mech Ind 18(2):222. https://doi.org/10.1051/meca/2016047

Sharma SK, Kumar A (2018a) Ride comfort of a higher speed rail vehicle using a magnetorheological suspension system. Proc Inst Mech Eng Part K J Multi-body Dyn 232:32–48. https://doi.org/10.1177/1464419317706873

Sharma SK, Kumar A (2018b) Disturbance rejection and force-tracking controller of nonlinear lateral vibrations in passenger rail vehicle using magnetorheological fluid damper. J Intell Mater Syst Struct 29(2):279–297. https://doi.org/10.1177/1045389X17721051

Sharma SK, Kumar A (2018c) Impact of longitudinal train dynamics on train operations: a simulation-based study. J Vib Eng Technol 6(3):197–203. https://doi.org/10.1007/s42417-018-0033-4

Sharma SK, Lee J (2020a) Finite element analysis of a fishplate rail joint in extreme environment condition. Int J Veh Struct Syst 1. https://doi.org/10.4273/ijvss.12.5.03

Sharma SK, Lee J (2020b) Design and development of smart semi active suspension for nonlinear rail vehicle vibration reduction. Int J Struct Stab Dyn 20:2050120. https://doi.org/10.1142/S0219455420501205

Sharma SK, Lee J (2021) Crashworthiness analysis for structural stability and dynamics. Int J Struct Stab Dyn 21:2150039. https://doi.org/10.1142/S0219455421500395

Sharma RC, Palli S (2016) Analysis of creep force and its sensitivity on stability and vertical-lateral ride for railway vehicle. Int J Veh Noise Vib 12:60. https://doi.org/10.1504/IJVNV.2016.077474

Sharma SK, Sharma RC (2018a) An investigation of a locomotive structural crashworthiness using finite element simulation. SAE Int J Commerc Veh 11(4):235–244. https://doi.org/10.4271/02-11-04-0019

Sharma SK, Sharma RC (2018b) Simulation of quarter-car model with magnetorheological dampers for ride quality improvement. Int J Veh Struct Syst 10(3):169–173. https://doi.org/10.4273/ijvss.10.3.03

Sharma RC, Sharma SK (2018c) Sensitivity analysis of three-wheel vehicle's suspension parameters influencing ride behavior. Noise Vib Worldwide 49(7–8):272–280. https://doi.org/10.1177/0957456518796846

Sharma RC, Sharma SK (2022) Ride analysis of road surface-three-wheeled vehicle-human subject interactions subjected to random excitation. SAE Int J Commerc Veh 15(3):02-15–03-0017. https://doi.org/10.4271/02-15-03-0017

Sharma SK, Sharma RC (2019) Pothole detection and warning system for Indian roads. In: Advances in interdisciplinary engineering . lecture notes in mechanical engineering. Springer, Singapore, pp 511–519. https://doi.org/10.1007/978-981-13-6577-5_48

Sharma SK, Sharma RC (2021) Multi-objective design optimization of locomotive nose. SAE Technical PapeR, 10. https://doi.org/10.4271/2021-01-5053

Sharma SK, Sharma RC, Kumar A, Palli S (2015) Challenges in rail vehicle-track modeling and simulation. Int J Veh Struct Syst 7(1):1–9. https://doi.org/10.4273/ijvss.7.1.01

Sharma RC, Palli S, Sharma SK, Roy M (2017) Modernization of railway track with composite sleepers. Int J Veh Struct Syst 9(5):321–329. https://doi.org/10.4273/ijvss.9.5.10

Sharma RC, Sharma SK, Palli S (2018) Rail vehicle modelling and simulation using Lagrangian method. Int J Veh Struct Syst 10(3):188–194. https://doi.org/10.4273/ijvss.10.3.07

Sharma SK, Saini U, Kumar A (2019) Semi-active control to reduce lateral vibration of passenger rail vehicle using disturbance rejection and continuous state damper controllers. J Vib Eng Technol 7(2):117–129. https://doi.org/10.1007/s42417-019-00088-2

Sharma S, Sharma RC, Sharma SK, Sharma N, Palli S, Bhardawaj S (2020a) Vibration isolation of the quarter car model of road vehicle system using dynamic vibration absorber. Int J Veh Struct Syst 12. https://doi.org/10.4273/ijvss.12.5.05

Sharma RC, Sharma SK, Palli S (2020b) Linear and non-linear stability analysis of a constrained railway wheelaxle. Int J Veh Struct Syst 12. https://doi.org/10.4273/ijvss.12.2.04

Sharma SK, Sharma RC, Sharma N (2020c) Combined multi-body-system and finite element analysis of a rail locomotive crashworthiness. Int J Veh Struct Syst 12. https://doi.org/10.4273/ijvss.12.4.15

Sharma RC, Sharma SK, Sharma N, Sharma S (2020d) Analysis of ride and stability of an ICF railway coach. Int J Veh Noise Vib 16:127. https://doi.org/10.1504/IJVNV.2020.117820

Sharma RC, Sharma S, Sharma SK, Sharma N (2020e) Analysis of generalized force and its influence on ride and stability of railway vehicle. Noise Vib Worldw 51:95–109. https://doi.org/10.1177/0957456520923125

Sharma RC, Palli S, Sharma N, Sharma SK (2021a) Ride behaviour of a four-wheel vehicle using H infinity semi-active suspension control under deterministic and random inputs. Int J Veh Struct Syst 13. https://doi.org/10.4273/ijvss.13.2.18

Sharma RC, Sharma S, Sharma N, Sharma SK (2021b) Linear and nonlinear analysis of ride and stability of a three-wheeled vehicle subjected to random and bump inputs using bond graph and simulink methodology. SAE Int J Commer Veh 14:02-15-01–0001. https://doi.org/10.4271/02-15-01-0001

Sharma SK, Sharma RC, Lee J (2021c) Effect of rail vehicle-track coupled dynamics on fatigue failure of coil spring in a suspension system. Appl Sci 11:2650. https://doi.org/10.3390/app11062650

Sharma RC, Sharma S, Sharma SK, Sharma N, Singh G (2021d) Analysis of bio-dynamic model of seated human subject and optimization of the passenger ride comfort for three-wheel vehicle using random search technique. Proc Inst Mech Eng Part K J Multi-body Dyn 235:106–121. https://doi.org/10.1177/1464419320983711

Sharma SK, Mohapatra S, Sharma RC, Alturjman S, Altrjman C, Mostarda L, Stephan T (2022a) Retrofitting existing buildings to improve energy performance. Sustainability 14:666. https://doi.org/10.3390/su14020666

Sharma SK, Sharma RC, Lee J, Jang H-L (2022b) Numerical and experimental analysis of DVA on the flexible-rigid rail vehicle carbody resonant vibration. Sensors 22:1922. https://doi.org/10.3390/s22051922

Sharma SK, Lee J, Jang H-L (2022c) Mathematical modeling and simulation of suspended equipment impact on car body modes. Machines 10:192. https://doi.org/10.3390/machines10030192

Sharma SK, Sharma RC, Lee J (2022d) In situ and experimental analysis of longitudinal load on carbody fatigue life using nonlinear damage accumulation. Int J Damage Mech 31:605–622. https://doi.org/10.1177/10567895211046043

Sharma RC, Gopala Rao LVV, Sharma SK, Palli S, Satyanarayana VSV (2022e) Analysis of lateral stability and ride of an indian railway constrained dual-axle bogie frame. SAE Int J Commerc Veh 16(2):02-16–02-0014. https://doi.org/10.4271/02-16-02-0014

Sharma RC, Palli S, Gopala Rao LVV, Duppala A, Sharma SK (2022f) Four-wheel vehicle response under bump, Pothole, Harmonic, and Random Excitations Using Bond Graph/Simulink Technique. SAE Int J Commerc Veh 16(2):02-16-02–0008. https://doi.org/10.4271/02-16-02-0008

Sharma RC, Palli S, Jha AK, Bhardawaj S, Sharma SK (2022g) Vibration and ride comfort analysis of railway vehicle system subjected to deterministic inputs. Resmilitaris 12(5):1345–1355

Sharma RC, Palli S, Bhardawaj S (2023a) A review of wheel-rail dynamic behavior on curved tracks. Int J Veh Struct Syst 15. https://doi.org/10.4273/ijvss.15.1.04

Sharma RC, Palli S, Sharma SK (2023b) Ride analysis of railway vehicle considering rigidity and flexibility of the carbody. J Chin Inst Eng 46:355–366. https://doi.org/10.1080/02533839.2023.2194918

Sharma RC, Bhardawaj S, Sharma SK, Bahl S (2023c) Ride behaviour of three-wheel vehicle using semi-active suspension using skyhook control strategy. Int J Veh Struct Syst 15(1). https://doi.org/10.4273/ijvss.15.1.05

Sharma SK, Sharma RC, Choi Y, Lee J (2023d) Experimental and mathematical study of flexible–rigid rail vehicle riding comfort and safety. Appl Sci 13(9):5252

Sharma SK, Sharma RC, Choi Y, Lee J (2023e) Modelling and dynamic analysis of adaptive neuro-fuzzy inference system-based intelligent control suspension system for passenger rail vehicles using magnetorheological damper for improving ride index. Sustainability 15(16):12529

Sharma SK, Sharma RC, Lavania S, Palli S, Avesh M (2023f) Industry-driven approach for ANFIS-based intelligent control suspension system with MR damper for enhanced ride quality in passenger rail vehicles for technological investigations. Nat Acad Sci Lett 1–4

Sharma SK, Sharma RC, Upadhyay RK, Lee J (2023g) Coupled torsional–transverse vibration reduction in marine propulsion dynamics with novel approach using magnetorheological damping and adaptive control system. J Vib Eng Technol 1–11

Sharma SK, Upadhyay RK, Kumar V, Valera H (2023h) Introduction to sustainable transportation system. In: Transportation energy and dynamics. Springer, pp 3–6

Sharma RC, Palli S, Dharmana L, Sharma SK, Sharma A, Kharola A, Sharma N (2024a) Simulation of transient dynamic response of a locomotive considering its flexibility and rigidity. Noise and Vibrat Worldwide 55(1–2):33–47

Sharma RC, Palli S, Duppala A, Bhardawaj S, Sharma SK, Sharma N, Sharma A (2024b) Rigid body analysis of ride behaviour of roadway vehicle using Lagrangian dynamics. Int J Veh Struct Syst (IJVSS), 16(1)

Sharma RC, Sharma N, Palli S, Sharma SK, Joshi S, Rallabandi SR, Sharma A (2024c) Dynamic behavior of three-wheel vehicle with inerter-based dynamic absorber suspension system. Noise & Vib Worldwide 09574565241243398

Sharma SK, Kumar N, Avesh M, Sharma RC, Siddiqui MIH, Lee J (2024d) Navigating tranquillity with H∞ controller to mitigate ship propeller shaft vibration. J Vib Eng Technol 1–11

Sharma SK, Sharma RC, Lee J (2024e) Modelling and simulation of MR damper characterization and uncertainty assessment in marine engine vibration isolation with FLPID control. Int J Modell Simul 1–11

Sharma SK, Sharma RC, Lee J (2024f) Propelling precision of longitudinal vibration mitigation in ship propeller shafts through advanced nonlinear intelligent semi-active control leveraging adaptive neuro-fuzzy inference system with linear quadratic regulator. J Vib Cont 10775463241244836

Sharma SK, Sharma RC, Upadhyay RK, Lee J (2024g) Enhanced ship engine vibration reduction using magnetorheological dampers and adaptive neuro-fuzzy control system. Ships and Offshore Struct 1–17

Sinha AK, Sengupta A, Gandhi H, Bansal P, Agarwal KM, Sharma SK, Sharma RC, Sharma SK (2019) Performance enhancement of an all-terrain vehicle by optimizing steering, powertrain and brakes. In: Advances in engineering design. Lecture notes in mechanical engineering. Springer, Singapore, pp 207–215. https://doi.org/10.1007/978-981-13-6469-3_19

Swapna Rekha H, Nayak J, Behera HS (2020) Pi-Sigma neural network: survey of a decade progress. In: Computational intelligence in pattern recognition. Advances in intelligent systems and computing, vol 1120. Springer, Singapore, pp 429–441. https://doi.org/10.1007/978-981-15-2449-3_37

Vishwakarma PN, Mishra P, Sharma SK (2022a) Characterization of a magnetorheological fluid damper a review. Mater Today: Proc 56:2988–2994. https://doi.org/10.1016/j.matpr.2021.11.143

Vishwakarma PN, Mishra P, Sharma SK (2022b) Formulation of semi-active suspension system and controls in rail vehicle. SSRN Electron J. https://doi.org/10.2139/ssrn.4159616

Upadhyay RK, Sharma SK, Kumar V, Valera H (2023a) Introduction to transportation systems technology and integrated management. In: Transportation systems technology and integrated management. Springer, pp 1–4. https://doi.org/10.1007/978-981-99-1517-0_1

Upadhyay RK, Sharma SK, Kumar V, Valera H (2023b) Transportation systems technology and integrated management. In: Upadhyay RK, Sharma SK, Kumar V, Valera H, (eds) Springer Nature Singapore. https://doi.org/10.1007/978-981-99-1517-0

Upadhyay RK, Sharma SK, Kumar V (Eds) (2024a) Intelligent transportation system and advanced technology. Springer Nature Singapore. https://doi.org/10.1007/978-981-97-0515-3

Upadhyay RK, Sharma SK, Kumar V (2024b) Introduction to intelligent transportation system and advanced technology. In: Intelligent transportation system and advanced technology. Energy, Environment, and Sustainability. pp 3–6. Springer, Singapore. https://doi.org/10.1007/978-981-97-0515-3_1

Wu Q, Cole C, Spiryagin M, Chang C, Wei W, Ursulyak L, Shvets A, Murtaza MA, Mirza IM, Zhelieznov K, Mohammadi S, Serajian H, Schick B, Berg M, Sharma RC, Aboubakr A, Sharma SK, Melzi S, Di Gialleonardo E, Cantone L et al (2023) Freight train air brake models. Int J Rail Transp 11(1):1–49. https://doi.org/10.1080/23248378.2021.2006808

Chapter 9
Improving On-Time Performance: Predicting Train Delays with Machine Learning Techniques

Rakesh Chandmal Sharma, Ismail Hossain, and Amit Kumar

9.1 Introduction

The Indian railway system ranks as the fourth-largest rail network globally, boasting a comprehensive route network spanning 67,415 km. It comprises 12,147 locomotives and encompasses 7,321 stations, facilitating transportation for a staggering 8.439 billion passengers in the year 2019 (Adeoye et al. 2023a). However, a notable challenge plaguing the Indian railway network is the issue of train delays. In 2019, the punctuality index stood at 69.23%, according to data from the Indian Railways Yearbook 2018–19 (Badgujar et al. 2024; Kokkinos et al. 2024; Raju Rallabandi et al. 2024; Rallabandi et al. 2024; Sharma et al. 2024a, b, c, d, e, f, g; Upadhyay et al. 2024a, b). These delays can vary widely, ranging from minor ten-minute setbacks to significant delays lasting several hours (Sharma et al. 2020c). The extent of the delay often depends on various factors, including the specific train, time of day, and the distance of the journey. This paper aims to delve deeper into the analysis and prediction of train delays, employing machine learning techniques. The motivation for this research is to develop an "N-Order Markov Late Minutes Prediction Framework" using a dataset spanning the years 2016 to 2018, which included information on 135

R. C. Sharma
Mechanical Engineering Department, Graphic Era (Deemed to Be University), Dehradun 248002, India

Mechanical Engineering Department, Graphic Era Hill University, Dehradun 248002, India

I. Hossain
School of Natural Sciences and Mathematics, Ural Federal University, Yekaterinburg 620000, Russia

A. Kumar (✉)
SSRAS safety and risks services Pvt. Ltd, Roorkee 247667, India
e-mail: amitk88.kumar@gmail.com

S. K. Sharma et al. (eds.), *Dynamics of Transportation Ecosystem, Modeling, and Control*, Energy, Environment, and Sustainability,
https://doi.org/10.1007/978-981-97-0437-8_9

trains passing through Mugalsarai station. Their system exhibited the capability to predict delay minutes at stations with a commendable level of accuracy.

9.2 Literature Review

The initial research paper examined the application of an N-order Markov Model. The primary aim of this paper was to forecast train delays, measured in minutes, based on specific input parameters, including train particulars and route details (Mohapatra et al. 2016). The researchers collected train delay data for Mugalsarai station for a period of two years, from 2016 to 2018. They selected this station based on the frequency of trains passing through it (Rao et al. 2022). Subsequently, they performed data pre-processing, data analysis, data segregation, and data preparation. The proposed model relied on N-order Markov process regression models. They later applied a two-step k-nearest neighbor search to identify the station semantically closest to the given station. Finally, they employed an N-order Markov model in conjunction with a random forest regression model. The model achieved significant accuracy. The dataset consisted of 135 trains over the two-year period with 52 known trains and 83 unknown trains. The dataset was divided into two parts: 52 known trains for training (80%) and testing/validation (20%). The model was then used to predict delays for the 83 unknown trains, which it had never been trained on (Mohapatra et al. 2016). The independent variables used in the model included actarr_date, station_code, latemin, distance, month, and weekday. These variables represented the actual arrival date of the train at a station, the station code name, the delay in minutes, the distance of a station from the source station in kilometers, the month name, and the weekday name. Two separate regression models, namely, Random Forest Regression and Ridge Regression, were employed to construct the model and analyze and compare the results (Kohli et al. 2022).

The model achieved a significant accuracy rate of 95%. One of the key advantages of this model was its scalability. It was constructed using data from 52 known trains and tested on 83 unknown trains, demonstrating its ability to handle large train datasets. For future work, there are several potential avenues for improvement (Bhardawaj et al. 2023; Hossain et al. 2023; Lee and Sharma 2023; Sharma et al. 2023c, d, e, f, g, h; Upadhyay et al. 2023a, b; Wu et al. 2023). First, the dataset could be expanded by incorporating data from additional stations and trains. Moreover, new machine learning models, such as artificial neural networks and recurrent neural networks, could be explored to enhance prediction accuracy. Additionally, factors like weather conditions, incidents such as riots or train strikes, could be integrated into the model to provide more comprehensive predictions. Finally, an online model that continuously learns from train delay data throughout its lifecycle could be developed. Another research paper aims to predict train delays using big data analytics. The methods employed to achieve this objective include shallow learning machines and deep extreme learning machines. The paper utilizes various models, such as extreme learning machines, shallow learning machines, and deep learning machines, to fulfill its research objectives (Sharma and Kumar 2017a).

To understand the techniques discussed above, it's essential to grasp the concept of feedforward neural networks, as all the mentioned models are a type of feedforward neural network (Sharma and Kumar 2018c). A feedforward neural network consists of several layers, with the first layer being the input layer and the last layer being the output layer. Between these two layers, there are hidden layers, each equipped with an activation function to process the inputs effectively (Sharma et al. 2021b). These layers consist of numerous neurons, and each neuron is interconnected with neurons in other layers. Through the adjustment of these connections, the network learns and adapts itself to minimize the squared error in its output. In both of the research papers mentioned, a common challenge encountered throughout the process was the quality of the data (Bhardawaj et al. 2020a). Data collection was a time-consuming task, involving both manual collection and data retrieval through APIs (Kumar et al. 2022; Sharma et al. 2022e, f, g; Sharma and Sharma 2022; Vishwakarma et al. 2022a, b). Regarding neural networks, they are a type of machine learning architecture designed to mimic the human brain's structure. In essence, neural networks take in data, undergo training to identify patterns within it, and subsequently make predictions based on new data sets that exhibit similar patterns (Sharma et al. 2022a).

A neural network is essentially a network of interconnected mathematical formulas. It comprises an input layer with multiple hidden layers, where the input layer contains various attributes and independent variables denoted as X_1 and X_2. The hidden layer consists of one or more assumptions, and the corresponding layer is similar, featuring one or more specified outputs (Bandekar et al. 2017). It's important to note that each layer can have any number of nodes, and a neural network can have as many layers as necessary (Hossain et al. 2017). To illustrate this concept with a case study, consider a scenario where we have images of different dog breeds and want to build a neural network capable of identifying the breed of a given dog from an image. To achieve this, we feed the neural network a dataset containing images of various dog breeds. The network learns to make accurate predictions based on subjective images (Mohapatra et al. 2017a).

Activation in a neural network can be likened to a function that determines whether a particular node is activated or not, resulting in a numerical value between 0 and 1. An input image is processed by each node using a linear function, generating a value represented as "z." This "z" value is then passed through an activation function, which determines whether the node's feature is activated or not (Sharma and Sharma 2019). Eventually, each node in the corresponding layer recognizes whether it's activated before producing an output (Swapna Rekha et al. 2020). Neural networks consist of nodes interconnected in a manner inspired by the neural networks found in human brains. In artificial neural networks (ANNs), the individual neurons are loosely modeled after their organic counterparts in the human brain. Each connection between neurons, akin to synapses in the organic brain, can transmit signals to multiple other neurons (Sinha et al. 2019). When a neuron receives a signal, it processes it and can subsequently send a signal to nearby neurons. These "signals" in neural networks are numerical values, resulting from each neuron's non-linear computation of the sum of its inputs. The connections are referred to as "edges," and their weights typically change as the network learns, affecting the strength of the signal transmitted along a connection. Some neurons have a threshold and only send a signal if the cumulative signal surpasses a certain threshold (Goswami et al. 2019).

Neural networks commonly consist of layers, and different layers can serve various purposes. Signals flow from the first layer (the input layer) to the last layer (the output layer), potentially passing through intermediate layers multiple times (Sharma et al. 2022b). One way to visualize artificial neural networks is as a network of interconnected nodes, with neural neurons forming the nodes, and weighted edges representing connections between them. Artificial Neural Networks receive input signals from external sources, converting them into vectors (Mohanty et al. 2016). These input signals, denoted as x(n), are then mathematically processed based on various inputs. Different types of Artificial Neural Networks (ANNs) exist, each designed for specific tasks. While most artificial neural networks share similarities with their biological counterparts, they can be highly effective in their intended functions (Samantaray et al. 2014; Pattnaik et al. 2022; Nibedita et al. 2021; Biswas et al. 2021). For instance, ANNs can be used for tasks like classification or clustering. The output from an ANN is often used to make informed decisions or provide predictions based on the data it has been trained on.

Train delay prediction is a critical aspect of modern railway management, with significant implications for both passengers and operators (Sharma et al. 2022c, 2020b, 2021b; Pundir et al. 2023a; Das et al. 2023). The ability to accurately forecast delays allows for better planning and resource allocation, ultimately improving service reliability and customer satisfaction. Over the years, various approaches have been employed to address this challenge, ranging from traditional statistical methods to more advanced machine learning techniques (Agrahari et al. 2021a; Sharma et al. 2022d; Maity et al. 2020; Misra et al. 2022; Sharma and Lee 2020a). In the traditional methods, time series analysis has long been a foundational approach to train delay prediction. Techniques such as Autoregressive Integrated Moving Average (ARIMA) and Exponential Smoothing have been applied to historical delay data to model and predict future delays. While these methods provide a good starting point, they often struggle to capture the complex, non-linear relationships inherent in train delay data (Goyal et al. 2019; Sharma 2019; Sharma et al. 2019; Bhardawaj et al. 2020b, 2021; Lee and Sharma 2020; Choi et al. 2021; Lee et al. 2021; Sharma and Sharma 2021).

Regression models have also been extensively used in train delay prediction. These models attempt to establish a relationship between delay and various independent variables, such as time of day, weather conditions, and station-specific factors. Linear regression and its variants, including multiple linear regression and ridge regression, have been applied (Mohapatra et al. 2017b; Kohli et al. 2020; Devadarshini et al. 2022; Sharma et al. 2021a; Sarkar et al. 2018). While these models offer interpretability and transparency, they may not effectively capture non-linear relationships. In recent years, machine learning techniques have gained prominence in the field of train delay prediction. These methods leverage the power of data to learn complex patterns and make predictions. Artificial Neural Networks (ANNs) have particularly garnered attention due to their capacity to handle non-linearity effectively. Researchers have applied deep learning techniques, including Recurrent Neural Networks (RNNs) and Long Short-Term Memory networks (LSTMs), to time-series train delay data with promising results. These models can capture intricate temporal dependencies and patterns in the data, improving prediction accuracy (Palli et al. 2022; Bhardawaj et al. 2020b, 2019a).

Ensemble learning methods like Random Forest and Gradient Boosting have also found application in train delay prediction. These techniques combine the predictions of multiple models to enhance overall accuracy (Sharma and Lee 2020b, 2021; Acharya et al. 2020). They are particularly adept at handling feature importance analysis, which can provide insights into the factors contributing to delays. Recent advancements in the field have focused on addressing key challenges and improving prediction accuracy. One notable advancement involves the integration of weather data into prediction models (Pundir et al. 2023b; COVID-19 2020). Weather conditions, such as heavy rainfall or snowfall, can significantly impact train operations. By incorporating weather data, models can better account for external factors contributing to delays. Additionally, the availability of real-time data from sensors and Global Positioning System (GPS) devices has transformed the landscape of train delay prediction. Real-time data allows for the continuous monitoring of train movements and the adjustment of prediction models in response to changing conditions. This adaptability improves the accuracy of predictions and enhances the ability to proactively manage delays (Pattnaik et al. 2020; Mohapatra et al. 2020).

Hybrid models have also gained traction, combining the strengths of different machine learning techniques. For instance, a hybrid model might blend a time series model's ability to capture temporal patterns with a regression model's capability to account for external factors (Sharma et al. 2020e; Mohapatra et al. 2022, 2017c; Adeoye et al. 2023b; Pati et al. 2022, 2020). This approach can result in more robust prediction models. Furthermore, researchers have been working on improving the interpretability of machine learning models in the context of train delay prediction (Chatterjee et al. 2022; Adeoye et al. 2023c; Nayak et al. 2022; Mohapatra et al. 2017d). While complex models like deep learning can offer high accuracy, they are often seen as "black boxes" that lack transparency. Recent efforts aim to develop models that provide not only accurate predictions but also insights into why a particular prediction was made. This interpretability can be crucial, especially in safety–critical applications where decisions are based on model outputs (Adeoye et al. 2022; Sharma et al. 2020a, 2020b, 2021d; Bhardawaj et al. 2020c; Palli et al. 2021).

Despite the progress made in train delay prediction using machine learning techniques, several challenges persist. Data quality remains a critical issue, as the accuracy of predictions heavily relies on the quality of historical data. Incomplete or inaccurate data can lead to unreliable predictions and hinder the effectiveness of machine learning models (Sharma et al. 2021d; Palli and Koona 2015). Moreover, the volume of data required for training machine learning models can be a challenge, particularly for less-frequented train routes or newly introduced services. Gathering a sufficient volume of high-quality data is essential for model training and validation. Another challenge is the generalization of models across different railway systems and regions (Kanth Choppara et al. 2019; Mohapatra et al. 2015; Bhardawaj et al. 2019b). Models trained on data from one region may not generalize well to another region with different operating conditions and infrastructure. Developing models that can adapt and generalize effectively is an ongoing research area. In conclusion, train delay prediction using machine learning techniques holds immense promise for the transportation industry (Mohapatra et al. 2021; Bhattarai et al. 2022; Agrahari et al. 2021b; Agarwal et al. 2020; PanduRanga Vital et al. 2019). The ability to forecast delays accurately can lead to improved service reliability and customer satisfaction. While traditional methods have laid the foundation, recent advancements in machine learning, including the use of artificial neural networks, ensemble learning, and the integration of real-time data, have propelled the field forward (Sharma et al. 2023a, 2023b; Satyanarayana et al. 2022; Sharma and Palli 2016; Avesh and Srivastava 2020). However, challenges such as data quality, model transparency, and generalization remain. Future research should focus on addressing these challenges and exploring new avenues for enhancing train delay prediction models, ultimately benefiting both operators and passengers in the railway transportation (Avesh and Srivastava 2019, 2012; Avesh et al. 2019a, 2019b; Mohd and Srivastava 2019).

9.3 Methodology

In this section, a data collection, preparation, partition, training, and testing have been discussed.

9.3.1 Data Collection

This section offers a detailed account of the dataset employed in this analysis, which aligns with the dataset previously utilized by Gaurav and Srivastava (Gaurav and Srivastava 2018) in their earlier study. The dataset under consideration comprises a total of 135 trains passing through the designated station. Within this dataset, 52 trains are classified as known entities, while 83 trains remain unclassified and unidentified in terms of historical data. A comprehensive summary of data statistics pertaining to the entire set of 135 trains is thoughtfully presented in Table 9.1.

In the predictive modeling, the selection and inclusion of variables are of paramount importance. The endeavor to identify variables that wield substantial influence over train delay assumes a central role in the modeling process (Kohli et al. 2022). The significance of this undertaking lies in its potential to inform the strategic utilization of these variables within the prediction model. In light of a comprehensive analysis of the dataset and the intricacies inherent to the problem domain, a discerning examination has revealed specific data fields that exert notable impact upon the target

Table 9.1 Train dataset analysis

Train count	135
Unique station count	819
Train journey count (Maximum)	334
Train average journey count	48
Station count in train's route (Maximum)	129
Average station count in train's route	30
Maximum station distance from source (km)	

Table 9.2 Description of data fields

station_code	Station code
scharr	Scheduled arrival time of the train on that station
latemin	Delay period of train in minutes
distance	Distance of the station from the source
weekday	Distance of the station from the source
month	Month
delay_15	Delay time is more than or equal to 15 min

variable, namely, "schrr," "schdep," "distance," "weekday," and "month." These identified variables have subsequently been thoughtfully integrated into the development of the prediction model (Table 9.2).

9.3.2 Data Preparation

This section provides a comprehensive overview of the dataset employed in this study. Initially, the dataset encompassed 16 distinct data fields for each train, as elucidated in Table II. To streamline the dataset and facilitate the modeling process, a deliberate reduction was undertaken, resulting in a dataset comprising 8 key data fields. Subsequently, an additional boolean data field was introduced, denoting delays exceeding 15 min. This boolean variable assumes a "true" value when the delay duration equals or exceeds 15 min, and "false" when the delay is less than 15 min. Two columns within the dataset, namely, "scharr" and "schdep," originally featured values denoting "source" and "destination." These values signified whether the train journey initiated or concluded at the station under consideration. To enhance data uniformity and simplify model creation, all instances of "source" and "destination" were standardized to "00:00."

The dataset preparation process involved addressing data types, particularly converting data objects into floating-point values. This conversion was primarily executed through the utilization of Python libraries, with most data objects being converted automatically. However, certain data objects necessitated manual conversion to ensure consistency (Palli et al. 2018; Sharma et al. 2015, 2017, 2018; Sharma and Sharma 2018a, b, c; Sharma and Kumar 2017b, 2018a, b). For the "weekday" variable, a transformation was implemented whereby weekdays were converted into their respective numeric counterparts. For instance, "Monday" was represented as "1," "Tuesday" as "2," and so forth. Similarly, the "month" variable was transformed, assigning numerical values to each month, with "January" corresponding to "1," "February" to "2," and so on. The final step in dataset preparation involved the creation of dummy variables to accommodate station codes. As machine learning models are incapable of processing string values such as station codes, this conversion was imperative. Python's Pandas library facilitated the creation of individual data fields for each unique station code. For each row, the corresponding station code was assigned a value of "1," while all other station code fields received "0," ensuring that each observation was suitably represented. This meticulous dataset preparation process not only ensures the uniformity and compatibility of the data but also lays the groundwork for accurate and effective predictive modeling, ultimately enhancing the model's capacity to generate meaningful insights and predictions related to train delays.

9.3.3 Data Partition Testing and Training

The process of developing a train delay prediction model involves two critical phases: data partitioning and model evaluation. In the initial phase, known as data partitioning, the dataset is systematically divided into two distinct subsets: training data and testing data. This partitioning strategy is a fundamental step in machine learning model development. Specifically, 80% of the dataset is allocated to the training data subset, while the remaining 20% constitutes the testing data subset. In the subsequent phase, the training data plays a pivotal role. It serves as the foundation upon which the train delay prediction model is constructed. During this phase, the model learns the intricate patterns and relationships inherent in the data, thereby acquiring the ability to make accurate predictions. The testing data subset, distinct from the training data, serves as an essential tool for model validation and evaluation. This subset contains data that the model has not encountered during its training phase. Consequently, applying the model to the testing data allows for an assessment of its performance and the determination of its accuracy in predicting train delays.

In practical implementation, Python, a widely used programming language, offers a diverse range of libraries and tools that streamline the data partitioning process. By employing these Python libraries, the dataset is partitioned into training and testing subsets in accordance with the specified 80–20 split. This meticulous partitioning approach ensures that the model is trained on a substantial portion of the data, while retaining an independent, unseen dataset for rigorous evaluation. This evaluation step is essential for gauging the model's reliability and accuracy in forecasting train delays, thereby contributing to the robustness of the model's predictive capabilities.

9.3.4 *Random Forest.*

The machine learning model employed for train delay prediction in this study is the Random Forest Classifier (RFC). Random Forest is a versatile algorithm capable of handling both regression and classification tasks, as detailed by Breiman in 2001. The essence of the Random Forest lies in the amalgamation of multiple decision trees, where the value predicted by each tree depends on a random vector sampled independently and following the same distribution across all trees in the forest, as described by Freund and Schapire, along with insights provided by Stergiou and Siganos. To implement the Random Forest Classifier, the popular Scikit-learn Python library was utilized, with default parameters being applied. At its core, the model operates by evaluating the importance of each node within the decision trees. For instance, in the scenario where a decision tree possesses only two child nodes, a mathematical equation can be formulated to determine the feature importance of a given node as follows:

$$n_{ij} = W_j * \left(C_j - W_{left(j)} * C_{left(j)} - W_{right(j)} * C_{right(j)}\right) \tag{9.1}$$

Here, n_{ij} represents the importance of node j's feature, W_j signifies the weight of samples reaching node j, C_j denotes the impurity value of node j, left(j) designates the left split child node of node j, and right(j) corresponds to the right split child node of node j. This calculation of node importance is fundamental to the Random Forest algorithm's predictive power and contributes significantly to its effectiveness in train delay prediction.

9.4 Results

Figure 9.1 provides a detailed visualization of the decision tree used within the Random Forest Classifier. Each node, represented by small dots, is shown, allowing for a comprehensive examination of the model's decision-making process. This visualization not only illustrates the structure of the tree but also highlights key decision points and feature importance, aiding in model interpretability. Upon closer inspection, this detailed view of the decision tree provides insights into how the model processes and makes decisions based on various input features.

9.4.1 *Multilayer Deep Neural Network Architecture*

Figure 9.2 showcases the architecture of the Multilayer Deep Neural Network employed in this study. This architecture comprises multiple layers of interconnected neurons, allowing for the modeling of complex non-linear relationships in

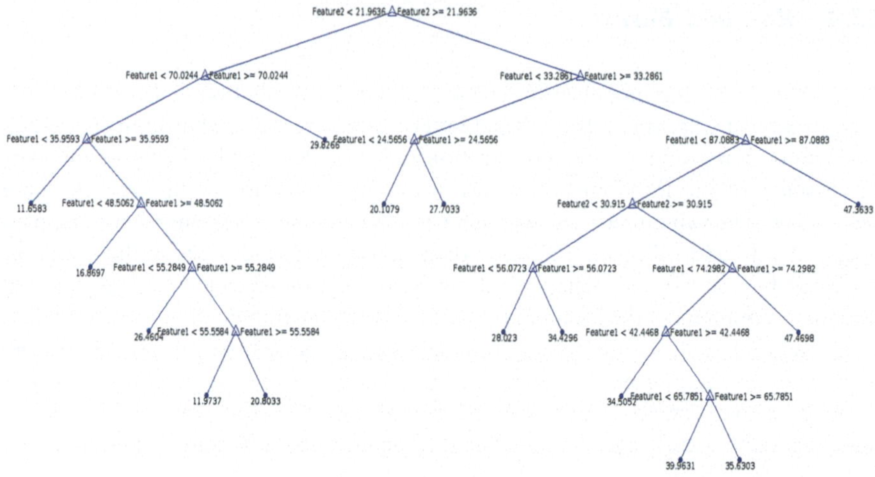

Fig. 9.1 Visualization of decision tree

the data. This neural network architecture serves as an alternative or complementary approach to the Random Forest model. The deep neural network architecture offers the advantage of capturing intricate patterns in the data and may potentially improve predictive accuracy in certain scenarios.

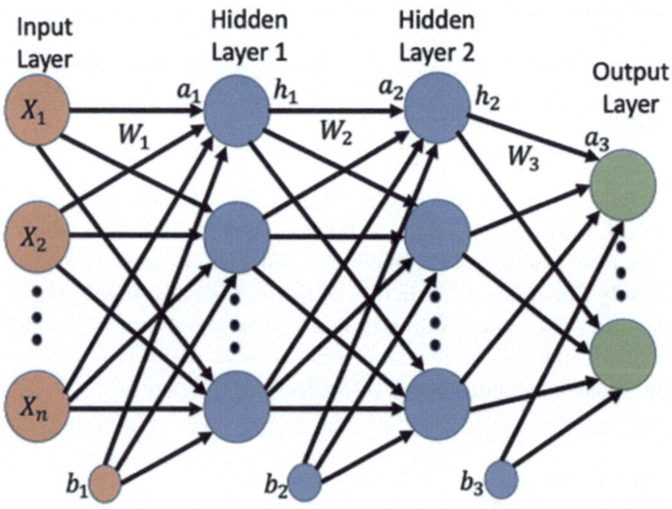

Fig. 9.2 A multilayer deep neural network architecture

9.4.2 Prediction Accuracy

The Random Forest Classification model yielded a remarkable prediction accuracy of 97.85%. This high accuracy percentage is a noteworthy outcome and underscores the effectiveness of the model in predicting train delays accurately. Achieving such a high level of accuracy is indicative of the model's ability to generalize well to unseen data and make reliable predictions.

It's important to note that this level of accuracy is particularly valuable in the context of train delay prediction, where timely and accurate information can significantly enhance operational efficiency and passenger satisfaction. The model's proficiency in classifying train delays speaks to its potential real-world applicability in improving rail transportation services.

9.4.3 Implications and Future Directions

The exceptional accuracy achieved by the Random Forest Classification model has several implications. Firstly, it underscores the feasibility of machine learning techniques, particularly ensemble methods like Random Forest, in addressing complex prediction tasks in the transportation sector. Furthermore, it highlights the potential for data-driven approaches to enhance decision-making and operational planning in railway systems.

Moving forward, there are several avenues for future research. The use of a Multilayer Deep Neural Network, as depicted in Fig. 9.2, opens up possibilities for further exploration. Comparing the performance of different models, such as Random Forest and deep neural networks, on large and diverse datasets could yield insights into the strengths and weaknesses of each approach.

Additionally, model interpretability remains an important area of consideration. While high accuracy is valuable, understanding how the model arrives at its predictions is crucial, especially in safety–critical domains like transportation. Developing techniques to interpret the decision-making processes of complex models like deep neural networks is a promising direction.

9.4.4 Limitations

It's essential to acknowledge certain limitations of the study. Firstly, the results are specific to the dataset and context used in this research. Generalizing the model's performance to different railway systems or under varying conditions may require further investigation. Additionally, the model's high accuracy could potentially be influenced by the quality and representativeness of the data used for training and testing. Ensuring the integrity and completeness of data sources is an ongoing challenge in data-driven modeling.

9.5 Conclusion

The culmination of this comprehensive study on train delay prediction through the application of machine learning techniques has unveiled profound insights into the potential transformation of the railway industry. The exceptional predictive precision achieved by the Random Forest Classification (RFC) model, registering an impressive 97.85% accuracy, carries profound implications for railway operations, passenger satisfaction, and the broader arena of data-driven decision-making. Of paramount significance is the RFC model's achievement in predicting train delays with remarkable accuracy. In the context of rail transport, where timeliness is of paramount importance, this level of precision signifies a potential paradigm shift. Railway authorities can now leverage advanced machine learning models to forecast delays with precision, promising improved operational efficiency and enhanced passenger experiences. This high degree of accuracy equips decision-makers with the tools necessary to proactively address potential disruptions, ensuring smoother and more dependable journeys for passengers.

The research underscores the transformative potential of machine learning within the realm of rail transport. Historically, railway operations have adhered to pre-established schedules, often necessitating manual intervention to tackle delays. However, the findings underscore the ability of machine learning models, particularly ensemble methods like Random Forest, to analyze historical data, discern intricate patterns, and furnish precise predictions. This transition from a reactive to a proactive operational approach holds the promise of revolutionizing rail transport by curbing delays, optimizing resource allocation, and elevating overall service quality. The study adopted a versatile approach, embracing multiple modeling techniques. Beyond the Random Forest, a Multilayer Deep Neural Network Architecture was introduced into the framework. This inclusion underscores the adaptability of machine learning methodologies to diverse problem domains. The neural network's proficiency in capturing complex non-linear relationships complements the Random Forest's effectiveness in handling structured data. This synergistic interplay among models broadens the scope of the research, presenting avenues for future exploration in model selection and hybrid methodologies.

The implications of this research extend far beyond the confines of the study. Railway operators, whether in the public or private sector, are presented with a compelling case for the integration of machine learning models into their operations. The ability to predict train delays with remarkable precision empowers operators to mitigate disruptions, optimize maintenance schedules, and allocate resources judiciously. This translates into improved reliability and reduced travel uncertainties for passengers, fostering higher satisfaction levels and potentially stimulating increased ridership. Looking ahead, numerous avenues for future research beckon. The application of machine learning models to diverse railway systems under varying operational conditions warrants investigation. Each railway network possesses unique challenges, necessitating tailored machine learning solutions for enhanced predictive accuracy.

Additionally, enhancing model interpretability remains a pressing concern. While high accuracy is commendable, understanding the rationale behind specific model predictions, especially in safety–critical domains, is imperative. Research endeavors should prioritize the development of techniques for transparent decision-making within complex models such as deep neural networks. Furthermore, the incorporation of external data sources, including weather data, passenger load metrics, and real-time updates, offers the potential to further augment prediction model accuracy and utility. The fusion of diverse data streams and the refinement of adaptable models beckon as exciting prospects for future research.

In conclusion, this research illuminates the path toward a data-driven future for rail transport. The exceptional predictive accuracy of the Random Forest Classification model and the versatility of machine learning methodologies underscore the transformative potential of data-driven decision-making. Railway operators stand on the threshold of leveraging machine learning's capabilities to revolutionize their services, enhance passenger satisfaction, and optimize operational efficiency. As the intersection of data science and rail transportation unfolds, the horizon brims with possibilities for innovation and positive industry impact. The adoption of data-driven solutions transcends choice; it stands as an imperative for the modern railway sector. This research constitutes a significant stride toward realizing this vision, heralding a future where rail travel unfolds as a more dependable, efficient, and enjoyable experience for passengers worldwide.

Acknowledgements The research funding from the Ministry of Science and Higher Education of the Russian Federation (Ural Federal University Program of Development within the Priority-2030 Program) is gratefully acknowledged.

References

Acharya A, Gahlaut U, Sharma K, Sharma SK, Vishwakarma PN, Phanden RK (2020) Crashworthiness analysis of a thin-walled structure in the frontal part of automotive chassis. Int J Veh Struct Syst 12. https://doi.org/10.4273/ijvss.12.5.06

Adeoye AO, Quadri RO, Lawal OS (2022) Assessment of biofuel potential of tenera palm kernel shell via fixed bed pyrolysis and thermal characterization. Results Surf Interf 9:100091. https://doi.org/10.1016/j.rsurfi.2022.100091

Adeoye AO, Quadri RO, Lawal OS (2023a) Wet synthesis, characterization of goethite nanoparticles and its application in catalytic pyrolysis of palm kernel shell in TGA. Results in Surfaces and Interfaces. 11:100118. https://doi.org/10.1016/j.rsurfi.2023.100118

Adeoye AO, Quadri RO, Lawal OS (2023b) Biomass waste pyrolysis in the context of Nigeria's oil and gas industry: an overview. Int J Ambient Energy 44:2525–2541. https://doi.org/10.1080/01430750.2023.2256325

Adeoye AO, Lawal OS, Quadri RO, Malomo D, Aliyu MT, Dang GE, Emojevu EO, Maikato MJ, Yahaya MG, Omonije OO, Edidem VG, Abubakar YK, Offor OF, Sochima EH, Peter BE, Hikon BN (2023) Sustainable energy via thermochemical and biochemical conversion of biomass wastes for biofuel production. In: Transportation energy and dynamics. energy, environment, and sustainability. Springer, Singapore, pp 245–306. https://doi.org/10.1007/978-981-99-2150-8_11

Agarwal KM, Mohapatra S, Sharma P, Sharma S, Bhatia D, Mishra A (2020) Study and overview of the novel corona virus disease (COVID-19). Sensors Int 1:100037. https://doi.org/10.1016/j.sintl.2020.100037

Agrahari R, Mohanty S, Vishwakarma K, Nayak SK, Samantaray D, Mohapatra S (2021b) Update vision on COVID-19: Structure, immune pathogenesis, treatment and safety assessment. Sensors Int 2:100073. https://doi.org/10.1016/j.sintl.2020.100073

Agrahari R, Sarraf G, Joshi NC, Mohapatra S, Varma A (2021) Insight of biopolymers and applications of polyhydroxyalkanoates. In Microbial polymers, pp 177–191. Springer Singapore, Singapore (2021). https://doi.org/10.1007/978-981-16-0045-6_8

Avesh M, Srivastava R (2012) Modeling simulation and control of active suspension system in Matlab Simulink environment. In 2012 Students conference on engineering and systems, pp 1–6. IEEE. https://doi.org/10.1109/SCES.2012.6199124

Avesh M, Srivastava R (2019) Parametric optimization to design a passenger car suspension system for better dynamic performance. Eur Transp—Trasp Eur

Avesh M, Srivastava R, Chandmal Sharma R, Sharma N (2019) Effective design of active suspension system using fuzzy logic control approach. Int J Veh Struct Syst 11. https://doi.org/10.4273/ijvss.11.5.13

Avesh M, Srivastava R, Sharma RC (2019b) Parametric optimization for the design of passenger vehicle suspension system with the application of genetic algorithm. Int J Veh Struct Syst 11. https://doi.org/10.4273/ijvss.11.2.08

Avesh M, Srivastava R (2020) Passenger car active suspension system model for better dynamic characteristics. Natl Acad Sci Lett 43:37–41. https://doi.org/10.1007/s40009-019-00807-z

Badgujar P, Sharma S, Rahul D, Sharma SK, Upadhyay RK (2024) Feasibility study of Mumbai-Pune hyperloop corridor: challenges and implementation. Res Transp Bus Manage 52:101089

Bandekar D, Chouhan OP, Mohapatra S, Hazra M, Hazra S, Biswas S (2017) Putative protein VC0395_0300 from Vibrio cholerae is a diguanylate cyclase with a role in biofilm formation. Microbiol Res 202:61–70. https://doi.org/10.1016/j.micres.2017.05.003

Bhardawaj S, Chandmal Sharma R, Kumar Sharma S (2019a) Development and advancement in the wheel-rail rolling contact mechanics. IOP Conf Ser Mater Sci Eng 691:012034. https://doi.org/10.1088/1757-899X/691/1/012034

Bhardawaj S, Chandmal Sharma R, Kumar Sharma S (2019b) A survey of railway track modelling. Int J Veh Struct Syst 11. https://doi.org/10.4273/ijvss.11.5.08

Bhardawaj S, Sharma RC, Sharma SK (2020a) Ride analysis of track-vehicle-human body interaction subjected to random excitation. J Chin Soc Mech Eng Trans Chinese Inst Eng Ser C/ Chung-Kuo Chi Hsueh K. Ch'eng Hsuebo Pao 41:229–238

Bhardawaj S, Sharma RC, Sharma SK (2020b) Development in the modeling of rail vehicle system for the analysis of lateral stability. Mater Today Proc 25:610–619. https://doi.org/10.1016/j.matpr.2019.07.376

Bhardawaj S, Sharma RC, Sharma SK (2020c) Analysis of frontal car crash characteristics using ANSYS. Mater Today Proc 25:898–902. https://doi.org/10.1016/j.matpr.2019.12.358

Bhardawaj S, Sharma RC, Sharma SK, Sharma N (2021) On the planning and construction of railway curved track. Int J Veh Struct Syst 13(2):151–159. https://doi.org/10.4273/ijvss.13.2.04

Bhardawaj S, Sharma RC, Sharma SK, Gopala Rao L, Vashist A (2023) Modeling of biomechanical human body model for seat to head transmissibility analysis. Noise Vib Worldwide 54(2–3):62–74. https://doi.org/10.1177/09574565231154243

Bhattarai S, Mhamdi A, Hossain I, Raoui Y, Pandey R, Madan J, Bouazizi A, Maiti M, Gogoi D, Sharma A (2022) A detailed review of perovskite solar cells: Introduction, working principle, modelling, fabrication techniques, future challenges. Micro Nanostruct 172:207450. https://doi.org/10.1016/j.micrna.2022.207450

Biswas N, Saha S, Biswas BK, Chowdhury M, Rahaman A, Junghare V, Mohapatra S, Hazra S, Zangrando E, Roy Choudhury R, Roy Choudhury C (2021) The DNA- and protein-binding properties and cytotoxicity of a new copper(II) hydrazone Schiff base complex. J Coord Chem 74:1482–1504. https://doi.org/10.1080/00958972.2021.1913128

Chatterjee A, Nirwan S, Mohapatra S, Sharma P, Agnihotri A, Shrivastava N (2022) Biochemical aspects of pathogenic variability in white rust infected Indian mustard. Mycologia 114:757–768. https://doi.org/10.1080/00275514.2022.2060007

Choi S, Lee J, Sharma SK (2021) A study on the performance evaluation of hydraulic tank injectors. In: Joshi P, Gupta SS, Shukla AK, Gautam SS (eds) Advances in engineering design. Lecture notes in mechanical engineering, pp 183–190. https://doi.org/10.1007/978-981-33-4684-0_19

COVID-19: Analysing the legal nuances of the lockdown order. Indian J Forensic Med Toxicol (2020). https://doi.org/10.37506/ijfmt.v14i4.11553

Das S, Swapnarekha H, Vimal S (2023) Integration of blockchain technology with renewable energy for sustainable development: issues, challenges and future direction. Presented at the (2023). https://doi.org/10.1007/978-981-99-3734-9_48

Devadarshini D, Mohapatra S, Pati S, Maity S, Rath CC, Jena PK, Samantaray D (2022) Evaluation of PHAs production by mixed bacterial culture under submerged fermentation. Biologia (bratisl). 78:895–902. https://doi.org/10.1007/s11756-022-01302-5

Gaurav R, Srivastava B (2018) Estimating train delays in a large rail network using a zero shot markov model. In 2018 21st International Conference on Intelligent Transportation Systems (ITSC). pp 1221–1226. IEEE (2018). https://doi.org/10.1109/ITSC.2018.8570014

Gopala Rao L, Sharma RC, Satyanarayana V, Palli S (2022) Stochastic optimal preview control response of a quarter-car nonlinear suspension model using spectral decomposition method. Noise Vib Worldw 53:225–232. https://doi.org/10.1177/09574565221093234

Goswami B, Rathi A, Sayeed S, Das P, Sharma RC, Sharma SK (2019) Optimization design for aerodynamic elements of Indian locomotive of passenger train. In: Advances in engineering design. Lecture notes in mechanical engineering. Springer, Singapore, pp 663–673. https://doi.org/10.1007/978-981-13-6469-3_61

Goyal S, Anand CS, Sharma SK, Sharma RC (2019) Crashworthiness analysis of foam filled star shape polygon of thin-walled structure. Thin-Walled Struct 144:106312. https://doi.org/10.1016/j.tws.2019.106312

Hossain I, Ahmed YE, Sharma SK (2023) The use of both metallic-insert and TiO_2/water nanofluid to improve the performance of photovoltaic modules; experimental study. Therm Sci Eng Prog 42:101944. https://doi.org/10.1016/j.tsep.2023.101944

Hossain I, Velkin VI, Shcheklein SE (2017) The study of passive vibration dampers in pipelines using PIV-methodology for single phase flow. In WIT transactions on ecology and the environment, pp 565–570. https://doi.org/10.2495/ESUS170521

Kanth Choppara R, Chandmal Sharma R, Kumar Sharma S, Gupta T (2019) Aero dynamic cross wind analysis of locomotive. IOP Conf Ser Mater Sci Eng 691:012035. https://doi.org/10.1088/1757-899X/691/1/012035

Kohli I, Joshi NC, Mohapatra S, Varma A (2020) Extremophile—an adaptive strategy for extreme conditions and applications. Curr Genomics 21:96–110. https://doi.org/10.2174/1389202921666200401105908

Kohli I, Mohapatra S, Kumar P, Goel A, Varma A, Joshi NC (2022) Role of bacterial endophytes in the promotion of plant growth. In: Bacterial endophytes for sustainable agriculture and environmental management, pp 235–253. Springer Singapore, Singapore (2022). https://doi.org/10.1007/978-981-16-4497-9_12

Kokkinos NC, Emmanouilidou E, Sharma SK (2024) Waste-to-biofuel production for the transportation sector. In: Intelligent transportation system and advanced technology. Springer, pp 99–122

Kumar U, Kasvekar R, Sharma SK, Upadhyay RK (2022) Wear of wheels and axle in locomotive and measures taken by Indian railway. In: Advances in engine tribology. Springer, pp 77–96. https://doi.org/10.1007/978-981-16-8337-4_5

Lee J, Sharma SK (2020) Numerical investigation of critical speed analysis of high-speed rail vehicle. 한국정밀공학회학술발표대회 논문집 (Korean Society of Precision Engineering), p 696

Lee J, Sharma SK (2023) Intelligent systems for railway infrastructure. In: Applied sciences, vol 13, issue 20. MDPI, p 11274

Lee J, Han J, Sharma SK (2021) Structural analysis on the separated and integrated differential gear case for the weight reduction. In: Advances in engineering design: select proceedings of FLAME 2020, p 175

Maity S, Das S, Mohapatra S, Tripathi AD, Akthar J, Pati S, Pattnaik S, Samantaray DP (2020) Growth associated polyhydroxybutyrate production by the novel Zobellellae tiwanensis strain DD5 from banana peels under submerged fermentation. Int J Biol Macromol 153:461–469. https://doi.org/10.1016/j.ijbiomac.2020.03.004

Misra S, Mohanty D, Mohapatra S (2022) Food-based probiotics: Functional dietary ingredients. In Probiotics, pp 257–275. Elsevier. https://doi.org/10.1016/B978-0-323-85170-1.00006-3

Mohanty DP, Mohapatra S, Misra S, Sahu PS (2016) Milk derived bioactive peptides and their impact on human health—A review. Saudi J Biol Sci 23:577–583. https://doi.org/10.1016/j.sjbs.2015.06.005

Mohapatra S, Rath SN, Pradhan SK, Samantaray DP, Rath CC (2016) Secondary structural models (16S rRNA) of polyhydroxyalkanoates producing bacillus species isolated from different rhizospheric soil: Phylogenetics and chemical analysis. Int J Bioautomation 20:329–338

Mohapatra S, Mishra R, Roy P, Yadav KL, Satapathi S (2017b) Systematic investigation and in vitro biocompatibility studies on implantable magnetic nanocomposites for hyperthermia treatment of osteoarthritic knee joints. J Mater Sci 52:9262–9268. https://doi.org/10.1007/s10853-017-1136-0

Mohapatra S, Sarkar B, Samantaray DP, Daware A, Maity S, Pattnaik S, Bhattacharjee S (2017c) Bioconversion of fish solid waste into PHB using Bacillus subtilis based submerged fermentation process. Environ Technol 38:3201–3208. https://doi.org/10.1080/09593330.2017.1291759

Mohapatra S, Maity S, Dash HR, Das S, Pattnaik S, Rath CC, Samantaray D (2017d) Bacillus and biopolymer: prospects and challenges. Biochem Biophys Reports 12:206–213. https://doi.org/10.1016/j.bbrep.2017.10.001

Mohapatra S, Pattnaik S, Maity S, Mohapatra S, Sharma S, Akhtar J, Pati S, Samantaray DP, Varma A (2020) Comparative analysis of PHAs production by *Bacillus megaterium* OUAT 016 under submerged and solid-state fermentation. Saudi J Biol Sci 27:1242–1250. https://doi.org/10.1016/j.sjbs.2020.02.001

Mohapatra S, Samantaray DP, Samantaray SM (2015) Study on polyhydroxyalkanoates production using rhizospheric soil bacterial isolates of sweet potato. Indian J Sci Technol 8:57. https://doi.org/10.17485/ijst/2015/v8iS7/64027

Mohapatra S, Samantaray DP, Samantaray SM, Mishra BB, Das S, Majumdar S, Pradhan SK, Rath SN, Rath CC, Akthar J, Achary KG (2016) Structural and thermal characterization of PHAs produced by Lysinibacillus sp. through submerged fermentation process. Int J Biol Macromol 93:1161–1167. https://doi.org/10.1016/j.ijbiomac.2016.09.077

Mohapatra S, Mohanta PR, Sarkar B, Daware A, Kumar C, Samantaray DP (2017a) Production of polyhydroxyalkanoates (PHAs) by bacillus strain isolated from waste water and its biochemical characterization. Proc Natl Acad Sci India Sect B Biol Sci 87:459–466. https://doi.org/10.1007/s40011-015-0626-6

Mohapatra S, Vishwakarma K, Joshi NC, Maity S, Kumar R, Ramchander M, Pattnaik S, Samantaray DP (2021) A review on PHAs: the future biopolymer. In: Environmental and agricultural microbiology, pp 83–100. Wiley. https://doi.org/10.1002/9781119525899.ch5

Mohapatra S, Mohapatra S, Han H, Ariza-Montes A, López-Martín M del C (2022) Climate change and vulnerability of agribusiness: assessment of climate change impact on agricultural productivity. Front Psychol 13. https://doi.org/10.3389/fpsyg.2022.955622

Mohd A, Srivastava R (2019) Parametric optimization and experimental validation for nonlinear characteristics of passenger car suspension system. Period Polytech Transp Eng 49:103–113. https://doi.org/10.3311/PPtr.12999

Nayak J, Swapnarekha H, Routray AR, Nayak SR, Behera HS (2022) Big data in education: present and future. In Computational intelligence in data mining. smart innovation, systems and technologies, 281. Springer, Singapore, pp 721–739. https://doi.org/10.1007/978-981-16-9447-9_54

Nibedita S, Swati P, Pattnaik M, Mohapatra S (2021) Methane emission and strategies for mitigation in livestock. In: Environmental and agricultural microbiology, pp 257–274. Wiley (2021). https://doi.org/10.1002/9781119525899.ch12

Palli S, Dontikurti R, Sharma RC, Sharma N (2021) Analysis of dynamic response of a railway locomotive using ANSYS. Int J Veh Struct Syst 13. https://doi.org/10.4273/ijvss.13.2.21

Palli S, Duppala A, Sharma RC, Gopala Rao LVV (2022) Dynamic simulation of automotive vehicle suspension using MATLAB simulink. Int J Veh Struct Syst 14. https://doi.org/10.4273/ijvss.14.3.04

Palli S, Koona R (2015) Analyses of dynamic response of a railway bogie. Int J Veh Noise Vib 11:103. https://doi.org/10.1504/IJVNV.2015.070015

Palli S, Koona R, Sharma SK, Sharma RC (2018) A review on dynamic analysis of rail vehicle coach. Int J Veh Struct Syst 10(3):204–211. https://doi.org/10.4273/ijvss.10.3.10

PanduRanga Vital T, Lakshmi BG, Swapna Rekha H, DhanaLakshmi M (2019) Student performance analysis with using statistical and cluster studies. In Soft computing in data analytics. Advances in intelligent systems and computing, 758. Springer, Singapore, pp 743–757. https://doi.org/10.1007/978-981-13-0514-6_71

Pati S, Maity S, Dash A, Jema S, Mohapatra S, Das S, Samantaray DP (2020) Biocompatible PHB production from bacillus species under submerged and solid-state fermentation and extraction through different downstream processing. Curr Microbiol 77:1203–1209. https://doi.org/10.1007/s00284-020-01922-7

Pati S, Mohanty MK, Mohapatra S, Samantaray D (2022) Bioethanol production by Enterobacter hormaechei through carbon monoxide-rich syngas fermentation. Int J Energy Res 46:20096–20106. https://doi.org/10.1002/er.8678

Pattnaik S, Dash D, Mohapatra S, Pattnaik M, Marandi AK, Das S, Samantaray DP (2020) Corrigendum to "Improvement of rice plant productivity by native Cr(VI) reducing and plant growth promoting soil bacteria Enterobacter cloacae" [Chemosphere 240 (2020) 124895]. Chemosphere 238:125011. https://doi.org/10.1016/j.chemosphere.2019.125011

Pattnaik S, Mohapatra S, Pati S, Dash D, Devadarshini D, Tanaya K, Mishra BB, Samantaray D (2022) Microbial bioremediation of Cr(VI)-contaminated soil for sustainable agriculture. In: Microbial biodegradation and bioremediation, pp 395–407. Elsevier (2022). https://doi.org/10.1016/B978-0-323-85455-9.00008-4

Pundir S, Upadhyay S, Kumar N, Chandra Joshi N, Priya R, Ahmad Mir R, Hossain I, Pandey OP (2023a) Interconnected CuO nanoplates as a highly efficient electrocatalyst for oxygen evolution reaction. Mater Lett 336:133921. https://doi.org/10.1016/j.matlet.2023.133921

Pundir S, Upadhyay S, Priya R, Kumar N, Chetana S, Hossain I, Joshi NC, Pandey OP (2023b) Synthesis of 1D β-MnO_2 for high-performance supercapacitor application. J Solid State Electrochem 27:531–538. https://doi.org/10.1007/s10008-022-05347-z

Raju Rallabandi S, Pilla DP, Dowluru S, Palli S, Sharma N, Sharma SK, Sharma RC (2024) Critical evaluation of epoxy-hemp-pineapple-palm fiber composites using hybrid AHM-TOPSIS technique for sustainable structural applications. J Chin Inst Eng 1–12

Rallabandi SR, Srinivas L, Palli S, Sharma RC, Sharma N, Sharma A, Sharma SK (2024) A contrastive characterization of pure Mg and AZ91D alloy based on the testing of mechanical, corrosion, wear, and erosion properties. Eng Res Express 6(1):015017

Samantaray D, Mohapatra S, Mishra BB (2014) Microbial bioremediation of industrial effluents. In: Microbial biodegradation and bioremediation, pp 325–339. Elsevier. https://doi.org/10.1016/B978-0-12-800021-2.00014-5

Sarkar MS, Segu H, Bhaskar JV, Jakher R, Mohapatra S, Shalini K, Shivaji S, Reddy PA (2018) Ecological preferences of large carnivores in remote, high-altitude protected areas: insights from Buxa Tiger Reserve, India. Oryx 52:66–77. https://doi.org/10.1017/S0030605317000060

Satyanarayana VSV, Sharma RC, Sateesh B, Gopala Rao LVV, Mohan Rao N, Palli S (2022) Optimum response of a nonlinear passive vehicle suspension system under random road excitations. SAE Int J Commer Veh 16:02-16-01-0004. https://doi.org/10.4271/02-16-01-0004

Sharma SK (2019) Multibody analysis of longitudinal train dynamics on the passenger ride performance due to brake application. Proc Inst Mech Eng Part K J Multi-Body Dyn 233(2):266–279. https://doi.org/10.1177/1464419318788775

Sharma SK, Kumar A (2017a) Ride performance of a high speed rail vehicle using controlled semi active suspension system. Smart Mater Struct 26:055026. https://doi.org/10.1088/1361-665X/aa68f7

Sharma SK, Kumar A (2017b) Impact of electric locomotive traction of the passenger vehicle ride quality in longitudinal train dynamics in the context of Indian railways. Mech Ind 18(2):222. https://doi.org/10.1051/meca/2016047

Sharma SK, Kumar A (2018a) Disturbance rejection and force-tracking controller of nonlinear lateral vibrations in passenger rail vehicle using magnetorheological fluid damper. J Intell Mater Syst Struct 29(2):279–297. https://doi.org/10.1177/1045389X17721051

Sharma SK, Kumar A (2018b) Impact of longitudinal train dynamics on train operations: a simulation-based study. J Vib Eng Technol 6(3):197–203. https://doi.org/10.1007/s42417-018-0033-4

Sharma SK, Kumar A (2018c) Ride comfort of a higher speed rail vehicle using a magnetorheological suspension system. Proc Inst Mech Eng Part K. J Multi-body Dyn 232:32–48. https://doi.org/10.1177/1464419317706873

Sharma SK, Lee J (2020a) Finite element analysis of a fishplate rail joint in extreme environment condition. Int J Veh Struct Syst 12. https://doi.org/10.4273/ijvss.12.5.03

Sharma SK, Lee J (2020b) Design and development of smart semi active suspension for nonlinear rail vehicle vibration reduction. Int J Struct Stab Dyn 20:2050120. https://doi.org/10.1142/S0219455420501205

Sharma SK, Lee J (2021) Crashworthiness analysis for structural stability and dynamics. Int J Struct Stab Dyn 21:2150039. https://doi.org/10.1142/S0219455421500395

Sharma RC, Palli S (2016) Analysis of creep force and its sensitivity on stability and vertical-lateral ride for railway vehicle. Int J Veh Noise Vib 12:60. https://doi.org/10.1504/IJVNV.2016.077474

Sharma SK, Sharma RC (2018a) An investigation of a locomotive structural crashworthiness using finite element simulation. SAE Int J Commercial Veh 11(4):235–244. https://doi.org/10.4271/02-11-04-0019

Sharma SK, Sharma RC (2018b) Simulation of quarter-car model with magnetorheological dampers for ride quality improvement. Int J Veh Struct Syst 10(3):169–173. https://doi.org/10.4273/ijvss.10.3.03

Sharma RC, Sharma SK (2018c) Sensitivity analysis of three-wheel vehicle's suspension parameters influencing ride behavior. Noise Vib Worldwide 49(7–8), 272–280. https://doi.org/10.1177/0957456518796846

Sharma SK, Sharma RC (2019) Pothole detection and warning system for Indian roads. In Advances in interdisciplinary engineering. Lecture notes in mechanical engineering. Springer, Singapore, pp 511–519. https://doi.org/10.1007/978-981-13-6577-5_48

Sharma SK, Sharma RC (2021) Multi-objective design optimization of locomotive nose. SAE Technical Paper, p 10. https://doi.org/10.4271/2021-01-5053

Sharma RC, Sharma SK (2022) Ride analysis of road surface-three-wheeled vehicle-human subject interactions subjected to random excitation. SAE Int J Commercial Veh 15(3):299–312, 02-15-03-0017. https://doi.org/10.4271/02-15-03-0017

Sharma SK, Sharma RC, Kumar A, Palli S (2015) Challenges in rail vehicle-track modeling and simulation. Int J Veh Struct Syst 7(1):1–9. https://doi.org/10.4273/ijvss.7.1.01
Sharma RC, Palli S, Sharma SK, Roy M (2017) Modernization of railway track with composite sleepers. Int J Veh Struct Syst 9(5):321–329. https://doi.org/10.4273/ijvss.9.5.10
Sharma RC, Sharma SK, Palli S (2018) Rail vehicle modelling and simulation using Lagrangian method. Int J Veh Struct Syst 10(3):188–194. https://doi.org/10.4273/ijvss.10.3.07
Sharma SK, Saini U, Kumar A (2019) Semi-active control to reduce lateral vibration of passenger rail vehicle using disturbance rejection and continuous state damper controllers. J Vib Eng Technol 7(2):117–129. https://doi.org/10.1007/s42417-019-00088-2
Sharma RC, Sharma SK, Sharma N, Sharma S (2020a) Analysis of ride and stability of an ICF railway coach. Int J Veh Noise Vib 16:127. https://doi.org/10.1504/IJVNV.2020.117820
Sharma RC, Sharma S, Sharma SK, Sharma N (2020b) Analysis of generalized force and its influence on ride and stability of railway vehicle. Noise Vib Worldw 51:95–109. https://doi.org/10.1177/0957456520923125
Sharma S, Sharma RC, Sharma SK, Sharma N, Palli S, Bhardawaj S (2020c) Vibration isolation of the quarter car model of road vehicle system using dynamic vibration absorber. Int J Veh Struct Syst 12. https://doi.org/10.4273/ijvss.12.5.05
Sharma RC, Sharma SK, Palli S (2020d) Linear and non-linear stability analysis of a constrained railway wheelaxle. Int J Veh Struct Syst 12. https://doi.org/10.4273/ijvss.12.2.04
Sharma SK, Sharma RC, Sharma N (2020e) Combined multi-body-system and finite element analysis of a rail locomotive crashworthiness. Int J Veh Struct Syst 12. https://doi.org/10.4273/ijvss.12.4.15
Sharma SK, Sharma RC, Lee J (2021a) Effect of rail vehicle-track coupled dynamics on fatigue failure of coil spring in a suspension system. Appl Sci 11:2650. https://doi.org/10.3390/app11062650
Sharma RC, Palli S, Sharma N, Sharma SK (2021b) Ride behaviour of a four-wheel vehicle using H infinity semi-active suspension control under deterministic and random inputs. Int J Veh Struct Syst 13. https://doi.org/10.4273/ijvss.13.2.18
Sharma RC, Sharma S, Sharma N, Sharma SK (2021c) Linear and nonlinear analysis of ride and stability of a three-wheeled vehicle subjected to random and bump inputs using bond graph and simulink methodology. SAE Int J Commer Veh 14:02-15-01-0001. https://doi.org/10.4271/02-15-01-0001
Sharma RC, Sharma S, Sharma SK, Sharma N, Singh G (2021d) Analysis of bio-dynamic model of seated human subject and optimization of the passenger ride comfort for three-wheel vehicle using random search technique. Proc Inst Mech Eng Part K J Multi-body Dyn 235:106–121. https://doi.org/10.1177/1464419320983711
Sharma SK, Mohapatra S, Sharma RC, Alturjman S, Altrjman C, Mostarda L, Stephan T (2022a) Retrofitting existing buildings to improve energy performance. Sustainability 14:666. https://doi.org/10.3390/su14020666
Sharma SK, Sharma RC, Lee J, Jang H-L (2022b) Numerical and experimental analysis of DVA on the flexible-rigid rail vehicle carbody resonant vibration. Sensors 22:1922. https://doi.org/10.3390/s22051922
Sharma SK, Lee J, Jang H-L (2022c) Mathematical modeling and simulation of suspended equipment impact on car body modes. Machines 10:192. https://doi.org/10.3390/machines10030192
Sharma SK, Sharma RC, Lee J (2022d) In situ and experimental analysis of longitudinal load on carbody fatigue life using nonlinear damage accumulation. Int J Damage Mech 31:605–622. https://doi.org/10.1177/10567895211046043
Sharma RC, Gopala Rao LVV, Sharma SK, Palli S, Satyanarayana VSV (2022e) Analysis of lateral stability and ride of an Indian railway constrained dual-axle bogie frame. SAE Int J Commercial Veh 16(2):213–228, 02-16-02-0014. https://doi.org/10.4271/02-16-02-0014
Sharma RC, Palli S, Gopala Rao LVV, Duppala A, Sharma SK (2022f) Four-wheel vehicle response under bump, pothole, harmonic, and random excitations using bond graph/simulink technique.

SAE Int J Commercial Veh 16(2):129–139, 02-16-02-0008. https://doi.org/10.4271/02-16-02-0008
Sharma RC, Palli S, Jha AK, Bhardawaj S, Sharma SK (2022g) Vibration and ride comfort analysis of railway vehicle system subjected to deterministic inputs. Resmilitaris 12(5):1345–1355
Sharma RC, Palli S, Bhardawaj S (2023a) A review of wheel-rail dynamic behavior on curved tracks. Int J Veh Struct Syst 15. https://doi.org/10.4273/ijvss.15.1.04
Sharma RC, Palli S, Sharma SK (2023b) Ride analysis of railway vehicle considering rigidity and flexibility of the carbody. J Chinese Inst Eng 46:355–366. https://doi.org/10.1080/02533839.2023.2194918
Sharma RC, Bhardawaj S, Sharma SK, Bahl S (2023c) Ride behaviour of three-wheel vehicle using semi-active suspension using skyhook control strategy. Int J Veh Struct Syst 15(1). https://doi.org/10.4273/ijvss.15.1.05
Sharma SK, Sharma RC, Choi Y, Lee J (2023d) Experimental and mathematical study of flexible–rigid rail vehicle riding comfort and safety. Appl Sci 13(9):5252
Sharma SK, Sharma RC, Choi Y, Lee J (2023e) Modelling and dynamic analysis of adaptive neuro-fuzzy inference system-based intelligent control suspension system for passenger rail vehicles using magnetorheological damper for improving ride index. Sustainability 15(16):12529
Sharma SK, Sharma RC, Lavania S, Palli S, Avesh M (2023f) Industry-driven approach for ANFIS-based intelligent control suspension system with MR damper for enhanced ride quality in passenger rail vehicles for technological investigations. Natl Acad Sci Lett 1–4
Sharma SK, Sharma RC, Upadhyay RK, Lee J (2023g) Coupled torsional–transverse vibration reduction in marine propulsion dynamics with novel approach using magnetorheological damping and adaptive control system. J Vib Eng Technol 1–11
Sharma SK, Upadhyay RK, Kumar V, Valera H (2023h) Introduction to sustainable transportation system. In: Transportation energy and dynamics. Springer, pp 3–6
Sharma RC, Palli S, Dharmana L, Sharma SK, Sharma A, Kharola A, Sharma N (2024a) Simulation of transient dynamic response of a locomotive considering its flexibility and rigidity. Noise Vib Worldwide 55(1–2):33–47
Sharma RC, Palli S, Duppala A, Bhardawaj S, Sharma SK, Sharma N, Sharma A (2024b) Rigid body analysis of ride behaviour of roadway vehicle using Lagrangian dynamics. Int J Veh Struct Syst (IJVSS) 16(1)
Sharma RC, Sharma N, Palli S, Sharma SK, Joshi S, Rallabandi SR, Sharma A (2024c) Dynamic behavior of three-wheel vehicle with inerter-based dynamic absorber suspension system. Noise Vib Worldwide 55(4–5):253–263. https://doi.org/10.1177/09574565241243398
Sharma SK, Kumar N, Avesh M, Sharma RC, Siddiqui MIH, Lee J (2024d) Navigating tranquillity with H∞ controller to mitigate ship propeller shaft vibration. J Vib Eng Technol 1–11
Sharma SK, Sharma RC, Lee J (2024e) Modelling and simulation of MR damper characterization and uncertainty assessment in marine engine vibration isolation with FLPID control. Int J Model Simul 1–11
Sharma SK, Sharma RC, Lee J (2024f) Propelling precision of longitudinal vibration mitigation in ship propeller shafts through advanced nonlinear intelligent semi-active control leveraging adaptive neuro-fuzzy inference system with linear quadratic regulator. J Vib Control. https://doi.org/10.1177/10775463241244836
Sharma SK, Sharma RC, Upadhyay RK, Lee J (2024g) Enhanced ship engine vibration reduction using magnetorheological dampers and adaptive neuro-fuzzy control system. Ships Offshore Struct 1–17
Sinha AK, Sengupta A, Gandhi H, Bansal P, Agarwal KM, Sharma SK, Sharma RC, Sharma SK (2019) Performance enhancement of an all-terrain vehicle by optimizing steering, powertrain and brakes. In Advances in engineering design . lecture notes in mechanical engineering. Springer, Singapore, pp 207–215. https://doi.org/10.1007/978-981-13-6469-3_19
Swapna Rekha H, Nayak J, Behera HS (2020) Pi-sigma neural network: survey of a decade progress. In: Computational intelligence in pattern recognition. Advances in intelligent systems and

computing, 1120. Springer, Singapore, pp 429–441. https://doi.org/10.1007/978-981-15-2449-3_37

Upadhyay RK, Sharma SK, Kumar V, Valera H (2023a) Introduction to transportation systems technology and integrated management. In: Transportation systems technology and integrated management. Springer, pp 1–4. https://doi.org/10.1007/978-981-99-1517-0_1

Upadhyay RK, Sharma SK, Kumar V, Valera H (2023b) Transportation systems technology and integrated management. Edited by RK Upadhyay, SK Sharma, V Kumar, H Valera. Springer Nature, Singapore. https://doi.org/10.1007/978-981-99-1517-0

Upadhyay RK, Sharma SK, Kumar V (Eds) (2024a) Intelligent transportation system and advanced technology. Springer Nature Singapore. https://doi.org/10.1007/978-981-97-0515-3

Upadhyay RK, Sharma SK, Kumar V (2024b) Introduction to intelligent transportation system and advanced technology. In: Intelligent transportation system and advanced technology. Energy, Environment, and Sustainability. pp 3–6. Springer, Singapore. https://doi.org/10.1007/978-981-97-0515-3_1

Vishwakarma PN, Mishra P, Sharma SK (2022a) Characterization of a magnetorheological fluid damper a review. Mater Today: Proc 56:2988–2994. https://doi.org/10.1016/j.matpr.2021.11.143

Vishwakarma PN, Mishra P, Sharma SK (2022b) Formulation of semi-active suspension system and controls in rail vehicle. SSRN Electron J. https://doi.org/10.2139/ssrn.4159616

Wu Q, Cole C, Spiryagin M, Chang C, Wei W, Ursulyak L, Shvets A, Murtaza MA, Mirza IM, Zhelieznov K, Mohammadi S, Serajian H, Schick B, Berg M, Sharma RC, Aboubakr A, Sharma SK, Melzi S, Di Gialleonardo E, Cantone L et al (2023) Freight train air brake models. Int J Rail Transp 11(1):1–49. https://doi.org/10.1080/23248378.2021.2006808

Chapter 10
Enhancing Transparency and Efficiency in the Fruits and Vegetables Cold Chain Through Blockchain Integration

Abhinav Yash, Ashwani Shukla, and Rakesh Chandmal Sharma

10.1 Introduction

Perishable foods production has grown rapidly in India over the past decade due to its wide and diverse agricultural terrain. Agriculture and related industries have contributed significantly to India's Gross Value Added (GVA). In 2020–21, agricultural and associated industries made up 20.2% of India's GVA, according to the January 2022 Economic Survey. India has become a major producer of fruits, vegetables, dairy products, meat, and fish due to its good agricultural circumstances and abundant natural resources. India is a big manufacturer of perishable goods. India is the world's second-largest producer of fruits and vegetables, producing 300.26 million metric tons in 2020–21. India's dairy industry dominates the world, producing and consuming 20% of milk (Adeoye et al. 2023a; Sharma et al. 2020a; Mohapatra et al. 2016a, b; Gopala Rao et al. 2022).

In the meat and poultry sector, India annually produces approximately 5.3 million metric tons of meat and a remarkable 75 billion eggs. India is not only the largest producer of buffalo meat globally but also the second-largest producer of goat meat. The export of meat and poultry products from India reached a total value of USD 5 billion in the fiscal year 2014–15, with buffalo meat accounting for a significant portion of this figure (Kohli et al. 2022; Sharma and Kumar 2017, 2018; Sharma et al. 2021a). Furthermore, India holds the distinction of being the second-largest fish producer in the world, with a market value of around USD 15 billion. Despite

A. Yash · A. Shukla
School of Technology, Gati Shakti Vishwavidyalaya, Vadodara, Gujarat 390 004, India

R. C. Sharma (✉)
Mechanical Engineering Department, Graphic Era (Deemed to Be University), Dehradun 248002, India
e-mail: rcsharmaiitr@gmail.com

Mechanical Engineering Department, Graphic Era Hill University, Dehradun 248002, India

S. K. Sharma et al. (eds.), *Dynamics of Transportation Ecosystem, Modeling, and Control*, Energy, Environment, and Sustainability,
https://doi.org/10.1007/978-981-97-0437-8_10

these impressive production figures, India's contribution to the global farm trade remains relatively modest (Sharma et al. 2015; Sharma et al. 2017; Sharma and Kumar 2017; Palli et al. 2018; Sharma et al. 2018; Sharma and Sharma 2018; Sharma and Kumar 2018a, b; Sharma and Sharma 2018a, b). One of the primary impediments to expanding its footprint in the international market is the substantial waste incurred throughout the value chain of perishable goods. Perishable items such as fruits, vegetables, meats, and dairy products require controlled environments, encompassing temperature and humidity control, for their storage and seamless movement within the supply chain. These controlled environments are collectively referred to as the "cold chain." Within the cold chain, the processes of microbiological, chemical, or enzymatic deterioration are slowed down, preserving the quality and freshness of these products from the farm to the end consumer (Bhardawaj et al. 2020a; Sharma et al. 2022a; Bandekar et al. 2017; Hossain et al. 2017; Mohapatra et al. 2017a).

A well-established cold chain infrastructure is vital for preserving the quality and nutritional value of perishable goods, which form a significant part of the daily diet for a majority of the Indian population. However, India faces substantial challenges in this regard, which lead to considerable food loss and wastage. Inadequate cold storage and perishable food infrastructure contribute to losses of up to 40% of overall production, resulting in price fluctuations, hunger, and financial losses for farmers and fishermen. The deficiencies in cold storage facilities impact not only the economic well-being of farmers and fishermen but also the overall food supply chain, causing price volatility and inflation. This underscores the urgent need for a comprehensive solution to address the inadequacies in India's cold supply chain infrastructure (Sharma and Sharma 2019; Swapna Rekha et al. 2020; Sinha et al. 2019; Goswami et al. 2019; Sharma et al. 2022b).

The impact of these challenges extends beyond the economic sphere and affects the environment as well. The decomposition of wasted perishable goods leads to the emission of significant amounts of greenhouse gases, contributing to climate change. Consequently, the need for efficient and sustainable cold chain logistics becomes paramount to mitigate these challenges (Goyal et al. 2019; Sharma 2019; Sharma et al. 2019; Bhardawaj et al. 2020; Lee and Sharma 2020; Bhardawaj et al. 2021; Choi et al. 2021; Lee et al. 2021; Sharma and Sharma 2021). Despite the impressive growth in the production of perishable items and their crucial role in India's economy, there is a critical mismatch between supply and demand. The Indian population, which is rapidly growing and urbanizing, exhibits changing dietary preferences, an increased demand for convenience foods, and a rising inclination toward online food shopping. These trends necessitate a transformation in the cold supply chain to meet the evolving needs of consumers (Kumar et al. 2022; Sharma et al. 2022a; Sharma et al. 2022b; Sharma et al. 2022c; Sharma and Sharma 2022; Vishwakarma et al. 2022a, b). Moreover, India's vast climatic diversity poses unique challenges. The produce grown in one climatic region often needs to be transported to another region with entirely different weather conditions, exacerbating the risk of deterioration during transit. This underscores the importance of preserving the seasonal produce to meet year-round demand and support sustainable agriculture. Additionally, globalization has ushered in new opportunities and challenges (Mohanty et al. 2016; Samantaray et al. 2014; Pattnaik et al. 2022; Nibedita et al. 2021).

The global demand for locally grown food necessitates strict quality standards and infrastructure and process changes to satisfy international norms. India must prioritize quality management and production methods to address these issues (Bhardawaj et al. 2023; Hossain et al. 2023; Lee and Sharma 2023; Sharma et al. 2023a, b, c, d, e, f; Upadhyay et al. 2023a, b; Wu et al. 2023).

The rise of e-commerce and changing customer behavior have highlighted the importance of a cold chain. Online food sales are rising, making perishable items delivery essential to modern logistics. To meet demand for processed and fresh food, cold storage facilities must be larger and more modern. These facts show that India's cold supply chain faces significant problems and tremendous opportunity. India needs restructure and invest in cold chain infrastructure, technology, and qualified people to maximize its perishable goods sector (Biswas et al. 2021). The key to cold supply chain management is in Table 10.1.

This paper explores India's complex cold supply chain. It evaluates the cold chain infrastructure's current situation, difficulties, and urgent need for transformation. It also examines how blockchain could improve cold supply chain traceability, quality control, and efficiency. We hope this detailed research will help alter India's cold supply chain, benefiting farmers, consumers, and the economy.

10.2 Scope and Objective of the Study

The Indian F&V cold chain business ensures safe and efficient distribution of fresh goods from fields to consumers. This industry has many issues that affect its effectiveness, profitability, and the quality and shelf life of the produce it manages. Due to the urgency of these issues, this paper conducts a detailed investigation with particular goals. It begins with a comprehensive literature study of F&V cold chain publications, research papers, reports, and scholarly works. This evaluation identifies and highlights the sector's significant difficulties, with a focus on technological concerns.

Second, this paper examines global and Indian cold chain management technology. This exploration seeks innovative answers to India's F&V cold chain sector's problems. Technology advances include temperature control, monitoring systems, packaging, and shipping logistics.

The third purpose is to analyze how the listed difficulties affect F&V product quality and shelf life. The study examines these effects to determine how these challenges affect cold chain efficiency and profitability. It also seeks to illuminate how these issues may impair the sector's ability to fulfill India's rising fresh produce demand.

Based on the literature review and technical exploration, this study's fourth objective is to provide practical recommendations and important insights to improve F&V cold chain operations. These suggestions include technical solutions, efficiency improvements, and sector stakeholder profitability increases.

Table 10.1 Shows the key aspects of cold supply chain management

S.r.No	Aspect	Summary
	India's agricultural growth	India has witnessed significant growth in the production of perishable items, contributing to its economic development
	Prominence in perishable production	India ranks among the top producers globally in categories such as fruits, vegetables, dairy, meat, poultry, and seafood
	Perishable production statistics	–2nd largest producer of fruits and vegetables –20% share in global milk production –large meat and egg production –2nd largest fish producer
	Challenges in the cold supply chain	India faces challenges in preserving the quality and reducing waste in the cold supply chain
	Importance of cold chain infrastructure	Cold chain infrastructure is vital for maintaining the quality of perishable items, especially in a developing country like India
	Impact on farmers and the economy	Poor cold storage facilities result in economic losses for farmers and inflation in the food supply chain
	Role in preserving seasonal produce	Cold chain infrastructure helps meet year-round demand by preserving seasonal produce
	Globalization and quality standards	India needs to adhere to quality standards to meet the rising global demand for locally grown food
	Rise of E-commerce in perishable sales	Online food sales are increasing, boosting the demand for cold chain logistics and cold storage facilities
	Challenges in the cold supply chain	Challenges include inadequate infrastructure, temperature fluctuations, high energy costs, lack of technology adoption, and an unorganized industry
	Environmental impact	Inefficient cold chains contribute to food waste and greenhouse gas emissions, exacerbating climate change

Lastly, this research aspires to contribute meaningfully to the broader body of knowledge related to F&V cold chain management in India. By synthesizing and critically evaluating existing research and emerging technological solutions, this report aims to serve as a valuable reference for future studies and endeavors in this field. In essence, this study seeks to provide a holistic and well-informed assessment of the challenges faced by the F&V cold chain sector in India, while also proposing viable technological solutions to address these challenges. By doing so, it aims to empower stakeholders within the sector with the knowledge and insights needed to enhance their operations, improve product quality, and ultimately contribute to the efficiency and sustainability of the F&V cold chain in India.

10.3 Literature Review

The literature review serves as a comprehensive summary of relevant studies, consolidating key findings, assessing the strengths and weaknesses of prior research, identifying knowledge gaps, and presenting a theoretical framework for the adoption of blockchain technology within the Indian Fruits & Vegetables (F&V) sector's cold chain. Additionally, it offers an overview of the existing research on technological advancements in the F&V cold chain sector. Numerous studies, articles, and research efforts have explored technological advancements and improvements in the F&V cold chain industry (Sharma et al. 2020b, 2021b, 2022c, 2022d; Pundir et al. 2023a; Das et al. 2023; Agrahari et al. 2021a; Maity et al. 2020; Misra et al. 2022; Sharma and Lee 2020a). While previous research has highlighted the potential of various technologies to enhance the effectiveness and efficiency of the cold chain, most of these studies have predominantly centered on monitoring and controlling the physical parameters of the cold chain. This focus often overlooks the underlying issues contributing to inefficiencies, such as the lack of trust and transparency in the supply chain, representing a notable limitation of existing research efforts. The existing research primarily aims to bolster the infrastructure of the F&V sector without placing sufficient emphasis on technological advancements capable of introducing transparency and quality control. A significant research gap persists concerning the utilization of blockchain technology within the F&V cold chain industry to address these trust and transparency challenges (Mohapatra et al. 2017b; Kohli et al. 2020; Devadarshini et al. 2022; Sharma et al. 2021c; Sarkar et al. 2018; Palli et al. 2022).

Blockchain technology, by creating immutable and tamper-proof transaction records and cold chain data, possesses the potential to significantly enhance trust and transparency within the F&V cold chain industry. It can enable end-to-end traceability of F&V, from farm to consumer, ensuring transparent and reliable documentation of every transaction, including harvesting, transportation, storage, and distribution. Furthermore, blockchain technology has the capacity to enhance traceability and reduce the risk of fraud within the F&V sector (Bhardawaj et al. 2019a, 2020b; Sharma and Lee 2020b, 2021; Acharya et al. 2020; Pundir et al. 2023b). As blockchain records are unalterable, information pertaining to the cold chain remains accurate and cannot be tampered with. This not only enhances consumer confidence in the F&V supply chain but also provides a mechanism for rapid identification of any food safety issues (Badgujar et al. 2024; Kokkinos et al. 2024; Raju Rallabandi et al. 2024; Rallabandi et al. 2024; Sharma et al. 2024a, b, c, d, e, f, g, h; Upadhyay and Sharma Vikram 2024a, b). Nonetheless, it is essential to note that instead of addressing the larger systemic issues, such as the lack of trust and transparency within the cold supply chain, most research efforts have been narrowly focused on monitoring, improving, and controlling the physical aspects of the cold chain. Despite the considerable potential of blockchain technology to enhance trust and transparency within the F&V cold chain sector, a significant research gap remains in understanding its adoption and implementation in this context (COVID-19: Analysing the Legal Nuances of the Lockdown Order 2020; Pattnaik et al. 2020; Mohapatra et al. 2020, 2022; Sharma et al. 2020c).

10.4 Methodology

Supply chain management requires innovation for efficiency and reliability. Freshness and quality of perishable goods like fruits and vegetables (F&V) depend on the cold supply chain. Cold supply chain issues are specific to India, a large F&V producer. This article discusses a study on how blockchain integration can improve India's F&V cold supply chain. The qualitative study "Improvement in F&V cold supply chain of India" uses secondary data and a conceptual framework for blockchain integration in the cold chain. The case study examines India's cold fruit and vegetable supply chain (Adeoye et al. 2023b, c; Pati et al. 2022, 2020; Mohapatra et al. 2017c; Chatterjee et al. 2022). This research aims to identify difficulties in India's F&V cold supply chain and explore blockchain integration as a solution. This research used academic publications, research papers, reports, articles, and online databases. Microsoft Excel and Python, popular academic tools for data cleansing, coding, analysis, and visualization, were used to carefully analyze the acquired data. This technology accelerated research and ensured accuracy (Nayak et al. 2022; Mohapatra et al. 2017d; Adeoye et al. 2022; Sharma et al. 2020d).

Blockchain integration in India's F&V cold supply chain was a key finding. The study explored how blockchain technology improved facilities in this area. This required creating a conceptual model for blockchain, its functions, and its potential benefits to the cold supply chain. Programming and coding were done to incorporate blockchain technology in the Indian F&V business. Additionally, current startups in this industry were researched to determine their operations and compatibility with the proposed technology (Sharma et al. 2020e, 2021d; Bhardawaj et al. 2020c, 2019b; Palli et al. 2021; Palli and Koona 2015; Kanth Choppara et al. 2019; Mohapatra et al. 2015). This hands-on method assessed blockchain's potential. The research went well, but the lack of accurate and dependable cold supply chain data in India hampered it. The developing infrastructure, insufficient stakeholder cooperation, and low investment in new technology contributed to this difficulty. To overcome this challenge, the research triangulated data from multiple sources and extensively verified its credibility and authenticity. This study achieved its goals with its research method. The case study approach illuminated India's intricate cold supply chain. Thematic study revealed cold chain problems and blockchain integration benefits. What sets this research apart is its reliance on secondary data and the conceptual framework for blockchain integration. In a field where primary data collection through surveys and interviews is common, this study's use of secondary data is noteworthy (Mohapatra et al. 2021; Bhattarai et al. 2022; Agrahari et al. 2021b; Agarwal et al. 2020; PanduRanga Vital et al. 2019; Sharma et al. 2023a). The complexity of collecting primary data on the cold chain in India made this approach not only practical but also innovative. Furthermore, the conceptual framework for blockchain integration in the cold chain is a novel approach (Sharma et al. 2023b; Satyanarayana et al. 2022; Sharma and Palli 2016; Avesh and Srivastava 2020, 2019). The implementation of blockchain technology is still in its infancy in the Indian market, and there is limited research on its potential benefits in the cold supply chain. This approach paved the way for exploring the benefits and identifying the challenges that must be addressed before implementing blockchain in the cold chain (Avesh et al. 2019a, b; Avesh and Srivastava 2012; Mohd and Srivastava 2019).

This research on "Improvement in F&V cold supply chain of India" offers a fresh perspective on enhancing the efficiency and reliability of the cold supply chain in a nation with a significant stake in the global F&V market. By embracing innovative approaches and technologies like blockchain, India can overcome the challenges it faces and establish a more robust and secure cold supply chain system.

10.5 Results and Discussion

The literature reviews suggest that blockchain technology holds great promise for addressing challenges in the Fruits and Vegetables (F&V) cold supply chain in India. Here are the key findings.

The global blockchain market within the supply chain sector has experienced remarkable growth, surging from a valuation of USD 93.16 million in 2020 to a projected USD 3,485.25 million by 2028. This expansion underscores the substantial potential that blockchain technology holds for the supply chain industry. Deloitte's survey outcomes reinforce this sentiment, as 53% of executives and managers across diverse industries acknowledge the significant impact that blockchain can have on their supply chains. Furthermore, a noteworthy 33% of respondents have already begun integrating blockchain into their supply chain operations, indicative of the growing interest and adoption of this technology.

Accenture's research demonstrates the tangible financial advantages of blockchain implementation, highlighting the potential for a 0.5–1% increase in revenue and a corresponding reduction of 0.5–1% in supply chain expenses. These financial gains underscore the value of enhanced transparency and efficiency that blockchain offers. Ethical sourcing, a crucial aspect of modern supply chains, finds a practical solution in blockchain technology. Tesla's integration of blockchain for raw material sourcing, particularly cobalt for electric vehicle production, illustrates how this technology fosters accountability, transparency, and sustainability within the supply chain.

Real-time data sharing is a critical feature of blockchain-based platforms like Trade Lens, a collaborative effort by IBM and Maersk. This platform facilitates seamless, real-time data exchange among shippers, carriers, ports, and forwarders, significantly reducing processing time by up to 40% and mitigating errors and disputes. Blockchain's capacity for end-to-end product traceability is another key advantage, effectively reducing counterfeiting, enhancing quality control, and building consumer trust. Walmart's application of blockchain to trace the origins of its leafy greens exemplifies this capability.

The food and agriculture sector benefits substantially from blockchain by improving food safety through comprehensive traceability. Projections suggest that the blockchain industry in this domain will achieve a remarkable compound annual growth rate (CAGR) of 48.1% from 2021 to 2028, ultimately reaching a valuation of USD 1.4 billion. Furthermore, blockchain's transformative potential extends to the pharmaceutical industry, where enhanced supply chain traceability is anticipated to drive the healthcare blockchain market to a valuation of USD 1.636 billion by 2025. In essence, these insights collectively highlight the pivotal role that blockchain technology plays in revolutionizing the supply chain landscape and enhancing various facets of modern industries. To address challenges in the F&V cold supply chain, blockchain technology can provide benefits such as transparency, traceability, reduced waste, and enhanced quality control. However, there are challenges to adoption, including technological literacy, lack of access to technology, industry fragmentation, and resistance to change. Education, infrastructure development, and collaborative efforts are necessary to overcome these challenges. Blockchain technology can also transform the Indian apple market by providing end-to-end traceability, improving supply chain efficiency, and enhancing transparency. It can help farmers access quality inputs, demonstrate compliance with standards, and achieve fair prices for their produce. Blockchain technology has the potential to revolutionize the F&V cold supply chain in India and the apple market by addressing key challenges and improving various aspects of the supply chain. Tables 10.2, 10.3, and 10.4 provide a concise overview of the benefits and challenges associated with blockchain adoption in the supply chain and F&V cold chain sectors.

The data and insights presented here leave no doubt about the significant growth potential of blockchain in the global supply chain industry. With the market projected to reach USD 3,485.25 million by 2028, it is evident that blockchain is not merely a theoretical concept but a rapidly evolving reality with substantial economic implications. The survey results from Deloitte, involving over 1,200 industry executives and managers, further underline this point. A majority of respondents, 53%, acknowledge blockchain's potential to revolutionize their supply chains, indicating a growing awareness of its transformative capabilities. Even more striking is the fact that 33% of the respondents are already actively implementing blockchain technology in their supply chains, demonstrating that real-world adoption is well underway. These survey findings not only validate the technology's relevance but also emphasize the urgency for businesses to explore its potential benefits.

Moreover, the financial advantages of blockchain integration cannot be overlooked. The Accenture report's estimates of improving revenue by 0.5–1% and

Table 10.2 Benefits of blockchain in supply chain

Benefit	Description
Market potential	Rapid growth of the global blockchain in supply chain market, indicating substantial opportunities for adoption
Interest and adoption	A significant percentage of executives and managers express interest and are implementing blockchain in their supply chains
Financial gains	Blockchain technology can increase revenue and reduce supply chain expenses through improved transparency and efficiency
Ethical sourcing	Integration of blockchain ensures ethical sourcing of raw materials, enhances transparency, and aligns with sustainability goals
Real-time data sharing	Blockchain platforms like Trade Lens provide real-time data sharing, reducing processing time and disputes
Product traceability	Blockchain enables end-to-end product traceability, reducing counterfeiting, improving quality control, and boosting customer confidence
Food safety	Blockchain technology enhances food safety by enabling traceability in the food and agriculture industry
Pharmaceutical traceability	Blockchain enhances supply chain traceability in the pharmaceutical sector, supporting industry growth

Table 10.3 Challenges in blockchain adoption

Challenge	Description
Technological illiteracy	Lack of digital literacy among stakeholders, particularly in rural areas, poses a significant hurdle
Lack of access to technology	Small-scale farmers and workers may not have access to advanced technology required for blockchain adoption
Industry fragmentation	The F&V cold supply chain is highly fragmented, making it challenging to implement a blockchain system involving all stakeholders
Resistance to change	Resistance to adopting new technology is a common challenge that must be addressed through education and training
Unorganized industry	The disorganized nature of the F&V cold chain industry complicates the implementation of blockchain solutions

Table 10.4 Examples of blockchain adoption in supply chain

Company	Application
Tesla	Ethical sourcing of raw materials, transparency in the supply chain
IBM and Maersk	Real-time data sharing, reducing processing time and disputes
Walmart	Enhanced supply chain traceability, food safety

reducing supply chain expenses by 0.5–1% through enhanced transparency and efficiency make a compelling case for decision-makers and stakeholders in the F&V cold supply chain sector. These financial gains provide a tangible incentive for embracing blockchain technology. The manuscript further solidifies its argument by presenting real-world cases. Tesla's use of blockchain to ensure ethical sourcing of raw materials showcases how the technology can enhance accountability, transparency, and efficiency while aligning with sustainability and ethical sourcing goals. The Trade Lens platform, a collaboration between IBM and Maersk, demonstrates the practical benefits of blockchain through increased visibility, reduced processing time, and minimized disputes in the supply chain. Additionally, the manuscript highlights the profound implications of blockchain for traceability and quality control. Walmart's example of using blockchain to trace the journey of leafy greens from farm to store exemplifies how the technology can swiftly identify the source of foodborne diseases, ensuring product safety and bolstering consumer confidence.

Furthermore, the manuscript expands its scope by exploring blockchain's potential in other industries, such as pharmaceuticals, where it can enhance supply chain traceability and reduce drug fraud. The anticipated growth of the healthcare blockchain market to USD 1.636 billion by 2025 underscores the broader industry interest in blockchain adoption.

10.6 Conclusion

This manuscript highlights the tremendous potential of blockchain technology in addressing the complex challenges within the Fruits and Vegetables (F&V) cold supply chain in India. The results and insights presented throughout the manuscript paint a vivid picture of the transformative power of blockchain, not only in India but on a global scale. The projected market growth, the enthusiasm of industry leaders revealed in the Deloitte survey, and the tangible financial benefits outlined in the Accenture report all underscore the urgency for businesses and stakeholders to embrace this technology. Real-world cases from companies like Tesla, Walmart, and IBM-Maersk's Trade Lens platform demonstrate the practicality and benefits of blockchain implementation. These examples showcase how blockchain can enhance transparency, traceability, ethical sourcing, and efficiency across the supply chain. The technology's potential extends beyond the F&V sector, with promising implications in pharmaceuticals and healthcare as well. As India's F&V cold supply chain sector grapples with challenges related to fragmentation, lack of technology access, and the need for modernization, blockchain emerges as a potent solution. However, it is crucial to address issues of technological literacy, access, and resistance to change. Initiatives such as training programs, technology centers in rural areas, and incentives can facilitate blockchain adoption. In light of the evidence presented in this manuscript, it is imperative for India to lead the way in embracing blockchain technology. By doing so, India has the opportunity to revolutionize its F&V cold chain sector, bringing about increased efficiency, transparency, and sustainability.

This transformation will not only benefit businesses but also enhance consumer trust and ensure the safety and quality of food products, ultimately contributing to the well-being of the nation's population. The time is ripe for India to seize the blockchain revolution and reshape its F&V cold supply chain for the better.

References

Acharya A, Gahlaut U, Sharma K, Sharma SK, Vishwakarma PN, Phanden RK (2020) Crashworthiness analysis of a thin-walled structure in the frontal part of automotive chassis. Int J Veh Struct Syst 12. https://doi.org/10.4273/ijvss.12.5.06

Adeoye AO, Quadri RO, Lawal OS (2022) Assessment of biofuel potential of tenera palm kernel shell via fixed bed pyrolysis and thermal characterization. Results Surf Interfaces 9:100091. https://doi.org/10.1016/j.rsurfi.2022.100091

Adeoye AO, Quadri RO, Lawal OS (2023a) Wet synthesis, characterization of goethite nanoparticles and its application in catalytic pyrolysis of palm kernel shell in TGA. Results Surf Interfaces 11:100118. https://doi.org/10.1016/j.rsurfi.2023.100118

Adeoye AO, Quadri RO, Lawal OS (2023b) Biomass waste pyrolysis in the context of Nigeria's oil and gas industry: an overview. Int J Ambient Energy 44:2525–2541. https://doi.org/10.1080/01430750.2023.2256325

Adeoye AO, Lawal OS, Quadri RO, Malomo D, Aliyu MT, Dang GE, Emojevu EO, Maikato MJ, Yahaya MG, Omonije OO, Edidem VG, Abubakar YK, Offor OF, Sochima EH, Peter BE, Hikon BN (2023) Sustainable energy via thermochemical and biochemical conversion of biomass wastes for biofuel production. In: Transportation energy and dynamics. energy, environment, and sustainability. Springer, Singapore, pp 245–306. https://doi.org/10.1007/978-981-99-2150-8_11

Agarwal KM, Mohapatra S, Sharma P, Sharma S, Bhatia D, Mishra A (2020) Study and overview of the novel corona virus disease (COVID-19). Sensors Int 1:100037. https://doi.org/10.1016/j.sintl.2020.100037

Agrahari R, Mohanty S, Vishwakarma K, Nayak SK, Samantaray D, Mohapatra S (2021b) Update vision on COVID-19: Structure, immune pathogenesis, treatment and safety assessment. Sensors Int 2:100073. https://doi.org/10.1016/j.sintl.2020.100073

Agrahari R, Sarraf G, Joshi NC, Mohapatra S, Varma A (2021) Insight of biopolymers and applications of polyhydroxyalkanoates. In Microbial polymers, pp 177–191. Springer Singapore, Singapore (2021). https://doi.org/10.1007/978-981-16-0045-6_8

Avesh M, Srivastava R (2012) Modeling simulation and control of active suspension system in Matlab Simulink environment. In 2012 Students conference on engineering and systems, pp 1–6. IEEE. https://doi.org/10.1109/SCES.2012.6199124

Avesh M, Srivastava R (2019) Parametric optimization to design a passenger car suspension system for better dynamic performance. Eur Transp—Trasp Eur

Avesh M, Srivastava R, Chandmal Sharma R, Sharma N (2019) Effective design of active suspension system using fuzzy logic control approach. Int J Veh Struct Syst 11. https://doi.org/10.4273/ijvss.11.5.13

Avesh M, Srivastava R, Sharma RC (2019b) Parametric optimization for the design of passenger vehicle suspension system with the application of genetic algorithm. Int J Veh Struct Syst 11. https://doi.org/10.4273/ijvss.11.2.08

Avesh M, Srivastava R (2020) Passenger car active suspension system model for better dynamic characteristics. Natl Acad Sci Lett 43:37–41. https://doi.org/10.1007/s40009-019-00807-z

Bandekar D, Chouhan OP, Mohapatra S, Hazra M, Hazra S, Biswas S (2017) Putative protein VC0395_0300 from Vibrio cholerae is a diguanylate cyclase with a role in biofilm formation. Microbiol Res 202:61–70. https://doi.org/10.1016/j.micres.2017.05.003

Badgujar P, Sharma S, Rahul D, Sharma SK, Upadhyay RK (2024) Feasibility study of Mumbai-Pune Hyperloop corridor: Challenges and implementation. Res Transport Bus Manage 52:101089

Bhardawaj S, Chandmal Sharma R, Kumar Sharma S (2019a) Development and advancement in the wheel-rail rolling contact mechanics. IOP Conf Ser Mater Sci Eng 691:012034. https://doi.org/10.1088/1757-899X/691/1/012034

Bhardawaj S, Sharma RC, Sharma SK (2020b) Development in the modeling of rail vehicle system for the analysis of lateral stability. Mater Today Proc 25:610–619. https://doi.org/10.1016/j.matpr.2019.07.376

Bhardawaj S, Sharma RC, Sharma SK (2020c) Analysis of frontal car crash characteristics using ANSYS. Mater Today Proc 25:898–902. https://doi.org/10.1016/j.matpr.2019.12.358

Bhardawaj S, Sharma RC, Sharma SK (2020) Development in the modeling of rail vehicle system for the analysis of lateral stability. Mater Today: Proceed 25:610–619. https://doi.org/10.1016/j.matpr.2019.07.376

Bhardawaj S, Chandmal Sharma R, Kumar Sharma S (2019) A survey of railway track modelling. Int J Veh Struct Syst 11. https://doi.org/10.4273/ijvss.11.5.08

Bhardawaj S, Sharma RC, Sharma SK (2020a) Ride analysis of track-vehicle-human body interaction subjected to random excitation. J Chin Soc Mech Eng Trans Chinese Inst Eng Ser C/Chung-Kuo Chi Hsueh K. Ch'eng Hsuebo Pao 41:229–238

Bhardawaj S, Sharma RC, Sharma SK, Sharma N (2021) On the planning and construction of railway curved track. Int J Vehicle Struct Syst 13(2):151–159. https://doi.org/10.4273/ijvss.13.2.04

Bhardawaj S, Sharma RC, Sharma SK, Gopala Rao L, Vashist A (2023) Modeling of biomechanical human body model for seat to head transmissibility analysis. Noise Vibr Worldwide 54(2–3):62–74. https://doi.org/10.1177/09574565231154243

Bhattarai S, Mhamdi A, Hossain I, Raoui Y, Pandey R, Madan J, Bouazizi A, Maiti M, Gogoi D, Sharma A (2022) A detailed review of perovskite solar cells: Introduction, working principle, modelling, fabrication techniques, future challenges. Micro Nanostruct 172:207450. https://doi.org/10.1016/j.micrna.2022.207450

Biswas N, Saha S, Biswas BK, Chowdhury M, Rahaman A, Junghare V, Mohapatra S, Hazra S, Zangrando E, Roy Choudhury R, Roy Choudhury C (2021) The DNA- and protein-binding properties and cytotoxicity of a new copper(II) hydrazone Schiff base complex. J Coord Chem 74:1482–1504. https://doi.org/10.1080/00958972.2021.1913128

Chatterjee A, Nirwan S, Mohapatra S, Sharma P, Agnihotri A, Shrivastava N (2022) Biochemical aspects of pathogenic variability in white rust infected Indian mustard. Mycologia 114:757–768. https://doi.org/10.1080/00275514.2022.2060007

Choi S, Lee J, Sharma SK (2021) A study on the performance evaluation of hydraulic tank injectors. In: Joshi P, Gupta SS, Shukla AK, Gautam SS (eds) Advances in Engineering Design. Lecture Notes in Mechanical Engineering, pp 183–190. https://doi.org/10.1007/978-981-33-4684-0_19

COVID-19: Analysing the legal nuances of the lockdown order. Indian J Forensic Med Toxicol (2020). https://doi.org/10.37506/ijfmt.v14i4.11553

Das S, Swapnarekha H, Vimal S (2023) Integration of blockchain technology with renewable energy for sustainable development: issues, challenges and future direction. Presented at the (2023). https://doi.org/10.1007/978-981-99-3734-9_48

Devadarshini D, Mohapatra S, Pati S, Maity S, Rath CC, Jena PK, Samantaray D (2022) Evaluation of PHAs production by mixed bacterial culture under submerged fermentation. Biologia (bratisl). 78:895–902. https://doi.org/10.1007/s11756-022-01302-5

Gopala Rao L, Sharma RC, Satyanarayana V, Palli S (2022) Stochastic optimal preview control response of a quarter-car nonlinear suspension model using spectral decomposition method. Noise Vib Worldw 53:225–232. https://doi.org/10.1177/09574565221093234

Goswami B, Rathi A, Sayeed S, Das P, Sharma RC, Sharma SK (2019) Optimization design for aerodynamic elements of Indian locomotive of passenger train. In: Advances in engineering design. Lecture notes in mechanical engineering. Springer, Singapore, pp 663–673. https://doi.org/10.1007/978-981-13-6469-3_61

Goyal S, Anand CS, Sharma SK, Sharma RC (2019) Crashworthiness analysis of foam filled star shape polygon of thin-walled structure. Thin-Walled Struct 144:106312. https://doi.org/10.1016/j.tws.2019.106312

Hossain I, Velkin VI, Shcheklein SE (2017) The study of passive vibration dampers in pipelines using PIV-methodology for single phase flow. In WIT transactions on ecology and the environment, pp 565–570. https://doi.org/10.2495/ESUS170521

Hossain I, Ahmed YE, Sharma SK (2023) The use of both metallic-insert and TiO2/water nanofluid to improve the performance of photovoltaic modules; experimental study. Thermal Sci Eng Prog 42:101944. https://doi.org/10.1016/j.tsep.2023.101944

Kokkinos NC, Emmanouilidou E, Sharma SK (2024) Waste-to-biofuel production for the transportation sector. In: Intelligent transportation system and advanced technology, pp 99–122, Springer

Kanth Choppara R, Chandmal Sharma R, Kumar Sharma S, Gupta T (2019) Aero dynamic cross wind analysis of locomotive. IOP Conf Ser Mater Sci Eng 691:012035. https://doi.org/10.1088/1757-899X/691/1/012035

Kohli I, Joshi NC, Mohapatra S, Varma A (2020) Extremophile – an adaptive strategy for extreme conditions and applications. Curr Genomics 21:96–110. https://doi.org/10.2174/1389202921666200401105908

Kohli I, Mohapatra S, Kumar P, Goel A, Varma A, Joshi NC (2022) Role of bacterial endophytes in the promotion of plant growth. In: Bacterial endophytes for sustainable agriculture and environmental management, pp 235–253. Springer Singapore, Singapore (2022). https://doi.org/10.1007/978-981-16-4497-9_12

Kumar U, Kasvekar R, Sharma SK, Upadhyay RK (2022) Wear of wheels and axle in locomotive and measures taken by indian railway. In: Advances in Engine Tribology, pp. 77–96, Springer. https://doi.org/10.1007/978-981-16-8337-4_5

Lee J, Han J, Sharma SK (2021) Structural analysis on the separated and integrated differential gear case for the weight reduction. Adv Eng Des: Select Proceed FLAME 2020:175

Lee J, Sharma SK (2020) Numerical investigation of critical speed analysis of high-speed rail vehicle. 한국정밀공학회학술발표대회 논문집 (Korean Society of Precision Engineering), 696

Lee J, Sharma SK (2023) Intelligent systems for railway infrastructure. Appl Sci 13(20): 11274. MDPI

Maity S, Das S, Mohapatra S, Tripathi AD, Akthar J, Pati S, Pattnaik S, Samantaray DP (2020) Growth associated polyhydroxybutyrate production by the novel Zobellellae tiwanensis strain DD5 from banana peels under submerged fermentation. Int J Biol Macromol 153:461–469. https://doi.org/10.1016/j.ijbiomac.2020.03.004

Misra S, Mohanty D, Mohapatra S (2022) Food-based probiotics: Functional dietary ingredients. In Probiotics, pp 257–275. Elsevier. https://doi.org/10.1016/B978-0-323-85170-1.00006-3

Mohanty DP, Mohapatra S, Misra S, Sahu PS (2016) Milk derived bioactive peptides and their impact on human health—A review. Saudi J Biol Sci 23:577–583. https://doi.org/10.1016/j.sjbs.2015.06.005

Mohapatra S, Rath SN, Pradhan SK, Samantaray DP, Rath CC (2016b) Secondary structural models (16S rRNA) of polyhydroxyalkanoates producing bacillus species isolated from different rhizospheric soil: Phylogenetics and chemical analysis. Int J Bioautomation 20:329–338

Mohapatra S, Mishra R, Roy P, Yadav KL, Satapathi S (2017b) Systematic investigation and in vitro biocompatibility studies on implantable magnetic nanocomposites for hyperthermia treatment of osteoarthritic knee joints. J Mater Sci 52:9262–9268. https://doi.org/10.1007/s10853-017-1136-0

Mohapatra S, Sarkar B, Samantaray DP, Daware A, Maity S, Pattnaik S, Bhattacharjee S (2017c) Bioconversion of fish solid waste into PHB using *Bacillus subtilis* based submerged fermentation process. Environ Technol 38:3201–3208. https://doi.org/10.1080/09593330.2017.1291759

Mohapatra S, Maity S, Dash HR, Das S, Pattnaik S, Rath CC, Samantaray D (2017d) Bacillus and biopolymer: prospects and challenges. Biochem Biophys Reports 12:206–213. https://doi.org/10.1016/j.bbrep.2017.10.001

Mohapatra S, Pattnaik S, Maity S, Mohapatra S, Sharma S, Akhtar J, Pati S, Samantaray DP, Varma A (2020) Comparative analysis of PHAs production by *Bacillus megaterium* OUAT 016 under submerged and solid-state fermentation. Saudi J Biol Sci 27:1242–1250. https://doi.org/10.1016/j.sjbs.2020.02.001

Mohapatra S, Samantaray DP, Samantaray SM (2015) Study on polyhydroxyalkanoates production using rhizospheric soil bacterial isolates of sweet potato. Indian J Sci Technol 8:57. https://doi.org/10.17485/ijst/2015/v8iS7/64027

Mohapatra S, Samantaray DP, Samantaray SM, Mishra BB, Das S, Majumdar S, Pradhan SK, Rath SN, Rath CC, Akthar J, Achary KG (2016) Structural and thermal characterization of PHAs produced by Lysinibacillus sp. through submerged fermentation process. Int J Biol Macromol 93:1161–1167. https://doi.org/10.1016/j.ijbiomac.2016.09.077

Mohapatra S, Mohanta PR, Sarkar B, Daware A, Kumar C, Samantaray DP (2017a) Production of polyhydroxyalkanoates (PHAs) by bacillus strain isolated from waste water and its biochemical characterization. Proc Natl Acad Sci India Sect B Biol Sci 87:459–466. https://doi.org/10.1007/s40011-015-0626-6

Mohapatra S, Vishwakarma K, Joshi NC, Maity S, Kumar R, Ramchander M, Pattnaik S, Samantaray DP (2021) A review on PHAs: the future biopolymer. In: Environmental and agricultural microbiology, pp 83–100. Wiley. https://doi.org/10.1002/9781119525899.ch5

Mohapatra S, Mohapatra S, Han H, Ariza-Montes A, López-Martín M del C (2022) Climate change and vulnerability of agribusiness: assessment of climate change impact on agricultural productivity. Front Psychol 13. https://doi.org/10.3389/fpsyg.2022.955622

Mohd A, Srivastava R (2019) Parametric optimization and experimental validation for nonlinear characteristics of passenger car suspension system. Period Polytech Transp Eng 49:103–113. https://doi.org/10.3311/PPtr.12999

Nayak J, Swapnarekha H, Routray AR, Nayak SR, Behera HS (2022) Big data in education: present and future. In Computational intelligence in data mining. smart innovation, systems and technologies, 281. Springer, Singapore, pp 721–739. https://doi.org/10.1007/978-981-16-9447-9_54

Nibedita S, Swati P, Pattnaik M, Mohapatra S (2021) Methane emission and strategies for mitigation in livestock. In: Environmental and agricultural microbiology, pp 257–274. Wiley. https://doi.org/10.1002/9781119525899.ch12

Palli S, Dontikurti R, Sharma RC, Sharma N (2021) Analysis of dynamic response of a railway locomotive using ANSYS. Int J Veh Struct Syst 13. https://doi.org/10.4273/ijvss.13.2.21

Palli S, Duppala A, Sharma RC, Gopala Rao LVV (2022) Dynamic simulation of automotive vehicle suspension using MATLAB simulink. Int J Veh Struct Syst 14. https://doi.org/10.4273/ijvss.14.3.04

Palli S, Koona R (2015) Analyses of dynamic response of a railway bogie. Int J Veh Noise Vib 11:103. https://doi.org/10.1504/IJVNV.2015.070015

Palli S, Koona R, Sharma SK, Sharma RC (2018) A review on dynamic analysis of rail vehicle coach. Int J Vehicle Structures Syst 10(3):204–211. https://doi.org/10.4273/ijvss.10.3.10

PanduRanga Vital T, Lakshmi BG, Swapna Rekha H, DhanaLakshmi M (2019) Student performance analysis with using statistical and cluster studies. In Soft computing in data analytics. Advances in intelligent systems and computing, 758. Springer, Singapore, pp 743–757. https://doi.org/10.1007/978-981-13-0514-6_71

Pati S, Maity S, Dash A, Jema S, Mohapatra S, Das S, Samantaray DP (2020) Biocompatible PHB production from Bacillus species under submerged and solid-state fermentation and extraction through different downstream processing. Curr Microbiol 77:1203–1209. https://doi.org/10.1007/s00284-020-01922-7

Pati S, Mohanty MK, Mohapatra S, Samantaray D (2022) Bioethanol production by *Enterobacter hormaechei* through carbon monoxide-rich syngas fermentation. Int J Energy Res 46:20096–20106. https://doi.org/10.1002/er.8678

Pattnaik S, Dash D, Mohapatra S, Pattnaik M, Marandi AK, Das S, Samantaray DP (2020) Corrigendum to "Improvement of rice plant productivity by native Cr(VI) reducing and plant growth promoting soil bacteria Enterobacter cloacae" [Chemosphere 240 (2020) 124895]. Chemosphere 238:125011. https://doi.org/10.1016/j.chemosphere.2019.125011

Pattnaik S, Mohapatra S, Pati S, Dash D, Devadarshini D, Tanaya K, Mishra BB, Samantaray D (2022) Microbial bioremediation of Cr(VI)-contaminated soil for sustainable agriculture. In: Microbial biodegradation and bioremediation, pp 395–407. Elsevier (2022). https://doi.org/10.1016/B978-0-323-85455-9.00008-4

Pundir S, Upadhyay S, Kumar N, Chandra Joshi N, Priya R, Ahmad Mir R, Hossain I, Pandey OP (2023a) Interconnected CuO nanoplates as a highly efficient electrocatalyst for oxygen evolution reaction. Mater Lett 336:133921. https://doi.org/10.1016/j.matlet.2023.133921

Pundir S, Upadhyay S, Priya R, Kumar N, Chetana S, Hossain I, Joshi NC, Pandey OP (2023b) Synthesis of 1D β-MnO2 for high-performance supercapacitor application. J Solid State Electrochem 27:531–538. https://doi.org/10.1007/s10008-022-05347-z

Raju Rallabandi S, Pilla DP, Dowluru S, Palli S, Sharma N, Sharma SK, Sharma RC (2024) Critical evaluation of epoxy-hemp-pineapple-palm fiber composites using hybrid AHM-TOPSIS technique for sustainable structural applications. J Chin Inst Eng 1–12

Rallabandi SR, Srinivas L, Palli S, Sharma RC, Sharma N, Sharma A, Sharma SK (2024) A contrastive characterization of pure Mg and AZ91D alloy based on the testing of mechanical, corrosion, wear, and erosion properties. Eng Res Express 6(1):015017

Samantaray D, Mohapatra S, Mishra BB (2014) Microbial bioremediation of industrial effluents. In: Microbial biodegradation and bioremediation, pp 325–339. Elsevier. https://doi.org/10.1016/B978-0-12-800021-2.00014-5

Sarkar MS, Segu H, Bhaskar JV, Jakher R, Mohapatra S, Shalini K, Shivaji S, Reddy PA (2018) Ecological preferences of large carnivores in remote, high-altitude protected areas: insights from Buxa Tiger Reserve, India. Oryx 52:66–77. https://doi.org/10.1017/S0030605317000060

Satyanarayana VSV, Sharma RC, Sateesh B, Gopala Rao LVV, Mohan Rao N, Palli S (2022) Optimum response of a nonlinear passive vehicle suspension system under random road excitations. SAE Int J Commer Veh 16:02-16-01-0004. https://doi.org/10.4271/02-16-01-0004

Sharma SK (2019) Multibody analysis of longitudinal train dynamics on the passenger ride performance due to brake application. Proceedings of the Institution of Mechanical Engineers. Part K: J Multi-Body Dyn 233(2):266–279. https://doi.org/10.1177/1464419318788775

Sharma SK, Kumar A (2017) Ride performance of a high speed rail vehicle using controlled semi active suspension system. Smart Mater Struct 26:055026. https://doi.org/10.1088/1361-665X/aa68f7

Sharma RC, Sharma SK (2018) Sensitivity analysis of three-wheel vehicle's suspension parameters influencing ride behavior. Noise Vibr Worldwide 49(7–8):272–280. https://doi.org/10.1177/0957456518796846

Sharma SK, Sharma RC (2021) Multi-objective design optimization of locomotive nose. SAE Tech Paper 10. https://doi.org/10.4271/2021-01-5053

Sharma RC, Sharma SK, Palli S (2018) Rail vehicle modelling and simulation using lagrangian method. Int J Vehicle Struct Syst 10(3):188–194. https://doi.org/10.4273/ijvss.10.3.07

Sharma SK, Saini U, Kumar A (2019) Semi-active control to reduce lateral vibration of passenger rail vehicle using disturbance rejection and continuous state damper controllers. J Vibr Eng Technol 7(2):117–129. https://doi.org/10.1007/s42417-019-00088-2

Sharma SK, Lee J (2020b) Design and development of smart semi active suspension for nonlinear rail vehicle vibration reduction. Int J Struct Stab Dyn 20:2050120. https://doi.org/10.1142/S0219455420501205

Sharma SK, Lee J (2021) Crashworthiness analysis for structural stability and dynamics. Int J Struct Stab Dyn 21:2150039. https://doi.org/10.1142/S0219455421500395

Sharma RC, Palli S (2016) Analysis of creep force and its sensitivity on stability and vertical-lateral ride for railway vehicle. Int J Veh Noise Vib 12:60. https://doi.org/10.1504/IJVNV.2016.077474
Sharma RC, Sharma SK (2022) Ride analysis of road surface-three-wheeled vehicle-human subject interactions subjected to random excitation. SAE Int J Commer Vehicles 15(3):02-15-03–0017. https://doi.org/10.4271/02-15-03-0017
Sharma RC, Palli S, Sharma SK, Roy M (2017) Modernization of railway track with composite sleepers. Int J Vehicle Struct Syst 9(5):321–329. https://doi.org/10.4273/ijvss.9.5.10
Sharma RC, Sharma SK, Sharma N, Sharma S (2020d) Analysis of ride and stability of an ICF railway coach. Int J Veh Noise Vib 16:127. https://doi.org/10.1504/IJVNV.2020.117820
Sharma RC, Sharma S, Sharma SK, Sharma N (2020e) Analysis of generalized force and its influence on ride and stability of railway vehicle. Noise Vib Worldw 51:95–109. https://doi.org/10.1177/0957456520923125
Sharma SK, Sharma RC, Lee J (2021c) Effect of rail vehicle-track coupled dynamics on fatigue failure of coil spring in a suspension system. Appl Sci 11:2650. https://doi.org/10.3390/app11062650
Sharma SK, Mohapatra S, Sharma RC, Alturjman S, Altrjman C, Mostarda L, Stephan T (2022a) Retrofitting existing buildings to improve energy performance. Sustainability 14:666. https://doi.org/10.3390/su14020666
Sharma SK, Sharma RC, Lee J, Jang H-L (2022b) Numerical and experimental analysis of DVA on the flexible-rigid rail vehicle carbody resonant vibration. Sensors 22:1922. https://doi.org/10.3390/s22051922
Sharma SK, Lee J, Jang H-L (2022c) Mathematical modeling and simulation of suspended equipment impact on car body modes. Machines 10:192. https://doi.org/10.3390/machines10030192
Sharma SK, Sharma RC, Lee J (2022d) In situ and experimental analysis of longitudinal load on carbody fatigue life using nonlinear damage accumulation. Int J Damage Mech 31:605–622. https://doi.org/10.1177/10567895211046043
Sharma RC, Palli S, Sharma SK (2023b) Ride analysis of railway vehicle considering rigidity and flexibility of the carbody. J Chinese Inst Eng 46:355–366. https://doi.org/10.1080/02533839.2023.2194918
Sharma RC, Gopala Rao LVV, Sharma SK, Palli S, Satyanarayana VSV (2022a) Analysis of lateral stability and ride of an indian railway constrained dual-axle bogie frame. SAE Int J Commer Vehicles 16(2):02-16-02–0014. https://doi.org/10.4271/02-16-02-0014
Sharma RC, Palli S, Gopala Rao LVV, Duppala A, Sharma SK (2022b) Four-Wheel Vehicle Response under Bump, Pothole, Harmonic, and Random Excitations Using Bond Graph/Simulink Technique. SAE Int J Commer Vehicles 16(2):02-16-02–0008. https://doi.org/10.4271/02-16-02-0008
Sharma RC, Palli S, Jha AK, Bhardawaj S, Sharma SK (2022c) Vibration and ride comfort analysis of railway vehicle system subjected to deterministic inputs. Resmilitaris 12(5):1345–1355
Sharma RC, Sharma SK (2022c) Ride analysis of road surface-three-wheeled vehicle-human subject interactions subjected to random excitation. SAE Int J Commer Vehicles 15(3):02-15-03–0017. https://doi.org/10.4271/02-15-03-0017
Sharma SK, Kumar A (2017) Impact of electric locomotive traction of the passenger vehicle Ride quality in longitudinal train dynamics in the context of Indian railways. Mech Ind 18(2):222. https://doi.org/10.1051/meca/2016047
Sharma SK, Kumar A (2018) Ride comfort of a higher speed rail vehicle using a magnetorheological suspension system. Proc Inst Mech Eng Part K. J Multi-body Dyn 232:32–48. https://doi.org/10.1177/1464419317706873
Sharma SK, Kumar A (2018a) Disturbance rejection and force-tracking controller of nonlinear lateral vibrations in passenger rail vehicle using magnetorheological fluid damper. J Intel Mater Syst Struct 29(2):279–297. https://doi.org/10.1177/1045389X17721051

Sharma SK, Kumar A (2018b) Impact of longitudinal train dynamics on train operations: a simulation-based study. J Vibr Eng Technol 6(3):197–203. https://doi.org/10.1007/s42417-018-0033-4

Sharma SK, Sharma RC (2018a) An investigation of a locomotive structural crashworthiness using finite element simulation. SAE Int J Commer Vehicles11(4):235–244. https://doi.org/10.4271/02-11-04-0019

Sharma SK, Sharma RC (2018b) Simulation of quarter-car model with magnetorheological dampers for ride quality improvement. Int J Vehicle Struct Syst 10(3):169–173. https://doi.org/10.4273/ijvss.10.3.03

Sharma SK, Lee J (2020a) Finite element analysis of a fishplate rail joint in extreme environment condition. Int J Veh Struct Syst 12. https://doi.org/10.4273/ijvss.12.5.03

Sharma SK, Sharma RC (2019) Pothole detection and warning system for Indian roads. In Advances in interdisciplinary engineering. Lecture notes in mechanical engineering. Springer, Singapore, pp 511–519. https://doi.org/10.1007/978-981-13-6577-5_48

Sharma S, Sharma RC, Sharma SK, Sharma N, Palli S, Bhardawaj S (2020) Vibration isolation of the quarter car model of road vehicle system using dynamic vibration absorber. Int J Veh Struct Syst 12. https://doi.org/10.4273/ijvss.12.5.05

Sharma SK, Sharma RC, Kumar A, Palli S (2015) Challenges in rail vehicle-track modeling and simulation. Int J Veh Struct Syst 7(1):1–9. https://doi.org/10.4273/ijvss.7.1.01

Sharma RC, Sharma SK, Palli S (2020b) Linear and non-linear stability analysis of a constrained railway wheelaxle. Int J Veh Struct Syst 12. https://doi.org/10.4273/ijvss.12.2.04

Sharma SK, Sharma RC, Sharma N (2020c) Combined multi-body-system and finite element analysis of a rail locomotive crashworthiness. Int J Veh Struct Syst 12. https://doi.org/10.4273/ijvss.12.4.15

Sharma RC, Palli S, Sharma N, Sharma SK (2021) Ride behaviour of a four-wheel vehicle using H infinity semi-active suspension control under deterministic and random inputs. Int J Veh Struct Syst 13. https://doi.org/10.4273/ijvss.13.2.18

Sharma RC, Sharma S, Sharma N, Sharma SK (2021) Linear and nonlinear analysis of ride and stability of a three-wheeled vehicle subjected to random and bump inputs using bond graph and simulink methodology. SAE Int J Commer Veh 14:02-15-01-0001. https://doi.org/10.4271/02-15-01-0001

Sharma RC, Sharma S, Sharma SK, Sharma N, Singh G (2021d) Analysis of bio-dynamic model of seated human subject and optimization of the passenger ride comfort for three-wheel vehicle using random search technique. Proc Inst Mech Eng Part K J Multi-body Dyn 235:106–121. https://doi.org/10.1177/1464419320983711

Sharma RC, Palli S, Bhardawaj S (2023) A review of wheel-rail dynamic behavior on curved tracks. Int J Veh Struct Syst 15. https://doi.org/10.4273/ijvss.15.1.04

Sharma RC, Bhardawaj S, Sharma SK, Bahl S (2023a) Ride behaviour of three-wheel vehicle using semi-active suspension using skyhook control strategy. Int J Veh Struct Syst 15(1). https://doi.org/10.4273/ijvss.15.1.05

Sharma SK, Sharma RC, Choi Y, Lee J (2023b) Experimental and mathematical study of flexible–rigid rail vehicle riding comfort and safety. Appl Sci 13(9):5252

Sharma SK, Sharma RC, Choi Y, Lee J (2023c) Modelling and dynamic analysis of adaptive neuro-fuzzy inference system-based intelligent control suspension system for passenger rail vehicles using magnetorheological damper for improving ride index. Sustainability, 15(16): 12529

Sharma SK, Sharma RC, Lavania S, Palli S, Avesh M (2023d) Industry-driven approach for anfis-based intelligent control suspension system with mr damper for enhanced ride quality in passenger rail vehicles for technological investigations. Nat Acad Sci Let 1–4

Sharma SK, Sharma RC, Upadhyay RK, Lee J (2023e) Coupled torsional–transverse vibration reduction in marine propulsion dynamics with novel approach using magnetorheological damping and adaptive control system. J Vibrat Eng Technol 1–11

Sharma SK, Upadhyay RK, Kumar V, Valera H (2023f) Introduction to sustainable transportation system. In: Transportation Energy and Dynamics, pp. 3–6, Springer

Sharma RC, Palli S, Dharmana L, Sharma SK, Sharma A, Kharola A, Sharma N (2024b) Simulation of transient dynamic response of a locomotive considering its flexibility and rigidity. Noise Vibr Worldwide 55(1–2):33–47

Sharma RC, Palli S, Duppala A, Bhardawaj S, Sharma SK, Sharma N, Sharma A (2024c) Rigid body analysis of ride behaviour of roadway vehicle using lagrangian dynamics. Int J Veh Struct Syst (IJVSS) 16(1)

Sharma RC, Sharma N, Palli S, Sharma SK, Joshi S, Rallabandi SR, Sharma A (2024d) Dynamic behavior of three-wheel vehicle with inerter-based dynamic absorber suspension system. Noise Vibrat Worldwide 09574565241243398

Sharma SK, Kumar N, Avesh M, Sharma RC, Siddiqui MIH, Lee J (2024e) Navigating tranquillity with H∞ controller to mitigate ship propeller shaft vibration. J Vibrat Eng Technol 1–11

Sharma SK, Sharma RC, Lee J (2024f) Modelling and simulation of MR damper characterization and uncertainty assessment in marine engine vibration isolation with FLPID control. Int J Model Simul 1–11

Sharma SK, Sharma RC, Lee J (2024g) Propelling precision of longitudinal vibration mitigation in ship propeller shafts through advanced nonlinear intelligent semi-active control leveraging adaptive neuro-fuzzy inference system with linear quadratic regulator. J Vibrat Control 10775463241244836

Sharma SK, Sharma RC, Upadhyay RK, Lee J (2024h) Enhanced ship engine vibration reduction using magnetorheological dampers and Adaptive Neuro-Fuzzy Control System. Ships Offshore Struct 1–17

Sinha AK, Sengupta A, Gandhi H, Bansal P, Agarwal KM, Sharma SK, Sharma RC, Sharma SK (2019) Performance enhancement of an all-terrain vehicle by optimizing steering, powertrain and brakes. In Advances in engineering design. Lecture notes in mechanical engineering. Springer, Singapore, pp 207–215. https://doi.org/10.1007/978-981-13-6469-3_19

Swapna Rekha H, Nayak J, Behera HS (2020) Pi-sigma neural network: survey of a decade progress. In: Computational intelligence in pattern recognition. Advances in intelligent systems and computing, 1120. Springer, Singapore, pp 429–441. https://doi.org/10.1007/978-981-15-2449-3_37

Upadhyay RK, Sharma SK, Kumar V, Valera H (2023a) Introduction to transportation systems technology and integrated management. In: Transportation Systems Technology and Integrated Management, pp. 1–4, Springer. https://doi.org/10.1007/978-981-99-1517-0_1

Upadhyay RK, Sharma SK, Kumar V, Valera H (2023b) Transportation systems technology and integrated management. In: Upadhyay RK, Sharma SK, Kumar V, Valera H (eds) Springer Nature, Singapore, https://doi.org/10.1007/978-981-99-1517-0

Upadhyay RK, Sharma SK, Kumar V (Eds) (2024a) Intelligent transportation system and advanced technology. Springer Nature Singapore. https://doi.org/10.1007/978-981-97-0515-3

Upadhyay RK, Sharma SK, Kumar V (2024b) Introduction to intelligent transportation system and advanced technology. In: Intelligent transportation system and advanced technology. Energy, Environment, and Sustainability. pp 3–6. Springer, Singapore. https://doi.org/10.1007/978-981-97-0515-3_1

Vishwakarma PN, Mishra P, Sharma SK (2022a) Characterization of a magnetorheological fluid damper a review. Materials Today: Proceedings 56:2988–2994. https://doi.org/10.1016/j.matpr.2021.11.143

Vishwakarma PN, Mishra P, Sharma SK (2022b) Formulation of semi-active suspension system and controls in rail vehicle. SSRN Electron J. https://doi.org/10.2139/ssrn.4159616

Wu Q, Cole C, Spiryagin M, Chang C, Wei W, Ursulyak L, Shvets A, Murtaza MA, Mirza IM, Zhelieznov K, Mohammadi S, Serajian H, Schick B, Berg M, Sharma RC, Aboubakr, A., Sharma SK, Melzi S, Di Gialleonardo E, Cantone L (2023) Freight train air brake models. Int J Rail Transport 11(1):1–49. https://doi.org/10.1080/23248378.2021.2006808

Chapter 11
Pyrolysis of Plastic Wastes Towards Achieving a Circular Economy: An Advanced Chemistry and Technical Approach

Abiodun Oluwatosin Adeoye, Jibrin Muhammad Yelwa, Naziru Imam, Rukayat Oluwatobiloba Quadri, Olayide Samuel Lawal, Dosu Malomo, Samson Abiodun Aasa, Augustine Eyikwuojo Onakpa, Ruth Kooaka Dorgbaa, Busuyi Patrick Omoniyi, Nuhu Ahmad, and Husaina Anchau Garba

A. O. Adeoye (✉) · N. Imam · R. O. Quadri · O. S. Lawal · D. Malomo
Department of Chemistry, Federal University Oye-Ekiti, Oye-Ekiti, Ekiti State, Nigeria
e-mail: bioken2017@gmail.com

J. M. Yelwa
Department of Scientific and Industrial Research, National Research Institute for Chemical Technology, Zaria, Kaduna-State, Nigeria

N. Imam
Analytical Chemistry Division, Department of Chemistry, University of Ibadan, Ibadan, Nigeria

S. A. Aasa
Department of Mechanical Engineering, Olabisi Onabanjo University, Ibogun Campus, Nigeria

A. E. Onakpa
Green Chemistry and Processes for Biomass (Green CAP), Toulouse INP-ENSIACET, Toulouse, France

R. K. Dorgbaa
Department of Metallurgical and Materials Engineering, University of Nigeria, Nsukka, Nigeria

B. P. Omoniyi
WATTCCON Chemical & Equipment Marketing Company Nig. LTD Commercial, Kaduna, Nigeria

N. Ahmad
Department of Civil Engineering, Bayero University, Kano, Kano-State, Nigeria

H. A. Garba
National Research Institute for Chemical Technology, Zaria, Kaduna State, Nigeria

S. K. Sharma et al. (eds.), *Dynamics of Transportation Ecosystem, Modeling, and Control*, Energy, Environment, and Sustainability,
https://doi.org/10.1007/978-981-97-0437-8_11

11.1 Introduction

The complexity of plastic waste is addressed in this chapter, and it is emphasised how important it is for waste management plans to consider the varied properties of plastics. In more precise terminology, the term "plastic" encompasses a wide range of materials, each with distinct chemical compositions and characteristic properties that render them ideal for uses. Given the significant impact that variations in composition can have on the pyrolysis behaviour of plastics, a comprehensive methodology is necessary for the purpose of designing and optimising the process. Moreover, the presence of plastic waste in various forms poses a significant barrier, necessitating the exploration of innovative solutions that frequently rely on a comprehensive comprehension of pyrolysis mechanisms.

To have a clear understanding of the complexities of pyrolysis processes and to characterise the products that these reactions produce, advanced instrumental analytical methods are an absolute need. Researchers get a clearer understanding of plastic waste pyrolysis reaction processes, product distribution, and material qualities via the use of spectroscopy, chromatography, and microscopy. Researchers can establish a more in-depth knowledge of pyrolysis processes by making use of these methodologies, which subsequently enables them to adjust process parameters for improved efficiency and selectivity (Ajaj et al. 2022; Abu Jadayil 2011; Abu Jadayil et al. 2017). In addition to this, the study highlights the significance of a circular economy as a suitable principle for the efficient, robust and sustainable management of plastic waste combined with the use of resources. Pyrolysis, which involves thermal conversion of waste plastic into useful materials, is consistent with the goals of reducing the amount of waste and increasing the effectiveness with which resources are used. The idea of circular economy goes beyond just getting rid of plastic waste. It's about turning this waste into valuable resources that can spur the economy's growth without ecological toxicity. In this context, this research delves into the potential applications of products derived from plastic waste pyrolysis. It explores how these products can play a role in various industries, such as petrochemicals, energy, and manufacturing.

The focus of this study is to shed light on how pyrolysis can be a game-changer in managing plastic waste. This section offers an in-depth look at the role of pyrolysis in paving the way for a circular economy. By weaving together principles from chemistry, engineering, economics, and environmental studies, we get a holistic view, thanks to a synthesis approach. This article lays out a blueprint for researchers, scholars, businesspeople, policymakers, and entrepreneurs to collaborate in steering our society towards sustainable practices. Given the pressing need to handle plastic waste and the environmental fallout of its production, the potential of pyrolysis is noteworthy. With the help of cutting-edge chemical processes and modern tech, pyrolysis can turn a global problem into a chance for meaningful change.

As the title of this publication, "Pyrolysis of Plastic Waste as a Route to a Circular Economy: An Advanced Chemistry Approach," implies, the main aim of this study is to determine the complex relationship that exists between the principles of advanced chemistry and the process of pyrolyzing plastic waste. The scientific foundations of pyrolysis are investigated in depth during the course of the article by means of an in-depth examination that focuses on the roles that temperature, reaction atmosphere, and catalysts play in determining the reaction routes and product yields. This basic knowledge is the basis for optimising pyrolysis processes to achieve desired results, whether that be maximising fuel production or adapting products for particular industrial purposes. In other words, this understanding is the foundation for optimising pyrolysis processes to achieve desired outcomes.

This book presents an in-depth investigation of the economic viability of pyrolysis processes, taking into consideration issues such as the prices of feedstock, the values of the products, and the efficiency of the processes. In addition, the effects of pyrolysis on the natural environment are analysed, taking into account aspects such as the amount of energy used, the emissions produced, and the life cycle evaluations. A thorough knowledge of the potential advantages and constraints connected with pyrolysis-based circular economy solutions is made possible by this exhaustive examination. The legislative, regulatory, and industrial application contexts are discussed in the paper alongside the scientific and technical elements of the topic. Within the context of the waste management industry, the regulatory frameworks and policy incentives that are in place play a critical part in determining the course that pyrolysis technologies will take. By creating a climate that is amenable to the introduction of cutting-edge solutions, governments may play a significant role in hastening the shift towards a circular economy. This can be accomplished via the adoption of pro-innovation laws. The book emphasises the need of stakeholder engagement, which should include governments, companies, and research institutions, in order to assist the incorporation of pyrolysis technologies into preexisting waste management systems.

11.2 Plastic Waste Towards Circular Economy

Plastic is fundamentally a substance composed of complex organic molecules of large molecular weight. The term "plastic" derives from the Greek word "Plastikos," which translates to "to mould." These molecules are predominantly composed of hydrocarbons that are obtained from fossil fuel sources. The monomers, which are fundamental components, combine to form long carbon chains referred to as polymers. They can generally be classified into two main categories depending on their physical properties: thermoplastic and thermosetting. Thermoplastic materials, such as polyvinyl chloride (PVC), nylon, and polyethylene, exhibit the property of becoming easily deformed when subjected to heat. The aforementioned materials comprise linear polymers as well as a combination of linear and cross-linked polymers. On the other hand, thermosetting polymers, e.g., bakelite and melamine, exhibit a permanent loss of their capacity to undergo softening after the moulding process. The bulk of plastic structures primarily comprise densely interconnected polymers, rendering bakelite advantageous for the manufacturing of electrical switches. On the other hand, melamine finds widespread application in the production of floor tiles.

11.2.1 *Plastic Wastes as Pollutant*

Plastic Waste

The increase use of plastics in everyday activities has engendered much discourse over the effective management of the resultant waste. In contemporary times, municipal solid waste (MSW) encompasses a substantial amount of plastic waste. Upon analysis, it is evident that approximately 10% of the composition consists of polyethylene terephthalate (PET), 19% comprises high-density polyethylene (HDPE), 6% is attributed to polyvinyl chloride (PVC), 23% is constituted by low-density polyethylene (LDPE), 14% is composed of polypropylene (PP), and 9% is comprised of polystyrene (PS) (Bodzay and Bánhegyi 2016). Plastic waste is undeniably a global problem that's getting worse. The amount of plastic we produce has skyrocketed, reaching nearly 300 million tons in 2013 and going up to 367 million tons by 2020 (Wan 2015). Even more worrying is that 8 million tons of this ends up polluting our oceans every year. Shockingly, a meagre 9% of all the plastic we've ever made has been recycled. Most of it, i.e., 79%, to be precise, is either in landfills or littering the natural world. (Garside 2020). While plastic isn't scientifically categorised as hazardous waste, it's clear that it's causing lasting harm to our environment because it has very slow natural decomposition rate. Is it any wonder that we're scrambling to find better ways to dispose of or ideally, reuse, all this plastic? After all, it's a ticking time bomb for the planet.

Plastic waste has emerged as a formidable environmental challenge of the modern era, casting a shadow over the concept of sustainability and threatening ecosystems, human health, and the delicate balance of the planet's ecosystems. The majority of this waste, around 79%, ends up in landfills or pollutes our natural environment, including our oceans. This not only poses significant environmental risks but also has been linked to various health concerns for those living near landfill sites. The ubiquity of plastics in our daily lives, driven by their convenience, durability, and versatility, has led to an exponential increase in their production and consumption. However, the inherent properties that make plastics so advantageous—their resistance to degradation and persistence in the environment—have also become their Achilles' heel, culminating in an urgent need to rethink our approach and methodology to plastic waste management in our societies. The commonly advocated solutions have been to reduce, reuse, and recycle. Reusing plastics isn't always feasible because the material degrades and can contain additives that limit its reuse. A critical assessment of our waste showed that municipal solid waste is largely made up of plastic materials like polyethylene terephthalate, high-density polyethylene, and polyvinyl chloride, among others. So far, the main methods to dispose of this waste have been landfills, biological degradation, and thermal or chemical recycling. But these aren't solving the problem fast enough.

11.2.2 Circular Economy

In simple terms, circular economy negates the approach of linear economy in the sense that it is a better and more modern approach to sustaining a modern economy. This economic strategy has the tendency to foster environmental consciousness among citizens, leading to the protection of the environment through a reduction in the indiscriminate disposal of plastic waste. This, in turn, helps mitigate the potential for plastic waste to serve as a breeding ground for disease-causing bacteria such as those responsible for malaria and typhoid. The circular economy is characterised by greater profitability and sustainability in comparison to the linear economy. Its primary objectives include promoting sustainable economic growth, enhancing competitiveness, and fostering job creation. It is essential to acknowledge that recycling plays a crucial role in maintaining the circulation of scarce recyclable materials, making it a significant component of the circular economy.

The concept of the circular economy is characterised by the implementation of the 3R framework, which encompasses the reduction, reuse, and recycling of resources. This approach aims to achieve the dual objectives of energy generation and waste reduction. The phenomenon under discussion is sometimes referred to as a paradigm shift, when traditional methods of doing economic business are replaced by a closed-loop approach in order to attain sustainable business practises. There exist notable instances of the accumulation of substantial quantities of waste, volatility in the pricing of distillation products derived from crude oil, inadequate implementation of sustainable business practises by both production and consuming products, insufficient levels of recycling, and deficiencies in public procurement oversight. The objective of achieving the conversion of waste into valuable resources is a crucial aim within the framework of a circular economy. This economic model emphasises the reintroduction of recyclable or reusable materials back into the economic system, so enabling them to function as a primary source of new raw materials.

In this context, the concept of a circular economy, which seeks to minimize plastic waste generation and maximize its utilization, has gained researchers attention as a viable solution. The journey towards a circular economy necessitates a paradigm shift in the way we perceive and handle plastic waste. Traditional linear models of "take-make-dispose" have been shown to be unsustainable, which has led to a situation in which more than 380 million tonnes of plastic are created worldwide each year, of which a large amount eventually becomes wastes. This debris, whether as a result of poor recycling or incorrect disposal, often makes its way into natural ecosystems, where it wreaks havoc on marine life, affects the quality of the soil, and adds to the worrisome occurrence of microplastic contamination. In addition, the extraction of natural fossil resources for the production of domestically or industrially useful plastic exacerbates the environmental issues since it is an energy-intensive process, it adds extensively to the emission of greenhouse gases (GHG), and it results in the depletion of natural resources. It is vital to take on these difficulties with a strategy that is both broad and inventive.

The limited availability of resources for plastic production has prompted the pursuit of certain objectives outlined in the United Nations' Sustainable Development Goal 12, which focuses on Sustainable Consumption and Production. In order to address this issue, it is imperative to increase recycling efforts and embrace a sustainable production system. These measures will have a profound impact on various aspects of our lives, including our work practises, purchasing habits, thought processes, and overall lifestyle. In recent years, a select few African countries have implemented various green initiatives, namely in the areas of green technology, green business promotion, and the encouragement of green consumerism among their citizens. In order to establish a circular economy, it is imperative for the government to develop a legally binding framework, conduct extensive awareness campaigns, implement monitoring and reporting mechanisms, involve stakeholders, enforce extended producer responsibility, establish recycling infrastructure, enhance capacity building efforts, and foster market engagement, among various other essential considerations.

The promotion or implementation of legislation is a crucial factor in aiding the achievement of sustainable development. To facilitate the progress of pyrolysis-based plastic waste recycling, it is crucial for governmental entities and policymakers to develop and support legislation and initiatives that promote public knowledge and involvement in environmental issues. Furthermore, it is imperative to implement strategies that provide incentives for the retrieval and collection of recyclable plastic wastes. This approach would effectively contribute to the promotion of sustainable economic development. In light of the recent fall in crude oil prices per barrel and currency rates, nations have been pushed to investigate sustainable development policies aimed at cultivating an inclusive economy that facilitates job creation, poverty reduction, and hunger alleviation.

11.2.3 Approach of Plastic Waste Management

Although plastics haven't been around for a very long time, recycling methods like pyrolysis offer promise. Projects that focus on turning plastics back into their basic components or liquid fuel are on the rise, and there's also emerging research on co-pyrolyzing biomass and plastic waste together. The technology of pyrolysis, which is a sophisticated chemistry-based process, has gained attention as a possible solution to solve the dilemma of plastic waste and to comply with the ideals of a circular economy. This is because pyrolysis can convert waste plastic into energy. The process of pyrolysis entails exposing plastic materials to high temperatures in the absence of oxygen. It's a process that breaks down long-chain polymers into simpler molecules by applying heat and pressure. This causes the plastic materials to undergo thermal breakdown, which results in the production of a variety of useful products which exist in the three states of matter, such as liquid fuels, gases, and solid chars (Chen et al. 2022; Ouda et al. 2016). What makes pyrolysis appealing is that it can produce a significant amount of liquid oil, sometimes up to 80 wt%, and it works at about

500 °C. Although the oil produced (paraffins, olefins, isoalkanes, and aromatics) is not immediately suitable for direct use as fuel in diesel engines because it contains unsaturated compounds but these can be upgraded into saturated compounds through a process called hydrogenation, making the oil more compatible with existing diesel standards (Mibei et al. 2023). This technique has a dual benefit: first, it prevents waste plastic from being buried in landfills or burned, landfills have health hazards, causing air and water pollution, living near landfills also increases the likelihood of problems with health issues and low birth rate.; second, it creates usable goods that can be incorporated into current industrial processes, which lessens the need for virgin fossil resources (Shareefdeen et al. 2020). The acceptance of pyrolysis as a better plastic waste management as compared to landfills and other existing methods will be based firstly on economically viability and also on environmental factors.

11.3 Plastic Wastes Management

11.3.1 Plastic Collection and Sorting

The process of gathering and classifying various types of plastic waste materials in order to recycle or properly dispose of them in an environmentally responsible way is known as plastic waste collection and sorting. Consumers typically start the process by separating plastic waste from other waste streams, such organic or metal waste, to start the process (Mihai et al. 2021). Due to the potential harmful effects of mixed wastes on the recycling process and the ensuing rise in associated costs, the practise of segregating waste is essential (Johnson 2019). The plastic waste is collected and then sent to material recovery facilities (MRFs), where it goes through extra sorting. Plastic identification codes (PICs), which identify different types of plastic resin, are frequently used at these facilities to categorise plastics (Lange 2021). In order to improve the accuracy and efficiency of the sorting process, there has been a rising use of automated sorting technologies, such as infrared sensors and conveyor belts, in recent years (Gundupalli et al. 2017).

Following sorting, recyclable plastics are subjected to extra processing, during which they are either made into new plastic products or used as raw materials in other industries. Utilising a lifecycle plan not only makes it easier to keep less wastes out of landfills, but it also helps to protect natural resources and reduce greenhouse gas emissions (Davis 2020).

The sorting of plastic waste is an essential step in the waste management process, particularly in relation to the recycling of plastics. Plastics are categorised into distinct classes according to their basic elements and composition. The implementation of precise categorization ensures the effective recycling or proper disposal of plastic waste, as the diverse forms of plastic often necessitate distinct processing techniques (Hopewell et al. 2009).

11.3.2 Methods of Plastic Waste Sorting

The process of thermal-adhesion sorting capitalises on the variation in softening temperatures shown by different materials (Lim et al. 2003). During its softening temperature, a plastic exhibits adhesive property towards surfaces, even other polymers that do not undergo softening. The observed variation in response can be attributed to the systematic manipulation of surface temperature, which enables the successive segregation of polymers. The thermal-adhesion sorting technique has the benefit of effectively segregating a specific kind of plastic from mixed plastic waste. However, it is limited in its ability to separate thermosetting polymers.

The float-sink sorting method capitalises on the variations in the specific gravity of different types of plastics (Bauer et al. 2018). Plastic waste is introduced into a liquid solution, wherein its specific gravity is subsequently manipulated. The float-sink sorting method is capable of separating plastic from waste plastic mixtures. However, it is not feasible to effectively sort plastics that have a specific gravity difference of less than 0.1.

Dry zig-zag separation method likewise capitalises on the variations in plastics' density (Howell 1992), but it employs an upward airflow to achieve the separation. A heterogeneous stream is introduced into a sorting chamber where the air is ascending. Those with a mass below a certain threshold, denoted as θ, are propelled upward and depart through the top, whereas those with a mass beyond θ descend and escape through the bottom. The dry zig-zag sorting method is capable of segregating plastic materials from combinations including both plastic and wastes. However, it is limited in its ability to separate polymers that have a specific gravity difference of less than 0.1.

Electrostatic sorting is a technique that capitalises on the disparity in the work function of conducting polymers, as demonstrated by Tilmatine et al. (2009). Initially, plastic flakes acquire an electric charge when coming into touch with a tribo-charging apparatus. Then, the discrete particles are suspended inside an electrostatic field and then segregated based on the relative magnitude of their electric charge. One of the notable advantages of electrostatic sorting is its ability to effectively sort materials at a very low operating cost, without the need for water or additives. However, it is important to note that this method is limited in its capacity to select just one kind of plastic from a mixture of two polymers.

The process of froth flotation sorting capitalises on the disparity in hydrophobicity among various polymers (Fraunholcz 2004). Initially, the plastic waste is introduced into the froth flotation machine, wherein it is immersed in a liquid solution including sodium sulfonate dissolved in water. The rate at which sinking occurs is regulated by the degree of hydrophobicity. Froth flotation is a technique employed for the purpose of separating plastics based on their specific gravity disparity, provided that the difference is below 0.1. However, in order to effectively sort the plastic, it is necessary to subject the surface of the waste plastic to appropriate treatment. The plastic waste collection issues are public awareness and infrastructural limitations,

Manual labour costs and potential for human error, Technological costs and the need for skilled operators.

11.3.3 Recycling System

The plastic that has undergone sorting is thereafter subjected to recycling processes, with the primary objective of producing utilitarian goods or harnessing energy. This research examines four distinct recycling methods. There exist four distinct methodologies that can be employed to address the issue of plastic waste: the production of pyrolysis oil via chemical recycling, the generation of waste plastic fuel through thermal recycling, the recovery of energy through incineration, and the reduction of plastic production by means such as material recycling.

To reduce plastic production, we often turn to recycling used polymers, keeping their chemical properties intact. This is made possible through various steps like separation, sorting, packing, washing, crushing, and mixing (Khair and Matsana 2021). Different kinds of plastic waste are first separated, then thermally melt, an essential part of the degradation cycle. Once melted, this plastic is then repurposed into a more basic, yet useful form.

A critical analysis of plastic waste conversion, pyrolysis oil emerges as a green fuel option. Detailed in study (Gopinath and Devan 2020), the pyrolysis process involves tools such as a, which entails the utilisation of a pyrolysis reactor, an oil separator, and a condenser. The reactor function is to thermally degrade varied plastic wastes at high temperature. The first stage of the procedure entails the condensation of oil within the condenser, which is then separated from the condensate in the oil separator.

Study (Shaha et al. 2020) discussed how Refuse-derived Fuel (RPF) production is about repurposing plastic waste. The procedure entails the fragmentation of various types of plastic wastes, followed by the conversion of the degraded plastic into pallets that may be utilised as a sustainable fuel source.

The incineration process includes a closed combustion of plastic waste, resulting in the generation of thermal energy. Plastic waste possesses a high calorific value, so enabling the release of sufficient thermal energy through the process of combustion. A study (Verma et al. 2016) showed that incineration of plastic waste results in the emission of a range of toxic gaseous pollutants, including HCl, HCN, NOx, SOx, volatile organic compounds, as well as greenhouse gases such as CO and CO_2. The presence of these pollutants leads to the occurrence of air pollution, which subsequently results in adverse impacts on both environmental and human health.

The value of recycled plastic varies depends on the end product. It is important to acknowledge that the capital and operational expenses associated with sorting and recycling procedures can vary greatly depending on the volume of plastic waste processed by each distinct recycling method. Hence, it is imperative to concurrently employ sorting and recycling methodologies to enhance the efficacy of the procedure. This would facilitate the utilisation of plastic waste as a more environmentally sustainable energy resource, concurrently mitigating the ecological hazards associated with plastic landfilling and the atmospheric contamination resulting from incineration.

11.3.4 The Conventional Methods

Landfilling

The practise of landfilling, often known as waste disposal in landfills, is a common method used for the management of solid waste. Landfilling is widely used as the predominant conventional approach of waste disposal in several countries. Nevertheless, the availability of landfill space has become limited due to the processes of urbanisation and population expansion. The disposal of plastic waste by landfilling has become more unfavourable as a waste management option, primarily due to growing concerns over its adverse impacts on the environment and public health. One of the primary focal points in the field of public health is on the various categories, volumes, and probabilities associated with the infiltration of hazardous compounds into groundwater at waste disposal locations (Teuten et al. 2009).

The pervasive use of plastics across various sectors has led to an alarming increase in plastic waste generation. Plastics, due to their recalcitrant nature, remain in the environment for extended periods, causing ecological harm and disrupting ecosystems. Plastic pollution in oceans, microplastics in soils, and the impact on wildlife have ignited a global call for innovative waste management strategies (See Fig. 11.1). Traditional methods such as landfilling have proven inadequate to address the scale of the issue, necessitating novel approaches like pyrolysis (Abu Jadayil et al. 2017; Linden 2013) & (Doria et al. 2009).

Direct Combustion or Incineration

The incineration of plastics within the context of municipal solid waste (MSW) is a well-established tradition that remains prevalent in several communities globally, particularly in underdeveloped and economically disadvantaged countries. The incineration of plastic waste in uncontrolled environments is a substantial contributor to atmospheric pollution. According to a study cited in reference (Hopewell et al. 2009), around 40% of global wastes is incinerated, whereas the predominant composition of municipal solid waste consists of approximately 12% plastics of diverse origins.

A combustion plastic is harmful for many reasons, impacting the environment and the health of all living beings, from plants to animals and humans. When plastics are combusted, they give off toxic gases which includes; dioxins, furans, mercury, and polychlorinated biphenyls (BCPs) (Verma et al. 2016). Dumping plastic waste in landfills isn't any better as it means we're wasting energy and precious raw materials. PVC, for instance, releases harmful substances like dioxins, carbon black, and certain aromatics when combusted. Additionally, if polyethylene (PE), polypropylene (PP), and polystyrene (PS) is combusted without doing it properly, it releases

Fig. 11.1 Plastic waste blocking drains a case of Nigeria

high amounts of carbon monoxide (CO) and other pollutants in the air. Furthermore, the combustion of plastic results in the formation of solid residual ash, characterised by a dark carbonaceous hue, as well as the release of soot, both of which contribute to the emission of airborne particles. The presence of soot and solid residue ash poses a significant risk to both human health and the environment. Studies (Verma et al. 2016, Valavanidis et al. 2008) suggest that inhaling or being exposed to this soot and ash can seriously affect our health and harm the environment. This risk is particularly pronounced in relation to volatile organic compounds (VOCs), semi-volatile organic compounds (semi-VOCs), smoke (particulate matter), particulate-bound heavy metals (PMHMs), polycyclic aromatic hydrocarbons (PAHs), poly-chlorinated dibenzofurans (PCDFs), and dioxins (Verma et al. 2016, Valavanidis et al. 2008). The need to be informed on the hazards associated with these conventional approaches on plastic waste reduction, along with its potential impacts on human health and the environment in both immediate and prolonged period of time, necessitates a focused and intentional research. This is particularly crucial given the prevailing peril posed by landfills and dumpsites in a significant number of urban areas worldwide.

Mechanical Recycling Approach

Mechanical recycling involves the process of melting and remoulding polymers that have been previously used or discarded, resulting in the creation of new materials (Schyns and Shaver 2020). This implies that the recycling process described may only be applied to waste generated from thermoplastic items, while the thermoset part will persistently be disposed of in landfills. In contrast, it is worth noting that pyrolysis processes may utilise both thermoset and thermoplastic materials as feedstock. The use of the pyrolysis method has the potential to effectively manage waste derived from a diverse range of advanced materials, such as composites. This is particularly relevant in the dynamic field of engineering, where polymer composites are progressively supplanting traditional materials across various applications. The process of mechanical recycling of plastic waste entails converting the waste into secondary raw materials or products, while minimising any significant changes to the chemical composition of the material (Schyns and Shaver 2020). The feasibility of mechanically recycling different thermoplastics depends on the number of recycling cycles they have through, as this factor defines the level of quality deterioration they have suffered. The potential improvement in the quality of the new product might be attained by the use of pristine plastic materials and the integration of further additives. Currently, it comprises more than 99% of the aggregate recycled materials and is commonly employed as the primary recycling method in most nations. Numerous plastic manufacturing enterprises choose to use virgin materials rather than recycled resources, mostly due to the economic inefficiencies linked to the collection, sorting, transportation, and recycling procedures. Occasionally, this decision is made at the expense of taking into account environmental factors. Downcycling and upcycling are two distinct classifications within the realm of mechanical recycling for plastic waste.

Downcycling of Plastic Waste

The term "mechanical downcycling" pertains to the practise of recycling plastic waste in a manner that progressively diminishes its quality and functionality. In contrast to the process of upcycling, which yields a final product that possesses equivalent or higher quality compared to its original state, downcycling involves the production of materials that exhibit diminished quality. Downcycling is a phenomenon that takes place during the recycling of plastic waste, wherein the resulting product is of worse quality and functionality compared to the original product (Eze et al. 2021). Nevertheless, unlike glass and metal, recovered plastics are unable to retain their original functionality during the recycling process. This phenomenon is commonly seen as plastics may experience a loss of essential characteristics, such as mechanical strength, optical transparency, and other properties, rendering them unfit for their initial intended uses. Presented below are a selection of interesting examples of downcycling (Zhao et al. 2022):

Outdoor furniture can be produced from high-quality plastics through the process of downcycling, resulting in the creation of benches or picnic tables. Nevertheless, these objects exhibit reduced durability and may exhibit diminished resistance to harsh climatic condition when compared to their unaltered counterparts (Eze et al. 2021).

The utilisation of recycled plastics in the production of synthetic fibres for garments is a viable option. Although this practise can be categorised as a kind of recycling, it is important to note that the fibres utilised in this process may not possess the same level of strength as those derived from non-recycled sources. Consequently, this could result in reduced durability of the clothing items produced.

Construction Materials: Downcycled plastics are frequently employed within the construction industry as non-structural components, such as insulation or carpet backing. These things possess a finite duration of usefulness and frequently lack the capacity for a subsequent recycling process.

Plastic timber, which is utilised for fencing or decking purposes, is manufactured using downcycled plastics. Although these products have the advantage of reducing waste in landfills, they often lack the ability to be recycled further and have limited functional applications when compared to conventional wood materials.

Road Materials: In certain nations, the practise of utilising downcycled plastics for the construction of sub-layers in road infrastructure is observed. Although this fulfils an immediate requirement, the durability of these roadways is sometimes shorter compared to those constructed using conventional materials.

The utilisation of downcycled plastic in packaging materials, particularly trays, is a prevalent application with notable drawbacks including a reduced lifespan and limited potential for reuse.

Disposable cutlery can also be utilised in the production of disposable plates, cups, and cutlery, which frequently cannot undergo additional recycling processes and may ultimately be deposited in landfills.

Footwear: The utilisation of recycled plastic material presents a viable option for the production of shoe soles. Nevertheless, it is possible that the durability of the soles could be compromised, resulting in a faster rate of wear compared to those constructed from authentic materials (Zhao et al. 2022).

Floor mats can be produced from downcycled plastics, which find application in both automotive and residential settings. Mats crafted from high-quality, virgin materials exhibit superior durability in comparison to the ones mentioned.

Pet Accessories: It is possible to manufacture pet dishes and toys using downcycled plastics. The lifespan of these products is comparatively shorter in comparison to those manufactured with newly produced plastics.

Upcycling

The process of upcycling plastic waste involves the conversion of wasted plastic materials into products that possess equivalent or enhanced value (Mantia 2004). In contrast to the process of downcycling, which leads to the production of end goods of reduced quality, upcycling endeavours to enhance the value and prolong the lifespan of plastic materials. Presented below are a selection of interesting examples (Zhao et al. 2022; Mantia 2004; Hundertmark et al. 2023).

Art Installations: Artists have ingeniously employed discarded plastic materials to fabricate visually captivating sculptures and art installations. These artistic creations not only fulfil aesthetic functions but also contribute to the promotion of awareness regarding the problem of plastic pollution.

The conversion of high-quality plastic waste into designer furniture pieces is a process that combines aesthetics and durability, resulting in items that possess greater value compared to the original material.

The practise of transforming plastic wastes into eco-bricks has gained traction in certain areas, wherein these eco-bricks are afterwards utilised in the construction of cost-effective dwelling structures. These bricks frequently exhibit higher levels of insulation compared to conventional bricks.

Plastic bottles possess the potential to undergo a conversion process, resulting in the creation of robust and reusable shopping bags. This transformation serves the purpose of substituting single-use plastic bags and thereby diminishing the overall accumulation of plastic waste.

The utilisation of plastic bottles in the production of premium textiles for apparel and accessories is on the rise, presenting a direct competition to textiles derived from primary sources.

The process of 3D printing involves the utilisation of specific categories of plastic waste that can be transformed into filament, hence enabling the production of a diverse array of items spanning from toys to tools.

Paving Stones: Several forward-thinking enterprises are employing the practise of repurposing plastics to fabricate durable and enduring paving stones. These stones provide comparable, if not superior, durability when compared to conventional materials.

The utilisation of recycled plastics in the manufacturing of kayaks, surfboards, and skateboards has resulted in the creation of resilient materials that exhibit both durability and exceptional performance.

Jewellery: Artisans are utilising discarded plastic materials to craft distinctive pieces of jewellery, so concurrently expressing a fashion-forward aesthetic and advocating for sustainable practises.

Home Decor: The use of plastic waste materials in the creation of home decor items, such as lampshades and wall clocks, is on the rise. This practise involves the process of upcycling, which transforms discarded materials into new products that combine both functionality and aesthetic appeal. The process of upcycling plastic waste entails the conversion of wasted plastic materials into products that possess either equivalent or enhanced value. In contrast to the process of downcycling, which leads to the production of end goods of reduced quality, upcycling endeavours to enhance the value and prolong the lifespan of plastic materials. Presented below are few striking examples:

Art Installations: Artists have ingeniously employed discarded plastic materials to fabricate visually captivating sculptures and art installations. These artistic works not only fulfil aesthetic functions but also contribute to the promotion of awareness regarding the issue of plastic pollution.

The conversion of high-quality plastic waste into designer furniture pieces is a process that combines aesthetics and durability, resulting in items that possess greater value than the original material.

Certain towns have adopted the practise of transforming plastic wastes into eco-bricks, which subsequently serve as building materials for the construction of cost-effective houses. These bricks frequently exhibit higher levels of insulation compared to conventional bricks.

The conversion of plastic bottles into durable, reusable shopping bags is a viable solution for the replacement of single-use plastic bags and the subsequent reduction of overall plastic waste.

The utilisation of plastic bottles in the production of premium textiles for apparel and accessories is on the rise, presenting a direct competition to textiles derived from primary resources.

The process of 3D printing involves the utilisation of specific categories of plastic waste that can be transformed into filament, which serves as the raw material for the production of diverse items, ranging from toys to tools.

Paving Stones: Several forward-thinking enterprises are employing the practise of repurposing plastics to fabricate durable and enduring paving stones. The durability of these stones is comparable, if not superior, to that of conventional materials.

Sporting goods: The utilisation of recycled plastics has facilitated the creation of sturdy materials employed in the manufacturing of kayaks, surfboards, and skateboards, hence providing enhanced durability and superior performance.

Jewellery: Artisans are employing discarded plastic materials to craft distinctive pieces of jewellery, so simultaneously creating a fashion statement and advancing the cause of sustainability.

The utilisation of plastic waste materials in the creation of home design objects, such as lampshades and wall clocks, is becoming more prevalent. This practise involves upcycling these materials, resulting in products that not only serve a practical purpose but also possess aesthetic appeal.

11.4 Pyrolysis of Plastic Wastes

This section provides a comprehensive examination of the need for pyrolysis technology, fundamental aspects of pyrolysis, encompassing many types and processes. This section provides a comprehensive review of thermal plastic degradation through pyrolysis, discussing key parameters for optimal liquid oil production and improving its quality. It also sheds light on the role of selective catalysts in affecting product yields and offer insights into how to optimize the liquid oil yield and its potential as a fuel. Additionally, it addresses the difficulties encountered when attempting to scale up pyrolysis technologies, namely the hurdles involved in transitioning from small-scale laboratory testing to large-scale commercial deployment. When pyrolysis technology is implemented on a large industrial scale, some issues, including process optimisation, feedstock variability, and reactor design, become more pronounced. This entails the advancement of state-of-the-art technology solutions, along with a thorough understanding of the fundamental principles of modern chemistry. This chapter offers a complete overview of different reactor designs, catalytic systems, and process control approaches in order to overcome these constraints and maximise the potential of pyrolysis for plastic waste management.

11.4.1 Need for Plastic Waste Pyrolysis

According to information currently available, 348 million tonnes of plastic were produced globally in 2017. According to projections, this need will quadruple by the year 2050 (www.plasticeurope.org factsheet). Globally, an estimated 6300 million tonnes of plastic waste have been produced since the introduction of mass manufacture of plastics in the 1950s. Approximately 79% of this sum has accumulated in landfills and the environment. Recently, a more pragmatic approach has come to the fore: converting plastic waste into energy. Since plastics are petroleum-derived, they possess a high calorific value, making them a promising source for energy generation. Various thermal treatment methods, including gasification, incineration, and plasma processes, offer potential solutions. Among them, pyrolysis stands out for its efficiency and low cost. Figure 11.2 showed the global plastic production, plastic waste and plastic waste treatment (Hundertmark et al. 2023).

To mitigate the negative consequences associated with plastic waste, the European Union has recently updated Directive 2018/851. This revision aims to encourage the retrieval of plastic waste for the purpose of recycling, hence discouraging its disposal in landfills. However, it is important to note that the quantity of waste deposited in landfills remains significantly high. Based on a recently released data, it has been observed that in Spain, the predominant method employed for the disposal of plastic waste generated after consumption is landfill, accounting for 46% of the total waste management approach (www.plasticeurope.org factsheet). The task of increasing the recycling rate and minimising landfill disposal solely through conventional mechanical recycling methods can be challenging and economically impossible. This is due to the presence of numerous plastic waste streams that consist of a diverse and mixed assortment of materials, particularly those originating from industrial recovery processes (López et al. 2010) Hence, there is a need for novel recycling options, and pyrolysis, which has just been classified as TRL (technology readiness level), appears to be a viable and promising choice.

The rising environmental problems linked with plastic waste have encouraged academics and companies alike to hunt for novel solutions that solve the issues presented by the buildup of non-biodegradable polymers in the environment. The

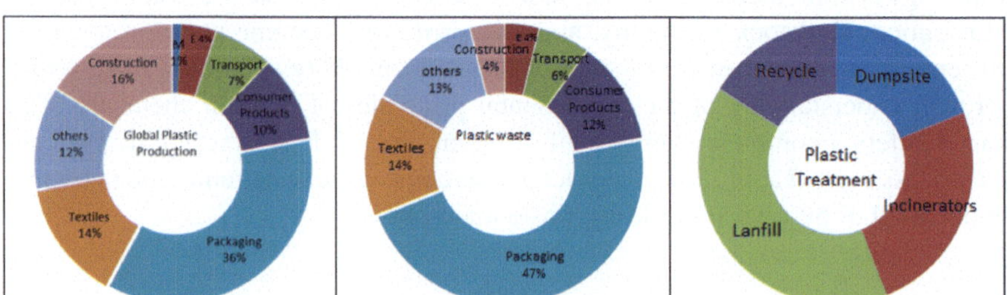

Fig. 11.2 Global plastic production, plastic generation and waste management (Hundertmark et al. 2023)

idea of a circular economy, which is characterised by resource efficiency and low waste creation, has attracted substantial attention as a strategy to alleviate the environmental effect of plastic waste. Circular economies are characterised by resource efficiency and minimal waste generation. Pyrolysis, which is a thermochemical process that involves the controlled thermal degradation of plastic materials, has emerged as a viable route within this framework to turn plastic waste into useful goods and contribute to the goals of the circular economy. Pyrolysis is a process that involves the controlled decomposition of plastic materials. One plausible consequence of technological advancements in waste management is the optimisation of efficiency and the amplification of the utilisation of recycled plastics in the manufacturing of novel products. The current limitations of the available technology need a recognition that solely relies on mechanical recycling of plastics which is inadequate for satisfying the projected demand of markets for recovered plastics. Consequently, there is an increasing endeavour towards the development of feedstock recycling technologies, which encompass the transformation of materials into their constituent monomers, as a feasible substitute. By reverting back to individual monomers and commencing future processing at that particular step, the preservation of historical importance compounds is guaranteed, since they are not transmitted to new products. This approach facilitates the establishment of a safe and circular economy.

11.4.2 Concept of Pyrolysis

Pyrolysis refers to the thermal degradation of plastic waste within a temperature range of 300–900 °C, in the absence of oxygen, resulting in the production of liquid and gaseous fuel (Karak et al. 2012; Adeoye et al. 2022, 2023a, b, 2024). Pyrolysis is essentially a thermal process that breaks down long-chain polymers into simpler molecules at higher heating rate for short durations, all in the absence of oxygen. The key outputs from this process—oil, gas, and char—have significant industrial value, particularly for energy production and refining processes (Karak et al. 2012; Adeoye et al. 2022, 2023a, b, 2024). Pyrolysis conditions can be highly varied, as its parameters can be manipulated to yield the desired products (Karak et al. 2012; Adeoye et al. 2022, 2023a, b, 2024). Notably, up to 80 wt% of liquid oil can be generated at moderate temperatures around 500 °C. This oil can be directly used in furnaces, boilers, turbines, and diesel engines, with no need for additional treatment. Compared to traditional recycling, pyrolysis is less labor-intensive and offers more flexibility, as it doesn't require meticulous sorting. When a feedstock, such as old plastic, undergoes combustion in an environment with limited or no oxygen, it undergoes pyrolysis instead of undergoing complete combustion, resulting in the formation of a mixture comprising of less complex hydrocarbons. Plastics are formed from minuscule constituent units, referred to as monomers, by a chemical reaction called polymerization. Specifically, this reaction involves the use of heat to facilitate the depolymerization process. The process of pyrolysis is employed to

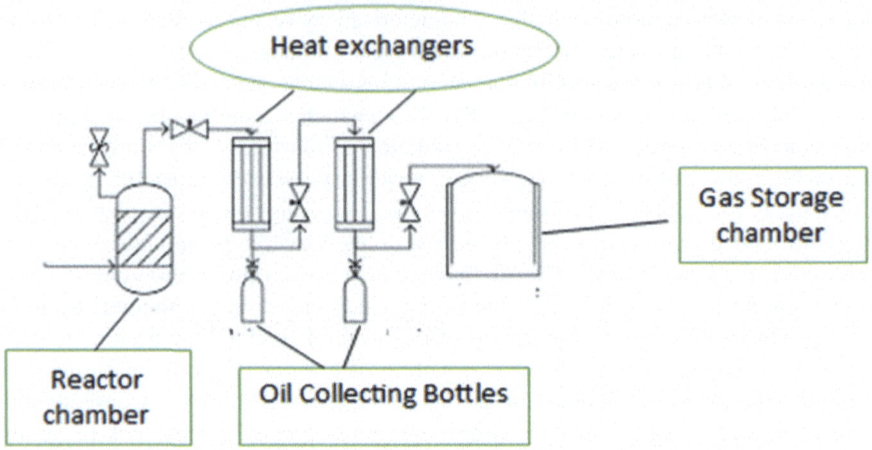

Fig. 11.3 Pyrolysis process

convert used and discarded plastics into useful resources, such as fuel, monomers, and other materials that may be effectively utilised as described by Fig. 11.3.

Pyrolysis has several advantages when compared to alternative traditional methods for managing waste plastic. For example, a significant portion of discarded plastic waste undergoes downcycling in the recycling process, wherein the plastic is transformed into goods of worse quality and limited suitability. This phenomenon can be attributed to the progressive degradation of plastic's inherent properties, such as clarity, strength, and flexibility, through many recycling processes. Furthermore, the pyrolysis method completely removes the expenses associated with the sorting, washing, and blending of waste plastics before engaging in mechanical recycling.

11.4.3 Reaction Mechanisms of Pyrolysis

Typically, the pyrolysis of plastic waste involves the fragmentation of its macromolecules, leading to a chain of chemical reactions. The free-radical route comprises three primary stages, namely initiation, propagation, and termination, which are frequently followed by chain scission reaction. Initially, the generation of free radicals occurs upon the application of heat. During the propagation stage, the generation of new radicals occurs alongside the formation of stable small molecules. Termination reactions are influenced by both radical coupling and radical disproportionation, when the transfer of hydrogen atoms occurs between radicals.

Polyethylene is largely used in the plastic world, with more than 80 million tons/yr of it being produced globally (www.plasticeurope.org factsheet). Produced from ethylene, a chemical byproduct of petroleum, polyethylene comes in different forms.

High-density, low-density, and linear low-density polyethylene are the types encountered most often (www.plasticeurope.org factsheet). Poly(ethene) can be produced using various techniques including the addition polymerization of ethene. Ethene, the primary precursor, is predominantly obtained through the cracking process of ethane, propane, naphtha, and gas oil (Poly(ethene) (Polyethylene) (n.d.)). The special features of polyethylene types are discussed below:

High-density polyethylene (HDPE): The production of high-density polyethylene (HDPE) primarily involves the use of two types of catalysts. One of these catalysts is the Ziegler–Natta organometallic catalyst, which consists of titanium compounds combined with an aluminium alkyl. The component in question is an inorganic substance that is commonly referred to as a Phillips-type catalyst. One notable illustration involves the use of chromium(VI) oxide on silica, a compound that is obtained through the process of heating a chromium(III) compound at around 1000 K in the presence of oxygen, followed by subsequent storage under a nitrogen atmosphere prior to its application (Poly(ethene) (Polyethylene) (n.d.)). High-density polyethylene (HDPE) is manufactured using three distinct production processes. All reactions occur under relatively low pressures, typically ranging from 10 to 80 atmospheres, in the presence of either a Ziegler–Natta or an inorganic catalyst. The temperature range often seen falls between the range of 350 and 420 K. Hydrogen is introduced into the ethene during all three steps in order to regulate the polymer's chain length. The utilisation of this material extends to other domains, such as food packaging and shopping bags, which are produced through the film-making process. Additionally, it finds application in the manufacturing of dustbins and crates through injection moulding, detergent bottles and drums through blow moulding, and water pipes through extrusion (Poly(ethene) (Polyethylene).

Low-density polyethylene (LDPE): The process is conducted at elevated pressures ranging from 1000 to 3000 atm, and at moderate temperatures ranging from 420 to 570 K, as can be inferred from the reaction equation standard enthalpy ΔH of −92 $kJmol^{-1}$ (Poly(ethene) (Polyethylene).

$$nC_2H_4 \rightarrow [-CH_2 - CH_2-]\cdot_n$$

The procedure employed in this context is known as radical polymerization, which involves the utilisation of an initiator, such as a minute quantity of oxygen or an organic peroxide. A highly pure form of ethene, with a purity exceeding 99.9%, is subjected to compression and afterwards introduced into a reactor with the initiator. The liquefied polyethylene is extracted, subjected to extrusion, and subsequently divided into granular form. The unreacted ethene is subjected to a recycling process. On average, polymer molecules exhibit a carbon atom count ranging from 4000 to 40,000, often accompanied by numerous shorter branches. The ratio of branches to carbon atoms is approximately 20 per 1000. The physical properties of low-density polyethylene (LDPE) are influenced by its relative molecular mass and branching. The presence of branching within a material has a direct impact on its degree of crystallinity, thereby influencing the overall density of the material. Low-density polyethylene (LDPE) is often characterised by its amorphous nature and clear appearance, exhibiting an approximate crystallinity level of 50%. The presence of branches inhibits the close packing of molecules, resulting in a lower density (Poly(ethene) (Polyethylene) (n.d.)). The utilisation of polyethylene can be observed in several industries. For instance, it finds application in the production of cling film and milk cartons through the film-making process. Additionally, it is employed in the manufacturing of buckets and bowls using injection moulding techniques. Moreover, polyethylene is utilised in the production of detergent squeezable bottles using blow moulding processes. Lastly, it is also employed in the production of flexible water pipes and cable jacketing through extrusion methods (www.plasticeurope.org factsheet). The utilisation of low density poly(ethene) encompasses a wide range of applications; however, the high pressure manufacturing technique employed for its production entails substantial capital expenditures. Nevertheless, a sophisticated methodology has been devised that combines Ziegler–Natta and inorganic catalysts to synthesise linear low density polyethylene (LLDPE), exhibiting enhanced characteristics compared to low density polyethylene (LDPE). The selection of a Ziegler–Natta catalyst allows for the utilisation of any of the three processes: slurry, solution, and gas phase. The utilisation of the gas phase method is employed in situations when an inorganic catalyst is utilised. Incorporation of minute quantities of a co-monomer, such as but-1-ene or hex-1-ene, is introduced into the primary material. The monomers undergo random polymerization, resulting in the formation of linear chains with occasional short branches composed of a few carbon atoms (Poly(ethene) (Polyethylene).(n.d.)). The utilisation of this technology is observed in the production of a stretch film through the film-making process, food boxes through injection moulding, and cable coating through extrusion (www.plasticeurope.org factsheet) & (Poly(ethene) (Polyethylene).(n.d.)).

When it comes to reducing this abundant plastic, pyrolysis is the best method. Pyrolysis is complex, though, and a lot depends on the conditions under which it is operated. Factors like temperature, the type of reactor used, and even the heating rate can all impact the product distribution (Sipra et al. 2018; Diaz Silvarrey and Phan 2016). Temperature, in particular, is key. Heating at a range between 300 and 500 °C, and you're more likely to get liquid products, which are often more useful and easier to handle (Demirbas 2004; Ciliz et al. 2004).

It's also worth noting that pyrolysis isn't too specific about the type of plastic it breaks down. But the process works best on polyolefins, producing liquid oils with low octane numbers that can substitute for conventional fuel (López et al. 2010) & (Ciliz et al. 2004).

One of the most common plastic types, polyethylene (PE) breaks down into waxes and oils when subjected to pyrolysis. The chemical equation governing this reaction is;

$$\mathrm{PE}(\mathrm{C_2\,H_4})_\mathrm{n} \rightarrow \mathrm{Waxes}(\mathrm{C_x\,H_y}) + \mathrm{Oils}(\mathrm{C_mH_n})$$

Polypropylene (PP) decomposes into olefins and waxes when pyrolysed:

$$PP(C_3H_6)_n \rightarrow Olefins(C_xH_y) + Waxes(C_mH_n)$$

Polyethylene Terephthalate (PET) pyrolysis typically yields terephthalic acid and ethylene glycol.

11.4.4 Pyrolysis Products and Applications

Pyrolysis plastic waste yields a spectrum of products with diverse applications. The liquid fraction, often referred to as pyrolysis oil or pyro-oil, can be further processed into chemicals, fuels, and specialty products. Gaseous products encompass hydrocarbons and syngas, which hold potential as energy sources or chemical raw materials. Solid residues, often containing carbonaceous materials, can be utilized as adsorbents or carbon sources. Advanced chemistry principles guide the design of downstream processes to upgrade and valorize these products.

11.4.5 Feedstock Selection for Plastic Waste Pyrolysis

The variety of plastics obtained from different sources of wastes exhibits a remarkable degree of diversity. There is significant variation in the composition of polymer blends utilised in packaging, waste electrical and electronic equipment (WEEE), and agricultural waste streams. Polyethylene terephthalate (PET) and polyvinyl chloride (PVC) are present in various streams, although it is noteworthy that polyolefins constitute the predominant portion of plastic goods. PET and PVC plastics provide challenges in the feed due to their limited presence, typically less than 5% and commonly around 1% to 2%, in feedstock combinations. On the other hand, polyolefins serve as an optimal feed for pyrolysis as they readily break down into useful hydrocarbon compounds. The thermal degradation byproducts of these polymers have detrimental effects on both the manufacturing process and the ultimate product. Polyvinyl chloride (PVC) undergoes heat degradation, resulting in the release of chlorinated hydrocarbons and hydrochloric acid (HCl). These byproducts can cause corrosion within the reactor and lead to the halogenation of the oil. Polyethylene terephthalate (PET) exhibits a tendency to undergo degradation, resulting in the formation of phthalic acids. This degradation process not only obstructs pipelines but also diminishes the overall quality of the oil. The co-pyrolysis of PET and PVC leads to an enhanced generation of organic chlorine (Bhaskar et al. 2004). Polyvinyl chloride (PVC) is predominantly utilised within the construction sector and does not make a substantial contribution to the packaging industry (www.plasticeurope.org factsheet) & (Lewandowski and Skórczewska, 2022). Nevertheless, the packaging industry extensively utilises polyethylene terephthalate (PET). The mechanical recycling of PET is widely recognised, however, chemical recycling methods such as glycolysis, methanolysis, and hydrolysis offer the capability to fully depolymerize

PET into its constituent monomers, namely terephthalic acid, dimethyl terephthalate (DMT), bis(hydroxylethylene) terephthalate (BHET), and ethylene glycol (Ragaert et al. 2017). Mechanically regenerated polyethylene terephthalate (PET) is commonly employed in cascaded applications due to its inferior material properties. The encouragement for the gathering of clean PET plastics stems from the existing limitations that restrict these methods only to clean PET feedstock. There is a limited number of methods available for the treatment of polyethylene terephthalate (PET) found in packaging waste by pyrolysis, while simultaneously inhibiting the formation of high boiling terephthalic acid (Chiu and Cheng 2000).

11.4.6 Pretreatment and Material Feeding for Pyrolysis

Prior to considering the conversion of plastic into fuel via pyrolysis, a fundamental stage must be undertaken, namely pretreatment. The removal of impurities such as metals or wood is necessary in order to ensure the optimal functionality of the process. The requirement at hand is not solely of a technical nature, but also holds significant economic implications. In order to achieve cost-effectiveness in a Plastic to Fuel (PTF) facility, it is important to implement a preliminary sorting process for the plastic waste.

The source of plastic wastes is also a significant factor to consider. The presence of diverse sources results in variations in shapes and sizes, hence rendering the process more intricate. Typically, it is advisable to achieve a consistent size for the waste material by subjecting it to crushing and sifting processes prior to its introduction into the pyrolysis reactor. Indeed, the inclusion of this element contributes to the cumulative expenditure, yet it is widely seen as indispensable. However, certain pyrolysis techniques exhibit less sensitivity towards the dimensions and configuration of the waste material (Adeoye et al. 2023a). Rotary kilns, for example, exhibit a notable degree of tolerance and are capable of accommodating a heterogeneous assortment of sizes. In the case of alternative reactor designs, such as fluidized bed reactors, the significance of uniform particle size becomes apparent due to its influence on the even distribution of heat throughout the process.

Regarding the introduction of plastic into these reactors, several factories employ distinct approaches. Certain individuals choose to utilize screw feeders, whereas others opt for hydraulic feeding devices. Plastic films provide challenges when used with screw feeders, as they have a tendency to entangle around the screw or adhere to it through melting, resulting in operational difficulties.

11.4.7 Factors Affecting Pyrolysis

Pyrolysis is a thermal degradation process that converts plastic waste into valuable products in the absence of oxygen. The process can be categorized into slow, fast, and intermediate pyrolysis, each differing in heating rates and product distributions. The fundamental chemistry behind pyrolysis involves the breakage of polymer chains into smaller molecular fragments, which subsequently undergo various reactions to form liquid hydrocarbons, gases, and solid residues. The efficiency and selectivity of pyrolysis depend on factors such as temperature, residence time, feedstock composition, and reaction atmosphere. Advanced chemistry knowledge is essential for tailoring these parameters to achieve specific product outcomes and minimize undesirable byproducts.

The composition of the feedstock, process variables (such as temperature, heating rate, and reaction gas), and the presence or absence of a catalyst (in the case of catalytic pyrolysis or thermal pyrolysis) can influence the content and yield of the final product. Temperature also plays a big role. Below 400 °C, more solids are obtained; between 400 and 600 °C, more aromatics; and above 600 °C, more gases like paraffins and olefins. Other factors include; type of reactors used, pressure, and residence time (Adeoye et al. 2023a).

11.4.8 Types of Pyrolysis Base on Heating Rates

Slow Pyrolysis (Non-Isothermal)

The process of gradually increasing the temperature of the feedstock in the absence of oxygen is commonly referred to as slow pyrolysis. The organic material undergoes partial evaporation of its volatile constituents, as opposed to combustion, resulting in the formation of a mostly carbonaceous residue known as char, often comprising 80% of its composition. In contrast to rapid pyrolysis, which places focus on the production of liquid byproducts, slow pyrolysis prioritises the generation of solid char as the primary output. The heating rate is maintained at 10 °C/s (Adeoye et al. 2023a).

Fast Pyrolysis (Isothermal)

Fast pyrolysis is a process that yields significant quantities of pyrolysis liquid fuel by rapid heating of the feedstock to moderate temperatures ranging from 400 to 600 °C, accompanied by a brief residence time of only a few seconds. Fast pyrolysis presents an opportunity to optimise the production of liquid products, with the reactor operating under isothermal conditions. The aforementioned technique is commonly employed in both research and practical applications of polymer pyrolysis. The heating rate is maintained at a constant value of 100 °C/s (Adeoye et al. 2023a).

Ultra-Fast/flash Pyrolysis

The primary outcomes of ultra-fast pyrolysis, also known as flash pyrolysis, characterised by rapid heating and swift thermal degradation, encompass gaseous substances and bio-oil. The heating rates encompass a range of 100 to 10,000 °C/s, while the duration of residency is rather short (Adeoye et al. 2023a).

11.4.9 Other Classification of Pyrolysis

Thermal Pyrolysis Technique

A degradation mechanism called thermal cracking, commonly referred to as pyrolysis, occurs when polymeric materials are subjected to high temperatures in an environment with little oxygen supply. Within the range of 350–900 °C, there is variation in the temperature. The end products are made up of a variety of materials, particularly gaseous, liquid, and carbonised char leftovers. It is a typical practise to keep the non-condensable residue, which has a high calorific value, and convert just the volatile portion of a product into liquid fuel.

The liquid output of thermal pyrolysis, a kind of thermochemical treatment (TCT), contains a wide range of paraffins, isoparaffins, olefins, naphthenes, and aromatics. It is necessary to subject the system to extremely high temperatures, up to a maximum of 900 °C, in order to start chain scission. Chemical bonds are broken during the aforementioned operation (Mantia 2004). According to a study (Scott et al. 1990), the reaction produced a solid byproduct as its primary byproduct at temperatures below 700 °C. Higher temperatures, on the other hand, resulted in the creation of a gaseous byproduct.

Microwave-Assisted Pyrolysis Technique

Microwave dielectric heating is used during microwave pyrolysis, sometimes referred to as microwave-assisted pyrolysis. Depending on the dielectric qualities of the material, microwaves can interact with it in one of three ways: via reflecting off conductors, transmitting through perfect insulators, or absorption and decaying inside the substance. By stirring up molecules with alternating electromagnetic fields, heat is produced in dielectric materials. The principle behind plastic microwave pyrolysis is the passage of thermal heat from an absorbent, which first absorbs microwave radiation, to the plastic. The uniformity of heating dispersion depends on the physical characteristics and volume ratio of the absorbent. Additionally, varying microwave strength results in an entirely different product dispersion.

Microwave-induced pyrolysis has been shown to have several benefits over thermal and catalytic pyrolysis, and may be utilised to produce chemicals and fuels with added value (Ludlow-Palafox and Chase 2001). The method offers the possibility of quick, volumetric, and selective heating of polymers for energy recovery. However, because microwave radiation has an extremely low dielectric loss factor, polymers cannot absorb it. As a result, the plastic needs to be combined with an absorbent to help in pyrolysis. For plastic pyrolysis, materials having a high dielectric loss factor, such as silicon carbide, carbon, and iron mesh, are suitable absorbents. The majority of single plastic types have been microwave pyrolyzed, according to published research. Polyethylene (PE), polypropylene (PP), polystyrene (PS), poly(ethylene terephthalate) (PET), and polyvinyl chloride (PVC) are a few of these polymers. Without the application of microwave absorbent, none of these polymers can absorb microwave.

Catalytic Pyrolysis Technique

In recent years, there has been a notable increase in interest surrounding the subject of waste plastic recycling. Pyrolysis emerges as a highly promising technology among the diverse range of technologies now under investigation. The aforementioned procedure entails the decomposition of both organic and synthetic substances under elevated temperatures, particularly inside an environment devoid of oxygen. Previous research has commonly utilised a temperature range spanning from 360 to 500 °C for the implementation of this procedure (López et al. 2010; Sarker et al. 2011).

Nevertheless, the practical difficulty arises from the high temperatures needed for pyrolysis, leading researchers to explore other methodologies. A technique that has been proposed involves the utilisation of catalysts to enhance the efficiency of the pyrolysis process and reduce the required operating temperatures (López et al. 2011; Serrano et al. 2001a).

The global environmental challenge of plastic wastes arises from the fact that polymers lack the inherent ability to undergo natural decomposition. Therefore, it is crucial to identify effective and environmentally friendly strategies for waste management. In the given setting, catalytic pyrolysis emerges as a more advantageous alternative in comparison to thermal pyrolysis. Based on current research conducted by Mibei et al. (2023), it has been observed that catalytic pyrolysis has the capability to attain the important pyrolysis products at reduced temperatures, hence enhancing its energy efficiency.

Catalytic pyrolysis, which employs solid catalysts to facilitate specific reactions, offers the potential to manipulate product distributions and improve the quality of pyrolysis liquids.

The characteristics of the resultant products obtained from the process of pyrolysis can exhibit substantial variations depending on whether thermal or catalytic pyrolysis is employed. According to a study (Miandad et al. 2016), thermal pyrolysis typically produces heavier oils that possess a lower octane number and a higher quantity of char or solid residue. Conversely, catalytic pyrolysis results in the production of lighter oils characterised by elevated octane values and reduced char content.

Catalysts play a crucial role in the pyrolysis process, serving many significant functions that contribute to the improvement of both the procedure's quality and efficiency. According to López et al. (2011), catalysts play a role in reducing the operating temperature, accelerating reaction rates, and so decreasing the overall energy demand of the process. The selection of a catalyst can also determine the varieties and quantities of the final products, including petrol, petrol fractions, heavy oils, and residue (Wan 2015).

One key benefit associated with the utilisation of catalysts lies in their ability to enhance the process of carbon chain reduction within the resulting products of pyrolysis. According to Yansaneh and Zein (2022), the reduction in melting temperatures of these plastic wastes results in enhanced manageability and perhaps increased value. A wide range of catalysts has been investigated in various studies, such as Fe_2O_3, $Ca(OH)_2$, FCC, natural and synthetic zeolites, red mud, ZSM-5, HZSM-5, and Y-zeolite, among others (Sarker et al. 2011; Lee 2009; Lee 2012; Syamsiro et al. 2014; López et al. 2011; Hernandez et al. 2007).

Upon critical analysis, it becomes apparent that the selection of a catalyst plays a crucial role in determining the effectiveness and final products of the pyrolysis process. The diverse available catalysts, each possessing distinct attributes and capacities, implies that they have potential for a particular use i.e. based on the needs of a pyrolysis process. However, due to the dynamic nature of the catalytic pyrolysis of plastic waste and the abundance of research conducted in the last decade, it is imperative to continue ongoing investigations to ascertain the most favourable parameters and catalysts for the efficient implementation of pyrolysis on a broad scale.

A good catalyst can make the pyrolysis process more efficient by lowering the temperature requirements, speeding up reactions, and reducing activation energy. One of its key benefits is its ability to break down long carbon chains, which essentially makes the end product easier to work with and more useful as a fuel. However, there's possibility that local materials like clay can alter the surface area and structure of plastic waste to further increase pyro-oil yield. When high-density polyethylene and polypropylene were subjected to pyrolysis at 300 °C with a 10 wt% concentration of the catalyst, we achieved higher pyro-oil yields of 87.23 wt% and 60.36 wt%, respectively. These results suggest that local clay can effectively replace imported catalysts. It's particularly noteworthy that these high yields were achieved at considerably lower temperatures (300–450 °C) compared to 600 °C needed for thermal pyrolysis.

Catalytic pyrolysis involves subjecting polymeric materials to elevated temperatures in the presence of a catalyst, under oxygen-free conditions, in order to induce their degradation. Catalysts are commonly employed in plastic pyrolysis to mitigate energy requirements, induce compositional alterations through cracking, and enhance the overall process efficiency. The catalysts often employed for the pyrolysis of plastic waste include silica alumina (Panda and Singh 2011), zeolites (Manos et al. 2001), fluid catalytic cracking (FCC) (Kumar and Srinivas 2020), and mobil classification of materials (MCM-41) (Ratnasari et al. 2016). Zeolite-based catalysts frequently exhibit enhanced conversion rates owing to their heightened acid strength in comparison to nonzeolitic catalysts, hence resulting in a greater generation of gaseous byproducts.

The study conducted by Uemichi et al. (1999) examined the process of catalytic degradation of polyethylene into gasoline-range fuels using silica-alumina and HZSM-5 zeolite in a two-stage system. Based on their research findings, it has been observed that the use of a two-stage strategy offers several advantages in comparison to the one-stage approach, which involves the gradual decline in performance of each catalyst. Notwithstanding the constrained quantity of aromatic compounds, the production of petrol was augmented, resulting in a notable increase in its octane rating. Significant results were obtained when a weight ratio of 9:1 of silica-alumina and HZSM-5 catalysts was employed in a flow reactor. The reversal of the catalysts' sequence did not yield any discernible benefits. The proposal suggested that the use of silica-alumina, which was loaded into the large pores of the top layer and possessed a moderate level of acidity, effectively served as a catalyst for the conversion of polyethylene into liquid hydrocarbons. The generation of oils of inferior quality occurred, resulted in a decrease in oil output. These low-quality oils were then transformed into high-grade petrol using the very acidic sites of the HZSM-5 catalyst placed in the lower layer.

The study conducted by Ratnasari et al. (2016) centred on the utilisation of staged catalysis as a unique approach for the pyrolysis of waste plastics. The primary objective of this research was to produce hydrocarbon oils that are well-suited for usage as fuel. The researchers utilised a distinctive experimental configuration that incorporated two distinct catalysts, namely mesoporous MCM-41 and microporous ZSM-5, placed in a predetermined sequence. Additionally, various ratios of these catalysts were tested in order to determine the optimal combination. In order to facilitate comparative analysis, individual testing of both catalysts, namely MCM-41 and ZSM-5, was conducted. Based on the findings, the use of a proportionate mixture of MCM-41 and ZSM-5 in a staged pyrolysis-catalysis procedure resulted in a substantial production of oil from high-density polyethylene, up to 83.15% in terms of weight. Regarding the gaseous by-products, the predominant gases comprised ethene, propene, butene, and butadiene. Remarkably, the oil that was generated had a substantial concentration of aromatic compounds, comprising 95.85% of its overall mass. Furthermore, an overwhelming majority of the oil, specifically 97.72%, consisted of hydrocarbons falling within the petrol range.

A study (Uddin et al. 1996) conducted a study that aimed to analyse the degradation mechanisms of different types of polyethylene, including high-density polyethylene (HDPE), low-density polyethylene (LDPE), linear low-density polyethylene (LLDPE), and cross-linked polyethylene (XLPE). The studies were conducted in batch forms, at a temperature of 430 °C. In the catalytic procedure, the study used selected silica-alumina as a solid acid catalyst. Additionally, experiments were done under heat settings without the presence of any catalyst. The findings showed both high-density polyethylene (HDPE) and cross-linked polyethylene (XLPE) exhibited a propensity to generate waxy compounds when exposed to thermal decomposition. In addition, it is worth noting that the liquid product yields obtained from both types exhibited a significant decrease, ranging from 58 to 63% by weight. On the other hand, it is worth noting that LDPE and LLDPE exhibited a comparatively elevated proportion of liquid products, ranging from 76 to 77% in terms of weight. The production of wax-like compounds by LDPE and LLDPE was found to be minimal. The product yields were influenced by the degradation of the polymers' structure. The liquid byproducts resulting from thermal degradation exhibited a broad boiling point range spanning from 36 to 405 °C. These byproducts were extensively distributed across the carbon fractions ranging from n-C_5 to n-C_{25}. The conversion of all polyethylenes was achieved by employing silica-alumina as a catalyst, resulting in the production of liquid products with high yields ranging from 77 to 83% by weight. Notably, no wax formation was observed during the process. The liquid products predominantly consisted of hydrocarbons with carbon chain lengths ranging from C_5 to C_{12} (n–C_5–n–C_{20}). The use of a solid acid catalyst resulted in an increased rate of unselective polyethylene breakdown, leading to the production of the light fuel oil. The user has provided a numerical range, specifically the numbers 52 and 53. In the studies investigating the impact of nonacidic catalysts on the pyrolysis of plastics, $MgCO_3$ and $BaCO_3$ were employed as catalysts at a temperature of 450 °C. The results revealed that the utilisation of $BaCO_3$ led to an oil production of 33.60%, while the use of $MgCO_3$ resulted in a somewhat lower oil output of 29.60%. Furthermore, the oil yield achieved a value of 32.20% when $CaCO_3$ was employed as a catalyst in the identical reaction conditions.

A study (Kumar and Srinivas 2020) investigated the pyrolysis of PS, PP, PE, and PET wastes in a batch reactor. The study examined the individual pyrolysis of each waste material, as well as their combined pyrolysis. Additionally, the effects of charcoal, activated carbon, and CaO catalysts on the pyrolysis process were explored. In the absence of catalysts, the obtained liquid yields from PS, PP, PP, and combination PS + PP + PE were 80%, 60.7%, 75%, and 66.86%, respectively. In contrast, the utilization of activated carbon results in an 82.43% increase in the liquid yield from PS + PP + PE. Similarly, the usage of charcoal leads to a 95.54% increase, while the combination of activated carbon and CaO as a catalyst yields a 75.50% increase. In the study conducted by Kassargy et al. (2017), a comparative analysis was performed on the thermal and catalytic pyrolysis of polyethylene (PE) and polypropylene (PP) utilizing a batch reactor. The findings indicated that the thermal pyrolysis process, conducted at a temperature of 450 °C, resulted in wax yields of 80% by weight and liquid yields of 85.5% by weight. When USY zeolite was employed in the process of catalytic pyrolysis, it was observed that the resulting liquid yields consisted of mixtures including C_5 to C_{39} and C_5 to C_{30}, which may be attributed to polyethylene (PE) and polypropylene (PP) respectively. These mixtures accounted for 71 and 82 wt% respectively. Following additional product separation via distillation, it is observed that 60.6% and 57% of gasoline with high octane values of 96 and 97, respectively, are produced for PP and PE. Observations have shown that from pyrolysis oil (PP) and pyrolysis gasoline (PE), diesel yields of 36.5% and 35.3% have been obtained, respectively. These diesel samples have been found to have cetane values of 52 and 53, respectively. Based on the findings of Demirbas (2004), it was observed that the polyolefin and PS blend, constituting about 46.6 wt% of the overall plastic waste, resulted in the generation of 35 wt% of solid waste and 2.2 wt% of gaseous byproducts. A study (Baiden 2018) conducted pyrolysis on a blend of several waste plastics, including HDPE, LDPE, PP, and thermoplastic polyurethane (TPU). According to their assertion, the utilisation of 80 V for a duration of 55 min, together with a plastic waste sample that initially weighed 89.5 wt%, yielded the highest quantity of liquid product. The maximum reaction temperature recorded for this yield was seen to be within the range of 497–500 °C. Study (Donaj et al. 2012) conducted a study on the pyrolysis of mixed polymers including polyolefins. The composition of the mixed plastics consists of 25% polypropylene (PP), 30% high-density polyethylene (HDPE), and 75% low-density polyethylene (LDPE). Two distinct sets of experiments were conducted using thermal pyrolysis at temperatures of 650 and 730 °C, and a Z–N catalyst in a fluidized quartz-bed reactor at temperatures of 500 and 650 °C. The mass fractions of gases, liquids, and solids resulting from thermal pyrolysis were found to be 36.9/48.4/15.7 wt% and 42.4/44.7/13.9 wt% at temperatures of 650 and 730 °C, respectively. In contrast, catalytic pyrolysis yielded mass fractions of 6.5/89.0/4.5 wt% and 54.3/41.9/3.8 wt% at temperatures of 500 and 650 °C, respectively. A research work (Pratama and Saptoadi 2014) was conducted on the pyrolysis of mixed waste plastics in batch reactors using a two-stage process, consisting of pyrolysis and catalytic reforming, at temperatures reaching 450 °C and 500 °C, respectively. The researchers employed discarded polyethylene (PE), polypropylene (PP), polystyrene (PS), and polyethylene terephthalate (PET) as their raw materials. Based on their research findings, it was determined that a composition of 50 wt% PE waste, 40 wt% PP waste, and 10 wt% PS waste resulted in the production of a liquid fuel comprising 45.13 wt%.

11.4.9.1 The Key Components of a Standard Pyrolysis Plant

Feedstock Storage Area: Picture a wide, airy space—either open or roofed—where we keep our raw material, be it plastic waste or biomass, ready for action.

Conveyor Belt: Imagine a continuous loop of rubber that carries this feedstock on a short journey to what we call the "Shredder".

Shredder: Here's where things get reduced. This machine takes the waste and grinds it down into smaller, more manageable chunks.

Second Conveyor Belt: The shredded bits then hop onto another conveyor, aimed straight for the Reactor, the plant's real nucleus.

Reactor: Envision a hefty, cylindrical chamber, built from robust steel that can handle extreme heat. This is where the primary pyrolysis reaction unfolds.

Heating Unit: This is our heat source—comprising burners or electric coils—that warms up the Reactor to the needed temperatures for pyrolysis.

Catalyst Feeder: If the process involves catalysts to speed things up, this part takes care of injecting them into the reactor at the right time.

Cooling System: Picture a series of condensers that bring down the temperature of the exiting pyrolytic liquid and gases, prepping them for storage.

Separator: A sophisticated device that draws a line between the liquid and gaseous outputs, directing them to their respective storage units.

Liquid Storage Tanks: Visualise large, sealed containers where we keep the liquid yield, often a form of pyrolytic oil.

Gas Storage Units: Gases that aren't channelled back into the plant for heating are stored here in dedicated units.

Control Room: Imagine a tech-savvy space, filled with computer systems and dashboards, from where the entire operation is orchestrated.

Exhaust Treatment: Last but crucially important, systems are in place to clean up exhaust gases, making them safe for environmental release.

11.4.9.2 Types of Pyrolysis Reactors

The selection of a reactor is a critical factor in determining the range of pyrolysis products. The economic worth of the products derived by pyrolysis is primarily contingent upon the composition of the feedstock. However, the utilisation of certain reactor technologies also holds significant influence in optimising the production of desired outputs. Each form of pyrolysis reactor has advantages and disadvantages, and the selection should be determined by the desired result and the ability to accommodate different feedstock variations. When choosing the right reactor, two main factors must be considered namely; heat and mass transfer efficiency (Grause et al. 2011). The assessment of reactors is conducted by evaluating many properties related to their operational and economic flexibility, drawing upon information gathered from multiple sources. However, a significant body of research has been derived from the study conducted by Arena and Mastellone (2006). It is important to acknowledge that this analysis focuses exclusively on the use of these pyrolysis reactors for plastic waste and may vary when applied to biomass waste or any other thermally convertible materials. Additionally, the critical analysis presented in this paper focuses on the features of the reactors as a single unit and does not consider any specific changes that may have been made to them. The reactor types are as shown in Fig. 11.4 and discussed in (i–iii).

Fluidized bed reactors, such as bubbling and circulating kinds (BFB and CFB), provide significant flexibility, particularly in relation to temperature regulation (Arena and Mastellone 2006). These reactors have gained recognition for their exceptional heat and mass transport capabilities, making them suitable for customization to optimize certain desired outcomes. In the context of the quick pyrolysis of polyolefins, the primary products obtained predominantly consist of waxes, which can subsequently undergo additional processing steps. Furthermore, these reactors exhibit compatibility with catalysts, enabling operation at reduced temperatures and facilitating targeted output. Catalysts have the capability to be included directly into the reactor bed or introduced into a secondary reactor in order to prolong their operational duration. Although there are advantages associated with these reactors,

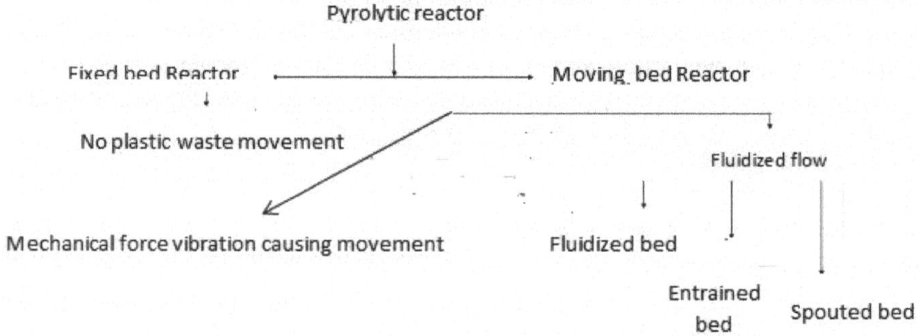

Fig. 11.4 Schematic representation of types of reactor

they can be expensive, particularly when implemented on a smaller scale. Additionally, ensuring proper feeding of the system can be challenging due to the specific particle size requirements (Lopez et al. 2017). The scalability of these systems, as demonstrated in commercial applications (Solantausta et al. 2012), is a significant advantage.

Rotary kilns exhibit a high level of simplicity and cost efficiency. They are frequently employed in the treatment of intricate waste materials, such as automobile shredder residue and industrial wastes. The provision of heat is commonly facilitated by the kiln walls. Previous examples of scaling up kilns include the Conrad and Toshiba Sapporo processes, which successfully converted plastics and tires into fuel and several other materials (Butler et al. 2011). The Faulkner technique for tires, characterized by its distinctive designs, facilitates the manipulation of heating conditions.

Screw kiln reactors are a derivative of rotary kilns. The kilns are equipped with a screw or auger mechanism that facilitates the movement of the feedstock within the heated chamber. This feature offers enhanced control over the residence time of the material in the reactor (Serrano et al. 2001b; Elmously et al. 2019; Kirby et al. 2017; Hense et al. 2015). The effectiveness of these reactors has been demonstrated in addressing intricate waste management challenges, such as the treatment of electronic scrap using the Haloclean process (Hornung et al. 1996), or the recycling of polystyrene under the ResolVe project. The simplicity of the design may give rise to potential challenges in heat transfer and temperature regulation, which in turn can result in the adhesion of materials to the inner surfaces of the reactors if not adequately addressed during the planning phase.

11.4.10 Comparing and Contrasting Reactors

In comparing and contrasting different types of reactors, we can consider four main categories: Fixed-Bed Reactors, Fluidized-Bed Reactors, Rotary Kiln Reactors, and Auger (Screw) Reactors. Fixed-Bed Reactors offer simplicity and lower upfront costs but have lower efficiency due to uneven heat transfer. They may also be prone to clogs and hotspots. Fluidized-Bed Reactors excel in heat transfer and breaking down plastics efficiently. They provide uniform temperature conditions and better overall efficiency. However, they tend to be operationally complex and come with higher costs.

Rotary Kiln Reactors offer excellent mixing capabilities but require more significant space. They excel in controlling residence time but can be constrained by space and cost considerations. Auger (Screw) Reactors are effective at handling larger volumes continuously. They offer continuous operation benefits but are more prone to wear and tear and can be harder to maintain.

11.5 Instrumental Assessment of Plastic Waste Pyrolysis Oil

11.5.1 Characterization Techniques

Pyrolysis of plastic waste serves as a promising approach to transform plastic into oils, gases, and char. However, the quality and efficiency of these byproducts depend on several factors, warranting thorough analysis through various instrumental methods.

The utilisation of instrumental analysis is of utmost importance in comprehending the intricate mechanisms involved in the process of plastic waste pyrolysis, including both the underlying reactions and the resultant products. Methods such as Fourier-transform infrared spectroscopy (FTIR) and nuclear magnetic resonance (NMR) spectroscopy provide significant insights on the chemical composition and functional groups found in the byproducts resulting from the pyrolysis of plastic wastes. Gas chromatography-mass spectrometry (GC–MS) and liquid chromatography-mass spectrometry (LC–MS) play a crucial role in the identification and quantification of volatile and liquid chemicals generated. To enhance morphology comprehension, the utilisation of scanning electron microscopy (SEM) and transmission electron microscopy (TEM) can be employed to monitor modifications in shape, morphology, and microstructures throughout the pyrolysis process.

11.6 Proximate and Ultimate Analysis

The purpose of doing elemental analysis on pyrolyzed plastic waste is to ascertain the ultimate parameters, namely carbon (C), hydrogen (H), nitrogen (N), oxygen (O), and sulphur (S), utilizing CHNSO analyser. According to Islam et al. (2004), the study of elemental composition in pyrolyzed plastic waste is frequently performed by the utilization of X-ray fluorescence (XRF) and inductively coupled plasma mass spectrometry (ICP-MS). The aforementioned methodologies demonstrate a notable degree of sensitivity and possess the capacity to identify and analyze minute constituents found in the char, gas, and liquid byproducts that are generated. The proximate parameters encompass moisture content, volatile matter, fixed carbon, and ash content. A study (Islam et al. 2004) reported that the moisture content of the pyrolyzed plastic waste was analyzed to be 0.41%. The volatile matter was discovered to be 96.88%, while the fixed carbon content was determined to be 0.28%. The ash content was measured to be 2.43%. The elemental composition of the pyroyzed plastic waste was determined to be as follows: carbon (83.93%), hydrogen (12.84%), and oxygen (0.80%). The ratio of carbon to hydrogen (C/H) was calculated to be 6.53. However, the presence of nitrogen and sulfur in the pyrolyzed plastic waste was not detected. According to a study (Yousef et al. 1763), the plastic waste samples exhibit a significant presence of C, H, and O, with average concentrations of 82.24 wt%, 14.07 wt%, and 3.21 wt%, respectively. The identification of these elements suggests that plastic waste samples possess a high concentration of carbon precursor, which effectively supports the conversion process. Furthermore, a little concentration of nitrogen (0.45 wt%) and sulfur (0.018 wt%) was detected, suggesting that the conversion procedure was accompanied by reduced emissions of NOx and SO_2, which are known to be environmentally harmful. The sample exhibits a significant proportion of volatile Matter, with a weight percentage of 90.652%. Additionally, it has a moderate amount of fixed carbon, with an average weight percentage of 7.49%, and a relatively low ash content, with an average weight percentage of 1.60%.

Understanding the elemental composition of plastic wastes is very important. Elemental analysis provides valuable insights into the composition and properties of many substances. A study (Wong and Broadbelt 2001) highlights the process of material recovery which involves the extraction of elements such as metals for use in industrial applications. To assess any harmful environmental impact, identifying the heavy metal is important. The determination of calorific value can be based on the elemental composition of the material e.g., biomass (C, H, N, S, & O) (Caballero et al. 2016).

Nuclear Magnetic Resonance (NMR) Spectroscopy

Nuclear magnetic resonance (NMR) spectroscopy is an advanced analytical instrumental used for the assessment of diverse materials, including waste plastic oil. The obtained information from NMR presents valuable knowledge of organic molecules, including their purity, chemistry, and physical composition. The application of NMR in plastic waste pyrolysis oil evaluation is to characterise its composition with the aim of identifying and detecting any potential contaminants or impurities that may be present within the pyro-oil. The analysis technique is non-destructive in nature making it a special primary advantage which allows for the examination of complex mixtures such as plastic waste oil without altering the sample. NMR spectra provides clear understanding into the various nuclei inside a given sample, such as hydrogen (^{1}H) and carbon-13 (^{13}C), together with their respective chemical environments. The identification of chemical groups and functional compositions present in pyrolysis oil can be determined by analysing the NMR spectra, specifically by critically analysing the chemical shifts, coupling patterns, and integration values (Adeoye et al. 2023a). Plastic waste pyrolysis oil via NMR analysis has been shown to consist of predominantly paraffinic and olefinic hydrocarbons (Das and Tiwari 2018). A study (Kayacan and M. 2008) researched to gain insight into the pyrolysis process of low-density polyethylene (LDPE) and high-density polyethylene (HDPE), done using a lab-scale batch-wise Pyrex® reactor enclosed within a vertical furnace. This facilitated the collection of the pyrolysis oil byproducts generated during the process of pyrolysis. In these experiments, nitrogen gas was employed as a carrier, while the temperature was systematically adjusted within the range of 293 to 823 K. In order to examine the structural characteristics of the pyrolysis products, the researchers employed Fourier Transform Infrared (FTIR) and Nuclear Magnetic Resonance (NMR) spectroscopy techniques, specifically utilizing ^{1}H and ^{13}C NMR spectroscopy (Kayacan 2008).

Gas Chromatography-Mass Spectrometry (GC–MS)

It has high accurate for qualitative and quantitative analysis, suitable for volatile compounds but not applicable for non-volatile and large molecular weight compounds. A recent study (Chin et al. 2014) analysed waste plastic pyrolysis oil using GC–MS which proved that hydrocarbon compounds exist within the C7–C40 range. A study (Yousef et al. 2020a) analysed pyrolyzed metallized food packaging plastics with different concentrations of ZSM-5 Zeolite catalyst. The GC–MS measurements showed Benzene, Toluene, Hexane, p-Xylene, etc. The study (Hernandez-Fernandez et al. 2023) investigated the thermal degradation and pyrolysis of hospital plastic waste, including polyethylene, polystyrene, and polypropylene. Molecular structures with aromatic rings were identified in the gas stream from pyrolysis and oxidation. The pyrolysis did not reveal polychlorinated dibenzo-p-dioxins or polychlorinated dibenzofurans, unlike classical incineration methods. Higher concentrations of CO, CO_2, phenol, acetic acid, and benzoic acid were found using GC–MS.

Fourier-Transform Infrared Spectroscopy (FTIR)

This technique is highly advantageous for identifying functional groups, and its notable advantage lies in its rapidity and non-destructive nature with respect to the samples. Nevertheless, this technique possesses several limitations that prevent it from detecting organic molecules.

The analysis of metallized food packaging polymers was conducted by Yousef et al. (1763) using TG-FTIR-GC–MS methodology. The findings indicated by the FTIR spectra and GC–MS data demonstrated the existence of several chemicals, such as propene, furan, 2,4-dimethyl-1-heptene, and isopropylcyclobutane, among other substances. The aforementioned chemicals had a key role in the generation of volatile products. Moreover, a study conducted by authors in Rajan et al. (2023) demonstrated that the main outcomes of the reaction between low-density polyethylene (LDPE) and polypropylene (PP) were alkanes and alkenes, as evidenced by Fourier-transform infrared (FTIR) analysis. In contrast, polyvinyl chloride (PVC) yields hydrochloric acid (HCl), alkenes, and a limited quantity of aromatic compounds as its primary byproducts.

Thermogravimetric Analysis (TGA)

Thermogravimetric analysis (TGA) is a commonly employed technique in both academic research and industrial applications. The process aids in quantifying the alterations in the mass of a given substance as a function of time and temperature. When examining low-density polyethylene (LDPE) using thermogravimetric analysis (TGA), certain similarities can be observed between LDPE and polypropylene (PP), whereas notable distinctions arise when comparing LDPE to polyvinyl chloride (PVC).

According to a study conducted by Rajan et al. (2023), rapid heating induces upward shifts in the initial, end, and peak temperatures, leading to elevated values. A study (Czajczyńska et al. 2017) investigated the pyrolysis characteristics of plastic waste, with a specific focus on the plastic casings found in television sets. TGA and mathematical models were employed to elucidate the degradation mechanism. The researchers also investigated the reaction mechanism and identified certain compensatory effects. The researchers discovered that the decomposition process of plastic wastes in a helium atmosphere may be categorized into three distinct phases: an initial phase characterized by little reduction in weight within the temperature range of 20–300 °C, an intermediate phase marked by substantial weight loss occurring between 300 and 500 °C, and a final phase where a consistent reduction in weight is observed between the temperature range of 500 and 1000 °C. In a separate investigation conducted by authors in Yousef et al. (2020b), it was observed that the predominant degradation of plastics used in food packaging occurred within the temperature range of 360–500 °C, leading to a substantial decrease in mass ranging from 67 to 96%.

Fourier Transform Infrared Spectroscopy (FTIR) and Gas Chromatography-Mass Spectrometry (GC–MS) are extensively employed analytical techniques across diverse scientific disciplines. In recent years, thermal gravimetric analysis (TGA) has garnered increasing attention due to its distinctive capacity to furnish significant insights into the thermal behavior of materials.

11.7 Advantages of Using Pyrolysis Technology in Plastic Waste Management, Current Challenges and Future Prospects

11.7.1 Advantages of Using Pyrolysis Technology in Plastic Waste Management

The global issue of non-biodegradable plastic waste demands urgent and effective solutions. One such promising approach is catalytic pyrolysis, which not only requires lower energy than its thermal counterpart but also produces valuable liquid fuel. Converting waste plastics into an alternative fuel source is feasible and effective using catalytic pyrolysis with indigenous clay as the catalyst. This fuel can be used for transportation, electricity generation, and heating. The heating value of liquid fuel derived from plastic waste is typically greater than that of conventional commercial grade fuel, with values exceeding 10,000 kcal/L (Khan et al. 2016). This characteristic offers advantages in terms of fuel utilisation for boilers, cement plants, steel mills, and glass producers.

Pyrolysis technique has the capability to handle a wide range of waste plastic, including both clean and unclean as well as unsorted varieties. In the context of a batch waste plastic pyrolysis factory, it is arguable that the shredding process may not be deemed essential. The waste pyrolysis reactor, together with its accompanying auxiliary system, encompasses all the necessary procedures involved in the conversion of plastic into fuel. This system is very advantageous due to its exceptional practicality and efficiency in reducing labour requirements. Additionally, there have been documented instances of the co-pyrolysis of plastics with waste paper (Sophonrat 2018), biomass (Supramono et al. 2018), and polymers derived from medical wastes (Baiden 2018). Unlike recycling, pyrolysis does not need the meticulous separation of different polymers. Due to the inherent incompatibility of most plastics, they are unable to undergo simultaneous processing in recycling procedures. As an illustration, the presence of a little quantity of PVC inside the PET recycling process might lead to the degradation of the entire PET resin, resulting in a yellowish appearance and increased fragility (Hopewell et al. 2009).

Pyrolysis is a method that exhibits relative simplicity and possesses the capacity to effectively handle a diverse array of plastic waste categories. This encompasses a wide range of waste streams, encompassing not only packaging waste, but also rubber, waste generated by electrical and electronic devices, medical waste, and end-of-life waste from automobiles. These materials frequently consist of hazardous or noxious compounds. In addition to producing liquid products, pyrolysis also generates a gaseous byproduct that is abundant in hydrocarbons. The heating value of this gas can vary between 25 and 45 MJ/kg, depending on the initial composition and prevailing conditions. Therefore, this characteristic renders it a very suitable contender for energy recuperation. A often employed and effective practise is the utilisation of recirculated pyrolysis-generated gas to provide the necessary heat for the operation. The process of recirculation substantially reduces the dependence on external heating sources.

Depending on the scale of operation, pyrolysis technology offers successful downscaling to microscopic plants, making them nearly equally transportable (Solantausta et al. 2012). These mobile devices may be used in waste concentration sites where feedstock is abundant but transportation costs are high. A distinguishing characteristic of pyrolysis, in comparison to alternative thermochemical conversion techniques, is its predominant generation of liquid oil and wax. These findings have practical benefits due to their ease of storage and versatility in their application as energy sources. The pyrolysis process offers flexibility in tailoring the final product through adjustments to the reactor design, operating temperatures, and utilisation of catalysts. From an economic standpoint, this treatment offers significant advantages. While the primary objective of the process often revolves on enhancing the production of pyrolysis liquid, it is possible to modify the technique to augment the formation of wax (Hopewell et al. 2009), monomers (Africa and Africa 2018), aromatics (Valavanidis et al. 2008; Qureshi et al. 2020), or particular compounds (Santaweesuk and Janyalertadun 2017) by employing a suitable catalyst (Bow et al. 2019).

11.7.2 Current Challenges and Future Prospects

One of the key challenges in pyrolysis of plastic waste is the inherent variability in feedstock composition, stemming from the wide range of plastic types and sources. Advanced chemistry insights into polymer structures and thermal behaviors are crucial for designing effective feedstock blends that can maximize product yields and quality. Researchers are working to develop strategies that account for this variability, including feedstock pre-processing techniques to standardize inputs and optimize conversion processes. Addressing this challenge is crucial for achieving consistent and efficient pyrolysis outcomes on both laboratory and industrial scales as shown in Fig. 11.5.

The economic viability of pyrolysis hinges on factors such as feedstock costs, product values, energy consumption, and capital investment. Life cycle assessments (LCAs) informed by advanced chemistry knowledge enable a comprehensive evaluation of the environmental impact of pyrolysis, encompassing greenhouse gas emissions, energy utilization, and resource depletion (Kalifa et al. 2021; Karak et al. 2012; Arafat and Jijakli 2013). Ongoing Research Efforts: Advanced Reactor Designs and Hybrid Processes. Research efforts are actively focused on developing advanced reactor designs that leverage advanced chemistry principles to improve pyrolysis efficiency and product selectivity. Fluidized bed reactors, rotary kilns, and microwave-assisted systems are examples of innovative designs being explored. Additionally, hybrid processes that combine pyrolysis with complementary technologies, such as gasification or incineration, hold promise for enhancing the utilization of waste resources. The integration of advanced chemistry-driven insights into these reactor designs and hybrid processes has the potential to revolutionize plastic waste management by achieving higher yields of valuable products and minimizing environmental impact (Rizwan et al. 2018; Al-Dabbagh 2021; Saif et al. 2022).

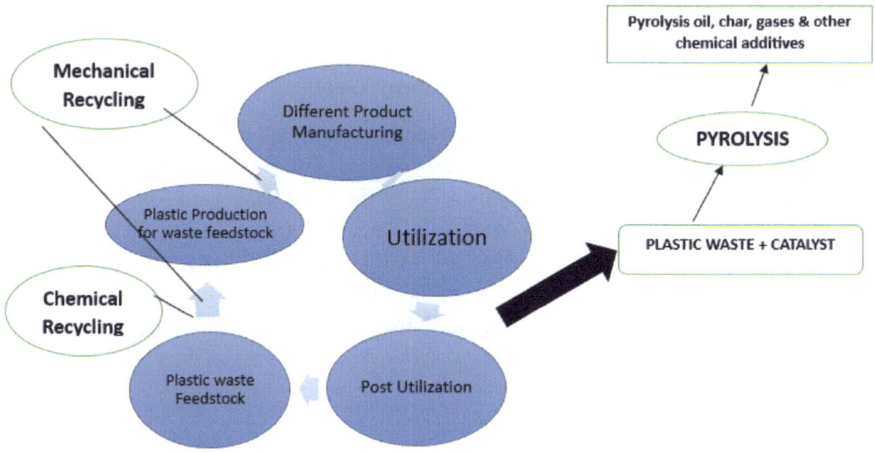

Fig. 11.5 Feedstock variability and reactor scalability

11.7.3 Collaborative Efforts in Innovating Pyrolysis Technology

Addressing the complex challenges posed by plastic waste and advancing pyrolysis technology requires collaborative efforts among researchers, industries, policymakers, and regulatory bodies. Interdisciplinary collaboration between chemists, engineers, environmental scientists, and economists is essential for developing holistic solutions. Furthermore, partnerships between academia and industry facilitate the translation of laboratory findings into practical applications, accelerating the adoption of advanced pyrolysis technologies in waste management systems. Collaborative initiatives ensure that the potential of pyrolysis is fully realized and that the technology aligns with circular economy principles. While pyrolysis holds immense potential, challenges such as feedstock variability, reactor scalability, and technology integration persist. Multidisciplinary collaboration, encompassing chemistry, engineering, and policy expertise, is crucial to overcoming these hurdles. Ongoing research efforts focus on the development of advanced reactor designs, improved catalytic systems, and hybrid processes that synergistically integrate pyrolysis with other waste-to-value technologies.

11.7.4 Advanced Chemistry's Role in Pyrolysis of Plastic Waste

The review underscores the pivotal role that advanced chemistry plays in shaping the pyrolysis of plastic waste as a route to a circular economy. From influencing feedstock selection and reaction pathways to enhancing product yields and quality, advanced chemistry insights are pivotal in optimizing pyrolysis processes. This approach transforms pyrolysis from a mere waste disposal method to a strategic waste management solution that aligns with circular economy goals.

Significance for Achieving Circular Economy Goals

The integration of advanced chemistry knowledge (catalysis) in pyrolysis contributes significantly to the broader objectives of the circular economy. By converting plastic waste into valuable resources, pyrolysis reduces the need for virgin fossil resources, minimizes waste generation, and promotes resource efficiency. The circular economy framework emphasizes the responsible use of resources and the reduction of environmental impacts, aligning perfectly with the advanced chemistry-guided pyrolysis approach (Chen et al. 2022; Elliott 1996; Sharma et al. 2022).

11.8 Conclusion and Recommendation

11.8.1 Conclusion

The use of advanced chemistry principles and the pyrolysis of plastic waste presents a transformative approach that holds significant promise in addressing the escalating global plastic waste crisis while advancing the ideals of a circular economy. This study has shed light on the essential part that cutting-edge chemistry plays in directing every stage of the pyrolysis process, from the selection of feedstocks and the optimisation of reactions to the characterisation of products and the subsequent valorisation of byproducts. Pyrolysis emerges not only as a solution for waste management but also as a catalyst for sustainable resource utilisation and environmental stewardship as a result of this combination. Previous studies are able to get insights into the complex relationships that regulate pyrolysis processes by using their extensive chemical expertise. These insights are unique and valuable. This level of knowledge makes it possible to alter the circumstances under which reactions take place in order to direct the conversion of plastic waste towards the production of certain goods. These products might include liquid fuels, valuable compounds, or even feedstocks for industrial operations. In addition, modern analytical methods, like as spectroscopy and chromatography, make it possible to conduct a full analysis of the products of pyrolysis. This opens up a window into the processes that are responsible for the transformation of plastics into useful resources. The relevance of modern chemistry goes far beyond the optimisation of processes; it also incorporates the question of whether or not pyrolysis is economically feasible and considers its effects on the environment. The incorporation of advanced chemistry into life cycle assessments enables the generation of a comprehensive assessment of the effect that pyrolysis has on the environment. This assessment takes into account energy consumption, emissions, and the depletion of resources. This nuanced understanding is essential for making informed decisions that contribute to a sustainable circular economy. In confronting the challenges of plastic waste management, advanced chemistry has guided innovations aimed at overcoming feedstock variability and reactor scalability. The pursuit of advanced reactor designs and the exploration of hybrid processes underscore the dynamic nature of pyrolysis research, as scientists and engineers collaborate to enhance efficiency, yield, and sustainability. Collaborative endeavors between academia, industry, and regulatory bodies are pivotal in translating laboratory findings into real-world solutions, ensuring that pyrolysis-driven circular economy practices become a tangible reality. As this review has illustrated, the role of advanced chemistry in the pyrolysis of plastic waste transcends its chemical underpinnings. It represents a paradigm shift, transforming the conversation surrounding plastic waste from one of crisis to one of opportunity. The promise of pyrolysis lies in its potential to decouple plastic production from fossil resource extraction and to redirect waste toward a new life as valuable resources. This alignment with circular economy principles resonates with the broader global agenda of sustainability, resource efficiency, and environmental resilience. In summary, when we combine advanced chemistry and pyrolysis, it presents a compelling narrative of innovation, green practices, and progress. This report shines a light on a pathway that can drive more research, get industries on board, and shape future policies that contribute to a world where plastic waste is no longer a burden but a catalyst for positive change. By embracing advanced chemistry principles in the context of pyrolysis, we embark on a journey toward a circular economy and sustainable environment that values not only the potential of our resources but also the well-being of our planet and future generations.

11.8.2 Recommendation

Pyrolysis clearly has a lot of advantages, but it's not without its challenges. As highlighted in this paper, continuous research efforts are vital for overcoming technical challenges, improving pyrolysis reactor designs, and enhancing the sustainability of pyrolysis processes. Governments, industries, and researchers are urged to continue collaborating to facilitate the widespread implementation of advanced pyrolysis technologies. By nurturing the fusion of advanced chemistry principles with waste management practices, we can transform the plastic waste crisis into an opportunity for sustainable resource utilization, fostering a more circular and resilient future. To achieve a successful integration of plastic waste pyrolysis into waste management systems there is a need to support policy frameworks and regulations that incentivize sustainable practices. Governments, in collaboration with plastic industries and researchers should play a pivotal role in fostering an enabling environment for pyrolysis-based circular economy solutions. In order to conduct a comprehensive analysis, it is important to acknowledge that catalytic pyrolysis presents certain benefits in relation to energy efficiency. However, it is crucial to highlight that the identification and accessibility of appropriate catalysts still necessitate additional investigation.

References

Abu Jadayil W (2011) Experimental investigation of solidification time effects on surface and subsurface aluminum casting defects. Int Rev Mech Eng 5:569. https://www.researchgate.net/journal/International-Review-of-Mechanical-Engineering-IREME-1970-8742

Abu Jadayil W, Khraisat W, Shakoor M (2017) Different strategies to improve the production to reach the optimum capacity in plastic company. Cogent Eng 4:1389831. https://doi.org/10.1080/23311916.2017.1389831

Adeoye AO, Quadri RO, Lawal OS (2022) Assessment of biofuel potential of Tenera Palm Kernel shell via fixed bed pyrolysis and thermal characterization. Results Surf Interface 9:100091. https://doi.org/10.1016/j.rsurfi.2022.100091

Adeoye AO, Quadri RO, Lawal OS (2023a) Biomass waste pyrolysis in the context of Nigeria's oil and gas industry: an overview. Int J Ambient Energy 44(1):2525–2541. https://doi.org/10.1080/01430750.2023.2256325

Adeoye AO, Quadri RO, Lawal OS (2023b) Wet synthesis, characterization of goethite nanoparticles and its application in catalytic pyrolysis of palm kernel shell in TGA. Results Surf Interfaces 11:100118. https://doi.org/10.1016/j.rsurfi.2023.100118

Adeoye AO, Quadri RO, Lawal OS (2024) Physicochemical characterization, valorization of lignocellulosic waste (Kola nut seed shell) via pyrolysis, and ultrasonication of its crude bio-oil for biofuel production. https://doi.org/10.1016/j.clwas.2024.100138

ESI Africa S. Africa (2018) 1.144 m tonnes of recyclable plastic dumped in landfills. Available from: https://www.esi-africa.com/top-stories/s-africa-1-144m-tonnes-of-recyclable-plastic-dumpedin-landfills/

Ajaj R, Abu-Jadayil W, Anver H, Aqil E (2022) A revision for the different reuses of polyethylene terephthalate (PET) water bottles. Sustainability 14:4583. https://doi.org/10.3390/su14084583

Al-Dabbagh R (2021) Waste management strategy and development in Ajman. UAE. Renew Energy Environ Sustain 6:14

Arafat HA, Jijakli K (2013) Modeling and comparative assessment of municipal solid waste gasification for energy production. Waste Manage 33:1704–1713

Arena U, Mastellone ML (2006) Fluidized bed pyrolysis of plastic wastes. Feedstock Recycling and Pyrolysis of Waste Plastics. John Wiley & Sons, Ltd. 435–474

Baiden T (2018) Pyrolysis of plastic waste from medical services facilities into potential fuel and/or fuel additives. Master's Theses, 3696 (2018). Available from: https://scholarworks.wmich.edu/masters_theses/3696

Bauer M, Lehner M, Schwabl D, Flachberger H, Kranzinger L, Pomberger R, Hofer W (2018) Sink–float density separation of post-consumer plastics for feedstock recycling. J Mater Cycles Waste Manag 20:1781–1791. https://doi.org/10.1007/s10163-018-0748-z

Bhaskar T, Kaneko J, Muto A, Sakata Y, Jakab E, Matsui T, Uddin A (2004) Pyrolysis studies of PP/PE/PS/PVC/HIPS-Br plastics mixed with PET and dehalogenation (Br, Cl) of the liquid products. J Anal Appl Pyrolysis 72:27–33

Bodzay B, Bánhegyi G (2016) Polymer waste: controlled breakdown or recycling? Int J Design Sci Technol 22(2):109–138

Bow Y, Rusdianasari Pujiastuti LS (2019) Pyrolysis of polypropylene plastic waste into liquid fuel. In: IOP Conf Ser: Earth Environ Sci 347:012128

Butler E, Devlin G, McDonnell K (2011) Waste polyolefins to liquid fuels via pyrolysis: review of commercial state-of-the-art and recent laboratory research. Waste Biomass Valor 2(3):227–255

Caballero B, de Marco I, Adrados A, López-Urionabarrenechea A, Solar J, Gastelu N (2016) Possibilities and limits of pyrolysis for recycling plastic rich waste streams rejected from phones recycling plants. Waste Manag 57:226–234. https://doi.org/10.1016/j.wasman.2016.01.002

Chen J, Mu X, Tu C, Hu G (2022) Influence of ecological cognition on consumers' willingness to participate in beverage packaging recycling. Resour Conserv Recyc Adv 14:200070

Chin BL, Yusup S, Al Shoaibi A, Kannan P, Srinivasakannan C, Sulaiman SA (2014) Kinetic studies of co-pyrolysis of rubber seed shell with high density polyethylene. Energy Convers Manag 87:746–753. https://doi.org/10.1016/j.enconman.2014.07.043

Chiu SJ, Cheng WH (2000) Promotional effect of copper (II) chloride on the thermal degradation of poly (ethylene terephthalate). J Anal Appl Pyrolysis 56(2):31–143

Ciliz NK, Ekinci E, Snape CE (2004) Pyrolysis of virgin and waste polypropylene and its mixtures with waste polyethylene and polystyrene. Waste Manage 24:173–181

Czajczyńska D, Anguilano L, Ghazal H, Krzyżyńska R, Reynolds A, Spencer N, Jouhara H (2017) Potential of pyrolysis processes in the waste management sector. Thermal Sci Eng Progress 3:171–197. https://doi.org/10.1016/j.tsep.2017.06.003

Das P, Tiwari P (2018) Valorization of packaging plastic waste by slow pyrolysis. Res Conserv Recyc 128:69–77. https://doi.org/10.1016/j.resconrec.2017.09.025

Davis M (2020) Environmental impacts of plastic recycling. J Sustain Develop 33(2):155–167

Demirbas A (2004) Pyrolysis of municipal plastic wastes for recovery of gasoline range hydrocarbons. J Anal Appl Pyrolysis 72:97–102

Diaz Silvarrey LS, Phan AN (2016) Kinetic study of municipal plastic waste. Int J Hydrogen Energy 41:16352–16364

Donaj PJ, Kaminsky W, Buzeto F, Yang W (2012) Pyrolysis of polyolefins for increasing the yield of monomers' recovery. Waste Manag 32(5):840–846. https://doi.org/10.1016/j.wasman.2011.10.009

Doria F, Pidgeon N, Hunter PR (2009) Perceptions of drinking water quality and risk and its effect on behaviour: a cross-national study. Sci Total Environ 407:5455–5464. https://doi.org/10.1016/j.scitotenv.2009.06.031

Elliott E (1996) Recycling: saving money and the environment. Text. World. 146:72–74. Eng 17:261–268. https://doi.org/10.1016/j.mineng.2003.10.028

Elmously MN, Jäger J, Neidel A, Apfelbacher R, Daschner A (2019) Hornung, Upscaling of thermo-catalytic reforming process from lab to pilot scale. Industrial Engineering Chemistry Research 58:15853–15862

Eze WU, Umunakwe R, Obasi HC, Ugbaja MI, Uche CC, Madufor IC (2021) Plastics waste management: A review of pyrolysis technology. Clean Technol Recycl 1(1):50–69. https://doi.org/10.3934/ctr.2021003

Fraunholcz N (2004) Separation of waste plastics by froth flotation—a review, part I. Miner Eng 17:261–268. https://doi.org/10.1016/j.mineng.2003.10.028

Garside M (2020) "Statista". http://www.statista.com/statistics/282732/global-production-of-plastics-since-1950/

Gopinath S, Devan PK (2020) Optimization and prediction of reaction parameters of plastic pyrolysis oil production using Taguchi method. Iran J Chem Eng 39:91–103. https://doi.org/10.30492/IJCCE.2020.33965

Grause G, Buekens A, Sakata Y, Okuwaki A, Yoshioka T (2011) Feedstock recycling of waste polymeric material. J Mater Cycles Waste Manag 265–282

Gundupalli SP, Hait S, Thakur A (2017) A review on automated sorting of source-separated municipal solid waste for recycling. Waste Manag 60:56–74. https://doi.org/10.1016/j.wasman.2016.09.015

Hense P, Reh K, Franke M, Aigner J, Hornung A, Contin A (2015) Pyrolysis of waste electrical and electronic equipment (WEEE) for recovering metals and energy: previous achievements and current approaches. Environ Eng Manag J 14:1637–1647

Hernandez M, Gomez A, Garcıa A, Agullo J, Marcilla A (2007) Effect of the temperature in the nature and extension of the primary and secondary reactions in the thermal and HZSM-5 catalytic pyrolysis of HDPE. Appl Catal A 317(2):183–194

Hernandez-Fernandez J, Lambis H, Reyes RV (2023) Application of pyrolysis for the evaluation of organic compounds in medical plastic waste generated in the city of cartagena-colombia applying TG-GC/MS. Int J Mol Sci 24(6):5397. https://doi.org/10.3390/ijms24065397

Hopewell J, Dvorak R, Kosior E (2009) Plastics recycling: challenges and opportunities. Philos Trans R Soc B 364(1526):2115–2126

Hornung A, Hornung U, Schoeneberger A, Weichmann J, Bockhorn H (1996) Staged pyrolysis, a process for chemical separation of waste plastics. In: Gestufte Pyrolyse Als Verfahrensprinzip Zur Chemischen Auftrennung Von Kunststoffgemischen

Howell SG (1992) A ten year review of plastics recycling. J Hazard Mater 29:143–164. https://doi.org/10.1016/0304-3894(92)85066-A

Hundertmark T, Mayer M, McNally C, Jan Simons T, Witte C (2023) How plastics waste recycling could transform the chemical industry, https://www.mckinsey.com/industries/chemicals/our-insights/how-plastics-waste-recycling-could-transform-the-chemical-industry. Accessed Sept 2023

Islam MN, Islam MN, MRA Beg (2004) Fixed bed pyrolysis of waste plastic for alternative fuel production. J Energy Environ 3:69–80

Johnson L (2019) Consumer behaviour and waste management. Waste Manag Rev 22(3):47–59

Kalifa A, Al-Maadid A, Koutiva I, Makropoulos C (2021) Individual water consumption behavior in relation to urban residential dynamics: the case of Qatar. Urban Water J 18:806–816

Karak T, Bhagat RM, Bhattacharyya P (2012) Municipal solid waste generation, composition, and management: the world scenario. Crit Rev Environ Sci Technol 42:1509–1630
Kassargy C, Awad S, Burnens G, Kahine K, Tazerout M (2017) Experimental study of catalytic pyrolysis of polyethylene and polypropylene over USY zeolite and separation to gasoline and diesel-like fuels. J Anal Appl Pyrolysis 127:31–37. https://doi.org/10.1016/j.jaap.2017.09.005
Kayacan, Doğan M (2008) Pyrolysis of low and high density polyethylene. Part I: non-isothermal pyrolysis kinetics. Energy Sources A: Recover Util, Environ Eff 30(5):385–391. https://doi.org/10.1080/15567030701457079
Khair F, Matsana J (2021) Sustainable plastic waste management strategy: optimization of plastic manufacturing plant waste (gas and product transition). IOP Conference Series Material Science Engineering 1041:012043. https://doi.org/10.1088/1757-899x/1041/1/012043
Khan MZ, Sultana M, Al-Mamun MR, Hasan MR (2016): Pyrolytic waste plastic oil and its diesel blend: fuel characterization. J Environ Public Health 1–6:2016. https://doi.org/10.1155/2016/7869080
Kirby M, Hornung A, Ouadi M, Theodorou M (2017) The role of thermo-catalytic reforming for energy recovery from food and drink supply chain wastes. Energy Proc 123:15–21
Kumar KP, Srinivas S (2020) Catalytic co-pyrolysis of biomass and plastics (polypropylene and polystyrene) using spent FCC catalyst. Energy Fuels 34:460–473
Lange JP (2021) Managing plastic waste–sorting, recycling, disposal, and product redesign. ACS Sustain Chem Eng 9(47):15722–15738. https://doi.org/10.1021/acssuschemeng.1c05013
Lee K (2009) Thermal and catalytic degradation of pyrolytic oil from pyrolysis of municipal plastic wastes. J Anal Appl Pyrol 85(1–2):372–379
Lee K (2012) Effects of the types of zeolites on catalytic upgrading of pyrolysis wax oil. J Anal Appl Pyrol 94:209–214
Lewandowski K, Skórczewska K (2022) A brief review of poly(vinyl chloride) (PVC) recycling. Polymers 14(15):3035. https://doi.org/10.3390/polym14153035
Lim S, Cho K, Chan H (2003) Separation of wasted plastics by thermal adhesion. J Kor Inst Resour Recycl 4:44–50
van der Linden S () Exploring beliefs about bottled water and intentions to reduce consumption. Environ Behav 47:526–550. https://doi.org/10.1177/0013916513515239
López A, de Marco I, Caballero BM, Laresgoiti MF, Adrados A (2010) Pyrolysis of municipal plastic wastes: influence of raw material composition. Waste Manage 30:620–627
Lopez G, Artetxe M, Amutio M, Bilbao J, Olazar M (2017) Thermochemical routes for the valorization of waste polyolefinic plastics to produce fuels and chemicals: a review. Renew Sustain Energy Rev 73:346–368
López A, de Marco I, Caballero B, Laresgoiti M, Adrados A (2011) Influence of time and temperature on pyrolysis of plastic wastes in a semi-batch reactor. Chem Eng J 173(1):62–71. https://doi.org/10.1016/j.cej.2011.07.037
Ludlow-Palafox C, Chase HA (2001) Microwave-induced pyrolysis of plastic wastes. Ind Eng Chem Res 40:4749–4756
Manos G, Yusof IY, Papayannakos N, Gangas NH (2001) Catalytic cracking of polyethylene over clay catalysts. Comparison with an Ultrastable Y zeolite. Ind Eng Chem Res 40(10):2220–2225. https://doi.org/10.1021/ie001048o
La Mantia FP (2004) Polymer mechanical recycling: downcycling or upcycling? In: Progress in rubber, plastics and recycling technology 20(1):11–24. https://doi.org/10.1177/147776060402000102
Miandad R, Barakat M, Aburiazaiza AS, Rehan M, Nizami A (2016) Catalytic pyrolysis of plastic waste: a review. Process safety and environmental protection 102:822–838, (2016). https://doi.org/10.1016/j.psep.2016.06.022
Mibei ZC, Kumar A, Talai SM (2023) Catalytic pyrolysis of plastic waste to liquid fuel using local clay catalyst. J Energy 1–11. https://doi.org/10.1155/2023/7862293

Mihai FC, Gündoğdu S, Markley LA, Olivelli A, Khan FR, Gwinnett C, Molinos-Senante M (2021) Plastic pollution, waste management issues, and circular economy opportunities in rural communities. Sustainability 14(1):20. https://doi.org/10.3390/su14010020

Ouda OR, Raza SA, Nizami AS, Rehan M, alWaked R, Korres NE (2016) Waste to energy potential: a case study of Saudi Arabia. Renew Sustain Energy 61:328–340

Panda AK, Singh RK (2011) Catalytic performances of kaoline and silica alumina in the thermal degradation of polypropylene. J Fuel Chem Technol 39:198–202

Poly(ethene) (Polyethylene). (n.d.). Retrieved from https://www.essentialchemicalindustry.org/polymers/polyethene.html#:~:text=Well%20over%2080%20million%20tonnes,the%20ethene%20manufactured%20each%20year. Accessed Sept 2023

Pratama NN, Saptoadi H (2014) Characteristics of waste plastics pyrolytic oil and its applications as alternative fuel on four-cylinder diesel engines. Int J Renew Energy Develop 3:13–20

Qureshi MS, Oasmaa A, Pihkola H, Deviatkin I, Tenhunen A, Mannila J, Laine-Ylijoki J (2020) Pyrolysis of plastic waste: opportunities and challenges. J Anal Appl Pyrolysis 152:104804. https://doi.org/10.1016/j.jaap.2020.104804

Ragaert K, Delva L, Van Geem K (2017) Mechanical and chemical recycling of solid plastic waste. Waste Manage 69:24–58

Rajan KP, Mustafa I, Gopanna A, Thomas SP (2023) Catalytic pyrolysis of waste low-density polyethylene (LDPE) carry bags to fuels: experimental and exergy analyses. Recycling 8(4):63. https://doi.org/10.3390/recycling8040063

Ratnasari DK, Nahil MA, Williams PT (2016) Catalytic pyrolysis of waste plastics using staged catalysis for production of gasoline range hydrocarbon oils. J Anal Appl Pyrol 124:631–637

Rizwan M, Saif Y, Almansoori A, Elkamel A (2018) Optimal processing route for the utilization and conversion of municipal solid waste into energy and valuable products. J Clean Prod 174:857–867. https://doi.org/10.1016/j.jclepro.2017.10.335

Saif Y, Griffiths S, Almansoori A (2022) Municipal solid waste supply chain management under an integrated optimization of sustainability targets. Comput Chem Eng 160:107725 (2022). https://doi.org/10.1016/j.compchemeng.2022.107725

Santaweesuk C, Janyalertadun A (2017) The production of fuel oil by conventional slow pyrolysis using plastic waste from a municipal landfill. Int J Environ Sci Develop 8:168–173

Sarker M, Kabir A, Rashid M, Molla M, Mohammad A (2011) Waste polyethylene terephthalate (PETE-1) conversion into liquid fuel. J Fund Renew Energy Appl 1:1–5

Schyns ZO, Shaver MP (2020) Mechanical recycling of packaging plastics: a review. Macromol Rapid Commun 42(3). https://doi.org/10.1002/marc.202000415

Scott DS, Czernik SR, Piskorz J, Radlein DS () Fast pyrolysis of plastic wastes. Energy Fuels 4(4):407–411, (1990). https://doi.org/10.1021/ef00022a013

Serrano DP, Aguado J, Escola JM, Garagorri E (2001a) Conversion of low-density polyethylene into petrochemical feedstocks using a continuous screw kiln reactor. J Anal Appl Pyrolysis 58:789–801

Serrano DP, Aguado J, Escola JM, Garagorri E (2001b) Conversion of low density polyethylene into petrochemical feedstocks using a continuous screw kiln reactor. J Anal Appl Pyrolysis 58:789–801

Shaha AP, Singamsetti MS, Tripathy BK, Srivastava G, Bilal M, Nkenyereye L (2020) Performance prediction and interpretation of a refuse plastic fuel fired boiler. IEEE Access 8:117467–117482. https://doi.org/10.1109/access.2020.3004156

Shareefdeen Z, Youssef N, Taha A, Masoud C (2020) Comments on waste to energy technologies in the United Arab Emirates (UAE). Environ Eng Res 25:129–134

Sharma SK, Mohapatra S, Sharma RC, Alturjman S, Altrjman C, Mostarda L, Stephan T (2022) Retrofitting existing buildings to improve energy performance. Sustainability 14:666. https://doi.org/10.3390/su14020666

Sipra AT, Gao N, Sarwar H (2018) Municipal solid waste (MSW) pyrolysis for bio-fuel production: a review of effects of MSW components and catalysts. Fuel Process Technol 175:131–147

Solantausta Y, Oasmaa A, Sipilä K, Lindfors C, Lehto J, Autio J, Jokela P, Alin J, Heiskanen J (2012) Bio-oil production from biomass: steps toward demonstration. Energy Fuels 26(1):233–240

Sophonrat N (2018) Pyrolysis of mixed plastics and paper to produce fuels and other chemicals. Doctoral Dissertation, KTH Royal Institute of Technology School of Industrial Engineering and Management Department of Materials Science and Engineering Unit of Processes, Sweden

Supramono D, Nabil M, Setiadi A, Nasikin M (2018) Effect of feed composition of co-pyrolysis of corncobs–polypropylene plastic on mass interaction between biomass particles and plastics. IOP Conf Series Earth Environ Sci 105:012049. https://doi.org/10.1088/1755-1315/105/1/012049

Syamsiro M, Saptoadi H, Norsujianto T, Noviasri CS, Alimuddin Z, Yoshikawa K (2014) Fuel oil production from municipal plastic wastes in sequential pyrolysis and catalytic reforming reactors. Energy Process 47:180–188

Teuten EL, Saquing JM, Knappe DR, Barlaz MA, Jonsson S, Björn A, Rowland SJ, Thompson RC, Galloway TS, Yamashita R, Ochi D, Watanuki Y, Moore C, Viet PH, Tana TS, Prudente M, Boonyatumanond R, Zakaria MP, Akkhavong K, Ogata Y, Hirai H, Iwasa S, Mizukawa K, Hagino Y, Imamura A, Saha M, Takada H (2009) Transport and release of chemicals from plastics to the environment and to wildlife. Philos Trans R Soc B: Biol Sci 364:2027–2045. https://doi.org/10.1098/rstb.2008.0284

Tilmatine A, Medles K, Bendimerad SE, Boukholda F, Dascalescu L (2009) Electrostatic separators of particles: Application to plastic/metal, metal/metal and plastic/plastic mixtures. Waste Manag 29:228–232. https://doi.org/10.1016/j.wasman.2008.06.008

Uddin A, Koizumi K, Sakata Y (1996) Thermal and catalytic degradation of high and low density polyethylene into fuel oil. United State. (1996) Retrieved from https://www.osti.gov/biblio/430479

Uemichi Y, Junko N, Toshihiro I, Masatoshi S, Arthur A, Garforth John D (1999) Conversion of polyethylene into gasoline-range fuels by two-stage catalytic degradation using silica−alumina and HZSM-5 zeolite. Ind Eng Chem Res 38(2):385–390. https://doi.org/10.1021/ie980341+

Valavanidis A, Iliopoulos N, Gotsis G, Fiotakis K (2008) Persistent free radicals, heavy metals and PAHs generated in particulate soot emissions and residue ash from controlled combustion of common types of plastic. J Hazard Mater 156(1–3):277–284. https://doi.org/10.1016/j.jhazmat.2007.12.019

Verma R, Vinoda KS, Papireddy M, Gowda AN (2016) Toxic pollutants from plastic waste—a review. Proc Environ Sci 35:701–708

Wan HM (2015) Waste plastics into fuel oil? Inst Sci Soc

Wong HW, Broadbelt LJ (2001) Tertiary resource recovery from waste polymers via pyrolysis: neat and binary mixture reactions of polypropylene and polystyrene. Ind Eng Chem Res 40(22):4716–4723. https://doi.org/10.1021/ie010171s

www.plasticeurope.org factsheet

Yansaneh O, Zein S (2022) Latest advances in waste plastic pyrolytic catalysis. Processes 10(4):683–684

Yousef S, Eimontas J, Striūgas N, Abdelnaby MA (2020a) Modeling of metalized food packaging plastics pyrolysis kinetics using an independent parallel reactions kinetic model. Polymers 12(8):1763. https://doi.org/10.3390/polym12081763

Yousef S, Eimontas J, Striūgas N, Zakarauskas K, Praspaliauskas M, Abdelnaby MA (2020b) Pyrolysis kinetic behaviour and TG-FTIR-GC–MS analysis of metallised food packaging plastics. Fuel 282:118737. https://doi.org/10.1016/j.fuel.2020.118737

Zhao X, Korey M, Li K, Copenhaver K, Tekinalp H, Celik S, Ozcan S (2022) Plastic waste upcycling toward a circular economy. Chem Eng J 428:131928. https://doi.org/10.1016/j.cej.2021.131928

Chapter 12
Revolutionizing Transportation: The Future Impact of Green Energy

Mohd Avesh, Ismail Hossain, and Rakesh Chandmal Sharma

12.1 Introduction

Indian Railways is a vital national asset for India, serving as a unified transportation network that connects the country's villages to major cities. Operating an impressive fleet of 19,000 passenger and freight trains, it ranks among the world's largest logistics and transportation networks. Each day, it manages the operation of 12,000 trains, facilitating the movement of approximately 23 million passengers across 8,500 stations throughout the subcontinent. To put this in perspective, it's akin to relocating the entire population of Australia. Furthermore, Indian Railways operates around 7,000 freight trains daily, carrying an astonishing 1.3 billion tons of cargo. Its network spans an impressive 65,000 path kilometers, surpassing half the circumference of the globe. With a freight load of 1008.09 million tons in 2012–13, it outperforms other regions such as China, Russia, and the United States in annual freight transport. Projections indicate that Indian Railways is poised to transport over 1.1 billion tons of cargo in the 2014–15 year (Adeoye et al. 2023a).

Transportation accounts for about 22.7% of global energy demand, with diesel and petrol being the most popular fuels. According to predictions, by 2035, the proportion would have increased to 60%, or half of overall global consumption (Sharma

M. Avesh (✉)
Mazedan International Research Academy, Okhla, New Delhi, Delhi 110025, India
e-mail: mail2avesh@gmail.com

I. Hossain
Department of Nuclear and Renewable Energy, Ural Federal University, Yekaterinburg 620002, Russia

R. C. Sharma
Mechanical Engineering Department, Graphic Era (Deemed to Be University), Dehradun 248002, India

Mechanical Engineering Department, Graphic Era Hill University, Dehradun 248002, India

S. K. Sharma et al. (eds.), *Dynamics of Transportation Ecosystem, Modeling, and Control*, Energy, Environment, and Sustainability,
https://doi.org/10.1007/978-981-97-0437-8_12

et al. 2020c). Furthermore, railways are regarded as a secure, energy-efficient, and low-carbon mode of transportation that plays a critical role not only in a country's economic development but is increasingly being viewed as a key player in the transition to a low-carbon, cohesive system by attracting passengers from other modes of transportation, optimizing electrification, incorporating technological innovations, and constructing infrastructure. Since 1975, the number of passenger-kilometers carried by rail has increased by 130 percent worldwide (Mohapatra et al. 2016a, 2016b; Gopala Rao et al. 2022). In this region, India is the leading contributor. The energy consumption in railways is rising to satisfy the demands of rail traffic. As a result, implementing energy-saving programs in transportation systems is a critical agenda item in numerous countries to achieve sustainable transportation infrastructure as rail traffic grows (Kohli et al. 2022; Sharma and Kumar 2017a, 2018; Sharma et al. 2021b).

Indian Railways (IR) operates one of the world's biggest rail networks, with 115,000 km (71,000 miles) of road, a 65,436-km (40,660-mile) path, and 8,495 stations. IR transported 8.397 billion passengers in 2014–15, or more than 23 million travelers a day, as well as 1050 million tons of shipping. By 2015, India's railway lines have been electrified for a total of 26,000 km, accounting for about 40% of the country's total railway network (Bhardawaj et al. 2020d; Sharma et al. 2022a; Bandekar et al. 2017; Hossain et al. 2017; Mohapatra et al. 2017a). All around the planet, transportation services are in high demand. Based on existing transportation market patterns, passenger and freight transportation activity is projected to double by 2050. This type of development indicates social and economic transition, but it also implies increasing power use (half of international coal consumption) and a quarter of regional carbon (CO_2) emissions. As a result, transportation market transitions are analogous to global oil transit. What are the possibilities for using renewable energy sources (RES) in airplanes, trains, cars, and other modes of transportation (Sharma and Sharma 2019; Swapna Rekha et al. 1120; Sinha et al. 2019; Goswami et al. 2019; Sharma et al. 2022b)

Energy stability, sustainability, deforestation, and the effects of climate change are some of the main threats that the current global economy faces. Some scholars and organizations have argued that transitioning to a 100 percent green economy is the only path to find a long-term answer to these problems. This decision is founded on the assumption that renewables are now developed innovations that are rapidly developing and will have a carbon footprint of nil (Mohanty et al. 2016; Samantaray et al. 2014; Pattnaik et al. 2022; Nibedita et al. 2021; Biswas et al. 2021; Sharma et al. 2022c). This last aspect makes them particularly well-suited to addressing climate change, which is perhaps the most pressing problem confronting humanity. Any shift to renewable energy in the medium term would almost certainly be aided by the wise use of carbon fuels, especially gas and oil, which are decreased dispatchable mediums that can supplement unreliable renewables. According to some of the studies listed above, our economy could become completely green in the long run. It will take between 40 and 50 years to complete a significant infrastructure restructuring and a globally concerted policy intervention (Sharma et al. 2020d, 2021c; Pundir et al. 2023a; Das et al. 2023). Despite the immediate need for such a transformation to avert drastic climate change, policymakers have failed to support such a concerted action effort. A feasible date for the start of a global green transition may be the end of all fossil fuel supply, which is estimated to occur between 2020 and 2036 by various researchers (Agrahari et al. 2021a; Sharma et al. 2022d; Maity et al. 2020; Misra et al. 2022), with completion in the second half of this decade. The 50-year timeframe is also consistent with the duration it takes for technical advances to spread across the economy in the past (Sharma and Lee 2020a; Mohapatra et al. 2017b; Kohli et al. 2020).

One of the most daunting facets of such a process continues to be the conversion of the current transportation infrastructure. Public transportation is also largely reliant on fossil fuels (mostly oil), which are predicted to diminish throughout the coming decades; additionally, global transportation emits a large amount of greenhouse emissions, pollutes urban environments, and is responsible for millions of incidents per year (Devadarshini et al. 2022; Sharma et al. 2021; Sarkar et al. 2018). This thesis examines existing global transportation trends, attitudes, and uses, as well as the major challenges that must be resolved in order for the global transportation sector to become completely decarbonized. We attempt to address the questions about how a 100 percent sustainable transportation infrastructure might be constructed using developed technology, what a 100 percent renewable economy's transportation system will look like, what the current system's capital and energy prices will be (Palli et al. 2022; Bhardawaj et al. 2020b, 2019a).

Transportation is essential in today's globalized environment because it allows for the trade of commodities and citizen connectivity, and it is one of the primary triggers of suburbanization of towns. The comprehensive studies on wear mitigation in locomotive components, lateral stability, ride comfort, and advanced suspension systems have significantly contributed to enhancing the performance and safety of rail and road vehicles (Kumar et al. 2022; Sharma et al. 2022e, f, g; Sharma and Sharma 2022; Vishwakarma et al. 2022a, 2022b). However, a heavy reliance on fossil fuels is one of the main issues resulting from the restructuring of the global transportation system. Fuel, in fact, is the most important energy source in the transportation energy mix, accounting for over 94% of overall energy production, followed by 3% for natural gas and other liquids, 2% for biofuels, and 1% for electricity. This broad number may be examined by looking at different means of transportation or how people utilize them (by passengers or freight) (Sharma and Lee 2020b, 2021c; Acharya et al. 2020; Pundir et al. 2023b). Light Commercial Vehicles (LDVs) absorbed about half of the overall transport resources in terms of passenger transportation. In 2009, freight transportation absorbed nearly 45 percent of overall transportation capacity, with Heavy Duty Vehicles (HDVs) accounting for more than half of it. The synthesis of all these fuels from batteries will consume a relatively high fraction of secondary resources if fuels were held as the primary energy carriers of energy in a 100% sustainable economy. Recent studies have significantly contributed to the understanding of Aerodynamic Optimization of rail vehicle and the improvement of railway systems. (Palli et al. 2018; Sharma et al. 2017; Sharma and Sharma 2018a; Sharma and Kumar 2018b, 2018c; Sharma and Sharma 2018b, 2018c). Furthermore, this demonstrates that a 100% renewable economy may face significant challenges in expanding beyond 12

TWa/a electricity generation. For the current scale of the petrochemical industry, the power source and the carbon that would be sustainably viable would not be enough to meet the electricity need, the demand for methane in biodiesel synthesis (for agriculture), and the synthetic production of gases and olefins. As a result, hydrocarbons will be limited and costly in such a situation, and the direct usage of the grid by the transportation industry, if necessary, will be the most cost-effective alternative (COVID-19 2020; Pattnaik et al. 2020; Mohapatra et al. 2020, 2022; Sharma et al. 2020e).

As a result, it is prudent to rationalize and reshape the transportation infrastructure and make direct and informed use of renewable energy. According to the study below, land transportation is largely feasible, but air and maritime transportation will have to depend primarily on renewable energy-derived fuels. Gilbert and Perl describe a series of revolutionary improvements that must be made to the existing transportation infrastructure in order to pursue transformation plans that will carry more individuals and freight without the use of oil until it is too late to avert a global energy shortage. These writers are likely underestimating the importance of intelligent fossil fuel usage in a transportation transformation (Adeoye et al. 2023b, c; Pati et al. 2022, 2020; Mohapatra et al. 2017c, d; Chatterjee et al. 2022; Nayak et al. 2022). They do, however, correctly emphasize the importance of fostering a transportation renaissance that refashions the current close links between mobility and oil-based energy sources. This isn't the only argument to favor a full transit reorganization. As Swenson points out, a direct substitution of today's fleet of internal combustion engines with an equally large fleet of electric motors would preserve the status quo in today's towns, which are clogged with automobiles and unsafe for pedestrians. The authors studied different scenarios for Energy system development under socioeconomic and ecological constraints as part of the MEDEAS project ("Modeling the Renewable Energy Transition in Europe," European Union's Horizon 2020 technology and development program, grant agreement No 691287EU of the Framework Program for Innovation actions, H2020 LCE-21-2015). This research, which is part of the effort, looks at some of the key developments that have been proven or are in the demonstration stage and may be used to replace conventional fossil fuel-based transportation. We determine which types of transportation are more compliant with content and energy restrictions. There has since been a calculation of the expense and resources needed to make such a move.

The study breaks down the overall framework of the global transportation system into and major carrier subsector, such as automobiles, aircraft, aviation, freight, urban transportation, and so on. We expect potential transportation to be primarily driven by (1) onboard batteries, (2) the electric grid, and (3) methane or methane to power fuel cells including internal combustion engines (ICE) of heavy-duty cars, ships, and aircraft, where the former options are unavailable. We assess the present and future costs of all three outlets in terms of energy efficiency, vital resource supply, and potential technical breakthroughs that might mitigate their shortcomings. The energy cost of aircraft fueled with hydrogen, liquid methane, and jet fuel. It gets better, faster, and in some instances, less expensive. Latest innovations, on the other hand, have not always proven useful to the planet. Transport absorbs 30% of global energy and has the lowest renewable energy share of any industry, according to a 2016 study by the International Renewable Energy Agency (IRENA). The country's energy demand is expected to increase by 1% each year until 2030. Transportation uses one-third of the world's oil and emits 14% of global greenhouse gas emissions. IRENA lays out a green roadmap to a more sustainable future in the transportation industry. They've described three places where development is required to bring Rebinding's findings to life: A Roadmap for the Future of Renewable Energy the Reconfigure is a global green energy roadmap that aims to reduce renewable geothermal share of the global energy mix by half by 2030 (Adeoye et al. 2022; Sharma et al. 2020a, b). The below are some of the action areas:

Use a system strategy approach that connects energy sectors as one of the three action areas, and increase electric vehicles in combination with sustainable electricity production.
Provide creative green energy pathways for all forms of transportation, including non-car models like freight, aircraft, and storage, that are both clean and cost-effective.
Look at innovative technology and innovations for evolving modes of transportation including aircraft, shipping, and military applications.

They also recommended taking steps to increase the usage of clean energy in transportation. In addition, IRENA has identified two major regions where transportation activities can be concentrated. Electricity is used to fuel these. In a liquid environment, vehicles and fossil fuels the global supply of electric vehicles (EVs) reached 1 million in 2015, and it surpassed 2 million by the end of 2016. However, for EVs to play a role in the global energy transformation, faster growth is needed, both in terms of reducing car pollution and rising the usage of renewable energy. Increased dependence on EV batteries and charging stations would gain higher shares of solar and wind power, the key variable renewable energy (VRE) sources projected to be widespread in potential power grids. Liquid biofuels are an alternative fuel for all internal combustion engines that operate on oil, petrol, or kerosene, such as passenger vehicles, vans, ships, and aircraft. Liquid biofuels, which comprise both conventional and modern forms of ethanol and biodiesel, are expected to account for 10% of transportation sector energy consumption by 2030, up from 3% in 2017. Liquid biofuels would be crucial if shipping and aviation are to become more competitive. These modes of transportation are the fastest-growing segments of the market, accounting for 20% of overall transportation energy production. Sustainable architecture is an important method for reducing road congestion and tackling pollution and environmental problems in large cities. The use of mass transportation and green travel is encouraged (Bhardawaj et al. 2020c).

A change from being vehicle-oriented to being person-oriented is needed for the sake of the environment and human health. Electric vehicles are those that operate on energy. The first electric car was built in 1838, 52 years before the first combustion engine vehicle. However, after 1913, the pervasive corporatization of the combustion engine led to a gradual fall in electric cars. Attempts to reintroduce electric cars in previous decades were often unsuccessful, and they remain a thin, niche market.

Electric vehicles are set to re-enter the industry on a large scale in the coming years, thanks to advances in battery technology. By 2050, electric cars may account for more than 60% of new vehicle purchases and up to 25% of the global car fleet, based on a modest growth scenario. According to a study published in 2012 by the Center for Automotive Research at the University of Duisburg-Essen in Germany, there is a global tendency for light-vehicle production to increasingly shift toward hybrid and electric cars. Combustion engines would account for 56% of automotive demand by 2030, while hybrid technology will account for 35% and electric technology will account for 9%. This suggests that 44% of small cars would not have a combustion engine to produce electricity to power the automobile. As a consequence, greenhouse gas (GHG) pollution would be significantly reduced (Palli et al. 2021).

Electric vehicles have significant environmental benefits, especially in urban areas. Many European countries, as well as the United States, Japan, China, and other countries, have also adopted electric cars. In India, the electric vehicle sector is also in its early stages. By 2017, electric vehicles are estimated to account for approximately 5% of the Indian automotive segment, or 1,75,000 vehicles, with the global market for the vehicles projected to reach about 20 million automobiles by 2020. An electric motor and an internal combustion engine combine to fuel a hybrid vehicle. From October 4 to December 2017, a hybrid vehicle with a traditional internal combustion engine and a fuel tank, as well as one or two electric motors and a battery pack, will be available. These are often confused with electric vehicles. Hybrids are usually gasoline-powered hybrids that store and conserve electricity that would otherwise be lost in traditional vehicles (Sharma et al. 2021d). Electric cars emit a fraction of the greenhouse gases that conventional automobiles do, varying from 26 to 90% less. These are a more environmentally friendly and fuel-efficient choice, resulting in lower operating costs and less environmental damage. Hydrogen-powered vehicles Hydrogen is used as a dominant form of electricity instead of gasoline. They haven't been a major hit yet, and only a few hydrogen cars have been made. There are two options to use hydrogen to drive an engine. In standard automobiles, hydrogen is burned in the same manner as gasoline is. Chemical-cell conversion fuel cells, which turn hydrogen into energy, power internal electric motors. Since hydrogen may be produced utilizing green energy sources, there will be no increase in carbon dioxide emissions. Commuting, trading cards, riding public transit (such as a bus), bicycling, walking, and running are also examples of transportation options. Alternative transportation requires more time, but it has a smaller environmental impact. Cleaner and greener transportation choices are a good alternative for commuting throughout the city (Palli and Koona 2015; Kanth Choppara et al. 2019; Mohapatra et al. 2015; Bhardawaj et al. 2019b). It reduces road congestion, noise, and other detrimental effects of driving alone. It saves cost as well as money on petrol and encourages social networking and group building. When freight and commuter transportation expand, caused by traffic congestion becomes more of a concern. The global need to reduce carbon pollutant emissions while still promoting human growth has prompted the creation of clean and affordable transportation networks. Transportation's environmental effect, which allows renewable energy more competitive and less expensive. While transportation has substantial economic advantages, it also has a negative effect on environmental systems. Transportation consumes a significant amount of electricity and emits significant amounts of greenhouse gases. Our travel has improved significantly in several respects over the past few decades. In terms of capability, comfort, and expense, transportation has increased. Technological advances, on the other hand, weren't always useful to the planet (Mohapatra et al. 2021; Bhattarai et al. 2022; Agrahari et al. 2021b).

12.2 Effect of Transportation on the Atmosphere

Passenger and freight stability are aided by transportation operations. Transportation activities, on the other hand, are linked to greater of environmental externalities. The value of transportation as a source of emissions and its myriad environmental consequences has grown as individual and freight movement has increased. More than a quarter of the world's energy is used by the automotive sector, which often releases a similar amount of CO_2 from combustion of fossil fuels. Direct effects: The actual influence of shipping operations on the world, typically with a well-defined cause-and-effect connection. Take, for example, noise and carbon dioxide. Direct harm has been shown to be caused by monoxide pollution.

The influence of transportation operations on secondary environmental systems is referred to as indirect impacts (or tertiary). They have more power than direct results, but the interactions they entail are often confused and challenging to establish. Particulates, among example, which are frequently generated as a consequence of combustion process in a combustion engine, are linked to cardiovascular and respiratory problems indirectly because they induce these diseases, among other factors. Transportation activities can have an additive, commutative, or synergistic impact. They take into account the varying consequences of overt and indirect environmental factors, which are also volatile. Global warming is the result of a variety of environmental and anthropogenic forces, including shipping, both of which have different sources and consequences. The shipping division is critical for 15% of all CO_2 pollution in the world (Agarwal et al. 2020; PanduRanga Vital et al. 2019; Sharma et al. 2023b, 2023g).

The transportation sector is linked to environmental issues. The most important results are as follows: Climate change has been a major environmental problem on a global scale. Greenhouse gas pollution triggers a long-term shift in the climate environment as a whole. Carbon dioxide is the most significant anthropogenic gas in the atmosphere. Methane, nitrous oxide, some reactive chlor and benzene residues, notably HFC, FC, and silicon use while (Sf6) and ozone are all significant greenhouse gases. Transportation is a significant contributor to the greenhouse change, responsible for about 27% of total emissions in 2015. "Asia's motorized transport pollution have been a significant contributor to the global problem of climate change," according to the International Financial Bank's report "Protecting the Environment in Transportation."

Green energy's importance and savings in the transportation market. Green transportation has several advantages, including fiscal, environmental, and personal budgetary advantages. With a 100% renewable economy, climate change, energy security, recycling, and pollution can also be tackled more efficiently. The conversion of the existing transportation infrastructure is also one of the most difficult facets of such a green transition. The innovations and structures that have been proposed or shown to be feasible alternatives to fossil-fuel transportation are examined in this report. Some of the major challenges to the modern global economy include energy stability, development, erosion, and the impact of climate change. As a long-term solution to these issues, some researchers and organizations have called for a transition to a 100% renewable climate. This judgment is based on the premise that renewables are already well-developed technologies that would have little carbon impact in the future. This last feature makes them especially well-suited to dealing with climate change, which is probably humanity's most urgent challenge.

In the medium term, any transition to renewable energy will almost certainly be helped by prudent use of fossil fuels, particularly natural gas, which is a low-carbon, dispatchable source that can replace unstable renewables. According to some of the studies mentioned above, our economy may eventually become fully green. A major infrastructure transformation and a nationally coordinated policy action would take about 40 and 50 years to complete. Despite the urgent need for such a revolution to avoid catastrophic climate change, lawmakers have been unable to rally around a coordinated policy plan. According to Garca-Olivares and Balibrera, the end of all fossil fuel supply, which is expected to occur between 2020 and 2036 by different authors, with completion in the second half of this century, may be a feasible date for the launch of a global green transition. The 50-year timeline is therefore in line with how long it has taken for technological advancements to propagate through the market in the past. The conversion of existing transportation facilities remains one of the most difficult aspects of such a process (Satyanarayana et al. 2022; Sharma and Palli 2016; Avesh and Srivastava 2020, 2019; Avesh et al. 2019a).

This research investigates current global transportation patterns, behaviors, and uses, as well as the big obstacles that must be overcome in order for the global transportation market to become carbon-free. We try to answer questions about how a 100% renewable transportation infrastructure might be built using developed technologies, what a 100% green economy's transport network will look like, and what the capital and energy costs of the current system would be.

Transportation is critical in today's globalized environment because it facilitates commodity trading, individual mobility, and is one of the driving forces behind community suburbanization. However, one of the major challenges stemming from the modernization of the international transport infrastructure is a strong dependence on fossil fuels. In reality, gasoline is the most critical energy source in the transport energy mix, accounting for over 94% of total energy demand, followed by 3% for fossil fuels and other liquids, 2% for biofuels, and 1% for electricity. This large figure may be investigated by looking at various modes of transportation or how citizens use them (by passengers or freight). In terms of travelers, light duty automobiles (LDVs) accounted for about half of all transport services. Freight transportation accounted for about 45% of total transportation power in 2009, with Heavy Duty Vehicles (HDVs) responsible for more than half of it. If oils were held as the primary energy carriers of transportation, the synthesis of these fuels from electricity will use a comparatively large fraction of secondary capital in a 100% green economy. Furthermore, this demonstrates that extending beyond 12 TWa/a electricity generation in a 100% renewable economy could pose major challenges. The power supply and biomass that would be sustainably feasible for the current size of the petrochemical industry would not be enough to satisfy the energy need, the demand for methane in ammonia production (through agriculture), and the synthetic production of hydrogen and olefins. As a consequence, hydrocarbons will be scarce and expensive in such a scenario, and the transportation industry's active use of the grid, if possible, will be the most cost-effective option.

As a consequence, it is wise to rationalize but restructure transportation systems, as well as to use green energies in a transparent and aware manner. Logistics sector is largely possible, according to the report below, but air and marine transportation would have to depend mainly on alternative energy fuels. Subsequently, policy initiatives were undertaken, including the formulation of the National Electric Mobility Mission Plan (NEMMP) 2020 in 2012, and the initiation of the initial phase of Faster Adoption and Manufacturing of (Hybrid &) Electric Vehicles (FAMHEV) in 2015 by the Ministry of Heavy Industries and Public Enterprises, Government of India. FAMHEV Phase-I, commonly referred to as FAME-I in India, was executed from April 2015 to March 2019, with a budget allocation of $140.7 million. Subsequently, in April 2019, FAMHEV Phase-II (or FAME-II) was launched for a three-year period to promote electric 2-wheelers, 3-wheelers, 4-wheelers, strong hybrid 4-wheelers, and e-buses. Additionally, provisions were made for the establishment of charging infrastructure, with a total allocation of $135.86 million. The financial assistance provided for electric vehicles is contingent on the battery capacity, as shown in Fig. 12.1.

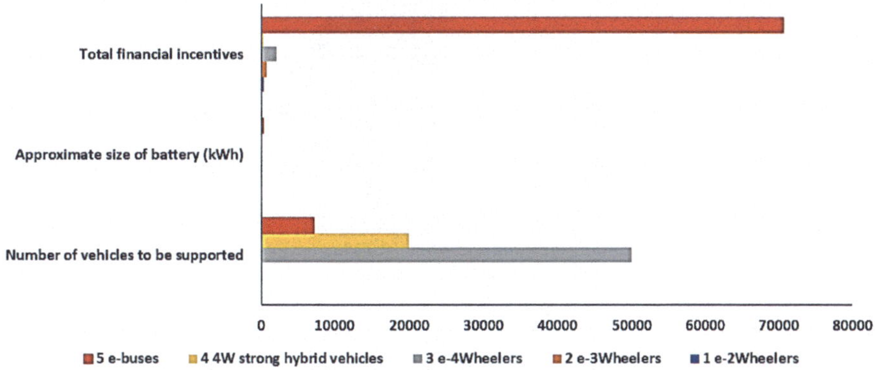

Fig. 12.1 Financial Incentives under FAMHEV-II for different electric vehicles

12.3 Energy Content of Combustibles

The energy intensity of different fuels differs, and it can be measured in terms of the same amount of energy released through combustion. And the higher the energy content, the greater the fuel output, fuel efficiency is in inverse proportion to chemical complexity. Gases become high-quality oils, solids are low-quality fats, and liquids are somewhere in the center. Hydrogen, being the most basic chemical element in existence, has the highest energy value of any fuel. Gasoline, which is generated by refining fuel oil, has much greater potential than coal (twice that of lower-grade coal combustion) or biomass (three times). Methane makes up almost half of the mass of liquid fossil fuels (LNG), and it also makes up about 85% of the mass of natural coal. Jet A-1 is the typical fuel for jetliner aircraft, and it's mostly made of kerosene with a few additives (denatured alcohol, antioxidant, and antistatic) to withstand extreme temperatures and altitudes. Bunker C gasoline, on the other side, is the most widely used marine fuel. In liquid shape, it is one of the least efficient oils, but it is suitable for big ship engines. While having a higher energy density than gasoline, methane and hydrogen are difficult to contain due to their gaseous existence. In addition, hydrogen must always be synthesized, which requires the use of materials. At a conversion rate of 100%, collecting the solar energy content of 1 kg of fuel on a substrate of one square meter would take 100 h. One of the most efficient energy storage devices, the lithium battery, can only hold around 0.5 MJ per kg, demonstrating the challenge of producing electric vehicles. The energy content density of fossil compared to alternative fuels in terms of density, mass (for gross and net heating value and volume basis are given in Figs. 12.2, 12.3, and 12.4 respectively.

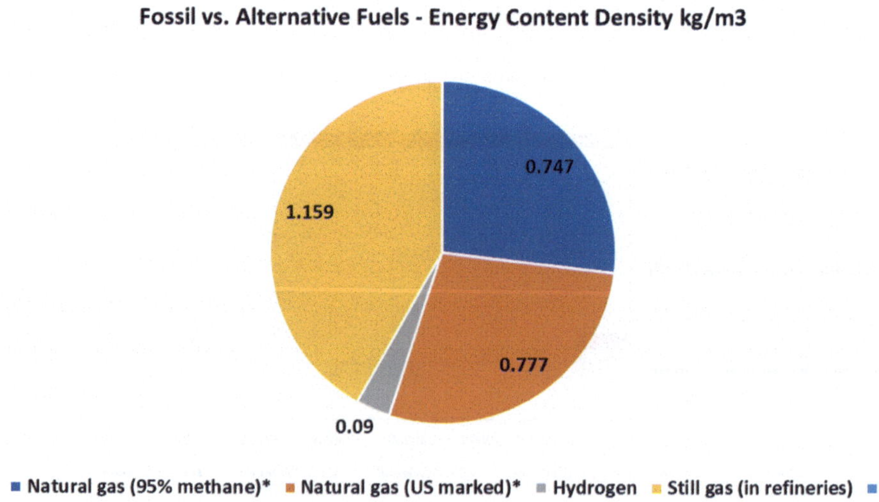

Fig. 12.2 Fossil compared to alternative fuels in terms of density

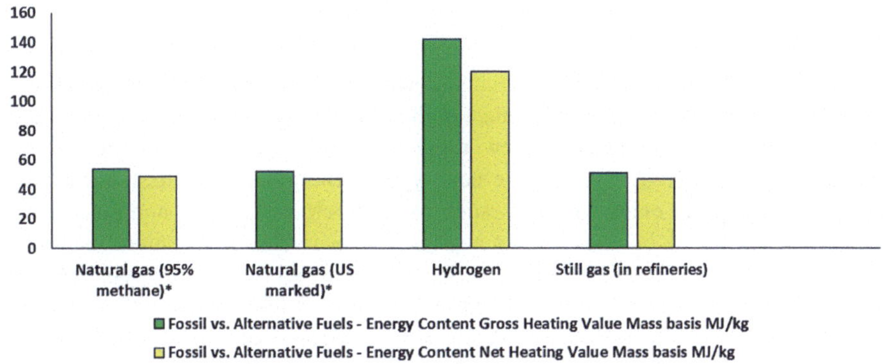

Fig. 12.3 Fossil compared to alternative fuels in terms of mass

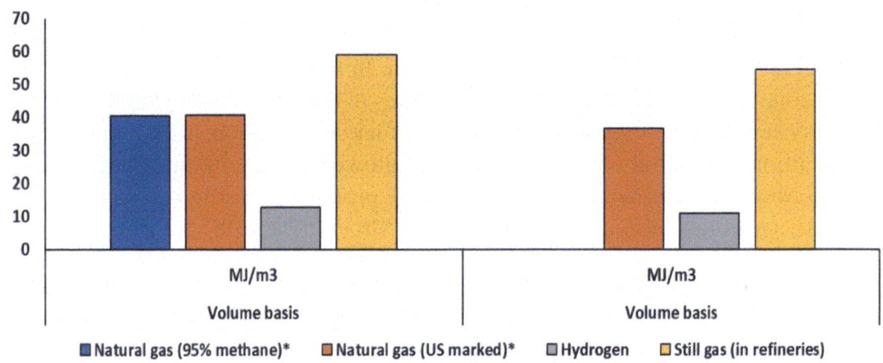

Fig. 12.4 Fossil compared to alternative fuels in terms of volume

12.4 Evolution of Energy

Shifts in energy supplies are tied to economic and technical advances. As shown by the transition from coal (solid) to oil (liquid) and natural gas (gas), the movement is toward the adoption of higher energy quality sources. This transition can be broken down into five main stages, one of which is speculative in nature: Energy was mostly derived from muscular and biomass sources prior to the industrial revolution (eighteenth century). The majority of the work was done by hand and cattle, with timber (primarily firewood) being used for home heating. Other fossil fuels, such as windmills and watermills, were available, but they only contributed a small amount of energy and were used for very particular purposes (e.g., milling flour).

The industrial revolution introduced a significant change in energy supplies with the use of gas, mostly for steam engines but also for power plants, by the mid-nineteenth century. This transition was fueled by the use of thermal energy to produce mechanical energy. When the twentieth century started, coal was the primary source of electricity, but a steady transition toward higher-energy sources such as oil began. Internal combustion engines and oil-powered ships were introduced during the second big transition. Petroleum products' dominance as the primary source of electricity achieved a high degree of reliance in the global economy in the late twentieth century. As scientific knowledge grew, more powerful fossil fuels, such as natural gas, became discovered, and a whole new type of energy, nuclear fission, became possible. Renewable energy options such as hydroelectric, wind, and solar have begun to be exploited, but they remain negligible. The twenty-first century would be marked by massive changes in energy supplies, with fossil fuels such as coal and oil gradually being phased out in favor of more competitive fossil fuels such as natural gas. There is also the possibility of developing a "clean coal" process (the term is more of an oxymoron). Biotechnology advancements highlight the growing promise of biomass-derived fuels, though wind and solar energy would also play a significant role in energy sources. Nuclear technology, particularly if experimental fusion power becomes a reality, may also play a major role, but this is also speculative. The use of hydrogen, mostly for fuel cells in cars, compact energy generators, and handheld computers, is expected to be a recent transition. In terms of final consumption, the leading countries showcase significant disparities in their utilization of both fuel and electricity resources. China leads the pack in fuel consumption, with a staggering 1,436 million tons of oil equivalent (Mtoe), followed closely by the United States at 1,106 Mtoe. Europe, with 982 Mtoe, holds a prominent position as well. Meanwhile, Africa's fuel consumption stands at 531 Mtoe, India at 487 Mtoe, and Russia at 369 Mtoe as shown in Fig. 12.5. On the other hand, when it comes to electricity consumption, China maintains its top position with 555 Mtoe, followed by the United States with 339 Mtoe and Europe with 309 Mtoe. These statistics highlight the vast differences in energy consumption patterns across the globe, reflecting the varying levels of industrialization, development, and population in these regions (Avesh and Srivastava 2012).

12.4.1 Green Transport: Conserving Energy for the Coming Future

- According to a 2016 report by the International Renewable Energy Agency (IRENA), transportation consumes 30% of global energy and has the lowest renewable energy share of any industry. Its fuel use is projected to increase by 1% each year before 2030. Transportation accounts for one-third of global electricity consumption and 14% of greenhouse gas emissions.
- In 2015, the global supply of hybrid cars (EVs) surpassed 1 million, and by the beginning of 2016, it had surpassed 2 million.

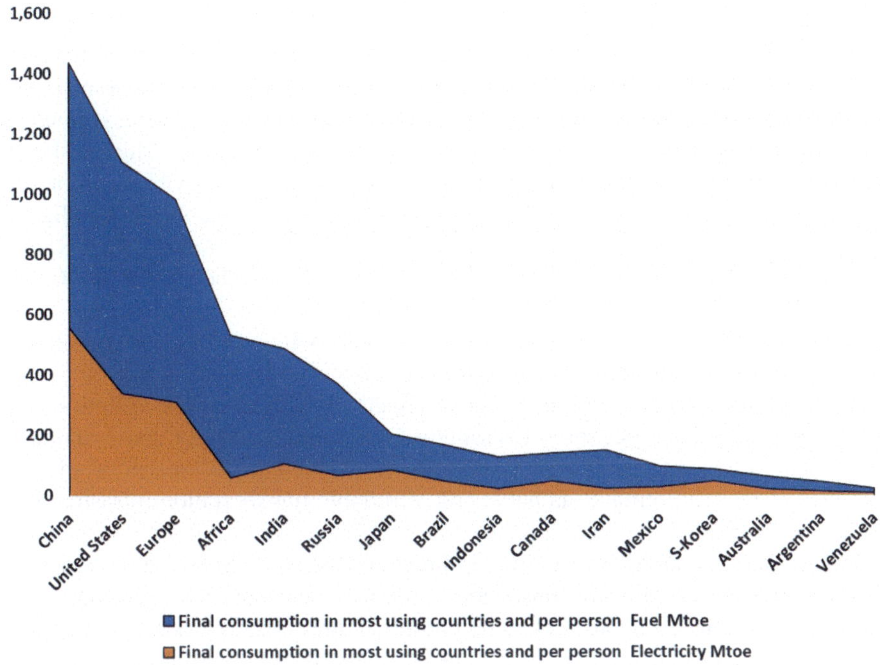

Fig. 12.5 World energy consumption country-wise

12.4.2 Environmental Impact—Transportation

Efforts toward cleaner and more cost-effective transportation solutions have gained momentum. Technological advancements, however, aren't always environmentally friendly. Climate change, deforestation, health concerns, and energy consumption have all entered the conversation. The transportation sector stands out, absorbing a substantial 30% of renewable energy while having the lowest proportion of renewables compared to other industries, as reported by the International Renewables Agency (IRENA) in 2016. Predictions show a steady annual increase of 1% in energy consumption before 2030. One-third of global energy demand and 14% of global carbon emissions can be attributed to transportation. IRENA has introduced a green roadmap, focusing on three key areas for the REmap's realization: increasing electric vehicles in tandem with renewable power output, implementing a device policy linking energy sectors, and developing economically viable pathways for various modes of transport. Policies promoting renewable technology in transportation are also in place. IRENA highlights two crucial aspects for transportation operations, namely, automobiles and liquid biofuels. The global production of electric vehicles (EVs) surpassed 2 million by the end of 2016, indicating the need for rapid EV development to support global energy transition, reduce vehicle emissions, and enhance green energy consumption. Fuel cells, both traditional and novel, including ethanol and biodiesel sources, are expected to compensate for 10% of transportation energy use by 2030, up from 3% in 2017. This becomes particularly significant for making shipping and aviation more sustainable, as they account for 20% of total transportation energy production, representing one of the industry's fastest-growing segments. The innovative approaches to transportation, from the feasibility of the Mumbai-Pune Hyperloop corridor and waste-to-biofuel production to advanced vibration control in marine and rail systems, highlight significant strides in sustainable and intelligent transportation technologies (Badgujar et al. 2024; Kokkinos et al. 2024; Raju Rallabandi et al. 2024a, b; Sharma et al. 2024a, b, c, d, e, f, g; Upadhyay et al. 2024a, b).

Alternative modes of transportation, such as carpooling, public transit, biking, and walking, offer greener and more environmentally friendly options, reducing road congestion, air emissions, and the negative effects of solo car travel. Pollution and traffic congestion have become increasingly concerning as freight and commuter transportation expand. The global imperative to reduce carbon emissions and air pollution while promoting sustainable and affordable transportation infrastructure has become evident.

Transportation operations also have indirect effects on secondary environmental systems, often with significant consequences. Fine particle emissions, linked to inefficient combustion in engines, indirectly contribute to respiratory and cardiac problems, among other health issues. The complex relationships between transportation operations and various environmental factors, such as climate change, are multifaceted and challenging to decipher. The shipping industry, for example, accounts for 15% of global CO_2 emissions. Climate change remains a significant environmental concern, with the release of greenhouse gases causing long-term climate alterations. Carbon dioxide is the most prominent anthropogenic gas in the atmosphere, alongside other key greenhouse gases like methane, nitrous oxide, and ozone. Transportation, responsible for approximately 27% of global emissions in 2015, plays a substantial role in climate change. According to the African Development Bank, Asia's motorized transport emissions significantly contribute to global greenhouse gas emissions.

Furthermore, transportation operations lead to congestion, air pollution, emissions of nitrous oxides (NOX), carbon monoxides (CO), and particulate matter. These factors often negatively impact both the economy and public health. Traffic expansion can reduce economic efficiency due to congestion and has adverse health effects on individuals exposed to air pollution. Transportation noise, a significant issue, has been associated with sleep disturbances, respiratory diseases, elevated cortisol levels, and even mortality, particularly among children.

To achieve global economic development and combat climate change effectively, a swift transition to clean energy technology is essential. Renewable energy sources are crucial for long-term energy security and pollution reduction. India, for instance, has established a renewable electricity supply route, encouraging the use of solar, wind, biofuel, and hydropower resources. Clean energy is not only environmentally friendly but often more cost-effective. India has ambitious goals, aiming to provide 175 GW of clean electricity by 2022, with substantial investments and pledges from various sectors. Solar and wind power capacities are projected to exceed 750 GW and 410 GW, respectively, by 2047. Meeting these targets will require a significant workforce and create numerous livelihood opportunities.

12.4.3 Green Transport: India's Policy.

According to reports, India's metropolitan population accounts for about 30% of both the country's total population. The history of the world has demonstrated that when economies develop, so does the pace of transition. It passes through urbanization until it begins to regenerate. From only 285 million in 2001, huge urban population is expected to rise to about 473 million by 2021 and 820 million by 2051. As a consequence, communities must have sufficient infrastructure to meet all passenger and freight transportation requirements. Transportation is vital in today's society, and it has a significant environmental impact. It is the world's largest oil producer, using the most of all the world largest petroleum. It pollutes the environment and by carbon dioxide pollution, leads greatly to global warming. By subsector, road transport is the most significant contributor to global warming. The transportation structure of a country determines its socioeconomic development. A more sustainable development policy is needed to promote overall "climate change" of the transportation industry and an environmental sustainability in the long run. The Ministries of Urban Development and Road Transport and Roads have announced a number of programs aimed at rendering India's transportation infrastructure more environmentally friendly.

12.4.4 Green Urban Transportation Scheme

The Ministry of Construction has announced a new initiative to improve green urban mobility with a central government expenditure of Rs. 25,000 crore. The name assigned to it is Green Urban Transport Scheme (GUTS). The program's mission is to develop and extend environmentally sustainable transportation systems in cities all over the world. The new law will encourage private investment in environmentally sustainable and effective public transit networks in major cities, including metro rail, nonmotorized transport, and other low-carbon emitting systems. The main aim is to reduce greenhouse gas pollution and build a long-term transportation system. The plan would help in the formation of a long-term arrangement for the government's body. Local mobility services may be financed with less emphasis on budgetary support if aggressive capital financing were promoted. The construction of nonmotorized transport networks will be a top priority for the nation. This Scheme will benefit the provision of non-motorized transit networks, increased public transportation access, the use of green energy, the introduction of Sensor Networks (ITS), and private sector participation in urban transportation initiatives. Cities and metro cities with communities of 5 lakh or more, as well as all capital cities, are being considered for the program. For the next five years, the federal government's initial pledge would encourage private sector interest in allocating funds for GUTS (Kohli et al. 2022).

12.4.5 Green Highways: A Long-Term Development Proposal

The Ministry of Road Transport and Highways (MoRTH) of the Government of India promulgated the Green Highways (Plantations, Transplantations, Beautification and Maintenance) Policy-2015 in September 2015 to construct green corridors along National Highways for a safe environment and inclusive growth. According to the policy, "development of environmentally friendly National Highways with private sector support." Economic production is influenced by communities, farmers, NGOs, the private community, institutions, government agencies, and forest departments. India seems to have a total road size of 46.99 lakh kilometers, with National Highways responsible for around 96,214 km, or 2% of total road length. Highways carry approximately 40% of the increased traffic. The data was collected and gathered by various studies (Bhardawaj et al. 2023; Hossain et al. 2023; Lee and Sharma 2023; Sharma et al. 2023h; Sharma et al. 2023a, b, d, e, f; Upadhyay et al. 2023a, b; Wu et al. 2023). Over the next three years, the Ministry has decided to convert all existing National Routes and 40,000 km of modern roads into Green Highways. The aim is to reduce the impact of air pollution and dust on embankment slopes by utilizing trees and plants along highways as natural carbon emissions sinks and to avoid soil erosion. Plants along highway center strips and around the edges limit oncoming vehicle glare, which can contribute to accidents. Local residents benefit directly from community tree planting when employment is created. Panchayats, NGOs, and some other Self-Help Categories (SHGs) will be involved in the plantation and maintenance process. Plant species will be selected based on variables such as precipitation, climate, soil type, and so on, and would be region-specific. For example, some soil conditions may be suitable for the planting of Jamun or mango trees, whereas others may be suitable for the cultivation of plants and grasses to produce biomass. Wherever possible, preference will be given to transplanting mature trees thus widening the routes (Mohanty et al. 2016).

12.4.6 Non-Motorized Transit Policy (NMT)

The most common non-motorized forms of transportation are walking, cycling, and riding a bike rickshaw. There are eco-friendly modes of transportation available. There is no local noise, limited oil consumption, and a low carbon footprint. They are less expensive than motorized vehicles and do not depend on fossil fuels. Many large-scale developments are underway on a national level to facilitate NMT as a primary mode of transportation in Indian cities. In a range of national and subnational policies,

NMT is either a critical consideration or a top priority. The following interventions and programs are examples of such initiatives and programs:

The National Urban Transportation Policy: NMT's function as a last-mile connection for urban transportation networks and as a stand-alone mode for short distances is described in the regulation. In municipalities, it can be implemented by local councils. As part of national Mission for Renewable Ecosystems: Prime Minister's National Strategy on Climate Change, a subcommittee on mass development has been created. It proposed eight principles for integrated urban mobility, with "hop" and "cycle" being the first two. The Ministry of Urban Development has established Service Level Benchmarks (SLBs) for urban transportation that can be used by all Indian cities. The task of the NMT scheme was specifically incorporated into this. Atal Mission for Rejuvenation and Urban Transformation (AMRUT) is an acronym for Atal Mission for Rejuvenation and Urban Transformation. One of the scheme's objectives is to reduce traffic pollution by switching to hybrid cars, building nonmotorized road networks, and taking public transportation. One of the characteristics of smart cities is the construction of walkable neighborhoods, which helps to reduce traffic, air pollution, and resource depletion while also improving local economies, fostering connections, and ensuring safety. The road network is planned or upgraded to accommodate pedestrians and cyclists as well as automobiles and mass transit, and administrative services are within walking or cycling distance. Scheme for Green Urban Mobility: The government is working on a campaign to make cities more eco-friendly and prosperous. The government is working on a mobility scheme that would promote the use of hybrid/electric vehicles and non-fossil fuels for mass transit in 103 cities with a population of more than 5 lakh residents, among other items. Bicycle lanes, walking paths, shared ride sharing, BRT networks, decentralized mobility systems, improved freight management, and creative financing for transportation programs are also part of the initiative, as is a phased shift to combination and non-fossil fuels for mass transit. In its estimation, the government to set aside Rs 70,000 crore. The government has developed a National Transmission Mobility Mission Plan 2020, which proposes incentivizing the deployment of green vehicles and promoting domestic automotive manufacturing capability. As part of its own Foreign Trade Policy, the government offers discounts in the form for duty waivers for some types of green vehicles (Sharma et al. 2020d).

12.4.7 India's Projected Primary Energy Use

Energy use is heavily influenced by the size and growth of a country's population. With 1.368 billion citizens as of January 2019, India is the nation's second most populous nation. It has a 1.18% average growth trend and represents 17.74% of the world population. The world's population is expected to reach 1.383 billion, 1.12 billion, 1.605 billion, and 1.58 billion people by the end of 2020, 2030, 2040, and 2050, respectively. India annually adds more people to the world than any other community, and the demographic of several of India's regions is comparable to that of some countries. India's energy consumption will grow at the highest pace of any global economy by 2040, with coal reaching the bulk of this need, with renewable energy leading the way. Renewables had replaced coal and instead oil as the single most effective source of domestic electricity by 2020. India's green energy prices will rise by 12% annually from 17 Mtoe throughout 2016 to 256 Mtoe in 2040.

12.5 Literature Review

12.5.1 Solar Energy at India Railways

Instead of using existing diesel, the massive amount of power required for the trains to run can be met with solar energy. The time has come, and there is an opportunity. There will be a lot of fuel scarcity in the next 5 to 10 decades, and there will be no fuel in the next 5 to 10 decades. As trains use the most electricity, the concept of equipping them with solar panels will be an excellent way to help mitigate fuel shortages while still promoting the usage of non-exhaustible and environmentally sustainable resources. While equipping the train with solar energy is not an easy job, we believe it is essential. While there are a few solar-powered trains on the market, they seem to be small and unspecialized. The track-crack detection safety feature will be the best implementation for this concept. This function would greatly contribute to rail protection and will help to prevent significant train incidents. The solar panels are available in a variety of sizes, shapes, and power outputs. The concept of a solar train may be applied by using the existing solar panels. The disadvantage is the large amount of power that must be supplied on a continuous basis for the train's engine to move about 21 compartments or perhaps more. As a result, regardless of the temperature, electricity supply, or solar energy strength, there is a need for continuous power to be generated. The risk of putting this latest train concept into action renders it an expensive proposition, because it lags economically. As a result, applying this concept for current trains would not be costly. Even though there is a significant investment, there are also significant returns and savings. The most significant point is that it renders the train safer, and would help us reduce the lack of usable gasoline. Solar trains are a cutting-edge method of educating citizens about solar resources and environmentally sustainable power generating systems. II. Data Collection: The installation of solar photovoltaic (PV) panels on established train roofs has proved to be one of the most feasible large-scale sources of renewable electricity for urban trains. There will be a train running from Sarai Rohilla in Delhi to Farukh Nagar in Haryana. It was the first solar-powered DEMU (diesel electrical multiple unit) train to leave Delhi's Safdarjung railway station on July 14, 2017. Six coaches are equipped with a total of 16 solar panels, each delivering 300Wp. These solar panels, which are made under the "Make in India" project, cost Rs 54 lakh.

This is the first-time solar panels have been used as a grid in Indian railways. The train has a battery backup which will operate for at least 72 h on battery charge. The solar panels produce about 17 units of electricity each day, which is enough to power the coach's lighting system. Solar panels can be installed on non-AC coaches only for the time being. In the coming days, the railways expect to add approximately 50 more of these buses. Solar electricity will be used in commuter trains first, and then in long-distance trains. The railways will save Rs 672 crore a year by saving an approximate 1.2 lakh kiloliters of fuel per year. Solar energy would also help to reduce carbon dioxide emissions by 2.7 lakh tons a year. The coaches of the second most famous train, the Kalka to Shimla toy train "The Himalayan Queen," have been converted to a solar-powered device in order to tap renewable energy and reduce carbon emissions. It is the country's first train with all of its coaches operated by solar energy. For the track that is on the UNESCO heritage list, the experiment would win carbon credit. Each coach costs Rs 1.25 lakh and comes with a 100-W solar panel. The fluorescent tube has been substituted by LED lamps, which are even lighter. Without the heat, the coach will run for two days and make two journeys. Solar power charging sockets have also been included in the buses. Trains traveling on the Pathankot-Joginder Nagar track have previously been used for trial runs. One coach on each of two trains has been driven by solar energy over the past year. The disadvantage is that the solar energy cannot be used to power the whole train; instead, only the illumination and other train equipment will use it. Diesel is needed in these trains in order for them to operate (Mohapatra et al. 2015).

Before 2030, the Indian Railways intends to be 100% electrified and a "direct zero carbon emitter." The plan is to utilize the rooftops of train and railway buildings, as well as the abandoned areas of Indian railways, which has a landpool of over 51,000 hectares. The Indian Railways reported that solar panels have been installed at 960 railway stations to satisfy their power needs. Orders for 198 MW of rooftop solar power for 550 stations have been placed and are currently being implemented. Some of the stations that have been solarized are New Delhi, Old Delhi, Jaipur, Secunderabad, Kolkata, Varanasi, Guwahati, Hyderabad, and Howrah. By 2030, 33 billion units would have been generated, which is 10 billion more than the existing demand. By 2030, it also intends to construct solar plants with a potential of 20 GW on its barren property. The Indian Railways has agreed to use solar energy to fulfill its traction power and lighting needs, making it a fully renewable form of transportation. The Railway Energy Management Limited (REMCL) was tasked with developing a GW of solar power projects on the Indian Railways' unused property. It has a total of 2 GW of solar projects planned on underutilized railway soil. It intends to issue a tender in July for the construction of 1 GW of solar projects along railway lines. Based on the information gathered, we have a plan to install solar energy capacity on the train roofs. We've also got the concept of renewable energy. The disadvantage to note is that only the coaches, not the whole train, including the engine, are driven by solar energy. But making the train work isn't feasible with little solar power since it takes about 22KW of energy only to start the motor, and running the train is difficult. There is, however, the option of operating the train entirely on solar electricity, which we propose in our concept.

We learned from the above data collection that the coaches will operate for two days on a single charge if the batteries are fully charged. As a result, the solar panels mounted on the train's roof-top, as seen in Fig. 12.1, are used to charge the batteries. Where solar panels with the best fabrication are mounted on the whole roof-top of the train compartments, taking up the least amount of room and producing the most power. Installing solar panels on the roof would not have enough electricity to operate the train. As a result, the frame with the compact solar panels, in addition to the roof-top panels, will be a perfect source of power production. The flexible solar panels should be mounted on the train's exterior walls so that even if the sun doesn't shine on the roof, electricity can be provided when the sun shines on the flexible solar panels. The locomotive engine generates electricity from all 21 compartments that are connected to it. As a result, the highest energy production from the panels can be harnessed and placed in the batteries mounted in the last compartment to store the energy. Furthermore, the energy contained in the batteries will be used to power the trains at night while there is no sunshine. The energy collected is split into two lines, one that assists the locomotive in running and the other that assists the batteries in charging. On the locomotive, control panels are being mounted so that the energy produced can be measured and the approximate distance to be covered from the train with the available power can be measured and shown so that the Loco-pilot can see how far he will go with the remaining energy. Considering solar resources during the offseasons, as there would be less hours during the day when sunshine is accessible, may be a disadvantage, although this can also be overcome. There is no need to prepare new trains with solar panels, but we should retrofit current trains with solar panels to provide an alternative mode of transportation. When there isn't enough sunshine or the battery is running low, trains should be used. The control panel displays the voltage of the panels, the charging rate of the batteries, the peak power production rate, the availability of power in the batteries, the approximate distance from the available batteries, the train's power usage, and other features such as track-crack detection. There will be a switching device that will enable the locomotive to convert from solar energy to diesel power. If there is sufficient energy in the batteries or available line power to run the train, the train will run on solar energy mode; if there is insufficient power to run the train to cover the required distance, the locomotive will switch to diesel mode, where diesel will be used to cover the distance, and so there will still be an alternative way to go. The Loco-pilot must keep his eyes on the control panels and determine when to swap between modes based on the amount of energy accessible. Another important function that this train would provide is track-crack identification, which would assist the train in detecting track cracks or disassembly such that the train will immediately halt and notify the loco-pilot through the control panel that there is a crack in the tracks on which the train is traveling. Since the tracks that are being mounted are always the same distance, height, duration, and intensity of the rail, we will use the sensor on the locomotive to maintain track of the track's width, height, length, and strength. This would cause the track to be tracked, because if some of the track's characteristics shift, such as distance, strength, height, or volume, or if the fish-plate is missed, an automated stop with minimal friction will occur, close to the stop that occurs when the chain is pulled. As a result, this protection mechanism would aid in the detection of track fractures and the reduction of injuries caused by train derailment.

12.6 Sustainable Energy Integration for Locomotives

In the realm of locomotive operations, the initial startup of a Diesel Locomotive consumes approximately 3–4 times more fuel than its standard consumption rate. To mitigate excessive fuel usage, railways often keep locomotive engines running, even during relatively short station stops lasting 30 min. Shutting down and restarting the engine within this time frame would expend around 350 L of diesel fuel for a typical large locomotive, which is roughly equivalent to the fuel required to cover a distance of 100 km or approximately 2 h of operation. This fuel consumption can vary depending on the locomotive model. The recent advancements in rail vehicle system modeling, structural analysis, and performance evaluation have been extensively studied in various contexts, including crashworthiness, lateral stability, critical speed analysis, and multi-objective optimization (Goyal et al. 2019; Sharma 2019; Sharma et al. 2019; Bhardawaj et al. 2020e; Lee and Sharma 2020; Bhardawaj et al. 2021; Choi et al. 2021; Lee et al. 2021; Sharma and Sharma 2021).

Given that an average train consumes 350 L of diesel to travel 100 km, the average power consumption for a train is approximately 8.2 kW per kilometer. Consequently, operating a locomotive for one kilometer necessitates 3.5 L of diesel. If the cost of one kilometer of diesel is about 177 rupees, then covering a distance of 100 km would cost roughly 17,700 rupees. In terms of power, one kilometer requires 8.2 kW, while 100 km would necessitate approximately 820 kW of power. By utilizing solar panels with a 300W output, it is feasible to generate the required electricity.

Each train compartment typically measures 23.54 m in length and 3.24 m in width, leaving a rooftop area of 76.29 square meters for a single compartment. Consequently, the installation of 300W solar panels, with dimensions of 1.95 m in length and 0.922 m in width, allows for approximately 42 panels per compartment, generating about 75% of their rated capacity. The total power generated by the solar panels on one compartment's roof amounts to 9.45 kW. Therefore, considering the 21 compartments, the combined rooftop panels would produce an estimated 1.98 MW of electricity. Additionally, if flexible solar panels were applied to the walls, further power generation could be achieved, contributing to the locomotive's smooth operation (Bhardawaj et al. 2019b).

12.6.1 Green Energy in Airways: Renewable Energy in the Wings

Energy is essential for economic growth and rising living standards. The burning of fossil fuels provides the bulk of the resources used for power, heating and cooling, and transportation.

Fossil fuels, like coal, gasoline, and natural gas, dominate global energy production. Nuclear and hydroelectric power also play significant roles, making up 11% and 17% of the global demand, respectively. In 2014, renewable sources provided

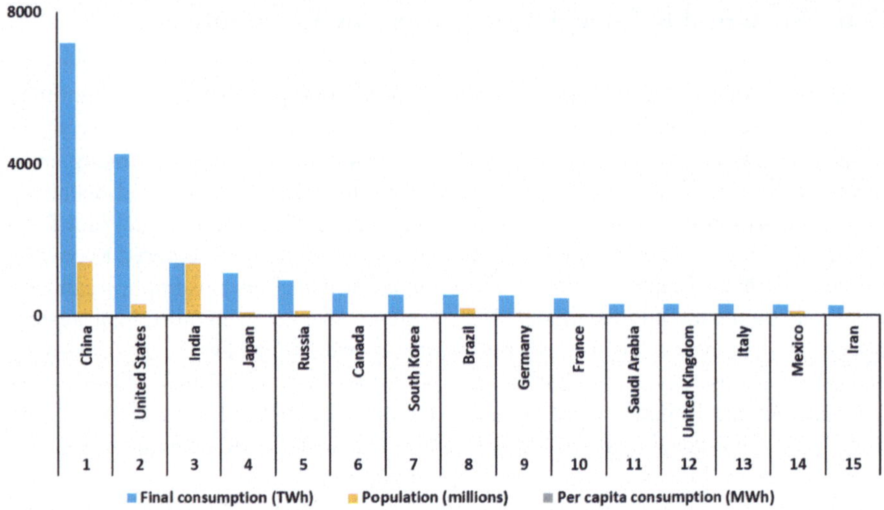

Fig. 12.6 Global electricity consumption by source for 2012

22% of global electricity demand, with hydropower contributing the most at 74%, followed by wind (13%), bioenergy (8%), solar (4%), and geothermal (1%). Non-hydro renewables made up 6% of this total. Solar technology saw the fastest growth between 2013 and 2014, expanding by 39%, followed by wind (12%), geothermal (7%), and bioenergy and hydropower (3%). To put it in perspective, the 5,294-terawatt hours of renewable energy in 2014 equated to significant benefits: it offset greenhouse gas emissions from 785,893,684 vehicles, powered 549,393,996 households for a year, saved 8,613,735,094 barrels of oil, and sequestered carbon equivalent to 3,521,837,479 acres of timberland (See Fig. 12.6).

In aviation, emissions from both domestic flights and airport activities are substantial. The International Civil Aviation Organization (ICAO) oversees emissions from international flights. Efforts to reduce jet fuel usage and CO_2 emissions include more fuel-efficient aircraft, direct flight paths, and electrification of on-ground functions. Renewable energy, gate electrification, and sustainable aviation fuels are also part of the solution (Agarwal et al. 2020). Collaboration among various airport stakeholders, including airlines and tourism companies, is crucial for effective pollution mitigation. While aviation relies heavily on liquid fuels, the industry is exploring renewable jet fuels from sources like camelina, jatropha, biomass, and animal oils. These efforts aim to reduce pollution and meet sustainability criteria, with the potential to cut emissions by up to 80% over the industry's lifecycle. In Scandinavia, there's a growing interest in renewable jet fuels, although large-scale production at affordable costs remains a challenge. Despite some airlines conducting flight trials, SAS, for instance, has refrained from purchasing these fuels due to their high prices and sustainability requirements.

12.6.2 Green Energy in Roadways

In conclusion, solar energy is poised to revolutionize the transportation sector, particularly on roadways, where it can play a pivotal role in achieving sustainability. Solar Roadways, a visionary concept, aims to harness the potential of green energy production on various surfaces, including roads, parking lots, cycle lanes, and more. India, a country blessed with abundant sunlight for approximately 300 days a year, stands to benefit significantly from solar-powered roadway projects. Initiatives like installing solar panels on highways, such as the proposed 205-km stretch of the Ahmedabad-Rajkot highway, hold the promise of generating substantial electricity (PanduRanga Vital et al. 2019). This clean energy can be utilized not only to power the highway infrastructure but also to address India's persistent power deficit and electrification challenges, potentially benefitting millions. Solar roads have the potential to usher in a transformative era by reducing our reliance on grid electricity, curbing fossil fuel emissions, and shrinking our carbon footprint. Additionally, they can eliminate the recurrent maintenance costs associated with traditional paved roads, contributing to long-term cost savings.

While solar roadways represent an innovative and sustainable solution, they come with their own set of challenges. The substantial upfront costs, as evidenced by projects in France and China, require careful consideration. Maximizing solar penetration on flat road surfaces and ensuring the durability and safety of the road materials are vital concerns. Engineers and manufacturers need to address these challenges as solar road technology evolves. Moreover, solar roads have the potential to redefine the solar industry and contribute to the development of a smart power grid. The possibilities for energy harvesting from road pavement, including PV sound barriers and solar panels integrated into parking lots, hold great promise for enhancing energy efficiency and sustainability. Solar road deployment may face hurdles; it presents a compelling opportunity to harness untapped solar energy resources and drive the transition toward a cleaner, more sustainable future for transportation and infrastructure. With ongoing research, innovation, and widespread adoption, solar roadways could play a pivotal role in shaping a greener world for generations to come (Mohd and Srivastava 2019).

12.7 Methodology

The methodology employed in this research aims to investigate critical factors influencing the future development of sustainable green energy, with a specific focus on the potential role of roadways in facilitating the adoption of green energy solutions. This approach integrates various strategies across interconnected sectors, offering insights into the energy industry, transportation modes, and innovative technologies.

Firstly, a systemic strategy is implemented, emphasizing the need to unite various energy sectors within a comprehensive plan. This includes a concerted effort to increase the deployment of electric vehicles alongside the generation of sustainable electricity, thereby creating clean and cost-effective energy pathways for all modes of transportation, encompassing not just traditional cars but also non-car alternatives such as freight transport, aircraft, and energy storage solutions. In addition to exploring technological advancements and innovations across multiple transportation sectors, the research takes into account the environmental consequences associated with traffic operations. This encompasses issues such as traffic congestion, air pollution, the emission of nitrous oxides (NOX) and carbon monoxides (CO), as well as the presence of dark smoke resulting from transportation-related activities. Furthermore, the research addresses the economic implications of transportation expansion, with a particular focus on the efficiency challenges posed by traffic congestion. In the aviation sector, where liquid fuels are the predominant energy source and no immediate alternatives are available, strategies are proposed to reduce carbon emissions while ensuring continued access to liquid fuels (Avesh et al. 2019b).

The sustainability criteria applied in this research underscore the importance of long-term production sustainability. This includes the need to avoid conflicts with food production or access to potable water, minimize habitat disruption, and employ land efficiently. A set of steps is presented to address these sustainability challenges, requiring endorsement and support from policymakers and civil society. Key strategies encompass a shift from traditional inter-urban land transportation to electric trains for both freight and commuter travel. The research advocates for the prioritization of electrified mass transportation and the selective use of fuel cells where their benefits are most evident. Furthermore, there is a call for reducing aviation fleets in favor of rail systems and marine transportation. Workflow and logistics optimization measures are proposed to reduce travel demand and prioritize transportation modes based on energy efficiency metrics. Efforts are also focused on improving energy quality by embracing advanced technologies and adapting metropolitan and public transportation infrastructure. The promotion of "Transport as a Service" (TaaS) is suggested as a means to reduce resource and material requirements in the road transportation sector.

However, it is essential to acknowledge the limitations of this research. These include uncertainties regarding the emergence of groundbreaking technologies, potential shifts in population distribution due to climate change and conflicts, economic disparities, and the necessity for political and business decisions that support the transition to electric transportation. Additionally, the research accounts for economic resources in hypothetical scenarios related to climate and environmental disasters. Moreover, the research recognizes that while a green transportation infrastructure is theoretically achievable, it may not align with the typical trajectory of accelerated resource demand growth. This raises concerns about the availability of essential metals such as copper (Cu), nickel (Ni), lithium (Li), platinum (Pt), and palladium (Pa). To tackle these challenges, the research proposes the adoption of geophysical and biodiversity metrics, a shift away from GDP as the primary measure of economic growth, and the introduction of innovative mechanisms to ensure resource stability without overburdening demand. Hence, the research methodology applied here offers a comprehensive, third-party perspective on the multifaceted challenges and opportunities associated with the development of sustainable green energy and transportation systems in the future.

12.8 Conclusion

The journey of human mobility has evolved significantly over time. From the humble beginnings of traveling on foot, the invention of the wheel revolutionized our ability to cover vast distances with speed and efficiency. What once took months or even years can now be accomplished in mere days and hours, thanks to the development of automobiles and the construction of extensive road networks. However, with progress comes the responsibility to adapt and innovate further. Our highways, the lifelines of modern transportation, need to evolve into smart highways to meet the challenges of the future. One pressing need is the efficient use of resources, especially when it comes to illuminating these vast stretches of road at night. Herein lies the importance of smart highways. Smart highways leverage cutting-edge technologies such as green energy sources like solar panels and windmills to generate power from the abundant sunlight and other green resources available. These highways incorporate elements of the Internet of Things (IoT), 5G connectivity, and cloud computing to enable rapid data communication and responsive actions when needed. India, with its favorable geography, ample sunlight, and a burgeoning population, is particularly well-positioned to embrace and implement the concept of smart highways. While the idea of smart technology has already permeated various aspects of our lives, from smart vehicles to smart homes, highways have remained relatively unchanged in their basic function of connecting points A and B. It is high time we shine a light on innovative solutions that can transform our highways into intelligent and sustainable transportation corridors.

One such innovation is the use of solar panels and wind turbines installed along the open land adjacent to highways, harnessing renewable energy to power various infrastructure needs. Additionally, advanced technologies like illuminating paints and sensors are being employed to replace traditional streetlights, making highways safer and more energy-efficient. Even the vibrations generated by passing vehicles can be converted into electric energy, highlighting the potential for sustainable solutions. In the context of an ever-growing concern for environmental conservation, it is crucial to recognize that conventional automobiles reliant on fossil fuels continue to contribute to pollution. As these resources become scarcer and environmental issues more pressing, hybrid vehicles represent a viable alternative to reduce emissions and mitigate their impact on the environment. In essence, the transformation of our highways into smart, sustainable, and energy-efficient corridors is not just a novel idea but a necessity. By embracing green technology, harnessing renewable energy sources, and adopting innovative solutions, we can pave the way for a cleaner, more efficient, and smarter future for our highways and, by extension, the world. It is a journey toward progress and environmental responsibility that holds great promise for generations to come.

Acknowledgements The research funding from the Ministry of Science and Higher Education of the Russian Federation (Ural Federal University Program of Development within the Priority-2030 Program) is gratefully acknowledged.

References

Acharya A, Gahlaut U, Sharma K, Sharma SK, Vishwakarma PN, Phanden RK (2020) Crashworthiness analysis of a thin-walled structure in the frontal part of automotive chassis. Int J Veh Struct Syst 12. https://doi.org/10.4273/ijvss.12.5.06

Adeoye AO, Quadri RO, Lawal OS (2022) Assessment of biofuel potential of tenera palm kernel shell via fixed bed pyrolysis and thermal characterization. Results Surf Interfaces 9:100091. https://doi.org/10.1016/j.rsurfi.2022.100091

Adeoye AO, Quadri RO, Lawal OS (2023a) Wet synthesis, characterization of goethite nanoparticles and its application in catalytic pyrolysis of palm kernel shell in TGA. Results Surf Interfaces 11:100118. https://doi.org/10.1016/j.rsurfi.2023.100118

Adeoye AO, Quadri RO, Lawal OS (2023b) Biomass waste pyrolysis in the context of Nigeria's oil and gas industry: an overview. Int J Ambient Energy 44:2525–2541. https://doi.org/10.1080/01430750.2023.2256325

Adeoye AO, Lawal OS, Quadri RO, Malomo D, Aliyu MT, Dang GE, Emojevu EO, Maikato MJ, Yahaya MG, Omonije OO, Edidem VG, Abubakar YK, Offor OF, Sochima EH, Peter BE, Hikon BN (2023) Sustainable energy via thermochemical and biochemical conversion of biomass wastes for biofuel production. In: Transportation energy and dynamics. energy, environment, and sustainability. Springer, Singapore, pp 245–306. https://doi.org/10.1007/978-981-99-2150-8_11

Agarwal KM, Mohapatra S, Sharma P, Sharma S, Bhatia D, Mishra A (2020) Study and overview of the novel corona virus disease (COVID-19) Sensors Int 1:100037. https://doi.org/10.1016/j.sintl.2020.100037

Agrahari R, Mohanty S, Vishwakarma K, Nayak SK, Samantaray D, Mohapatra S (2021b) Update vision on COVID-19: Structure, immune pathogenesis, treatment and safety assessment. Sensors Int 2:100073. https://doi.org/10.1016/j.sintl.2020.100073

Agrahari R, Sarraf G, Joshi NC, Mohapatra S, Varma A (2021) Insight of biopolymers and applications of polyhydroxyalkanoates. In Microbial polymers, pp 177–191. Springer Singapore, Singapore (2021). https://doi.org/10.1007/978-981-16-0045-6_8

Avesh M, Srivastava R (2012) Modeling simulation and control of active suspension system in Matlab Simulink environment. In 2012 Students conference on engineering and systems, pp 1–6. IEEE. https://doi.org/10.1109/SCES.2012.6199124

Avesh M, Srivastava R (2019) Parametric optimization to design a passenger car suspension system for better dynamic performance. Eur Transp—Trasp Eur

Avesh M, Srivastava R, Chandmal Sharma R, Sharma N (2019) Effective design of active suspension system using fuzzy logic control approach. Int J Veh Struct Syst 11. https://doi.org/10.4273/ijvss.11.5.13

Avesh M, Srivastava R, Sharma RC (2019b) Parametric optimization for the design of passenger vehicle suspension system with the application of genetic algorithm. Int J Veh Struct Syst 11. https://doi.org/10.4273/ijvss.11.2.08

Avesh M, Srivastava R (2020) Passenger car active suspension system model for better dynamic characteristics. Natl Acad Sci Lett 43:37–41. https://doi.org/10.1007/s40009-019-00807-z

Badgujar P, Sharma S, Rahul D, Sharma SK, Upadhyay RK (2024) Feasibility study of Mumbai-Pune Hyperloop corridor: challenges and implementation. Res Transp Bus Manage 52:101089

Bandekar D, Chouhan OP, Mohapatra S, Hazra M, Hazra S, Biswas S (2017) Putative protein VC0395_0300 from Vibrio cholerae is a diguanylate cyclase with a role in biofilm formation. Microbiol Res 202:61–70. https://doi.org/10.1016/j.micres.2017.05.003

Bhardawaj S, Chandmal Sharma R, Kumar Sharma S (2019) A survey of railway track modelling. Int J Veh Struct Syst 11. https://doi.org/10.4273/ijvss.11.5.08

Bhardawaj S, Chandmal Sharma R, Kumar Sharma S (2019a) Development and advancement in the wheel-rail rolling contact mechanics. IOP Conf Ser Mater Sci Eng 691:012034. https://doi.org/10.1088/1757-899X/691/1/012034

Bhardawaj S, Sharma RC, Sharma SK (2020b) Development in the modeling of rail vehicle system for the analysis of lateral stability. Mater Today Proc 25:610–619. https://doi.org/10.1016/j.matpr.2019.07.376

Bhardawaj S, Sharma RC, Sharma SK (2020c) Analysis of frontal car crash characteristics using ANSYS. Mater Today Proc 25:898–902. https://doi.org/10.1016/j.matpr.2019.12.358

Bhardawaj S, Sharma RC, Sharma SK (2020d) Ride analysis of track-vehicle-human body interaction subjected to random excitation. J Chin Soc Mech Eng Trans Chinese Inst Eng Ser C/ Chung-Kuo Chi Hsueh K. Ch'eng Hsuebo Pao 41:229–238

Bhardawaj S, Sharma RC, Sharma SK (2020e) Development in the modeling of rail vehicle system for the analysis of lateral stability. Mater Today: Proc 25:610–619. https://doi.org/10.1016/j.matpr.2019.07.376

Bhardawaj S, Sharma RC, Sharma SK, Sharma N (2021) On the planning and construction of railway curved track. Int J Veh Struct Syst 13(2):151–159. https://doi.org/10.4273/ijvss.13.2.04

Bhardawaj S, Sharma RC, Sharma SK, Gopala Rao L, Vashist A (2023) Modeling of biomechanical human body model for seat to head transmissibility analysis. Noise Vib Worldwide 54(2–3):62–74. https://doi.org/10.1177/09574565231154243

Bhattarai S, Mhamdi A, Hossain I, Raoui Y, Pandey R, Madan J, Bouazizi A, Maiti M, Gogoi D, Sharma A (2022) A detailed review of perovskite solar cells: Introduction, working principle, modelling, fabrication techniques, future challenges. Micro Nanostruct 172:207450. https://doi.org/10.1016/j.micrna.2022.207450

Biswas N, Saha S, Biswas BK, Chowdhury M, Rahaman A, Junghare V, Mohapatra S, Hazra S, Zangrando E, Roy Choudhury R, Roy Choudhury C (2021) The DNA- and protein-binding properties and cytotoxicity of a new copper(II) hydrazone Schiff base complex. J Coord Chem 74:1482–1504. https://doi.org/10.1080/00958972.2021.1913128

Chatterjee A, Nirwan S, Mohapatra S, Sharma P, Agnihotri A, Shrivastava N (2022) Biochemical aspects of pathogenic variability in white rust infected Indian mustard. Mycologia 114:757–768. https://doi.org/10.1080/00275514.2022.2060007

Choi S, Lee J, Sharma SK (2021) A study on the performance evaluation of hydraulic tank injectors. In: Joshi P, Gupta SS, Shukla AK, Gautam SS (eds) Advances in engineering design. Lecture notes in mechanical engineering, pp 183–190. https://doi.org/10.1007/978-981-33-4684-0_19

COVID-19: Analysing the legal nuances of the lockdown order. Indian J Forensic Med Toxicol (2020). https://doi.org/10.37506/ijfmt.v14i4.11553

Das S, Swapnarekha H, Vimal S (2023) Integration of blockchain technology with renewable energy for sustainable development: issues, challenges and future direction. Presented at the (2023). https://doi.org/10.1007/978-981-99-3734-9_48

Devadarshini D, Mohapatra S, Pati S, Maity S, Rath CC, Jena PK, Samantaray D (2022) Evaluation of PHAs production by mixed bacterial culture under submerged fermentation. Biologia (bratisl). 78:895–902. https://doi.org/10.1007/s11756-022-01302-5

Gopala Rao L, Sharma RC, Satyanarayana V, Palli S (2022) Stochastic optimal preview control response of a quarter-car nonlinear suspension model using spectral decomposition method. Noise Vib Worldw 53:225–232. https://doi.org/10.1177/09574565221093234

Goyal S, Anand CS, Sharma SK, Sharma RC (2019) Crashworthiness analysis of foam filled star shape polygon of thin-walled structure. Thin-Walled Struct 144:106312. https://doi.org/10.1016/j.tws.2019.106312

Goswami B, Rathi A, Sayeed S, Das P, Sharma RC, Sharma SK (2019) Optimization design for aerodynamic elements of Indian locomotive of passenger train. In: Advances in engineering design. Lecture notes in mechanical engineering. Springer, Singapore, pp 663–673. https://doi.org/10.1007/978-981-13-6469-3_61

Hossain I, Velkin VI, Shcheklein SE (2017) The study of passive vibration dampers in pipelines using PIV-methodology for single phase flow. In WIT transactions on ecology and the environment, pp 565–570. https://doi.org/10.2495/ESUS170521

Hossain I, Ahmed YE, Sharma SK (2023) The use of both metallic-insert and TiO_2/water nanofluid to improve the performance of photovoltaic modules; experimental study. Thermal Sci Eng Progress 42:101944. https://doi.org/10.1016/j.tsep.2023.101944

Kanth Choppara R, Chandmal Sharma R, Kumar Sharma S, Gupta T (2019) Aero dynamic cross wind analysis of locomotive. IOP Conf Ser Mater Sci Eng 691:012035. https://doi.org/10.1088/1757-899X/691/1/012035

Kohli I, Joshi NC, Mohapatra S, Varma A (2020) Extremophile – an adaptive strategy for extreme conditions and applications. Curr Genomics 21:96–110. https://doi.org/10.2174/1389202921666200401105908

Kohli I, Mohapatra S, Kumar P, Goel A, Varma A, Joshi NC (2022) Role of bacterial endophytes in the promotion of plant growth. In: Bacterial endophytes for sustainable agriculture and environmental management, pp 235–253. Springer Singapore, Singapore (2022). https://doi.org/10.1007/978-981-16-4497-9_12

Kokkinos NC, Emmanouilidou E, Sharma SK (2024) Waste-to-biofuel production for the transportation sector. In: intelligent transportation system and advanced technology. Springer, pp 99–122

Kumar U, Kasvekar, R, Sharma SK, Upadhyay RK (2022) Wear of wheels and axle in locomotive and measures taken by Indian railway. In: Advances in engine tribology. Springer, pp 77–96. https://doi.org/10.1007/978-981-16-8337-4_5

Lee J, Sharma SK (2020) Numerical investigation of critical speed analysis of high-speed rail vehicle. 한국정밀공학회 학술발표대회 논문집 (Korean Society of Precision Engineering), 696

Lee J, Han J, Sharma SK (2021) Structural analysis on the separated and integrated differential gear case for the weight reduction. Adv Eng Des: Select Proc of FLAME 2020, 175

Lee J, Sharma SK (2023) Intelligent systems for railway infrastructure. Appl Sci 13(20):11274. MDPI

Maity S, Das S, Mohapatra S, Tripathi AD, Akthar J, Pati S, Pattnaik S, Samantaray DP (2020) Growth associated polyhydroxybutyrate production by the novel Zobellellae tiwanensis strain DD5 from banana peels under submerged fermentation. Int J Biol Macromol 153:461–469. https://doi.org/10.1016/j.ijbiomac.2020.03.004

Misra S, Mohanty D, Mohapatra S (2022) Food-based probiotics: Functional dietary ingredients. In Probiotics, pp 257–275. Elsevier. https://doi.org/10.1016/B978-0-323-85170-1.00006-3

Mohanty DP, Mohapatra S, Misra S, Sahu PS (2016) Milk derived bioactive peptides and their impact on human health—A review. Saudi J Biol Sci 23:577–583. https://doi.org/10.1016/j.sjbs.2015.06.005

Mohapatra S, Rath SN, Pradhan SK, Samantaray DP, Rath CC (2016b) Secondary structural models (16S rRNA) of polyhydroxyalkanoates producing bacillus species isolated from different rhizospheric soil: Phylogenetics and chemical analysis. Int J Bioautomation 20:329–338

Mohapatra S, Mishra R, Roy P, Yadav KL, Satapathi S (2017b) Systematic investigation and in vitro biocompatibility studies on implantable magnetic nanocomposites for hyperthermia treatment of osteoarthritic knee joints. J Mater Sci 52:9262–9268. https://doi.org/10.1007/s10853-017-1136-0

Mohapatra S, Sarkar B, Samantaray DP, Daware A, Maity S, Pattnaik S, Bhattacharjee S (2017c) Bioconversion of fish solid waste into PHB using *Bacillus subtilis* based submerged fermentation process. Environ Technol 38:3201–3208. https://doi.org/10.1080/09593330.2017.1291759

Mohapatra S, Maity S, Dash HR, Das S, Pattnaik S, Rath CC, Samantaray D (2017d) Bacillus and biopolymer: prospects and challenges. Biochem Biophys Reports 12:206–213. https://doi.org/10.1016/j.bbrep.2017.10.001

Mohapatra S, Pattnaik S, Maity S, Mohapatra S, Sharma S, Akhtar J, Pati S, Samantaray DP, Varma A (2020) Comparative analysis of PHAs production by *Bacillus megaterium* OUAT 016 under submerged and solid-state fermentation. Saudi J Biol Sci 27:1242–1250. https://doi.org/10.1016/j.sjbs.2020.02.001

Mohapatra S, Samantaray DP, Samantaray SM (2015) Study on polyhydroxyalkanoates production using rhizospheric soil bacterial isolates of sweet potato. Indian J Sci Technol 8:57. https://doi.org/10.17485/ijst/2015/v8iS7/64027

Mohapatra S, Samantaray DP, Samantaray SM, Mishra BB, Das S, Majumdar S, Pradhan SK, Rath SN, Rath CC, Akthar J, Achary KG (2016a) Structural and thermal characterization of PHAs produced by Lysinibacillus sp. through submerged fermentation process. Int J Biol Macromol 93:1161–1167. https://doi.org/10.1016/j.ijbiomac.2016.09.077

Mohapatra S, Mohanta PR, Sarkar B, Daware A, Kumar C, Samantaray DP (2017a) Production of polyhydroxyalkanoates (PHAs) by bacillus strain isolated from waste water and its biochemical characterization. Proc Natl Acad Sci India Sect B Biol Sci 87:459–466. https://doi.org/10.1007/s40011-015-0626-6

Mohapatra S, Vishwakarma K, Joshi NC, Maity S, Kumar R, Ramchander M, Pattnaik S, Samantaray DP (2021) A review on PHAs: the future biopolymer. In: Environmental and agricultural microbiology, pp 83–100. Wiley. https://doi.org/10.1002/9781119525899.ch5

Mohapatra S, Mohapatra S, Han H, Ariza-Montes A, López-Martín M del C (2022) Climate change and vulnerability of agribusiness: assessment of climate change impact on agricultural productivity. Front Psychol 13. https://doi.org/10.3389/fpsyg.2022.955622

Mohd A, Srivastava R (2019) Parametric optimization and experimental validation for nonlinear characteristics of passenger car suspension system. Period Polytech Transp Eng 49:103–113. https://doi.org/10.3311/PPtr.12999

Nayak J, Swapnarekha H, Routray AR, Nayak SR, Behera HS (2022) Big data in education: present and future. In Computational intelligence in data mining. smart innovation, systems and technologies, 281. Springer, Singapore, pp 721–739. https://doi.org/10.1007/978-981-16-9447-9_54

Nibedita S, Swati P, Pattnaik M, Mohapatra S (2021) Methane emission and strategies for mitigation in livestock. In: Environmental and agricultural microbiology, pp 257–274. Wiley. https://doi.org/10.1002/9781119525899.ch12

Palli S, Koona R, Sharma SK, Sharma RC (2018) A review on dynamic analysis of rail vehicle coach. Int J Veh Struct Syst 10(3):204–211. https://doi.org/10.4273/ijvss.10.3.10

Palli S, Dontikurti R, Sharma RC, Sharma N (2021) Analysis of dynamic response of a railway locomotive using ANSYS. Int J Veh Struct Syst 13. https://doi.org/10.4273/ijvss.13.2.21

Palli S, Duppala A, Sharma RC, Gopala Rao LVV (2022) Dynamic simulation of automotive vehicle suspension using MATLAB simulink. Int J Veh Struct Syst 14. https://doi.org/10.4273/ijvss.14.3.04

Palli S, Koona R (2015) Analyses of dynamic response of a railway bogie. Int J Veh Noise Vib 11:103. https://doi.org/10.1504/IJVNV.2015.070015

PanduRanga Vital T, Lakshmi BG, Swapna Rekha H, DhanaLakshmi M (2019) Student performance analysis with using statistical and cluster studies. In Soft computing in data analytics. Advances in intelligent systems and computing, 758. Springer, Singapore, pp 743–757. https://doi.org/10.1007/978-981-13-0514-6_71

Pati S, Maity S, Dash A, Jema S, Mohapatra S, Das S, Samantaray DP (2020) Biocompatible PHB production from Bacillus species under submerged and solid-state fermentation and extraction through different downstream processing. Curr Microbiol 77:1203–1209. https://doi.org/10.1007/s00284-020-01922-7

Pati S, Mohanty MK, Mohapatra S, Samantaray D (2022) Bioethanol production by *Enterobacter hormaechei* through carbon monoxide-rich syngas fermentation. Int J Energy Res 46:20096–20106. https://doi.org/10.1002/er.8678

Pattnaik S, Dash D, Mohapatra S, Pattnaik M, Marandi AK, Das S, Samantaray DP (2020) Corrigendum to "Improvement of rice plant productivity by native Cr(VI) reducing and plant growth promoting soil bacteria Enterobacter cloacae" [Chemosphere 240 (2020) 124895]. Chemosphere 238:125011. https://doi.org/10.1016/j.chemosphere.2019.125011

Pattnaik S, Mohapatra S, Pati S, Dash D, Devadarshini D, Tanaya K, Mishra BB, Samantaray D (2022) Microbial bioremediation of Cr(VI)-contaminated soil for sustainable agriculture. In: Microbial biodegradation and bioremediation, pp 395–407. Elsevier (2022). https://doi.org/10.1016/B978-0-323-85455-9.00008-4

Pundir S, Upadhyay S, Kumar N, Chandra Joshi N, Priya R, Ahmad Mir R, Hossain I, Pandey OP (2023a) Interconnected CuO nanoplates as a highly efficient electrocatalyst for oxygen evolution reaction. Mater Lett 336:133921. https://doi.org/10.1016/j.matlet.2023.133921

Pundir S, Upadhyay S, Priya R, Kumar N, Chetana S, Hossain I, Joshi NC, Pandey OP (2023b) Synthesis of 1D β-MnO2 for high-performance supercapacitor application. J Solid State Electrochem 27:531–538. https://doi.org/10.1007/s10008-022-05347-z

Rallabandi SR, Pilla DP, Dowluru S, Palli S, Sharma N, Sharma SK, Sharma RC (2024a) Critical evaluation of epoxy-hemp-pineapple-palm fiber composites using hybrid AHM-TOPSIS technique for sustainable structural applications. J Chinese Inst Eng 1–12

Rallabandi SR, Srinivas L, Palli S, Sharma RC, Sharma N, Sharma A, Sharma SK (2024b) A contrastive characterization of pure Mg and AZ91D alloy based on the testing of mechanical, corrosion, wear, and erosion properties. Eng Res Express 6(1):015017

Samantaray D, Mohapatra S, Mishra BB (2014) Microbial bioremediation of industrial effluents. In: Microbial biodegradation and bioremediation, pp 325–339. Elsevier. https://doi.org/10.1016/B978-0-12-800021-2.00014-5

Sarkar MS, Segu H, Bhaskar JV, Jakher R, Mohapatra S, Shalini K, Shivaji S, Reddy PA (2018) Ecological preferences of large carnivores in remote, high-altitude protected areas: insights from Buxa Tiger Reserve, India. Oryx 52:66–77. https://doi.org/10.1017/S0030605317000060

Satyanarayana VSV, Sharma RC, Sateesh B, Gopala Rao LVV, Mohan Rao N, Palli S (2022) Optimum response of a nonlinear passive vehicle suspension system under random road excitations. SAE Int J Commer Veh 16:02-16-01-0004. https://doi.org/10.4271/02-16-01-0004

Sharma SK (2019) Multibody analysis of longitudinal train dynamics on the passenger ride performance due to brake application. Proc Inst Mech Eng, Part K: J Multi-Body Dynam 233(2):266–279. https://doi.org/10.1177/1464419318788775

Sharma SK, Kumar A (2017a) Ride performance of a high speed rail vehicle using controlled semi active suspension system. Smart Mater Struct 26:055026. https://doi.org/10.1088/1361-665X/aa68f7

Sharma SK, Kumar A (2017b) Impact of electric locomotive traction of the passenger vehicle Ride quality in longitudinal train dynamics in the context of Indian railways. Mechanics Ind 18(2):222. https://doi.org/10.1051/meca/2016047

Sharma SK, Kumar A (2018a) Ride comfort of a higher speed rail vehicle using a magnetorheological suspension system. Proc Inst Mech Eng Part K. J Multi-body Dyn 232:32–48. https://doi.org/10.1177/1464419317706873

Sharma SK, Kumar A (2018b) Disturbance rejection and force-tracking controller of nonlinear lateral vibrations in passenger rail vehicle using magnetorheological fluid damper. J Intell Mater Syst Struct 29(2):279–297. https://doi.org/10.1177/1045389X17721051

Sharma SK, Kumar A (2018c) Impact of longitudinal train dynamics on train operations: a simulation-based study. J Vibr Eng Technol 6(3):197–203. https://doi.org/10.1007/s42417-018-0033-4

Sharma SK, Lee J (2020a) Finite element analysis of a fishplate rail joint in extreme environment condition. Int J Veh Struct Syst 12. https://doi.org/10.4273/ijvss.12.5.03

Sharma SK, Lee J (2020b) Design and development of smart semi active suspension for nonlinear rail vehicle vibration reduction. Int J Struct Stab Dyn 20:2050120. https://doi.org/10.1142/S0219455420501205

Sharma SK, Lee J (2021c) Crashworthiness analysis for structural stability and dynamics. Int J Struct Stab Dyn 21:2150039. https://doi.org/10.1142/S0219455421500395

Sharma RC, Palli S (2016) Analysis of creep force and its sensitivity on stability and vertical-lateral ride for railway vehicle. Int J Veh Noise Vib 12:60. https://doi.org/10.1504/IJVNV.2016.077474

Sharma RC, Sharma SK (2018a) Sensitivity analysis of three-wheel vehicle's suspension parameters influencing ride behavior. Noise Vibr Worldwide 49(7–8):272–280. https://doi.org/10.1177/0957456518796846

Sharma SK, Sharma RC (2018b) An investigation of a locomotive structural crashworthiness using finite element simulation. SAE Int J Commercial Veh 11(4):235–244. https://doi.org/10.4271/02-11-04-0019

Sharma SK, Sharma RC (2018c) Simulation of quarter-car model with magnetorheological dampers for ride quality improvement. Int J Veh Struct Syst 10(3):169–173. https://doi.org/10.4273/ijvss.10.3.03

Sharma SK, Sharma RC (2019) Pothole detection and warning system for Indian roads. In Advances in interdisciplinary engineering. Lecture notes in mechanical engineering. Springer, Singapore, pp 511–519. https://doi.org/10.1007/978-981-13-6577-5_48

Sharma SK, Sharma RC (2021) Multi-objective design optimization of locomotive nose. SAE Tech PapeR, 10. https://doi.org/10.4271/2021-01-5053

Sharma RC, Sharma SK (2022) Ride analysis of road surface-three-wheeled vehicle-human subject interactions subjected to random excitation. SAE Int J Commercial Veh 15(3), 02-15-03–0017. https://doi.org/10.4271/02-15-03-0017

Sharma SK, Sharma RC, Kumar A, Palli S (2015) Challenges in rail vehicle-track modeling and simulation. Int J Veh Struct Syst 7(1):1–9. https://doi.org/10.4273/ijvss.7.1.01

Sharma RC, Palli S, Sharma SK, Roy M (2017) Modernization of railway track with composite sleepers. Int J Veh Struct Syst 9(5):321–329. https://doi.org/10.4273/ijvss.9.5.10

Sharma RC, Sharma SK, Palli S (2018) Rail vehicle modelling and simulation using lagrangian method. Int J Veh Struct Syst 10(3):188–194. https://doi.org/10.4273/ijvss.10.3.07

Sharma SK, Saini U, Kumar A (2019) Semi-active control to reduce lateral vibration of passenger rail vehicle using disturbance rejection and continuous state damper controllers. J Vib Eng Technol 7(2):117–129. https://doi.org/10.1007/s42417-019-00088-2

Sharma RC, Sharma SK, Sharma N, Sharma S (2020a) Analysis of ride and stability of an ICF railway coach. Int J Veh Noise Vib 16:127. https://doi.org/10.1504/IJVNV.2020.117820

Sharma RC, Sharma S, Sharma SK, Sharma N (2020b) Analysis of generalized force and its influence on ride and stability of railway vehicle. Noise Vib Worldw 51:95–109. https://doi.org/10.1177/0957456520923125

Sharma S, Sharma RC, Sharma SK, Sharma N, Palli S, Bhardawaj S (2020c) Vibration isolation of the quarter car model of road vehicle system using dynamic vibration absorber. Int J Veh Struct Syst 12. https://doi.org/10.4273/ijvss.12.5.05

Sharma RC, Sharma SK, Palli S (2020d) Linear and non-linear stability analysis of a constrained railway wheelaxle. Int J Veh Struct Syst 12. https://doi.org/10.4273/ijvss.12.2.04

Sharma SK, Sharma RC, Sharma N (2020e) Combined multi-body-system and finite element analysis of a rail locomotive crashworthiness. Int J Veh Struct Syst 12. https://doi.org/10.4273/ijvss.12.4.15

Sharma SK, Sharma RC, Lee J (2021) Effect of rail vehicle-track coupled dynamics on fatigue failure of coil spring in a suspension system. Appl Sci 11:2650. https://doi.org/10.3390/app11062650

Sharma RC, Palli S, Sharma N, Sharma SK (2021b) Ride behaviour of a four-wheel vehicle using H infinity semi-active suspension control under deterministic and random inputs. Int J Veh Struct Syst 13. https://doi.org/10.4273/ijvss.13.2.18

Sharma RC, Sharma S, Sharma N, Sharma SK (2021c) Linear and nonlinear analysis of ride and stability of a three-wheeled vehicle subjected to random and bump inputs using bond graph and simulink methodology. SAE Int J Commer Veh 14:02-15-01-0001. https://doi.org/10.4271/02-15-01-0001

Sharma RC, Sharma S, Sharma SK, Sharma N, Singh G (2021d) Analysis of bio-dynamic model of seated human subject and optimization of the passenger ride comfort for three-wheel vehicle using random search technique. Proc Inst Mech Eng Part K J Multi-body Dyn 235:106–121. https://doi.org/10.1177/1464419320983711

Sharma SK, Mohapatra S, Sharma RC, Alturjman S, Altrjman C, Mostarda L, Stephan T (2022a) Retrofitting existing buildings to improve energy performance. Sustainability 14:666. https://doi.org/10.3390/su14020666

Sharma SK, Sharma RC, Lee J, Jang H-L (2022b) Numerical and experimental analysis of DVA on the flexible-rigid rail vehicle carbody resonant vibration. Sensors 22:1922. https://doi.org/10.3390/s22051922

Sharma SK, Lee J, Jang H-L (2022c) Mathematical modeling and simulation of suspended equipment impact on car body modes. Machines 10:192. https://doi.org/10.3390/machines10030192

Sharma SK, Sharma RC, Lee J (2022d) In situ and experimental analysis of longitudinal load on carbody fatigue life using nonlinear damage accumulation. Int J Damage Mech 31:605–622. https://doi.org/10.1177/10567895211046043

Sharma RC, Gopala Rao LVV, Sharma SK, Palli S, Satyanarayana VSV (2022e) Analysis of lateral stability and ride of an Indian railway constrained dual-axle bogie frame. SAE Int J Commercial Veh 16(2), 02-16-02–0014. https://doi.org/10.4271/02-16-02-0014

Sharma RC, Palli S, Gopala Rao LVV, Duppala A, Sharma SK (2022f) Four-wheel vehicle response under bump, pothole, harmonic, and random excitations using bond graph/simulink technique. SAE Int J Commercial Veh 16(2), 02-16-02–0008. https://doi.org/10.4271/02-16-02-0008

Sharma RC, Palli S, Jha AK, Bhardawaj S, Sharma SK (2022g) Vibration and ride comfort analysis of railway vehicle system subjected to deterministic inputs. Resmilitaris 12(5):1345–1355

Sharma SK, Sharma RC, Lavania S, Palli S, Avesh M (2023a) Industry-driven approach for ANFIS-based intelligent control suspension system with MR damper for enhanced ride quality in passenger rail vehicles for technological investigations. Nat Acad Sci Lett 1–4

Sharma SK, Sharma RC, Upadhyay RK, Lee J (2023b) Coupled torsional–transverse vibration reduction in marine propulsion dynamics with novel approach using magnetorheological damping and adaptive control system. J Vib Eng Technol 1–11

Sharma RC, Palli S, Sharma SK (2023c) Ride analysis of railway vehicle considering rigidity and flexibility of the carbody. J Chinese Inst Eng 46:355–366. https://doi.org/10.1080/02533839.2023.2194918

Sharma SK, Sharma RC, Choi Y, Lee J (2023d) Experimental and mathematical study of flexible–rigid rail vehicle riding comfort and safety. Appl Sci 13(9):5252

Sharma SK, Sharma RC, Choi Y, Lee J (2023e) Modelling and dynamic analysis of adaptive neuro-fuzzy inference system-based intelligent control suspension system for passenger rail vehicles using magnetorheological damper for improving ride index. Sustainability 15(16):12529

Sharma SK, Upadhyay RK, Kumar V, Valera H (2023f) Introduction to sustainable transportation system. Transp Energy Dynam. Springer, pp 3–6

Sharma RC, Palli S, Bhardawaj S (2023g) A review of wheel-rail dynamic behavior on curved tracks. Int J Veh Struct Syst 15. https://doi.org/10.4273/ijvss.15.1.04

Sharma RC, Bhardawaj S, Sharma SK, Bahl S (2023h) Ride behaviour of three-wheel vehicle using semi-active suspension using skyhook control strategy. Int J Veh Struct Syst 15(1). https://doi.org/10.4273/ijvss.15.1.05

Sharma RC, Palli S, Dharmana L, Sharma SK, Sharma A, Kharola A, Sharma N (2024a) Simulation of transient dynamic response of a locomotive considering its flexibility and rigidity. Noise Vib Worldwide 55(1–2):33–47

Sharma RC, Palli S, Duppala A, Bhardawaj S, Sharma SK, Sharma N, Sharma A (2024b) Rigid body analysis of ride behaviour of roadway vehicle using lagrangian dynamics. Int J Veh Struct Syst (IJVSS) 16(1)

Sharma RC, Sharma N, Palli S, Sharma SK, Joshi S, Rallabandi SR, Sharma A (2024c) Dynamic behavior of three-wheel vehicle with inerter-based dynamic absorber suspension system. Noise Vibration Worldwide 09574565241243398

Sharma SK, Kumar N, Avesh M, Sharma RC, Siddiqui MIH, Lee J (2024d) Navigating tranquillity with H∞ controller to mitigate ship propeller shaft vibration. J Vib Eng Technol 1–11

Sharma SK, Sharma RC, Lee J (2024e) Modelling and simulation of MR damper characterization and uncertainty assessment in marine engine vibration isolation with FLPID control. Int J Modell Simul 1–11

Sharma SK, Sharma RC, Lee J (2024f) Propelling precision of longitudinal vibration mitigation in ship propeller shafts through advanced nonlinear intelligent semi-active control leveraging adaptive neuro-fuzzy inference system with linear quadratic regulator. J Vib Cont 10775463241244836

Sharma SK, Sharma RC, Upadhyay RK, Lee J (2024g) Enhanced ship engine vibration reduction using magnetorheological dampers and adaptive neuro-fuzzy control system. Ships Offshore Struct 1–17

Sinha AK, Sengupta A, Gandhi H, Bansal P, Agarwal KM, Sharma SK, Sharma RC, Sharma SK (2019) Performance enhancement of an all-terrain vehicle by optimizing steering, powertrain and brakes. In Advances in engineering design . lecture notes in mechanical engineering. Springer, Singapore, pp 207–215. https://doi.org/10.1007/978-981-13-6469-3_19

Swapna Rekha H, Nayak J, Behera HS (2020) Pi-sigma neural network: survey of a decade progress. In: Computational intelligence in pattern recognition. Advances in intelligent systems and computing, 1120. Springer, Singapore, pp 429–441. https://doi.org/10.1007/978-981-15-2449-3_37

Vishwakarma PN, Mishra P, Sharma SK (2022a) Characterization of a magnetorheological fluid damper a review. Mater Today: Proc 56:2988–2994. https://doi.org/10.1016/j.matpr.2021.11.143

Vishwakarma PN, Mishra P, Sharma SK (2022b) Formulation of semi-active suspension system and controls in rail vehicle. SSRN Electron J. https://doi.org/10.2139/ssrn.4159616

Upadhyay RK, Sharma SK, Kumar V, Valera H (2023a) Introduction to transportation systems technology and integrated management. In: Transportation systems technology and integrated management. Springer. pp 1–4. https://doi.org/10.1007/978-981-99-1517-0_1

Upadhyay RK, Sharma SK, Kumar V, Valera H (2023b) Transportation systems technology and integrated management. In: Upadhyay RK, Sharma SK, Kumar V, Valera H (eds) Springer Nature Singapore. https://doi.org/10.1007/978-981-99-1517-0

Upadhyay RK, Sharma SK, Kumar V (Eds) (2024a) Intelligent transportation system and advanced technology. Springer Nature Singapore. https://doi.org/10.1007/978-981-97-0515-3

Upadhyay RK, Sharma SK, Kumar V (2024b) Introduction to intelligent transportation system and advanced technology. In: Intelligent transportation system and advanced technology. Energy, Environment, and Sustainability. pp 3–6. Springer, Singapore. https://doi.org/10.1007/978-981-97-0515-3_1

Wu Q, Cole C, Spiryagin M, Chang C, Wei W, Ursulyak L, Shvets A, Murtaza MA, Mirza IM, Zhelieznov K, Mohammadi S, Serajian H, Schick B, Berg M, Sharma RC, Aboubakr A, Sharma SK, Melzi S, Di Gialleonardo E, Cantone L et al (2023) Freight train air brake models. Int J Rail Transp 11(1):1–49. https://doi.org/10.1080/23248378.2021.2006808

Chapter 13
Comparing Meta-Heuristic Algorithms for Transit Network Design

Obiora A. Nnene, Mark H. P. Zuidgeest, and Johan W. Joubert

13.1 Introduction

A public transport network comprises nodes that serve as stations joined in space by a links that satisfy people's travel demand. In terms of their function, public transport networks are essential in supporting socio-economic activities in any geographical area. However, the design of transit networks is a significantly complex endeavour (Nnene et al. 2019). This complexity is often due to the unique topography of each network in its unique context, the stochastic nature of travel demand and the relationship between the network the environment and different land-use systems (Rodrigue et al. 2013). These factors are crucial in shaping decisions around the type of service to be designed, the suitable technology to deploy for that service, and which segment of the population should be given priority in terms of service delivery. Designing public transit networks is a typical NP-hard optimisation problem (Newell 1979; Nnene et al. 2017), which is subject to some constraints like the configuration of network routes, passenger route choice and operational headway and other feasibility conditions for the length of route, capacity of vehicles and size of fleet must be accounted for.

The transit network design problem (TNDP) has been used to study transit network improvement, as seen in Cipriani et al. 2012 and Nnene et al. 2019), route network optimisation and frequency setting by Buba and Lee (2018) and Nikolić and Teodorović (2014). In addition, the bus transit network problem with variable transit demand was investigated by Fan and Mumford (2008) on multimodal transit network design Huang et al. (2018). Other TNDP related works have dwelt on transit

O. A. Nnene (✉) · M. H. P. Zuidgeest
Department of Civil Engineering, University of Cape Town, Cape Town, South Africa
e-mail: obiora.nnene@uct.ac.za

J. W. Joubert
Department of Industrial Systems Engineering, University of Pretoria, Pretoria, South Africa

S. K. Sharma et al. (eds.), *Dynamics of Transportation Ecosystem, Modeling, and Control*, Energy, Environment, and Sustainability,
https://doi.org/10.1007/978-981-97-0437-8_13

emissions management (Beltran et al. 2009), optimal network routing and intersection signal setting (Gallo et al. 2010) and rail frequency optimisation (Gallo et al. 2011). At the same time, (Johar et al. 2016; Ibarra-Rojas et al. 2015; Durán-Micco and Vansteenwegen 2022) undertook a detailed review of the TNDP and classified previous solution techniques used by other authors within the literature. To solve a TNDP, an attempt is made to identify a network solution and its accompanying service parameters, like frequency, amongst alternative solutions that best satisfy stakeholders' stated goal(s) like commuters, service operators and even the transit authorities in certain instances.

This chapter compares the performance of four meta-heuristic algorithms used within an innovative solution model developed by the authors for solving the TNDP. Similar comparison of meta-heuristic algorithms applied to transportation contexts like the airport gate assignment problem (Aktel et al. 2016), transport capacity fixed charge problem (Biswas et al. 2021) and road maintenance planning (Gerami Matin et al. 2017). Other use cases of similar comparative studies outside the transportation field may be found in Matnei Filho and Vergilio 2016; Ezugwu et al. 2020; Arıcı and Kaya 2019; Abd 2014; Silberholz and Golden 2010). A Simulation-based transit network design model (SBTNDM) is developed from the principles of Simulation-based optimisation (SBO). This discipline focuses on solving decision-based problems by integrating optimisation and simulation. In the SBTNDM, a discrete event activity-based simulation (ABS) is used to evaluate the network solutions, while multi-objective evolutionary meta-heuristics optimise. The model is a robust and versatile tool that allows for seamless integration with different meta-heuristics, and its technical details have been discussed in Nnene et al. 2023a. Its versatility is therefore harnessed in this work to allow a comparison of four bio-inspired algorithms, specifically the non-dominated sorting algorithm-II (NSGA-II) (Deb et al. 2000), strength Pareto evolutionary algorithm (SPEA2) (Zitzler et al. 2002), non-dominated sorting algorithm-III (NSGA-III) (Deb and Jain 2013) (Deb 2013) and the indicator-based evolutionary algorithm (IBEA) (Zitzler and Künzli 2004).

The named algorithms belong to the meta-heuristics class, which is known as Multi-objective Evolutionary Algorithms (MOEA). These algorithms operate by emulating biological processes, drawing inspiration from phenomena such as genetics. Therefore, they are often referred to as bio-inspired algorithms. Their function involves enabling the generation of improved and potentially better solutions from existing ones. Their operators' actions on the current population result in the creation of offspring solutions that are expected to be more fit or perform better than their parent solutions. Hence, this chapter presents a comparative study of how the earlier-mentioned algorithms perform within the SBTNDM. The chapter is structured as follows; the next section discusses SBO literature in the context of the TNDP. This is followed by a description of the earlier-mentioned meta-heuristic algorithms in Sect. 13.3. In Sect. 13.4, the actual performance evaluation is done, while the results are presented and discussed. Section 13.4 ends with a reflection on current trends in the transit network design while the chapter is concluded in Sect. 13.5.

13.2 Simulation-Based Optimisation of Transit Networks

The integration of optimisation algorithms with simulation to address complex optimisation problems such as the TNDP is referred to as Simulation-Based Optimisation (SBO). The discipline is fairly recent in the operations research specialisation but has gained recognition in the last twenty years (Gosavi 2015). In SBO, when presented with a set of alternative solutions within a problem's search space, optimisation algorithms are typically used to identify a unique or approximate solution, depending on whether the problem is convex or non-convex. Conversely, simulations are used to describe experimentally, the intricate interactions of an entity in a system. Simulations reproduce large-scale, real-life systems, providing insights into how the system performs under various scenarios. Essentially, optimisation and simulation models show distinct functional limitations (Nnene et al. 2023b). Optimisation algorithms excel at finding solutions but may struggle to comprehensively analyse large stochastic systems. On the other hand, simulations can adequately describe the behaviour of such systems but are not designed for optimisation tasks. As a result, SBO leverages the complementary nature of these two approaches to solve decision-based problems.

A detailed review of the SBO literature concerning transportation system design and planning may be found in Amaran et al. (2016). SBO problems and solution algorithms are categorised by the description of a problem's input variables and the number of objectives, among other indicators (Hachicha et al. 2010). Four main solution approaches are discussed in Arisha and Abo-Hamad (2010): gradient-based, statistical methods, meta-models and meta-heuristics. In SBO solutions based on meta-heuristics, the former is generally integrated with simulations of different kinds.

These solutions offer a notable advantage over the previously mentioned SBO methods, including meta-models. They excel in situations where the search space has a high-dimension, is non-continuous or involves decision variables that are qualitative (Arisha and Abo-Hamad 2010). An example of applying this technique in public transport systems can be found in Song et al. 2013, where researchers introduced an SBO for measuring and improving sustainable transportation systems. The primary goal of their work was to determine the optimal mix of transportation planning and operational strategies that minimise the generalised costs of travel on multimodal systems. These costs encompass factors like vehicle congestion pricing and public transport fares. Their model consisted of four major components: strategy, simulation, evaluation and optimisation. The researchers applied this model to analyse a small multimodal network in China, demonstrating its viability and effectiveness in practical transportation planning scenarios. Another work that uses SBO is (Dandl et al. 2020), where the authors propose a tri-level mathematical programming framework that combines Bayesian optimisation with activity-based simulation. The work aimed to model the decision dynamics between transit service operators and commuters as a basis for analysing and optimising transit policies. Their model was applied to the transit network of Munich, Germany, to determine the city's toll and parking cost requirements. Lastly, (Nnene et al. 2023c) uses a meta-heuristic SBO approach to develop the SBTNDM. The model integrated an Activity-based travel demand simulation and MOEA. The objective was to reduce travel time and operational costs for commuters and operators. The work was applied to a bus rapid transit network in Cape Town, South Africa.

In the SBTNDM, the search space encompasses different networks and their attendant operational parameters. Each network solution undergoes evaluation by simulating its travel demand using ABS. To initiate the optimisation framework, a set of public transit route networks is initialised. The primary task of the algorithm is to identify the most favourable solution from this initialised set. However, to accomplish this, these alternatives must undergo evaluation, which is jointly performed by the optimisation and simulation sub-routines. Consequently, the main interaction between these components of the SBTNDM occurs when networks are evaluated. The exchange of data between the elements is the basis for how the SBTNDM functions. To assess each network's performance, a transit schedule file containing the routes must be provided as input to the simulation which is called whenever a route network is to be evaluated. Commuters' travel demand is then simulated on the network, and the results analysed before being sent back to the optimisation sub-model. This iterative process continues until a suitable solution is identified. Following the completion of the simulation, the optimisation algorithm ranks the performance of each network using indicators obtained from the simulation results.

13.3 Description of Algorithms

13.3.1 Non-Dominated Sorting Algorithm-II

The NSGA-II was developed by Deb et al. (2000) as an enhancement to its predecessor, the non-dominated sorting genetic algorithm (NSGA). It begins by initialising the population of the network, after which these initial solutions are evaluated based on the specified objective functions. Subsequently, the solutions are sorted and grouped into distinct Pareto frontiers using a non-dominated sorting procedure. A solution earns the non-dominated classification when it surpasses other solutions in at least one objective while maintaining parity or superiority across other objectives. Consequently, all solutions that are non-dominated are temporarily removed and assigned a rank of one. The process repeats to identify and rank the subsequent sets of non-dominated solutions until all solutions are ranked. These ranks serve as fitness scores and indicate the dominance of solutions within their respective frontiers. Once the selection process concludes, genetic operators, namely, crossover or recombination and mutation, are deployed to produce a population of offspring or children equivalent in size to the original parent population. Following this, the generated offspring are merged with their parent solutions to create a population twice the size of the initial one. This combined population is again evaluated and ranked, with the superior half becoming the new parent population. This iterative process continues until a specified termination criterion is met. Furthermore, the algorithm incorporates elitism by preserving a small fraction of the best or "elite" solutions from the parent and offspring populations in all the generations. The elite solutions are reintroduced into the parent population in each subsequent generation, ensuring that the algorithm retains and builds on its best solutions as it progresses.

Algorithm 1 Non-dominated Sorting Genetic Algorithm II (adapted from Matnei Filho and Vergilio 2016).

```
procedure NSGA-II(P,f,G)
Initialize the population P
    Evaluate the population P with respect to the objectives f
    for g = 1,G do
    Create the offspring population O using crossover and mutation
    Evaluate the offspring population O with respect to the objectives f
    Combine the parent and offspring populations P = P ∪ O
    Sort the population P by non-dominance rank
    Select the best solutions
    end for
Return the best solutions P
end procedure
```

13.3.2 Non-Dominated Sorting Algorithm-III

NSGA-III, an advancement of the NSGA-II, employs a reference-point-based approach that focuses on non-dominated population members and reference points strategically placed within the search space. The process begins by establishing a set of reference points. Subsequently, the current parent is utilised to generate an offspring through the application of genetic operators. These parent and offspring populations are then merged. Following the merging step, the combined population is categorized into different non-domination levels using the same dominance ranking method found in the NSGA-II. Subsequently, the selection of solutions that will form the next generation occurs. In the NSGA-II, this selection is based on solutions with the biggest crowding distance values, signifying the widest diversity. However, this crowding distance mechanism isn't well-suited for problems with four or more objectives (Das and Pandit 2013). Consequently, the NSGA-III modifies the selection mechanism by introducing a set of reference points distributed along a normalised hyperplane within the solution space. This approach ensures that diversity among solutions is maintained, even in problems with many objectives, resulting in a more evenly distributed set of non-dominated solutions in the objective space. As a result, the preferred solutions tend to be closer to the identified reference points. In comparison to the NSGA-II, the primary drawback of the NSGA-III is its increased computational demands (Jain and Deb 2014).

Algorithm 2 Non-dominated Sorting Genetic Algorithm III (adapted from Matnei Filho and Vergilio 2016).

procedure NSGA-III(P,f,G)
Initialize the population P

	Evaluate the population P with respect to the objectives f
	for g = 1,…,G **do**
	Create the offspring population O using crossover and mutation
	Evaluate the offspring population O with respect to the objectives f
	Combine the parent and offspring populations P = P ∪ O
	Sort the population P by non-dominance rank
	Apply crowding distance to the population P
	Select the best solutions
	Calculate the reference points R
	Update the population P using the reference points R
	end for

Return the best solutions P
end procedure

13.3.3 Indicator-Based Evolutionary Algorithm

The main feature of indicator-based Multi-Objective Evolutionary Algorithms (MOEAs) is their fitness assignment operation, which depends on indicators such as hypervolume or additive epsilon indicators, in comparison with many other Pareto dominance-based MOEAs. The IBEA by Zitzler and Künzli (2004) is one of the most well-documented. It utilises the hypervolume indicator to rank solutions. The algorithm operates on the principle that a quality indicator effectively identifies solutions with desired qualities. A population is generated in the first generation. The algorithm employs a binary tournament selection operator to choose two individuals from the current population. These selected individuals undergo reproduction through crossover and mutation operations, resulting in the generation of two offspring. The children (offspring) are then merged with the parent population to create the next generation of individuals. Following this, the new population is iteratively evaluated, and in each successive iteration, the poorest-performing solutions are removed until the desired number of individuals for the next reproduction process is achieved. This process is referred to as environmental reproduction (Phan and Suzuki 2013).

However, an important drawback of this indicator-based method is that evaluating its performance becomes expensive in terms of computing requirements, especially when solving problems that involve a high number of objectives.

Algorithm 3 Indicator-Based Evolutionary Algorithm (adapted from Matnei Filho and Vergilio 2016).

	procedure IBEA(P,f,G)
	Initialize the population *P*.
	Evaluate the population P with respect to the objectives f.
	for $g = 1, G$ **do**
	Check the solution with the worst fitness value and remove it from *P*.
	Recalculate fitness values for *P*.
	If $g \geq G$, assign non-dominated solutions as the decision vector.
	Perform binary tournament selection to fill mating pool.
	Create the offspring population O using crossover and mutation.
	end for
	end procedure

13.3.4 Strength Pareto Evolutionary Algorithm 2

The SPEA2, and its predecessor, the SPEA, were developed by Zitzler et al. (2002). One key feature of SPEA2 is its utilisation of elitism through an external archive that maintains previously acquired non-dominated solutions. In each generation, non-dominated individuals from the current population are copied into this external non-dominated set. For each individual in this external set, a strength value is calculated, similar to the Multi-objective Genetic Algorithm (MOGA) ranking introduced by Fonseca and Fleming (1993). In the SPEA2 algorithm, the fitness of each member in the current population is evaluated based on the "strengths" of all external non-dominated solutions dominating it. This process considers the convergence and diversity of the approximated Pareto solution sets. This assignment strategy judges the number of individuals that a solution dominates and the number of solutions that dominate it simultaneously. Pareto dominance is then used to ensure effective distribution of the solutions along the Pareto front. It is noteworthy that as the size of the archived solutions grows, it can reduce the selection pressure in SPEA2, leading to a slowdown of the search process. To mitigate this issue, SPEA2 employs a clustering technique called the truncation operator. This operator trims the external non-dominated set, below a predefined threshold which effectively maintains an appropriate balance between exploring and exploiting which occurs during the evolutionary search for feasible solution sets.

Algorithm 4 Strength Pareto Evolutionary Algorithm (adapted from Matnei Filho and Vergilio 2016).

1. **procedure** SPEA2(P,f,G)
2. Initialize the population P

	Create empty external set Ø
	for g = 1, G **do**
	Evaluate fitness of P and Ø with respect to the objectives f
	Move non-dominated solutions to Ø
	Prune the Ø if its capacity is exceeded with the truncation operator
	If capacity of Ø is not exceeded, add dominated solutions
	Perform binary tournament selection
	Create the offspring population O using crossover and mutation
	end for

12. Return the best solutions P
13. **end procedure**

13.4 Performance Evaluation

In this section, we evaluate and compare the performance of the previously discussed algorithms using four Multi-Objective Evolutionary Algorithm (MOEA) quality indicators: hypervolume, generational distance, epsilon indicator and maximum Pareto front error. These indicators assess various aspects of the algorithms, including their convergence and diversity. For the evaluation, we run a population size of 30 networks for 50 generations each for all tests. Since all four algorithms under consideration are evolutionary algorithms, the main parameters shared across all of them are genetic operators, specifically crossover and mutation. In these tests, the crossover and mutation probabilities are set to 0.75 and 0.25, respectively, for all algorithms. (Siriwardene and Perera 2006) noted that a higher crossover probability promotes a better combination of parent solutions, while a lower mutation rate is recommended to enhance the algorithm's convergence. It's worth mentioning that the parameter values for the genetic operators, population size and generation count are determined through limited experimentation due to resource constraints. A more comprehensive sensitivity analysis may be carried out in the future to obtain optimal parameter values. However, the results presented here are based on the conducted experiments. The tests are conducted on the National Centre for High-Performance Computing's Lengau cluster, which is a Dell Linux High-Performance Computing cluster with 1368 nodes and 32,832 cores, allowing access to 240 nodes simultaneously. Each experiment takes approximately 15 min to run on this resource. The test duration for all four algorithms is limited to a 48-h window. The results are stored during the model run and are then later analysed.

13.4.1 Hypervolume

The hypervolume indicator is a metric that is widely used for assessing the quality of Multi-Objective Evolutionary Algorithm (MOEA) solutions. It quantifies the size of the search space covered by a Multi-Objective Problem's (MOP) approximation set (Bringmann and Friedrich 2013). The approximation set represents sub-optimal solutions also known as a local Pareto front, which serves as an estimate of the global Pareto front or reference set. This indicator is calculated in relation to a reference point called the nadir point (Hadka 2017), which typically represents the worst-case value for each objective function. A major advantage of the hypervolume indicator is its easy adaptability to problems with many objectives. However, one of its key limitations is that it can be computationally expensive. In Fig. 13.1, the plot illustrates the values of the hypervolume indicator for all four algorithms after 50 generations. Higher hypervolume values indicate better solutions because they dominate a larger portion of the search space. The figure demonstrates that the indicator's value consistently increases with the progression of generations, indicating that the SBTNDM's solutions are better in successive generations, in line with the typical behaviour of the hypervolume indicator.

Furthermore, the figure reveals that NSGA-II achieves the highest hypervolume indicator value among the four algorithms, followed by SPEA2, NSGA-III and IBEA. This implies that NSGA-II performs better, as a higher hypervolume indicator value signifies superior results. Finally, the indicator converges near 50 generations, suggesting that the number of generations required to get optimal network solutions with the SBTNDM is near its limit.

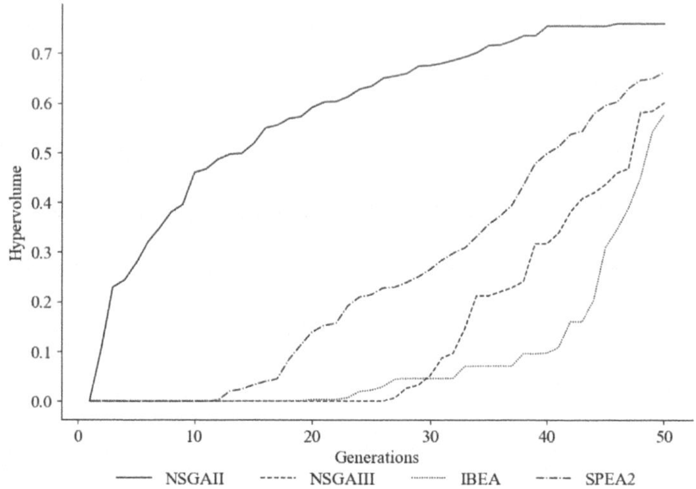

Fig. 13.1 The hypervolume indicator (Nnene 2020)

13.4.2 Generational Distance

The generational distance (GD) indicator evaluates the quality of solutions by calculating the mean distance between each solution in the approximation set and the nearest solution in the problem's reference set. As this indicator quantifies proximity to the reference set, it is well-suited for evaluating convergence in Multi-Objective Evolutionary Algorithms (MOEAs) (Liu et al. 2019). Smaller GD values are considered more favorable. When the approximation set is entirely contained within the reference set, the GD is zero. However, relying solely on GD may not be adequate for evaluating a MOEA's quality. This is because if the approximation set contains a single solution that is very close to the reference set when compared to other solutions, the GD measurements can be too low which may not reflect the reality. Therefore, GD is often used in conjunction with other quality or performance indicators. In Fig. 13.2, a plot of GD against the number of generations for the four algorithms is presented. Notably, NSGA-II achieves the lowest GD value, indicating superior performance compared to the other algorithms. SPEA2 comes in as the second-best performing algorithm. The values appear to converge after approximately 45 generations, suggesting that, in terms of the GD indicator, the SBTNDM is likely to yield good solutions after around 45 generations.

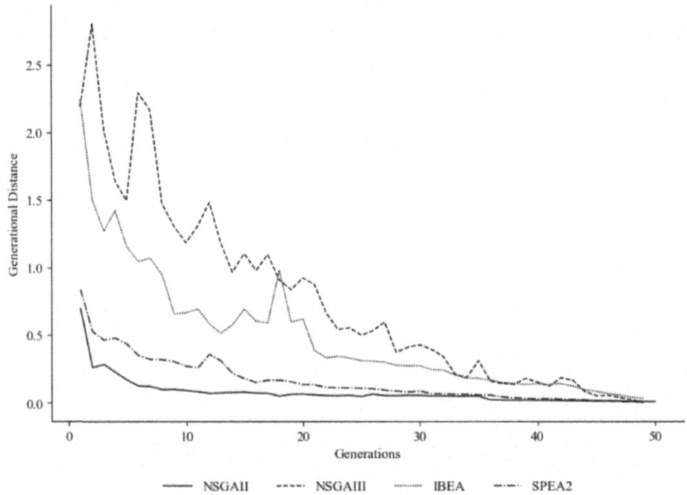

Fig. 13.2 The Generational distance indicator (Nnene 2020)

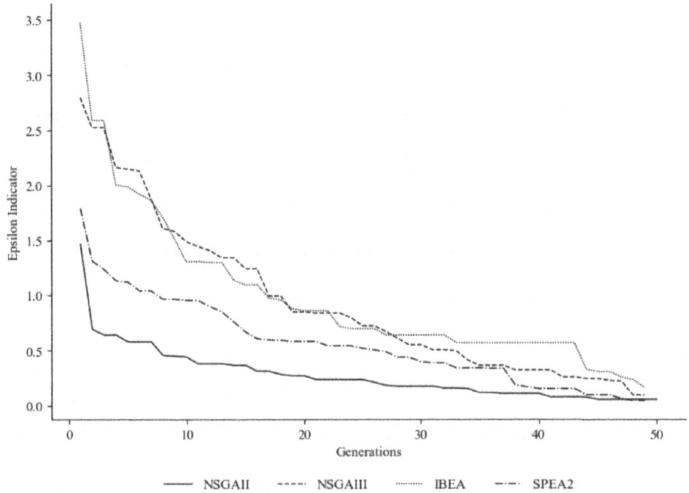

Fig. 13.3 The Epsilon indicator (Nnene 2020)

13.4.3 Epsilon Indicator

The epsilon or additive indicator quantifies the least distance required to transform an approximation set in a way that it completely dominates the reference set. In the context of this indicator, lower distance values needed to envelop the reference set imply better proximity and diversity within the approximation set. Conversely, if there are regions in the reference set that are poorly approximated by the approximation set, larger distances would be necessary. Hadka and Reed (2013) states that the additive indicator also measures the consistency of an approximation set. To achieve the best results with this indicator, the approximation set ideally should not have large gaps or areas of poor approximation. In Fig. 13.3, the plot displays the values of the epsilon or additive indicator against the generation parameter for the SBTNDM and the four algorithms. It is evident that the values show convergence after 50 generations, suggesting that, based on this indicator, the SBTNDM is likely to yield good solutions after approximately 50 generations.

13.4.4 Maximum Pareto Front Error

The maximum Pareto front error indicator evaluates how effectively one set of solutions compares to another, specifically measuring the largest minimum distance between solutions in the approximation set and its nearest matching solution in the reference set (Yen and He 2014). A well-spaced approximation set typically consists of solutions that are not densely clustered together or separated by large empty spaces.

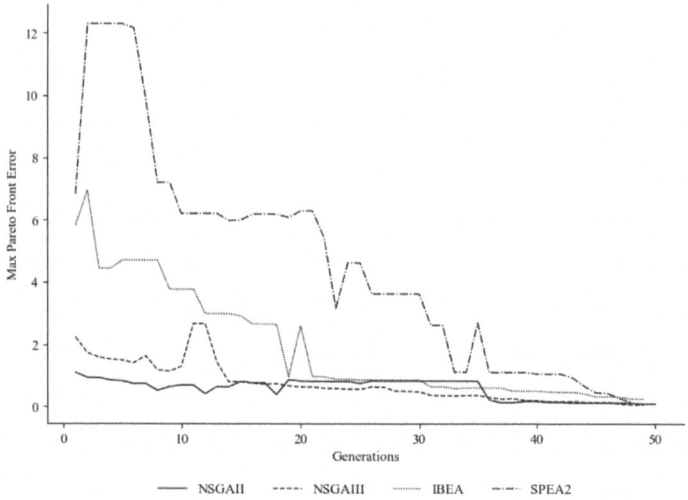

Fig. 13.4 The maximum Pareto error indicator (Nnene 2020)

It's important to note that this indicator is computed without a reference set, which means that an approximation set can exhibit good spacing even if its proximity to the reference set is poor. Therefore, it is often recommended to use the spacing indicator in conjunction with other performance metrics to gain a more comprehensive evaluation. In Fig. 13.4, the plot illustrates the values of the maximum Pareto front error indicator against the generation parameter of the SBTNDM and the four algorithms. The values also converge after 50 generations, indicating that, based on this indicator, the SBTNDM is expected to yield good solutions after approximately 50 generations.

Table 13.1 presents the relative performance of the algorithms using aggregated rank scores, with lower scores indicating better performance. The algorithms are ranked in ascending order of scores, where the best-performing algorithm has a rank of 1, and the worst has a rank of 4. It's important to note that for all indicators except the hypervolume, smaller values are more desirable. From the table, it's evident that NSGA-II consistently performs the best across all indicators, achieving the lowest aggregate score. SPEA2 is the next best performing algorithm, in three of the four indicators: hypervolume, generational distance and epsilon indicator. This suggests that SPEA2 could be a viable alternative for obtaining relatively acceptable solutions to this specific Transit Network Design Problem (TNDP) without relying on NSGA-II. In contrast, NSGA-III and IBEA exhibit the poorest performance among the four algorithms, according to the aggregated rank scores.

Table 13.1 Ranking values for the conducted tests

Algorithm	NSGA-II	NSGA-III	IBEA	SPEA2
Hypervolume	1	4	3	2
Generational distance	1	4	3	2
Epsilon indicator	1	3	4	2
Max Pareto Error	1	2	3	4
Total	4	13	13	10
Rank	1	3	3	2

13.4.5 Statistical Tests

The result of statistical analysis performed on the obtained results is discussed in this section. The utilisation of box plots and Mean Absolute Deviation (MAD) aids in conveying this information. Box plots offer insights into the distribution of algorithm results. As depicted in Fig. 13.5, the box plots illustrate the distributions of all four algorithms. Specifically, for the Epsilon, Generational distance and Max Pareto Error indicators, NSGA-II exhibits the smallest Interquartile Range (IQR), implying that its results have the least variability. The plots of the IBEA and NSGA-III algorithms for the Hypervolume indicator are of interest. The minimum values align with the median and lower quartile in these cases. This alignment indicates potential skewness in the distribution. Mean Absolute Deviation (MAD) is applied to verify these observed trends. MAD sheds light on data consistency and variability, particularly when comparing different datasets or groups within a dataset. Low MAD values suggest closely clustered data points around the mean, signifying reliability, stability or dataset homogeneity. Conversely, higher MAD values indicate greater dispersion of data points from the mean, hinting at possible noise, instability or dataset heterogeneity. Figure 13.6 shows MAD values for all four algorithms. Notably, NSGA-II presents smaller MAD values for all indicators, except for the Hypervolume, where larger values signify superior outcomes. Ultimately, the results of the statistical tests provide validation of the earlier discussed results.

13.4.6 Pareto Front Visualisation

When solving a multi-objective problem (MOP), the aim is to find a Pareto set of solutions when it exists. However, in challenging problems like the Transit Network Design Problem (TNDP), researchers often settle for locally optimal solution sets due to limited time and resources. These locally optimal sets provide the best possible approximation to the global Pareto optimal set, assuming it is known or available. The quality of Multi-Objective Evolutionary Algorithm (MOEA) solutions is typically assessed based on how closely the locally optimal solution set (approximation set)

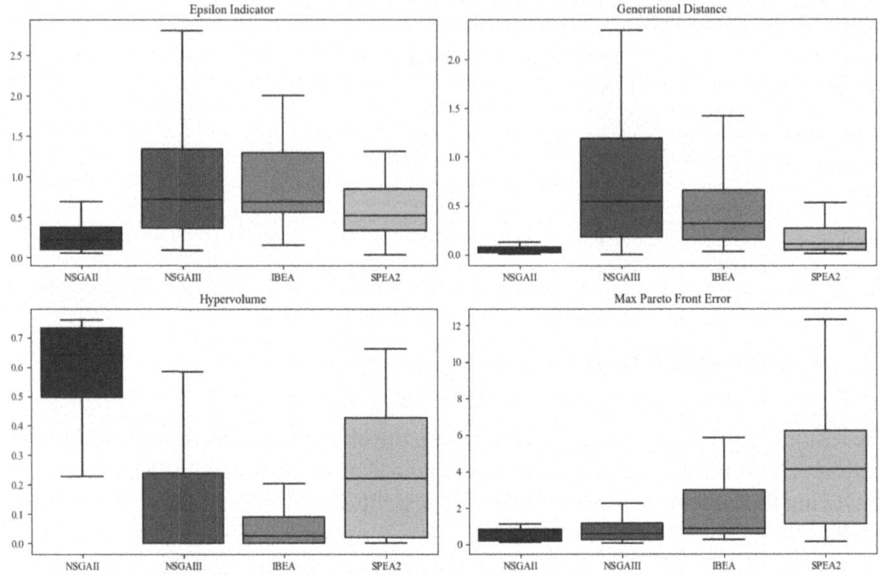

Fig. 13.5 Box plot of the indicators and algorithms

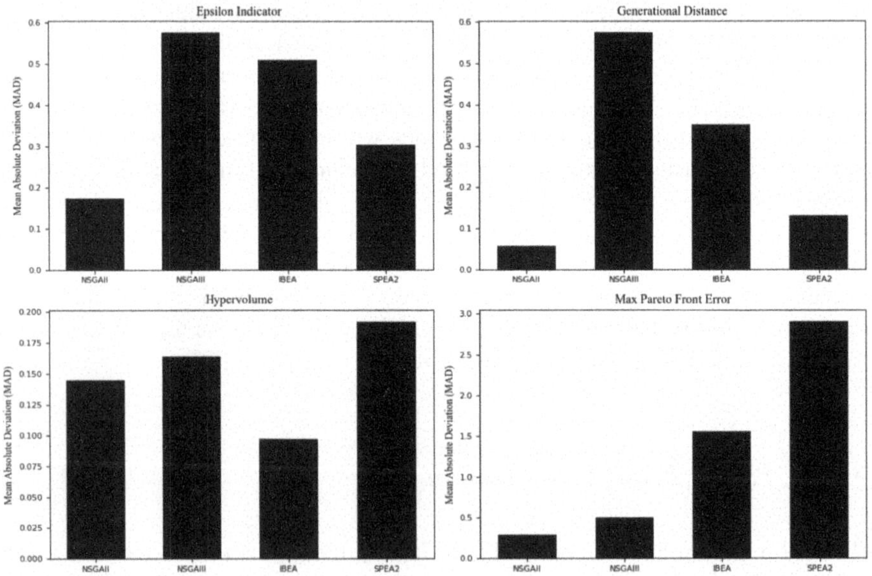

Fig. 13.6 Plot of mean absolute deviation for the indicators and algorithms

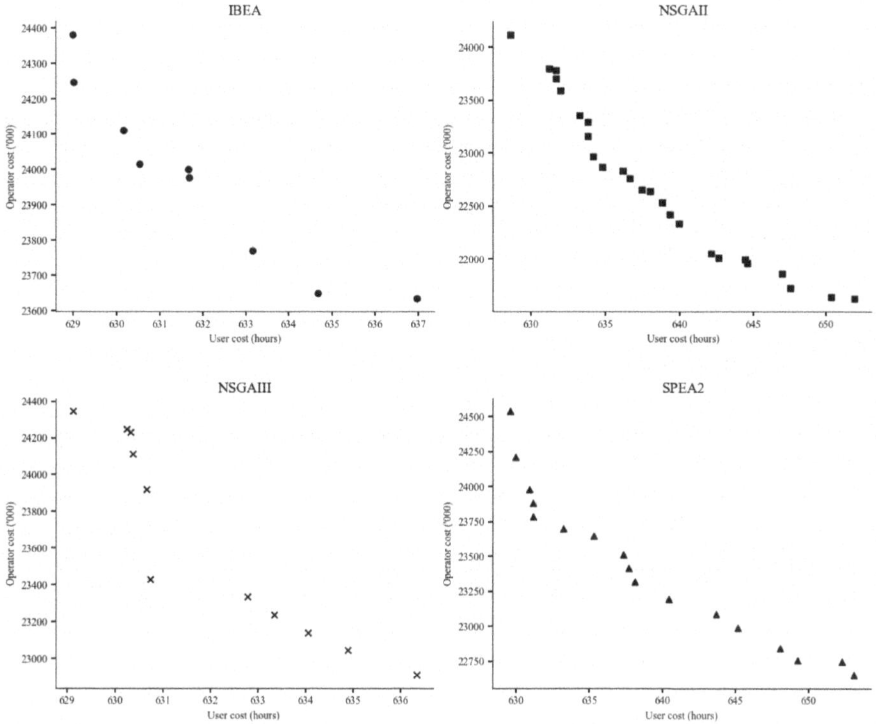

Fig. 13.7 Pareto front for the four algorithms (Nnene 2020)

approximates the global Pareto optimal set (reference set). To measure the quality of an MOEA's results, factors such as convergence and the spread of solutions across the Pareto front are considered. Figure 13.7 provides plots of the Pareto front for all four algorithms, offering valuable insights:

- In terms of the user objective, NSGA-II dominates the other algorithms, except for IBEA. This implies that, regarding the user cost objective, IBEA and NSGA-II are non-dominated relative to each other.
- Despite not dominating IBEA in terms of the user objective, NSGA-II generates solutions with a broader spread in the search space compared to IBEA and SPEA2.
- NSGA-III exhibits limited spread in its solutions when compared to NSGA-II and SPEA2, particularly concerning the operator's perspective, which represents the second objective.
- IBEA and SPEA2 are non-dominated relative to each other, as IBEA dominates SPEA2 in terms of the user objective, while SPEA2 dominates IBEA concerning the operator objective.
- NSGA-II and SPEA2 demonstrate an in-depth exploration of the solution space and generate significantly more solutions compared to NSGA-III and IBEA.

These observations align with the widely accepted definition of non-domination, where a solution is considered non-dominated if it performs better in at least one objective and does not necessarily perform worse in all other objectives. Therefore, it can be inferred that NSGA-II has produced non-dominated solutions compared to the other algorithms. These findings regarding the Pareto frontier of the four algorithms reinforce the results obtained with the quality indicators, indicating that the SBTNDM, when integrated with NSGA-II, outperforms the other algorithms, with SPEA2 being the second-best performer among the four.

13.4.7 Reflections on Practical Applications and Scalability

The improved performance of the NSGA-II, when compared with the other MOEA algorithms presented in this chapter, reflected by its smaller variability and more consistent results across various indicators, suggests that this algorithm has the potential to offer more reliable and stable solutions for transit network design when integrated with the SBTNDM. From a practical standpoint, the enhanced performance of NSGA-II implies that transit authorities and planners can rely on the model's results with a higher degree of confidence. This reliability translates into more informed decision-making and a greater likelihood of achieving the objective of improving the efficiency of the transit network for stakeholders like commuters and operators in terms of cost-effectiveness, service quality and resource utilisation. Furthermore, the findings highlight the potential for reducing the variability and susceptibility of the designed network solutions to extreme or unexpected changes. This characteristic can lead to smoother implementation and operation of transit systems, minimising the need for expensive adjustments or modifications after deployment.

Regarding its scalability, integrating NSGA-II with the SBTNDM shows promising attributes. The NSGA-II's ability to converge to well-distributed Pareto fronts suggests it can effectively handle larger, more diverse solution spaces. Its relatively consistent performance across different indicators also implies it can efficiently explore complex trade-offs in multi-objective transit network design problems. However, it is essential to note that scalability challenges may still arise, mainly when the algorithm is applied to large-scale and highly connected transit networks. This may be due to an increase in the dimensionality of the optimisation problem, potentially leading to longer convergence times and increased computational requirements, which might impact the practical feasibility of implementing the algorithm on large transit networks. An initial attempt to apply the SBTNDM to a real-life transit network may be found in Nnene et al. 2023c. Ultimately, the results underscore the practical significance of the SBTNDM as a decision-support tool for transit network design. The algorithm's ability to provide reliable, stable and well-balanced solutions can empower transit agencies to make more informed choices and achieve enhanced efficiency and effectiveness in their operations.

13.5 Conclusion

In this chapter, the result of a benchmarking exercise to determine the best-performing algorithm among four selected algorithms (NSGA-II, SPEA2, NSGA-III and IBEA) for the SBTNDM is done. The design of public transport networks is a complex endeavour aimed at realising networks that minimise the cost of accessing the network for commuters and of providing quality service by the operators. In the literature, meta-heuristic solutions have enjoyed greater acceptance by researchers and transportation experts searching for models that will give acceptable solutions to an otherwise intractable problem in a limited time. This work discusses a comparative study on the performance of different meta-heuristics used within a simulation-based optimisation solution model developed by the authors for solving the TNDP. The obtained results show that the NSGA-II outperforms the other listed algorithms. This will encourage the design of reliable transit networks that are more efficient and cost-effective. The comparison documented in this chapter has been limited to MOEAs. Future research by the authors will focus on extending the comparison to a broader range of optimisation algorithms, including traditional optimisation methods. Improvements in the computational efficiency of the SBO model will also be studied through a detailed sensitivity analysis of the parameters to obtain optimum test values.

References

Abd G, M. A, M. ES. A (2014) Comparative study of meta-heuristic algorithms for solving quadratic assignment problem. Int J Adv Comput Sci Appl 5(1):1–6

Aktel A, Betul Y, Özcan T, Yenisey M, Engin S (2016) The comparison of the metaheuristic algorithms performances on The comparison of the metaheuristic algorithms performances on airport gate assignment problem airport gate assignment problem. Transp Res Procedia [Internet] 22(2016):469–78. Available from: https://doi.org/10.1016/j.trpro.2017.03.061

Amaran S, Sahinidis NV, Sharda B, Bury SJ (2016) Simulation optimization: a review of algorithms and applications. Ann Oper Res 240(1):351–380

Arıcı F, Kaya E (2019) Comparison of meta-heuristic algorithms on benchmark functions. Acad Perspect Procedia. 2(3):508–517

Arisha A, Abo-Hamad W (2010) Optimisation methods in supply chain applications: a review. Irish J Manag 95–124

Beltran B, Carrese S, Cipriani E, Petrelli M (2009) Transit network design with allocation of green vehicles: a genetic algorithm approach. Transp Res Part C Emerg Technol 17(5):475–483

Biswas A, Pal T (2021) A comparison between metaheuristics for solving a capacitated fixed charge transportation problem with multiple objectives. Expert Syst Appl [Internet] 170(December 2020):114491. Available from: https://doi.org/10.1016/j.eswa.2020.114491

Bringmann K, Friedrich T (2013) Approximation quality of the hypervolume indicator. Artif Intell 195:265–290

Buba AT, Lee LS (2018) A differential evolution for simultaneous transit network design and frequency setting problem. Expert Syst Appl 106:277–289

Cipriani E, Gori S, Petrelli M. Transit network design: A procedure and an application to a large urban area. Transp Res Part C Emerg Technol [Internet]. 2012;20(1):3–14. Available from: https://doi.org/10.1016/j.trc.2010.09.003

Dandl F, Engelhardt R, Hyland M, Tilg G, Bogenberger K, Mahmassani HS (2021) Regulating mobility-on-demand services: Tri-level model and Bayesian optimization solution approach. Transp Res Part C Emerg Technol. 125(December 2020)

Das S, Pandit D (2013) Importance of user perception in evaluating level of service for bus transit for a developing country like India: a review. Transp Rev [Internet] 33(4):402–20. Available from: https://doi.org/10.1080/01441647.2013.789571

Deb K, Agrawal S, Pratap A, Meyarivan T (2000) A fast elitist non-dominated sorting genetic algorithm for multi-objective optimization: NSGA-II. In: Schoenauer M, Deb K, Rudolph G, Yao X, Lutton E, Julian J et al (eds) Parallel problem solving from nature PPSN VI PPSN 2000 lecture notes in computer science, vol 1917. Springer-Verlag, Berlin, pp 849–858

Deb K, Jain H. An Evolutionary many-objective optimization algorithm using reference-point based non-dominated sorting approach, Part I. IeeexploreIeeeOrg. 2013;18(c):1–1

Durán-Micco J, Vansteenwegen P (2022) A survey on the transit network design and frequency setting problem. Public Transp 14:155–190

Ezugwu AE, Adeleke OJ, Akinyelu AA, Viriri S (2020) A conceptual comparison of several metaheuristic algorithms on continuous optimisation problems [Internet]. Vol. 32, pp 6207–6251.Neural Computing and Applications. Springer London. Available from: https://doi.org/10.1007/s00521-019-04132-w

Fan L, Mumford CL (2008) A metaheuristic approach to the urban transit routing problem. J Heuristics [Internet] 16(3):353–72. Available from: https://doi.org/10.1007/s10732-008-9089-8

Fonseca CM, Fleming PJ (1993) Genetic algorithms for multiobjective optimization: formulation discussion and generalization. In: Proceedings of the 5th International Conference on Genetic Algorithms [Internet], pp 416–23. San Francisco: Morgan Kaufmann Publishers. Available from: http://dl.acm.org/citation.cfm?id=657757

Gallo M, D'Acierno L, Montella B (2010) A meta-heuristic approach for solving the Urban Network Design Problem. Eur J Oper Res 201(1):144–157

Gallo M, Montella B, D'Acierno L (2011) The transit network design problem with elastic demand and internalisation of external costs: an application to rail frequency optimisation. Transp Res Part C Emerg Technol. 19(6):1276–1305

Gerami Matin A, Vatani Nezafat R, Golroo A (2017) A comparative study on using meta-heuristic algorithms for road maintenance planning: Insights from field study in a developing country. J Traffic Transp Eng (English Ed [Internet] 4(5):477–86. Available from: https://doi.org/10.1016/j.jtte.2017.06.004

Gosavi A (2015) Simulation-based optimization : parametric optimization techniques and reinforcement learning. Second, p 554. Springer, New York

Hachicha W, Ammeri A, Masmoudi F, Chachoub H (2010) A comprehensive literature classification of simulation optimisation methods. In: international conference on multiple objective programming and goal programming—MOPGP10, p 13

Hadka D (2017) Beginner's Guide to the MOEA Framework. 2.12. Raleigh-Durham: CreateSpace Independent Publishing Platform (1833), p 214

Hadka D, Reed P (2013) Borg: an auto-adaptive many-objective evolutionary computing framework. Evol Comput 21(2):231–259

Huang D, Liu Z, Fu X, Blythe PT (2018) Multimodal transit network design in a hub-and-spoke network framework. Transp A Transp Sci 14(8):706–735

Ibarra-Rojas OJ, Delgado F, Giesen R, Muñoz JC (2015) Planning, operation, and control of bus transport systems: a literature review. Transp Res Part B Methodol 77(3):38–75

Jain H, Deb K (2014) An Evolutionary many-objective optimization algorithm using reference-point based nondominated sorting approach, Part II: handling constraints and extending to an adaptive approach. IEEE Trans Evol Comput 18(4):602–622

Johar A, Jain SS, Garg PK (2016) Transit network design and scheduling using genetic algorithm—a review. An Int J Optim Control Theor Appl 6(1):9–22

Liu Y, Wei J, Li X, Li M (2019) Generational distance indicator-based evolutionary algorithm with an improved niching method for many-objective optimization problems. IEEE Access. 7:63881–63891

Matnei Filho RA, Vergilio SR (2016) A multi-objective test data generation approach for mutation testing of feature models. J Softw Eng Res Dev [Internet] 4(1). Available from: https://doi.org/10.1186/s40411-016-0030-9

Newell GF (1979) Some issues related to the optimal design of bus routes. Transp Sci 13(1):20–35

Nikolić M, Teodorović D (2014) A simultaneous transit network design and frequency setting: Computing with bees. Expert Syst Appl 41(16):7200–7209

Nnene OA, Zuidgeest MHP, Beukes EA (2017) Application of metaheuristic algorithms to the improvement of the MyCiTi BRT network in Cape Town. J South African Inst Civ Eng. 59(4):56–63

Nnene OA, Joubert JW, Zuidgeest MHP (2019) Transit network design with meta-heuristic algorithms and agent based simulation. IFAC-PapersOnLine 52(3):13–18

Nnene OA, Joubert JW, Zuidgeest MHP (2019) An agent-based evaluation of transit network design. Proc Comput Sci 1(151):757–762

Nnene OA (2020) Simulation-based optimisation of public transport networks [Internet]. Faculty Eng Built Environ 2020. Available from: https://open.uct.ac.za/handle/11427/32308

Nnene OA, Joubert JW, Zuidgeest MHP (2023) A simulation-based optimization approach for designing transit networks. Public Transp

Nnene OA, Zuidgeest MHP, Joubert JW (2023) Optimising transit networks using simulation-based techniques. In: Upadhyay RK, Sharma SK, Kumar V, Valera H, editors. Transportation Systems Technology and Integrated Management [Internet], pp 317–45. Singapore: Springer Nature Singapore. Available from: https://doi.org/10.1007/978-981-99-1517-0_15

Nnene OA, Zuidgeest MHP, Joubert JW (2023) BRT network design for transit cost reduction in Cape Town, South Africa. J Public Transp [Internet]. 25(March):100042. Available from: https://doi.org/10.1016/j.jpubtr.2023.100042

Phan DH, Suzuki J (2013) R2-IBEA: R2 indicator based evolutionary algorithm for multiobjective optimization. 2013 IEEE Congr Evol Comput CEC 2013 (June):1836–45

Rodrigue J, Comtois C, Slack B (2013) The geography of transport systems. Third, p 432. United Kindom: Routledge

Silberholz J, Golden B. Comparison of Metaheuristics. In: Gendreau M, Potvin JY, editors (2010) Handbook of metaheuristics [Internet], pp 625–40. Boston, MA: Springer US. Available from: https://doi.org/10.1007/978-1-4419-1665-5_21

Siriwardene NR, Perera BJC (2006) Selection of genetic algorithm operators for urban drainage model parameter optimisation. Math Comput Model 44(5–6):415–429

Song M, Yin M, Chen X (Michael), Zhang L, Li M (2013) A simulation-based approach for sustainable transportation systems evaluation and optimization: theory, systematic framework and applications. Proc Soc Behav Sci 2013;96(Cictp):2274–86

Yen GG, He Z (2014) Performance metric ensemble for multiobjective evolutionary algorithms. IEEE Trans Evol Comput 18(1):131–144

Zitzler E, Künzli S (2004) Indicator-Based Selection in Multiobjective Search. In: International conference on parallel problem solving from nature, pp 832–42. Berlin, Heidelberg: Springer

Zitzler E, Laumanns M, Thiele L (2002) SPEA2: improving the strength pareto evolutionary algorithm for multiobjective optimizationations (Single Publication). In: Evolutionary methods for design, optimization and control with applications to industrial problems proceedings of the EUROGEN'2001, 95–100. Athens: CIMNE, Barcelona, Spain

Part IV
Sustainable Supply Chain and Environmental Impact

Chapter 14
Assessment and Mapping of Traffic Noise Levels in Coal Capital of India, Dhanbad: Implications for Noise Pollution Management

Rohit Patel, Prasoon Kumar Singh, Shivam Saw, Shashi Kant Tiwari, and Rakesh Kumar

14.1 Introduction

Noise pollution has become a rising problem in today's rapid and urbanized world. Noise pollution may originate from many different kinds of sources, including man-made and natural. Sources of noise mainly include transportation, industrial activities, construction and demolition noise, and household and domestic noise. Among all these sources, transportation noise has become a significant issue all around the globe, primarily due to the alarming increase in the number of vehicles and the excessive use of horns (Singh et al. 2021). India's transport sector is expanding rapidly, quickly increasing the number of motor vehicles on Indian roadways. As a result, highways are becoming increasingly congested, and noise pollution is rising. The industrial and transportation sectors combinedly contribute more than 75% to overall noise pollution in the urban environment. In the case of the transportation sector, a significant contribution is of road traffic noise which contributes nearly more than 70% of the total noise pollution (Fig. 14.1). Noise pollution can have a wide range of adverse effects on human life and the environment (Patel and Pandey 2017). A recent

R. Patel (✉) · P. K. Singh · S. Saw
Department of Environmental Science and Engineering, Indian Institute of Technology (Indian School of Mines), Dhanbad 826004, India
e-mail: rohitpatelp7@gmail.com

S. K. Tiwari
Balaji Institute of Modern Management, Sri Balaji University, Pune 411033, India

R. Kumar
School of Engineering, Department of Mechanical Engineering, P P Savani University, Surat 394125, India

S. K. Sharma et al. (eds.), *Dynamics of Transportation Ecosystem, Modeling, and Control*, Energy, Environment, and Sustainability,
https://doi.org/10.1007/978-981-97-0437-8_14

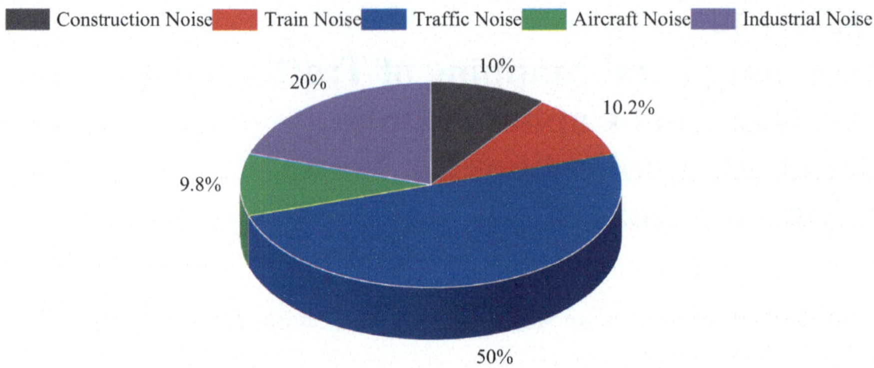

Fig. 14.1 Percentage distribution of sources of noise in an urban environment (Mishra et al. 2021)

study has found that regular noise exposure might cause disrupted sleep patterns, higher blood pressure, increased risk of cardiovascular disease, stroke, diabetes, irritability, and reduced cognitive performance (Okokon et al. 2018; Amoatey et al. 2020; Thacher et al. 2021).

In this regard, noise mapping is a crucial step in the procedure of identifying and visualizing the degrees of noise levels in the field of environmental studies. However, noise mapping represents computed or observed noise exposure over a specific geographical location (Murphy and King 2010). Noise pollution is a spatial feature that varies throughout geographic space based on factors such as the source and receiver's position and surrounding obstructions such as buildings, trees, and barriers (Fang and Ling 2003; Patel et al. 2022). A key component of strategic noise mapping is comprehending how it fluctuates over the study area, how many individuals it impacts, and how it may be controlled. Noise mapping facilitates town planners and regulatory bodies to detect locations that experience more significant noise pollution and determine if people who reside in those areas are negatively impacted by uncontrolled environmental noise exposure. Subsequently, suitable control measures and regulatory guidelines can be implemented to minimize noise levels to safeguard the community's health. Therefore, noise mapping is a vital tool to assess and find out the noise hotspot of any study location.

In the Indian context, many previous studies have been carried out to monitor noise pollution, identify the noisy environment, and prepare noise maps of various study areas. A detailed investigation in Asansol, West Bengal, was performed by noise level assessment and preparation of noise maps for the city (Banerjee et al. 2009). A noise pollution study with the help of noise maps was carried out in Mumbai City (Sharma and Joshi 2010). In another study, noise level assessment and GIS-based noise maps were prepared across Guwahati City in several areas (Alam 2011). Traffic noise mapping involving smartphone users and community-based assessment was conducted in Mandi, Punjab (Kumar et al. 2017). Noise monitoring and noise mapping employing the GIS technique were undertaken to find Surat City's noise hotspots (Tiwari et al. 2017). An identical study was also reported in Vellore City, where noise map preparation was done using SoundPLAN software (Manojkumar et al. 2019). Similarly, a study identified various noise sources and spatial distribution maps of traffic noise levels were prepared in Malda town (Das et al. 2019).

The Central Pollution Control Board (CPCB) of India has established standards for noise pollution levels throughout several zones. These restrictions are measured in decibels (dB) and vary according to the type of region or zone. The CPCB's noise pollution standards for different zones in India are presented in Table 14.1. According to CPCB, from 6 AM to 10 PM is considered day time while from 10 PM to 6 AM is night time.

This study's primary objective was to monitor and evaluate traffic noise levels at different sampling sites encompassing diverse land uses. The collected data was utilized to generate noise contour maps, enabling the identification of areas with high noise levels. These maps will provide valuable insights for town planners and highway engineers, allowing them to propose effective measures to alleviate noise pollution.

Noise pollution can have significant and far-reaching impacts on resident's daily lives, overall quality of life, and the environment. These impacts can be physical, psychological, and social in nature. The following are the potential adverse effects of noise pollution:

(i). Health Effects:

Physical Health: Prolonged exposure to high levels of noise can lead to various physical health issues such as hearing loss, elevated blood pressure, cardiovascular problems, and sleep disturbances. Chronic noise exposure is associated with an increased risk of heart disease and stroke.

Sleep Disruption: Noise pollution can disrupt sleep patterns, leading to reduced sleep quality, difficulty falling asleep, and frequent awakenings. Sleep deprivation can result in fatigue, reduced concentration, and decreased cognitive performance.

(ii). Psychological Effects:

Stress and Anxiety: Continuous exposure to loud and disruptive noise can cause stress and anxiety. It can trigger the body's "fight or flight" response, leading to elevated stress hormone levels, which may have long-term negative effects on mental health.

Table 14.1 Noise level standards prescribed by CPCB, India (Ministry of Environment and forest: The Noise Pollution (Regulation and Control) 2000)

Code of area	Category	Noise level as L_{eq} in dBA	
		Day time (6 AM-10 PM)	Night time (10 PM-6 AM)
A	Industrial area	75	70
B	Commercial area	65	55
C	Residential area	55	45
D	Silence zone	50	40

Irritability and Mood Changes: Noise pollution can contribute to irritability, mood swings, and decreased overall emotional well-being.

(iii). Quality of Life:

Annoyance: Persistent noise can be highly annoying and impact people's ability to enjoy their homes and outdoor spaces. Annoyance from noise pollution can reduce overall quality of life and lead to conflicts among residents.

Social Isolation: Noise pollution can limit outdoor activities and social interactions, leading to reduced community cohesion and a decreased sense of belonging.

(iv). Cognitive and Developmental Effects:

Cognitive Impairments: Children exposed to chronic noise may experience developmental delays in language acquisition and cognitive abilities. Noise can also interfere with concentration and learning in both children and adults.

Academic Performance: Noise pollution near schools can negatively impact students' ability to concentrate and learn, potentially leading to lower academic performance.

(v). Environmental Impact:

Wildlife Disturbance: Noise pollution can disrupt natural habitats and negatively affect wildlife behavior and communication. It can lead to altered migration patterns, reproductive disruptions, and habitat avoidance.

Ecosystem Disruption: Noise pollution can disrupt ecological balances and affect the health of ecosystems, potentially leading to shifts in species composition and ecosystem functions.

(vi). Occupational Impacts:

Workplace Productivity: Noise pollution in workplaces can reduce productivity, increase stress among employees, and contribute to communication difficulties.

14.2 Study Area

The objective of this study was to create noise contour maps for Dhanbad City, located in the Jharkhand state of India (Fig. 14.2). Dhanbad is situated between 85°5'E–86°30'E longitude and 23°32'N–24°5'N latitude in the eastern part of the country. Known as India's coal capital, Dhanbad has a population of 2,682,662 and a population density of 1,300 people per square kilometer, covering a total geographical area of 2,074 square kilometers (Report 2018). A total of 22 sampling locations were selected homogeneously to cover the entire city area. These locations were chosen from diverse land uses, including industrial, commercial, residential, and silent zones. The monitoring locations were classified based on significant intersection points, areas with high traffic volume, and marketplaces, primarily along the routes of three national highways (NH-19, NH-18, and NH-419) that traverse the city.

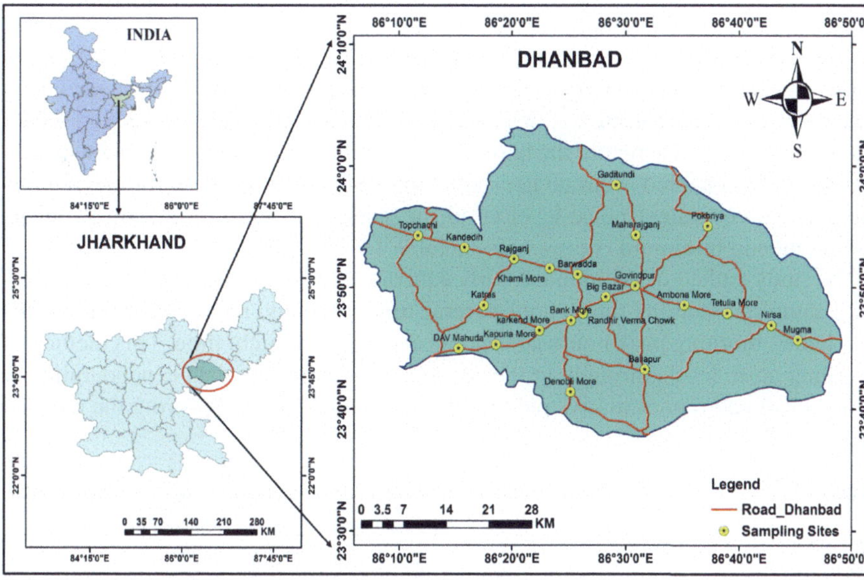

Fig. 14.2 Study area map showing all the sampling sites

14.3 Materials and Methodology

14.3.1 Data Collection

For data collection, various equipment, such as the Bruel & Kjaer 2238 sound level meter, were employed. The sound level meter was used to measure the continuous hourly A-weighted equivalent noise level (L_{eq}). It represents the average noise level for a given period of time and is generally used for quantifying noise values. Noise pollution monitoring was conducted for five-time slots, which include morning off-peak hours (6 AM to 9 AM), morning peak hours (9 AM to 12 PM), afternoon (2 PM to 5 PM), evening (5 PM to 8 PM), and night time (10 PM to 1 AM). The study examined noise pollution during different times of day and night. These time periods were chosen to observe the diurnal variations in noise levels. The study's temporal scope intended to investigate how noise pollution fluctuates during the day and night, taking into account potential effects on inhabitants' everyday life and well-being. This temporal study aids in the identification of peak noise hours, fluctuations in noise levels throughout the day, and potential areas of concern for noise abatement techniques. Moreover, to determine the periodic noise level in this study, the average equivalent noise level for a duration of 3 h is calculated using Eq. 14.1:

$$\text{Average } L_{eq} = 10xlog_{10}\left(10^{\left(\frac{Leq_1}{10}\right)} + 10^{\left(\frac{Leq_2}{10}\right)} + \ldots 10^{\left(\frac{Leq_n}{10}\right)}\right) \quad (14.1)$$

Furthermore, in this study, careful attention was given to correctly positioning the noise level meter on the tripod stand. It was placed at a height of 1.2 ± 0.1 m from the ground surface and maintained at a distance of 7.5 ± 0.2 m from the center of each lane to ensure accurate measurements (Patel et al. 2024). Moreover, a GPS meter was employed to record the latitude and longitude coordinates to obtain precise locations. This enabled accurate geospatial information for each sampling location. The average equivalent noise level at all the sampling sites in the form of mean, maximum, and minimum is presented in Table 14.2.

The study gathered continuous hourly equivalent noise level (Leq) measurements at 22 sampling stations throughout several time intervals, including morning off-peak hours, morning peak hours, afternoon, evening, and night time. This method of recording noise levels throughout the day and night improves data dependability by giving a full and representative picture of noise pollution trends during the day and

Table 14.2 Coordinates of selected sampling sites with their corresponding noise levels (in dBA)

S. No	Location	Lat	Long	Land Use	Average equivalent noise level in dBA		
					Mean	Maximum	Minimum
1	Topchachi	23.8096	86.5851	Commercial	73.4	75.9	67.9
2	Kandedih	23.7986	86.6491	Residential	73.2	75.7	67.8
3	Kharni More	23.7821	86.7132	Residential	73.4	76.0	68.0
4	Ambona More	23.7621	86.7524	Residential	73.7	75.7	67.9
5	Rajganj	23.8516	86.4293	Commercial	75.5	78.0	69.1
6	Barwadda	23.8603	86.3886	Commercial	75.0	77.1	68.8
7	Govindpur	23.8730	86.3362	Commercial	75.2	77.9	68.7
8	Mugma	23.8888	86.2632	Industrial	75.2	78.6	67.7
9	Nirsa	23.9051	86.1942	Commercial	75.1	78.0	67.8
10	Tetulia More	23.8361	86.5135	Residential	74.0	76.1	67.5
11	Big Bazar	23.8208	86.4704	Commercial	72.8	75.8	65.5
12	Randhir Verma Chowk	23.7983	86.4373	Silence	73.7	75.9	67.0
13	Bank More	23.7886	86.4196	Commercial	74.6	77.0	68.3
14	Kapuria More	23.7751	86.3734	Commercial	73.2	75.3	67.9
15	Karkend More	23.7504	86.2542	Commercial	73.6	75.8	67.8
16	DAV Mahuda	23.7558	86.3097	Silence	75.2	77.9	69.2
17	Denobli More	23.6909	86.4189	Silence	72.7	74.9	66.5
18	Katras	23.8095	86.2915	Industrial	74.9	77.5	69.5
19	Baliapur	23.7215	86.5274	Residential	73.8	75.9	68.7
20	Gaditundi	23.9745	86.4858	Residential	72.4	74.1	67.0
21	Maharajganj	23.9060	86.5144	Commercial	71.7	73.5	66.1
22	Pokhriya	23.9153	86.6271	Residential	72.7	74.6	67.9

night. Furthermore, the selection of sampling points scattered throughout varied land uses, including industrial, commercial, residential, and silent zones, improves the spatial representativeness of the data. The study includes a variety of noise pollution factors and situations inside the city by carefully selecting sampling locations from high-traffic volume zones, important junctions, and marketplaces. The sampling locations were spread across the city in order to cover the entire study region uniformly. This distribution ensures that noise data from heavily and sparsely inhabited areas, as well as other land use classifications, are included in the analysis. The study reduces the potential bias by focusing primarily on particular areas or types of sites by considering a diverse range of locations.

14.3.2 Noise Mapping Using GIS

Spatial distribution maps containing noise contours were generated for different time periods during workdays using ArcGIS version 10.5. The process involved in noise mapping in the present study is illustrated in Fig. 14.3. These maps were created utilizing the Inverse Distance Weighting (IDW) interpolation technique. IDW is a spatial interpolation method that estimates values at non-monitored locations based on measurements from known points. In ArcGIS, the IDW tool can be employed to create noise maps by interpolating measurements obtained from various sites. This tool allows users to specify the number of nearest neighbours to consider for interpolation and adjust the power of distance weighting. The resulting noise map is presented as a raster surface, where each cell represents the noise level at a specific location. In the context of noise mapping, IDW estimates noise levels at specified places within a study area employing noise measurements collected at known monitoring sites. The steps involved in IDW techniques are outlined below:

Noise Data Collection: Noise levels are measured at numerous monitoring stations around the study area using sound level meters. These measurements yield a dataset of noise values associated with certain geographic coordinates.

Neighborhood Selection: IDW finds a group of nearby monitoring points for each area where researcher want to estimate noise levels (interpolation point). These adjacent points are typically selected based on a defined radius or a predetermined number of the nearest spots.

Weighting: IDW weights neighbouring monitoring points based on their proximity to the interpolation point. The closer points are weighted more heavily, indicating that they have a greater influence on the predicted value at the interpolation point. Each point's weight is inversely proportional to its distance from the interpolation point.

Interpolation: The interpolation point's estimated noise level is calculated as a weighted average of the noise measurements at neighboring monitoring stations. The inverse of the distances between the interpolation point and the monitoring sites is used to calculate the weights.

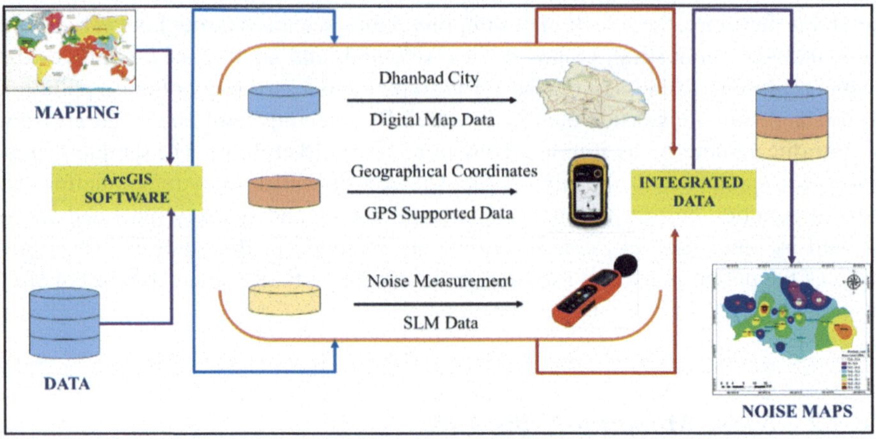

Fig. 14.3 Process involved in noise mapping (Modified from Esmeray and Eren (2021))

Contour Generation: After performing IDW interpolation for all needed interpolation locations, the estimated noise levels can generate a continuous noise contour map. Contours depict how noise distributes across the research region by representing bands of equal noise levels.

The mathematical expression used in the IDW technique is illustrated in Eq. 14.2:

$$Z(x_0) = \frac{\sum_{i=1}^{n} w_i(x_0).Z(x_i)}{\sum_{i=1}^{n} w_i(x_0)} \tag{14.2}$$

where

$Z\ (x_0)$ indicates the estimated value at the target location x_0.

$Z\ (x_i)$ represents the observed value at the known location x_i.

$w_i\ (x_0)$ is the weight assigned for each observed value depending on its closeness to the target location. It is determined employing the following formula (Eq. 14.3):

$$w_i(x_0) = \frac{1}{d(x_0, x_i)^p} \tag{14.3}$$

where

$d\ (x_0, x_i)$ signifies the Euclidean distance between the target location x_0 and the already-known location x_i .

p is the power parameter that is responsible for determining how the distance affects the weighting. Commonly, values of p fall between 1 and 3, with larger values assigning greater importance to observations that are closer in proximity.

In IDW, the estimated value at the target location is calculated as a weighted average of the observed values, with weights determined by the inverse of the distance raised to the power p. On the basis of this technique, 22 diversified land use sites were selected, and the noise contour maps for Dhanbad City, was prepared.

14.3.3 Significance of Noise Maps

The purpose of noise contour maps is to graphically display and aid in the comprehension of noise distribution patterns throughout a specified geographic area. These maps illustrate how noise levels fluctuate spatially, assisting researchers, policymakers, urban planners, and the general public in understanding the prevalence and intensity of noise pollution in various regions of the study area. Here's how noise contour maps help with this:

Visual Representation: Noise contour maps show noise levels at various places using graphical components such as contour lines and color gradients. This visual depiction enables viewers to identify locations with high and low noise levels, allowing for quick and easy interpretation of the spatial distribution of noise.

Patterns and Trends: Noise contour maps illustrate patterns and trends in noise distribution that raw data may not disclose. Viewers can identify noise hotspots, or regions where noise gradually increases or decreases.

Temporal Analysis: When noise contour maps for multiple time intervals are generated, as described in the study, viewers can examine noise distribution patterns throughout different periods. This temporal analysis aids in the identification of temporal fluctuations and leads to a thorough understanding of noise exposure throughout the day.

Decision-Making Support: Decision-makers and urban planners may gain information from noise contour maps. They can help prioritize sites for noise reduction measures, guide land use planning, recommend the design of noise barriers, and promote noise pollution policy creation.

14.4 Results and Discussions

14.4.1 Noise Contour Map for Morning Hours (6 AM to 9 AM, Off-Peak Hours).

In the time period 6 AM to 9 AM, the maximum noise intensity was reported near Nirsa and found to be 76.6 dBA. The region for these noise levels has been represented with the dark red color in the noise map (Fig. 14.4). Maharajganj recorded the lowest noise levels, and noise values obtained in this location was 72.4 dBA and indicated with white colors in the noise map. Additionally, the noise levels at Gaditundi, Pokhriya, and Denobli More exhibit a similar range of fluctuation, with purple color on the noise map designating noise levels between 73 and 73.4 dBA. A few areas, including Govindpur, Barwadda, and Rajganj, where the noise level was recorded to vary between 75.6 and 76.2 dBA, are highlighted with orange color in the noise map.

Similarly, the noise level in Kandedih, Kharni More, Karkend More, and Kapuria More was found to vary between 73.4 and 73.9 dBA and was shown with a dark

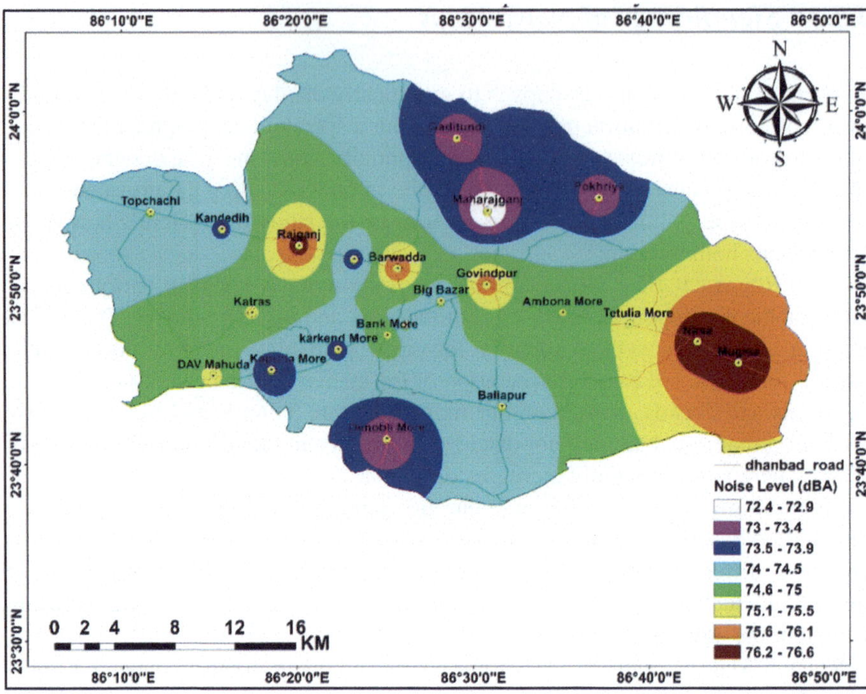

Fig. 14.4 Noise map for Dhanbad City during morning off-peak hours (6 to 9 AM)

blue color. Besides this, light blue, green, and yellow colors indicate all other monitored locations, including Baliapur, Topchachi, Big Bazaar, Bank More, Ambona More, Katras, and DAV Mahuda, where the noise level was recorded to be within 74 and 75.5 dBA. Concluding the discussion, we can say that all the locations in the present time duration violated the noise standards prescribed by CPCB.

Noise Contour Map for Morning Hours (9 AM to 12 PM, Peak Hours)

For the time period 9 AM to 12 PM, the noise level in the current research ranges from 73.2 to 78 dBA during peak morning hours. The highest noise level (78 dBA) was recorded at Nirsa, and the area surrounding this location is highlighted in brown color in the noise contour map (Fig. 14.5). Furthermore, the minimum noise level (73.2–73.8 dBA) displayed in white color was reported at Maharajganj and Gaditundi. Locations such as Rajganj, Barwadda, Govindpur, Bank More, Katras, and DAV Mahuda shows similar kind of noise variation, and noise level in these locations ranges from 76.3–77.4 dBA and are shown with yellow and orange color in the spatial distribution maps. However, the noise level at Topchachi, Kandedih, Big Bazar, Denobli More, Kapuria More, Baliapur, Ambona More, and Kharni More represents similar noise level which varies from 74.5 to 75.6 dBA and is shown with dark blue and light blue colors. Most of the area poses higher noise levels, ranging from 75 to 76.2 dBA. As a result, the noise level during the peak morning hours surpassed the CPCB's noise guidelines.

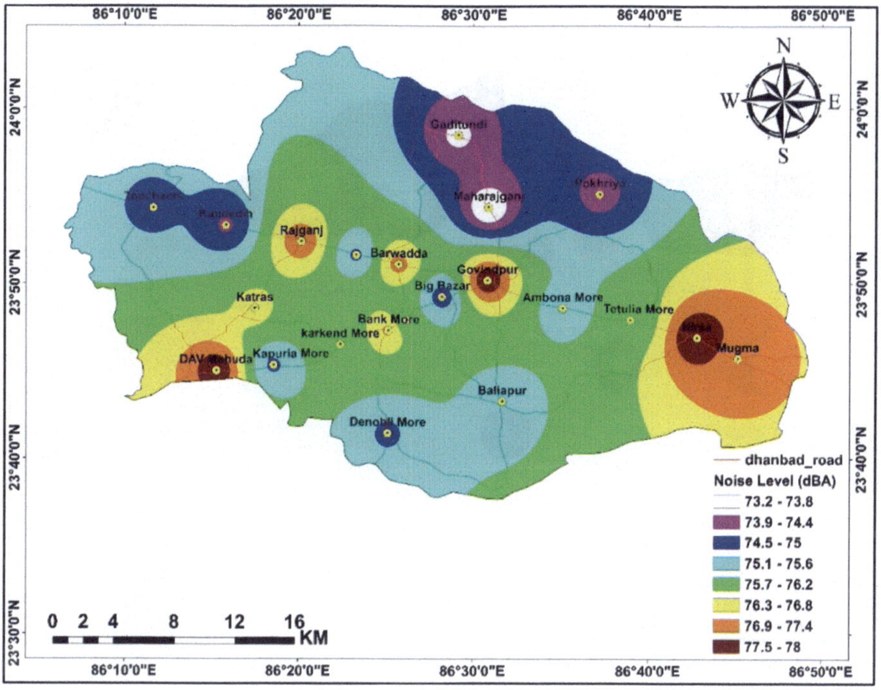

Fig. 14.5 Noise map for Dhanbad City during peak morning hours (9 AM to 12 PM)

14.4.2 Noise Contour Map for Afternoon Hours (2 PM to 5 PM, Off-Peak Hours)

The maximum (76.9 dBA) and minimum (73.5 dBA) noise level for this time period (2 PM to 5 PM) is observed at Rajganj and Maharajganj, respectively. The area with maximum noise is shown with brown color, whereas the area showing the minimum noise level is represented with white color in the noise map (Fig. 14.6). The spatial distribution map revealed that noise level at all locations for this time period varies from 73.5 to 76.9 dBA. Noise level at Barwadda, Govindpur, Katras, DAV Mahuda, Bank More and Tetulia More, Nirsa, and Mugma varies from 75.8 to 76.5 dBA, indicated with orange and yellow color. However, noise level near Topchachu, Kandedih, Kharni More, Big Bazar, Kapuria More, Denobli More, Gaditudi, and Pokhriya ranges from 74 to 74.8 dBA and are shown with purple and dark blue color. It can be depicted that noise exposure at all the sampling sites violated the noise limit set by CPCB, and it is comparatively lower than in the peak morning hours.

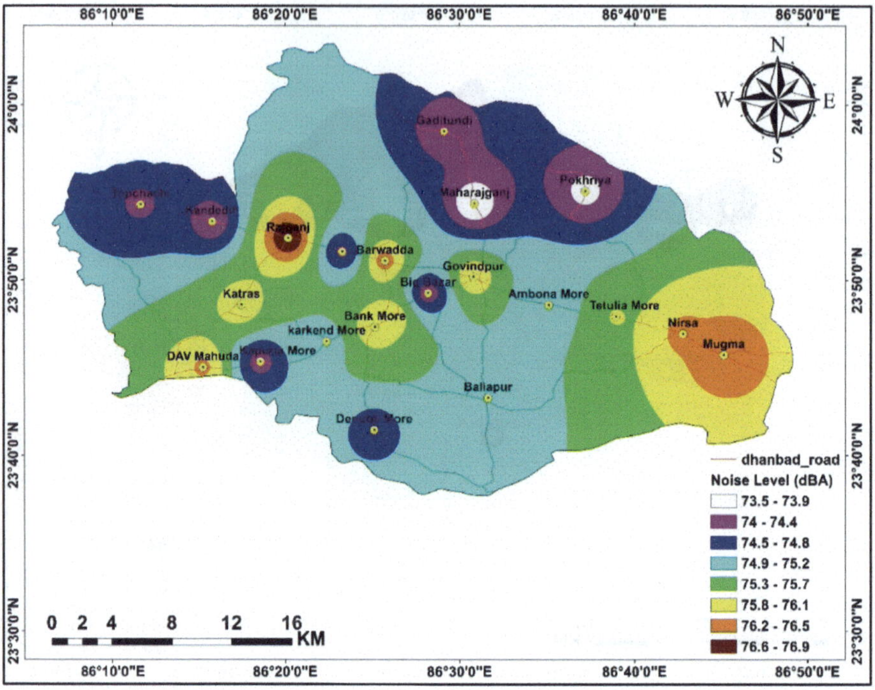

Fig. 14.6 Noise map for Dhanbad City during afternoon off-peak hours (2 PM to 5 PM)

14.4.3 Noise Contour Map for Evening Hours (5 PM to 8 PM, Peak Hours)

During the period of 5 PM to 8 PM, the highest recorded noise level (78.6 dBA) in the study area was observed at Mugma. Similarly, the Maharajganj area reports the minimum noise level (73.4 dBA) for this time period. The noise contour map (Fig. 14.7) for the present study area during 5 to 8 PM for a workday condition illustrated that noise levels range from 73.4 to 78.6 dBA. As a result, it can be concluded that the noise level exceeded the noise standard given by CPCB and was higher than in the morning hours.

14.4.4 Noise Map for Night Hours (10 PM to 1 AM, Off-Peak Hours)

For the time period 10 PM to 1 AM, the maximum noise level observed in the present research was 69.5 dBA at Katras because of the movement of more heavy vehicles in this region for coal transportation (see Fig. 14.8). The minimum noise level observed

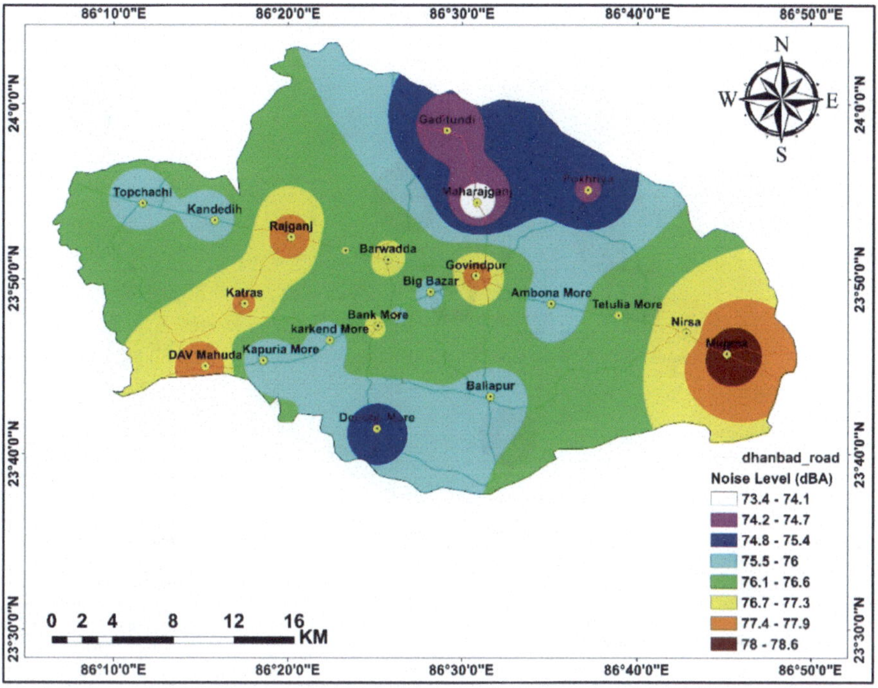

Fig. 14.7 Noise map for Dhanbad City during peak evening hours (5 PM to 8 PM)

in the current investigation was 65.5 dBA at Big Bazar. It may be quantified that all the sampling sites surpassed the suggested noise standards prescribed by CPCB in the night time.

The spatial analysis in Dhanbad City focused on measuring and visualizing the spatial distribution of noise pollution levels over various locations and time intervals. The investigation included a wide range of areas throughout the city to provide insights into noise exposure patterns and variations throughout the monitoring period. Therefore, according to the spatial distribution maps of L_{eq} for all time periods, the regions around Nirsa and Mugma in the eastern part of Dhanbad are noise hotspots during the day. However, during the night, the areas of Katras, DAV Mahuda, and Rajganj in the southwestern part of Dhanbad are identified to be the noisiest location. Furthermore, every sampling location exceeded the CPCB noise standards, with nearly 36% of the selected areas showing an average L_{eq} above 75 dBA and 100% of the chosen areas showing an average L_{eq} greater than 70 dBA.

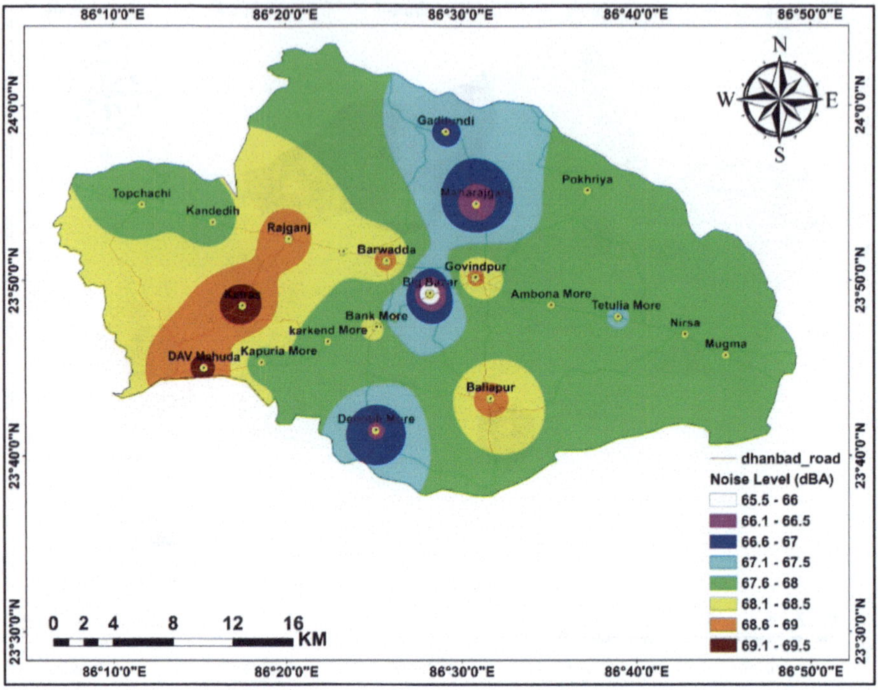

Fig. 14.8 Noise map for Dhanbad City during night off-peak hours (10 PM to 1 AM)

14.5 Recommendations and Strategies for Noise Pollution Management and Mitigation

The chapters' findings provide important implications for noise pollution control and urban planning by giving vital insights into the spatial distribution of noise pollution in Dhanbad City, India. Based on the study's findings, multiple recommendations and procedures for reducing noise pollution and improving urban design can be developed:

Zoning and Land Use Planning: Enforce strict zoning laws that distinguish noise-sensitive land uses (educational and healthcare) from noise-emitting factors (industrial, commercial, and transportation). Moreover, establish and enforce quiet zones, particularly in noise hotspots, to ensure that noise levels are managed and kept within acceptable ranges.

Transportation Planning and Traffic Management: Develop and execute effective traffic management approaches to alleviate congestion and encourage smoother traffic flow, which can assist in reducing vehicular traffic noise levels. In addition, encourage the use of public transit, cycling, and pedestrian-friendly infrastructure in order to minimize the overall number of motor vehicles on the road and consequently noise emissions.

Noise Barriers and Green Infrastructure: Construct noise barriers and sound-absorbing materials alongside highways and major roadways to block and absorb noise while safeguarding surrounding residential areas. Increase the amount of green space, parks, and vegetation along roadways and in noise-affected areas because the vegetation can act as a natural buffer to minimize noise transmission.

Building Design and Orientation: Encourage urban planners and architects to design buildings with suitable orientation and design to reduce resident noise exposure. Likewise, enforce building codes that require proper sound insulation in construction to promote a more peaceful indoor atmosphere.

Public Awareness and Education: Launch public awareness initiatives to educate communities about the health effects of noise pollution and how to decrease their exposure. Additionally, also involve schools, community organizations, and local media in raising awareness and promoting appropriate noise creation behavior.

Noise Mitigation Measures: Develop and implement noise-reduction laws for industry, construction sites, and commercial establishments during sensitive hours. Promote the adoption of electric or hybrid vehicles and impose noise restrictions on vehicle exhaust systems.

City Design and Urban Infrastructure: Incorporate noise pollution issues into urban planning by developing streetscapes, layouts, and infrastructures that prioritize noise reduction and a peaceful urban environment.

Continuous Monitoring and Evaluation: Implement a regular monitoring and assessment programme to evaluate the efficacy of noise pollution management techniques and make adjustments as needed over time.

14.6 Conclusions

The present work highlights the importance of noise pollution studies and methods to identify the noise-affected zones for further investigations. The key points concluded from the current research work are mentioned below.

An analysis of the noise maps indicated that noise pollution in Dhanbad City surpassed the noise standards prescribed by CPCB for both day and night.

During the day time, the noise hotspots were found in the eastern portion of the city, particularly around Nirsa and Mugma area. While at night time, the southwestern area of Dhanbad City, comprising Katras, DAV Mahuda, and Rajganj, recorded the highest noise levels.

The findings of this study add to an improved understanding of the acoustic environment of Dhanbad City and provide helpful information for decision-makers and town planners in developing strategies to manage and control noise pollution in the city.

For future research, there is an opportunity to increase the number of sampling sites and incorporate meteorological factors. Noise maps can be prepared with the help of dedicated software for noise mapping, such as CadnaA or SoundPLAN. This would enhance the precision and accuracy of noise map visualization.

References

Alam W (2011) GIS based assessment of noise pollution in Guwahati City of Assam, India. Int J Environ Sci

Amoatey P, Omidvarbona H, Baawain MS, Al-Mayahi A, Al-Mamun A, Al-Harthy I (2020) Exposure assessment to road traffic noise levels and health effects in an arid urban area. Environ Sci Pollut Res. https://doi.org/10.1007/s11356-020-09785-y

Banerjee D, Chakraborty SK, Bhattacharyya S, Gangopadhyay A (2009). Appraisal and Mapping the Spatial-Temporal Distribution of Urban Road Traffic Noise. https://doi.org/10.1007/BF03327636

Das P, Talukdar S, Ziaul S, Das S, Pal S (2019) Noise mapping and assessing vulnerability in meso level urban environment of Eastern India. Sustain Cities Soc. https://doi.org/10.1016/j.scs.2019.01.001

Esmeray E, Eren S (2021) GIS-based mapping and assessment of noise pollution in Safranbolu, Karabuk, Turkey. Environ Dev Sustain. https://doi.org/10.1007/s10668-021-01303-5

Fang CF, Ling DL (2003) Investigation of the noise reduction provided by tree belts. Landsc Urban Plan. https://doi.org/10.1016/S0169-2046(02)00190-1

Kumar R, Mukherjee A, Singh VP (2017) Traffic noise mapping of Indian roads through smartphone user community participation. Environ Monit Assess. https://doi.org/10.1007/s10661-016-5741-1

Manojkumar N, Basha K, Srimuruganandam B (2019) Assessment, prediction and mapping of noise levels in Vellore City, India. Noise Mapp. https://doi.org/10.1515/noise-2019-0004

Ministry of Environment and forest: The Noise Pollution (Regulation and Control) Rules (2000). 12311, 1088–1569

Mishra RK, Nair K, Kumar K, Shukla A (2021) Dynamic noise mapping of road traffic in an urban city. Arab J Geosci 14. https://doi.org/10.1007/s12517-020-06373-9

Murphy E, King EA (2010). Strategic environmental noise mapping: methodological issues concerning the implementation of the EU environmental noise directive and their policy implications. https://doi.org/10.1016/j.envint.2009.11.006

Okokon EO, Yli-Tuomi T, Turunen AW, Tiittanen P, Juutilainen J, Lanki T (2018) Traffic noise, noise annoyance and psychotropic medication use. Environ Int. https://doi.org/10.1016/j.envint.2018.06.034

Patel R, Pandey G (2017) A study of noise pollution at the campus of. Int Res J Eng Technol 1258–1263

Patel R, Kumar Singh P, Saw S (2022) Recent advancements in the challenges and strategies of globally used traffic noise prediction models. Environ Sci Pollut Res. https://doi.org/10.1007/s11356-022-20693-1

Patel R, Singh PK, Saw S (2024) Traffic noise modeling under mixed traffic condition in mid-sized Indian City: a linear regression and neural network-Based approach. Int J Math Eng Manag Sci. https://doi.org/10.33889/IJMEMS.2024.9.3.022

Report DS (2018) District Survey Report

Sharma V, Joshi BD (2010) Assessment of noise pollution during Deepawali festival in a small township of Haridwar City of Uttarakhand, India. Environmentalist. https://doi.org/10.1007/s10669-010-9265-x

Singh D, Francavilla AB, Mancini S, Guarnaccia C (2021) Application of machine learning to include honking effect in vehicular traffic noise prediction. Appl Sci 11. https://doi.org/10.3390/app11136030

Thacher JD, Poulsen AH, Hvidtfeldt UA, Raaschou-Nielsen O, Brandt J, Geels C, Khan J, Münzel T, Sørensen M (2021) Long-term exposure to transportation noise and risk for type 2 diabetes in a nationwide cohort study from denmark. Environ Health Perspect 129. https://doi.org/10.1289/EHP9146

Tiwari K, Tandel B, Vallabhbhai S, Jha SK, Tandel BN (2017) GIS based approach for noise mapping of urban road traffic. Nap-2017 S.V.N.I.T.

Chapter 15
Assessing Environmental Impacts and Mitigation Strategies for the DFCCIL Rail Projects

Mohd Avesh, Naziru Imam, and Abiodun Oluwatosin Adeoye

15.1 Introduction

The introduction to the dedicated freight corridors in India highlights a pivotal development in the country's transportation infrastructure similar to a landmark achievement of 6.57 million passengers transported in Nigeria railway in 7 years between Abuja-Kaduna due to the newly constructed rail corridor by Chinese conglomerate, CCECC. Specifically, the Eastern and Western dedicated freight corridors stand out as vital solutions for addressing both environmental and economic challenges. The significance of these corridors is underscored by their potential to save a substantial 457 million tons of CO_2 emissions over a 30-year period, as projected by DFCCIL (2020) (Dedicated Freight Corridor Projects 2020). Greenhouse gases have CO_2 as one of its most important components which are known to be responsible for global warming. Several technical approaches to include: pyrolysis of biomass waste, torrefaction of biomass, biodiesel production are new approaches to curb the emission of CO_2 (Adeoye et al. 2023b). Central to their environmental impact is the complete electrification of these corridors, a strategic move aimed at curbing harmful gas emissions within the regions they serve. This reduction in harmful gases is expected to materialize as freight transportation transitions from roadways and conventional diesel-powered rail locomotives to these advanced corridors. The scope of this report is focused on the projects within the Western Dedicated Freight Corridor (WDFC) and the APL 1 section of the Eastern Dedicated Freight Corridor (EDFC). Specifically, the APL 1 sections encompass Khurja–Dadri, Bhaupur–Khurja, and Kaurara–Chamraula. Given the ongoing impact of the COVID-19 pandemic, the study relies

M. Avesh
Mazedan International Research Academy, Okhla, New Delhi, Delhi 110025, India

N. Imam · A. O. Adeoye (✉)
Department of Chemistry, Federal University, Oye-Ekiti, Ekiti State, Nigeria
e-mail: bioken2017@gmail.com

S. K. Sharma et al. (eds.), *Dynamics of Transportation Ecosystem, Modeling, and Control*, Energy, Environment, and Sustainability,
https://doi.org/10.1007/978-981-97-0437-8_15

on the analysis of secondary data to comprehensively evaluate the environmental implications of these projects and estimate their cost–benefit analyses (Adeoye et al. 2023a).

The primary aim of this study is to conduct an extensive assessment of the environmental impacts of the projects. Additionally, the study aims to estimate the cost–benefit analysis associated with these projects, providing valuable insights into their economic and environmental implications. The objectives include a detailed examination of the WDFC and APL 1 projects within the EDFC, including an analysis of the utilized Environmental Management Frameworks, identification of significant environmental impacts, and the corresponding mitigation plans. Furthermore, the study involves a comprehensive calculation of the environmental costs incurred throughout the project, culminating in the determination of the resulting environmental cost benefits in monetary value. Map of DFCCIL in Fig. 15.1, Eastern and Western DFC in Fig. 15.2, EDFC section map in Fig. 15.3 (Dedicated Freight Corridor Projects 2020).

15.2 Literature Review

15.2.1 Study of EDFC Projects

India's Eastern Dedicated Freight Corridor (EDFC) projects are major railway infrastructure developments (Dedicated Freight Corridor Projects 2020; Adeoye et al. 2023a). These projects create dedicated rail routes to efficiently carry products and commodities between the east and north. The main EDFC project goals:

(a) The EDFC initiatives aim to reduce congestion and improve freight transportation efficiency by separating freight traffic from passenger traffic on the rail network.
(b) EDFC projects enhance connectivity between various industrial and commercial areas in eastern and northern India, fostering economic growth and trade.
(c) The dedicated corridors in India are designed to handle a large volume of freight, hence enhancing rail capacity.
(d) The projects incorporate environmental assessments and mitigation strategies to minimize their impact on the environment and local residents.

India's railway modernization and logistics and transportation sector improvement depend on the EDFC. It entails building new rail lines, bridges, tunnels, and other infrastructure to carry commodities efficiently (Sharma et al. 2020a). The Western Dedicated Freight Corridor (WDFC) and the EDFC projects are part of a wider strategy to build dedicated freight corridors across India. Table 15.1 shows EDFC Rail Project Environmental Impacts.

Table 15.1 summarizes the key environmental impacts associated with three sections of the Eastern Dedicated Freight Corridor (EDFC) rail projects, namely

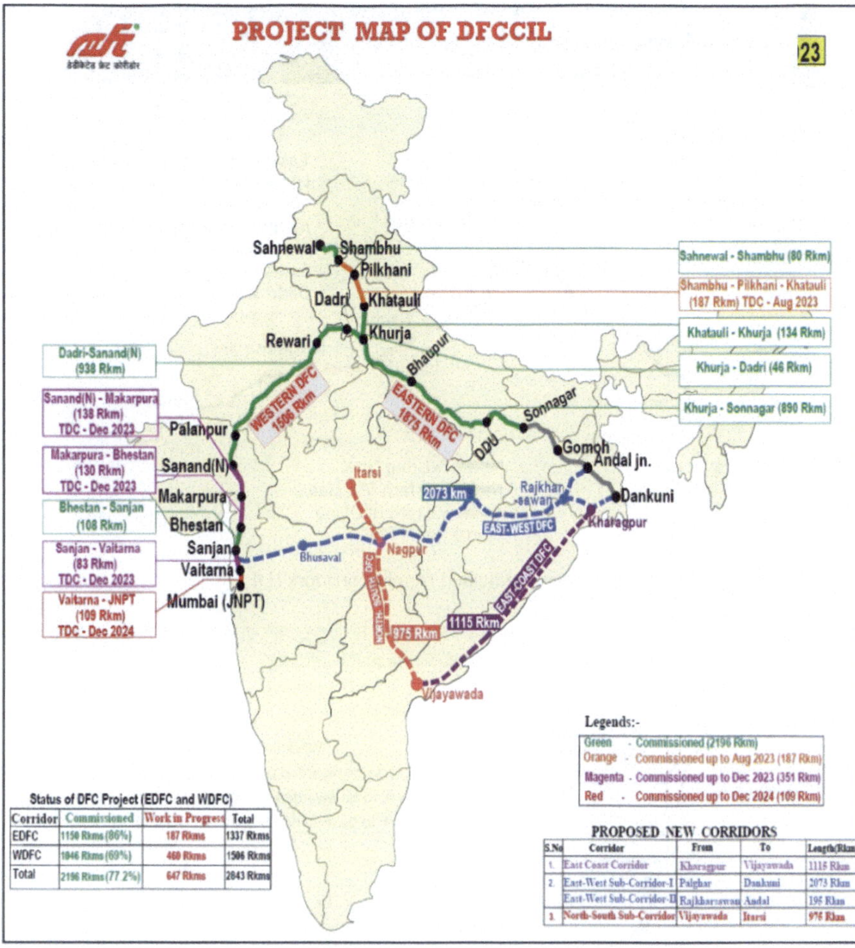

Fig. 15.1 Map of DFCCIL

Khurja–Dadri, Bhaupur–Khurja, and Kaurara–Chamraula. The environmental impacts encompass diverse aspects such as forest diversion, tree cutting, earthwork, noise and vibration levels, land acquisition, and the number of Project-Affected Families (PAFs) and Project-Affected Persons (PAPs) (Mohapatra et al. 2017a; Sharma and Sharma 2019; Swapna Rekha et al. 2020; Sinha et al. 2019; Goswami et al. 2019; Sharma et al. 2022b). These sections of the EDFC projects require careful environmental management, encompassing aspects such as land use, borrow area management, air and water quality, noise levels, and protection of flora and fauna. The table provides a concise overview of these environmental impacts, aiding in the understanding of the challenges and mitigation measures associated with each project section.

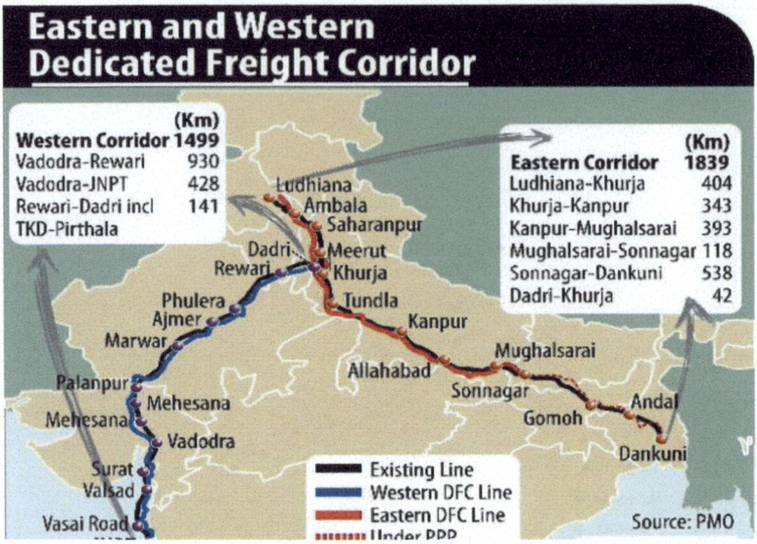

Fig. 15.2 Map of eastern and western dedicated freight corridors (DFC)

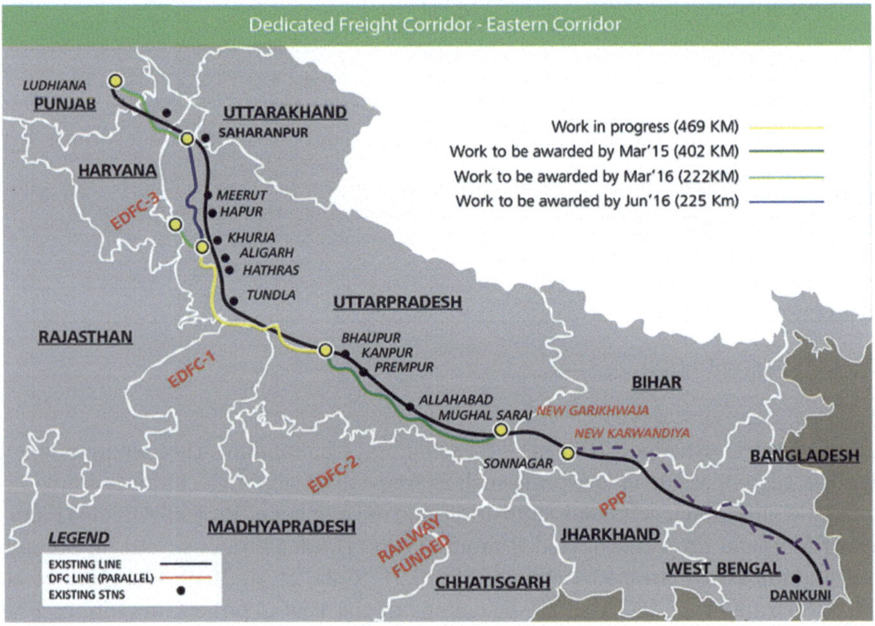

Fig. 15.3 Map of eastern dedicated freight corridor (EDFC) section

Table 15.1 Environmental Impacts of EDFC Rail Projects (Dedicated Freight Corridor Projects 2020)

S.r.no	Project name	Location and length	Environmental impact
1	Khurja–Dadri	49.6-km section in Uttar Pradesh	The diversion of 3.49 hectares of reserved forest, the felling of 2,550 trees, extensive earthwork, and the construction of railway materials. It also involves increased noise and vibration levels near Sensitive Receptors (SRs), affecting 14 Common Property Resources (CPRs) and posing health and safety concerns. Furthermore, this section requires significant land acquisition, impacting 81 PAFs and 5,841 PAPs (Mohapatra et al. 2016a, 2016b; Gopala Rao et al. 2022)
2	Bhaupur–Khurja	272 km	Environmental concerns similar to the Khurja-Dadri segment, including tree cutting and impacts on residential areas. Notably, it requires the cutting of 1,966 trees, with detours proposed to circumvent densely populated regions (Kohli et al. 2022; Sharma and Kumar 2017, 2018; Sharma et al. 2021a)
3	Kaurara–Chamraula	72 km	This segment traverses through reserved forest land and poses minimal wildlife or wetland impact. It involves cutting 4,352 trees and influences an archaeological monument, Budhiya-ki-Tal. Additionally, it falls within the Taj Trapezium Zone (TTZ), subject to pollution control regulations. Vibrations from the project can affect nearby buildings, and mitigation measures are in place, including underpasses, fencing, and vibration reduction techniques (Bhardawaj et al. 2020a; Sharma et al. 2022a; Bandekar et al. 2017; Hossain et al. 2017)

15.2.2 Study of WDFC Projects

The WDFC projects, or Western Dedicated Freight Corridor projects, are a significant railway infrastructure development initiative in India (Adeoye et al. 2023a). These projects are focused on establishing dedicated rail corridors for the efficient transportation of freight (goods and commodities) between the western and northern regions of the country. The primary objectives of the WDFC projects include:

(a) Decongesting Railways: By segregating freight traffic from passenger traffic on the existing rail network, the WDFC projects aim to alleviate congestion issues and enhance the overall efficiency of freight transportation.
(b) Facilitating Trade: The WDFC projects connect key industrial and commercial hubs in western and northern India, facilitating trade and economic growth in these regions.
(c) Increasing Rail Capacity: These dedicated corridors are designed to handle a substantial volume of freight, thereby increasing the overall rail capacity in India.
(d) Environmental Considerations: The projects include environmental impact assessments and mitigation measures to minimize their ecological footprint and mitigate any adverse effects on local communities and ecosystems.

The WDFC is an integral part of India's broader efforts to modernize and upgrade its railway infrastructure, with the goal of enhancing the logistics and transportation sector. The initiative involves the construction of new rail lines, bridges, tunnels, and other critical infrastructure components to support the efficient flow of goods (Mohanty et al. 2016; Samantaray et al. 2014; Pattnaik et al. 2022; Nibedita et al. 2021). The WDFC projects complement the Eastern Dedicated Freight Corridor (EDFC) in creating a comprehensive network of dedicated freight corridors across India. The WDFC projects play a crucial role in bolstering India's transportation and logistics sector. They contribute to economic development, reduce congestion on the existing rail network, and align with the country's sustainable development goals by incorporating environmental considerations into their planning and execution (Biswas et al. 2021; Sharma et al. 2022a, 2020b, 2020c). Table 15.2 summarizing the environmental impacts of the Western Dedicated Freight Corridor (WDFC) rail projects.

Table 15.2 provides an overview of the environmental impacts associated with various sections of the Western Dedicated Freight Corridor (WDFC) rail projects (Pundir et al. 2023a; Das et al. 2023; Agrahari et al. 2021a; Sharma et al. 2022d; Maity et al. 2020; Misra et al. 2022). These projects aim to create dedicated freight corridors in western India, enhancing the efficiency of freight transportation. The environmental impacts are detailed as follows:

(a) Rewari–Palanpur Section: This 641-km stretch is characterized by forest diversion, tree cutting, significant earthwork, and construction material usage. It also involves increased noise and vibration near Sensitive Receptors (SRs), impacts

Table 15.2 Environmental Impacts of WDFC Rail Projects (Adeoye et al. 2023a)

S.r.no	Project name	Location and length	Environmental impact
1	Rewari–Palanpur Section	Approximately 641 km	Forest diversion and habitat disruption—Tree cutting—Earthwork and construction materials—Noise and vibration near Sensitive Receptors (SRs)—Impact on water bodies—Land acquisition—Project-Affected Families (PAFs) and Project-Affected Persons (PAPs)
2	Palanpur–Makarpura Section	Approximately 345 km	–Forest diversion and habitat disruption—Tree cutting—Earthwork and construction materials—Noise and vibration near SRs—Impact on water bodies—Land acquisition—PAFs and PAPs
3	Makarpura–Vadodara Section	Approximately 68 km	–Tree cutting and land acquisition—Noise and vibration near SRs—Impact on water bodies—PAFs and PAPs
4	Vadodara–Ankleshwar Section	Approximately 72 km	–Tree cutting and land acquisition—Noise and vibration near SRs—Impact on water bodies—PAFs and PAPs
5	Ankleshwar–Palanpur Section	Approximately 232 km	–Tree cutting and land acquisition—Noise and vibration near SRs—Impact on water bodies—PAFs and PAPs

on water bodies, land acquisition, and affects Project-Affected Families (PAFs) and Project-Affected Persons (PAPs).

(b) Palanpur–Makarpura Section: Covering approximately 345 km, this section shares environmental impacts similar to the previous segment. It includes forest diversion, tree cutting, earthwork, construction materials, noise, and vibration near SRs, water body impacts, land acquisition, and affects PAFs and PAPs.

(c) Makarpura–Vadodara Section: Spanning around 68 km, this section involves tree cutting, land acquisition, noise and vibration near SRs, and water body impacts, affecting PAFs and PAPs.

(d) Vadodara–Ankleshwar Section: Covering approximately 72 km, this segment is associated with tree cutting, land acquisition, noise and vibration near SRs, and water body impacts, impacting PAFs and PAPs.

(e) Ankleshwar–Palanpur Section: This 232-km section involves tree cutting, land acquisition, noise and vibration near SRs, and water body impacts, affecting PAFs and PAPs.

The WDFC projects are critical for enhancing freight transportation in western India, but they also pose environmental challenges such as habitat disruption, noise, and water pollution (Sharma and Lee 2020a; Mohapatra et al. 2017b; Kohli et al. 2020; Devadarshini et al. 2022; Sharma et al. 2021c). Mitigation measures and environmental safeguards are typically implemented to address these concerns and minimize their impact on local ecosystems and communities.

15.2.3 Environment Management Framework

The Environment Management Frameworks (EMFs) for both the Eastern Dedicated Freight Corridor (EDFC) and Western Dedicated Freight Corridor (WDFC) rail projects: EMF for EDFC: EDFC's EMF consists of three components: EMRP identifies environmental specifications and clearances, CBP builds professionals' expertise, and ECoC offers guidance on responsible practices (Sarkar et al. 2018; Palli et al. 2022; Bhardawaj et al. 2020b, 2019a; Sharma and Lee 2020b; Acharya et al. 2020). EMF for WDFC: WDFC's EMF mirrors EDFC's, encompassing EMRP for clearances, CBP for capacity building, and ECoC for responsible actions. Table 15.3 highlights the key components comparison of the Environment Management Framework (EMF) for both the Eastern Dedicated Freight Corridor (EDFC) and Western Dedicated Freight Corridor (WDFC) rail projects (Sharma and Lee 2020b; Pundir et al. 2023b; COVID-19: Analysing the Legal Nuances of the Lockdown Order 2020).

15.2.4 Environmental Impacts and Mitigation Measures

The development of dedicated freight corridors in India, comprising the Eastern Dedicated Freight Corridor (EDFC) and Western Dedicated Freight Corridor (WDFC), brings about transformative changes in the country's transportation infrastructure. However, it also raises environmental concerns due to factors like land acquisition, tree cutting, noise, vibration, and more (Palli et al. 2021; Sharma et al. 2021d; Palli and Koona 2015; Kanth Choppara et al. 2019). In this section, we delve into the environmental impacts associated with these projects and explore the mitigation measures implemented to minimize their ecological footprint.

Table 15.4 provides a comparative overview of the environmental impacts and corresponding mitigation measures for the Eastern Dedicated Freight Corridor (EDFC) and Western Dedicated Freight Corridor (WDFC) rail projects (PanduRanga Vital et al. 2019; Sharma et al. 2023a, 2023b; Satyanarayana et al. 2022; Sharma and Palli 2016). It highlights their shared commitment to minimizing adverse effects on

Table 15.3 Comparison of the environment management framework of EDFC and WDFC projects

S.r.no	Component	EDFC rail projects	WDFC rail projects
1	Environmental Management Regulatory Procedure (EMRP)	Identifies environmental specifications and clearances for different project sections. Specifies assessment stages and procedures (Pattnaik et al. 2020; Mohapatra et al. 2020, 2022; Sharma et al. 2020c)	Similar to EDFC, identifies environmental specifications and clearances, detailing assessment stages and procedures (Adeoye et al. 2023b; Pati et al. 2022, 2020; Mohapatra et al. 2017c)
2	Capacity Building Plan (CBP)	Equips professionals with expertise in environmental mitigation, legal aspects, implementation plans, entitlement frameworks, monitoring, and inter-agency cooperation	Similar to EDFC, focuses on developing professionals' expertise in addressing environmental concerns (Chatterjee et al. 2022; Adeoye et al. 2023c; Nayak et al. 2022; Mohapatra et al. 2017d)
3	Environmental Code of Conduct (ECoC)	Offers recommendations and measures for addressing environmental concerns during assessment and management, ensuring compliance with environmental regulations	Similar to EDFC, provides guidance on responsible practices and actions to mitigate adverse environmental impacts (Adeoye et al. 2022; Sharma et al. 2020d, 2020e; Bhardawaj et al. 2020c)

the environment and communities. While the projects exhibit similar approaches to addressing environmental challenges, they tailor mitigation strategies to suit specific project requirements and geographic contexts, ensuring responsible and sustainable development along both corridors.

15.2.5 Policy Sustainability Initiatives and Tools

Sustainable large-scale infrastructure initiatives are important nowadays. Policy sustainability strategies and tools help integrate development with environmental, social, and economic issues for the Eastern Dedicated Freight Corridor (EDFC) and Western Dedicated Freight Corridor (WDFC) rail projects. Both corridors use sustainability policies, strategies, and tools to build responsible and durable infrastructure.

The Eastern Dedicated Freight Corridor (EDFC) and Western Dedicated Freight Corridor (WDFC) rail projects' policy sustainability measures and tools are compared in Table 15.5. These projects cover environmental protection, social welfare, energy efficiency, green construction, and economic sustainability. Both corridors adhere to national and international criteria for responsible and sustainable development to benefit the environment and communities they serve (Avesh and Srivastava 2020, 2019; Avesh et al. 2019a).

Table 15.4 Environmental impacts and mitigation measures for EDFC and WDFC Rail Projects

S.r. no	Environmental impact	EDFC	WDFC
1	Land Environment	Mitigation measures include minimizing land requirement, proper debris disposal, embankment protection, and hazardous waste management	Similar measures to minimize land use and address soil erosion (Mohapatra et al. 2015)
2	Borrow Area Management	Non-cultivable lands used, borrowing depths limited, and distance from settlements maintained	Borrowing from non-cultivable lands and distance from settlements ensured (Bhardawaj et al. 2019b)
3	Ambient Air Quality	Dust control with water sprinkling, covered storage of loose earth, and emission control on vehicles and machinery	Dust control, covered storage, and emission control applied consistently (Mohapatra et al. 2021)
4	Water Quality	Measures include silt fencing, groundwater recharge, continuous drainage, and hazardous waste management	Similar measures for water body protection and pollution prevention
5	Noise Level	Noise reduction with equipment adherence to noise standards, silencers, and worker protection	Noise control with equipment standards and protective barriers (Bhattarai et al. 2022)
6	Flora & Fauna	Compensation for trees felled, tree planting, and underpasses for animals	Similar measures for tree compensation and wildlife protection (Agrahari et al. 2021b)
7	Vibration	Enhanced track maintenance, engineering solutions, and protection for sensitive structures	Similar vibration control measures and structural protection (Agarwal et al. 2020)

15.2.6 Environmental Cost Analysis

In today's sustainability-driven society, massive infrastructure projects' environmental costs must be assessed. Environmental cost evaluations for the Eastern Dedicated Freight Corridor (EDFC) and Western Dedicated Freight Corridor (WDFC) rail projects reveal the financial costs and long-term benefits of environmental mitigation strategies. We examine both routes' environmental cost analyses to show how economic factors affect responsible and eco-friendly development (Avesh and Srivastava 2012).

The Eastern Dedicated Freight Corridor (EDFC) and Western Dedicated Freight Corridor (WDFC) rail projects' environmental cost analysis components are compared in Table 15.6. These costs include land acquisition and compensation, tree

Table 15.5 Policy Sustainability Initiatives and Tools for EDFC and WDFC Rail Projects

S.r.no	Sustainability aspect	Eastern dedicated freight corridor (EDFC)	Western dedicated freight corridor (WDFC)
1	Environmental policy	EDFC follows robust environmental policies, adhering to national regulations and international standards for eco-friendly development	WDFC mirrors EDFC's environmental policy, emphasizing compliance with ecological norms and standards
2	Social impact assessment (SIA)	Comprehensive SIA conducted, focusing on the well-being of affected communities, and aligning with inclusive development goals	WDFC carries out an extensive SIA, prioritizing community welfare and equitable growth
3	Green building initiatives	EDFC incorporates green building practices and energy-efficient designs in infrastructure development to reduce environmental impact	WDFC embraces green building principles, promoting sustainable construction methods and eco-friendly infrastructure
4	Energy efficiency measures	EDFC implements energy-efficient technologies to reduce power consumption and greenhouse gas emissions along the corridor	WDFC similarly adopts energy-efficient measures, contributing to reduced carbon footprint
5	Economic sustainability	EDFC emphasizes economic sustainability through cost-effective project management and long-term economic benefits for the region	WDFC shares a commitment to economic sustainability, focusing on efficient resource utilization and regional development

cutting and reforestation, pollution reduction, noise and vibration mitigation, and wildlife preservation. Both corridors encourage responsible development, balancing environmental mitigation costs against long-term financial and ecological benefits for sustainability and community well-being.

15.3 Environmental Assessment and Sustainability Analysis

In this analysis section, a critical aspects of environmental assessment and sustainability for both the Eastern Dedicated Freight Corridor (EDFC) and Western Dedicated Freight Corridor (WDFC) rail projects. This encompasses the assessment of CO_2 emissions, estimation of CO_2 emission factors, projection of future CO_2 emissions, the formulation of comprehensive Environmental Management Plans (EMPs), and conducting Environmental Cost–Benefit Analyses (Mohd and Srivastava 2019).

Table 15.6 Environmental cost analysis for EDFC and WDFC rail projects

S.r.no	Environmental cost component	Eastern dedicated freight corridor (EDFC)	Western dedicated freight corridor (WDFC)
1	Land acquisition and compensation	EDFC allocates substantial funds for land acquisition and compensation, factoring in ecological restoration costs	WDFC's budget also accounts for land acquisition and compensation, including provisions for ecological restoration
2	Tree cutting and reforestation	EDFC budgets for tree cutting expenses, offset by reforestation investments, ensuring minimal net loss of green cover	WDFC similarly budgets for tree cutting and reforestation, preserving ecological balance within the corridor
3	Emission reduction measures	EDFC invests in emission reduction technologies, weighing costs against long-term savings and environmental benefits	WDFC allocates resources to emission reduction measures, considering environmental and financial sustainability
4	Noise and vibration mitigation	EDFC accounts for noise and vibration mitigation costs, recognizing their importance for community well-being and compliance	WDFC's budget includes provisions for noise and vibration mitigation, prioritizing both financial prudence and community welfare
5	Wildlife protection measures	EDFC allocates funds for wildlife protection initiatives, ensuring responsible coexistence with natural habitats	WDFC also dedicates resources to wildlife protection, promoting ecological balance and sustainable development

These components play a pivotal role in evaluating the environmental impact, long-term sustainability, and economic viability of these monumental rail infrastructure projects.

Table 15.7 provides a comparative overview of the environmental assessment and sustainability components for both the Eastern Dedicated Freight Corridor (EDFC) and Western Dedicated Freight Corridor (WDFC) rail projects. These components encompass the assessment of CO_2 emissions, estimation of CO_2 emission factors, projection of future emissions, the formulation of comprehensive Environmental Management Plans (EMPs), and conducting Environmental Cost–Benefit Analyses. Both corridors adhere to internationally recognized standards and methodologies to ensure responsible and sustainable development, emphasizing environmental stewardship and long-term economic viability (Avesh et al. 2019b).

Table 15.7 Environmental Assessment and Sustainability Analysis

S.r.no	Aspect	Eastern dedicated freight corridor (EDFC)	Western dedicated freight corridor (WDFC)
1	Assessment of CO_2 emissions	EDFC conducts an in-depth assessment of CO_2 emissions throughout project phases, employing emission factors and calculation methodologies aligned with international standards	WDFC follows a rigorous CO_2 emission assessment, employing internationally recognized factors and methodologies to quantify project-related emissions
2	Estimation of CO_2 emission factors	EDFC estimates CO_2 emission factors specific to rail transportation, considering fuel consumption, electrification, and energy-efficient practices	WDFC similarly calculates CO_2 emission factors, accounting for rail-specific variables, energy sources, and efficiency measures
3	Future CO_2 emissions projection	EDFC projects future CO_2 emissions based on infrastructure development, traffic growth, and mitigation measures, enabling long-term environmental planning	WDFC forecasts future CO_2 emissions, incorporating infrastructure expansion and strategies to minimize environmental impact
4	Environmental management plan (EMP)	EDFC's EMP outlines comprehensive strategies for environmental protection, including pollution control, habitat conservation, and sustainable practices, ensuring responsible development	WDFC's EMP encompasses a holistic approach to environmental stewardship, emphasizing pollution prevention, biodiversity preservation, and sustainable resource management
5	Environmental cost–benefit analysis	EDFC conducts a thorough cost–benefit analysis, assessing the economic value of environmental initiatives and their contribution to sustainability, aligning with national and global goals	WDFC employs a robust cost–benefit analysis, quantifying the economic advantages of environmental investments and their role in fostering long-term sustainability

15.4 Results

15.4.1 Presenting Results

The total cost incurred in all three sections of the EDFC project is INR 82,763,700 or INR 82.76 million. The summary of the environmental cost–benefit analysis is presented in Table 15.8 for a 30-year period and Table 15.9 for an annual basis.

The ratio is 1:79, indicating that over a period of 30 years, the environmental cost will return 79 times more benefit. The annual ratio is 1:7, demonstrating that the return is 7 times more annually. This analysis illustrates that although the cost of

Table 15.8 Environmental Benefits for 30 Years

Positive impacts	Benefits (INR) over 30 years
Reduction in CO_2	1,087,575,993.20
Health cost savings	5,028,509,000
CA benefit in the future	384,830,640
Total environmental benefit	6,500,915,633.20

Table 15.9 Annual environmental benefits

Positive impacts	Benefits (INR) annually
Reduction in CO_2	36,250,533.11
Health cost savings	167,616,966.67
CA benefit in the future	384,830,640
Total environmental benefit	588,698,139.78

environmental management is initially high, it becomes highly beneficial in the long term.

15.4.2 Findings for WDFC

15.4.2.1 Reduction of CO_2 Emissions from the Western DFC

According to this study, the Delhi-Mumbai DFC project exhibits a substantial potential for reducing CO_2 emissions compared to expected emissions without the DFC project. Furthermore, implementing a low-carbon development path, including a stringent national carbon tax, can further reduce CO_2 emissions.

15.4.3 Potential for Climate Financing

The Clean Development Mechanism (CDM), which allows for the earning of tradable carbon credits (CERs), serves as the primary source of climate finance for infrastructure projects in developing countries. Moreover, innovative climate funds are anticipated to provide additional opportunities for climate financing in the future.

15.4.4 Impacts on Regional Development

The Western DFC's connectivity is poised to give rise to an intra-regional transportation system within the city as seen in the case of Lagos State, Nigeria where the corridors were designed to achieve streamlined supply chain process via a network of ports, industrial parks, and other logistics hubs. Simultaneously, the DFC project stands to benefit from Delhi-Mumbai Industrial Corridor (DMIC) industrial development. The challenge lies in constructing environmentally sustainable, low-carbon, and long-term regional transportation networks. The results unequivocally demonstrate that the long-term impact of the Dedicated Freight Corridor Corporation of India Limited (DFCCIL) is not only financially profitable but also has a positive environmental impact. The environmental cost–benefit analysis underscores that the measures taken by DFCCIL to mitigate the project's environmental impact yield substantial returns. This study holds significance, especially as carbon credit sales have faced challenges in the Indian market, which is still adapting to this mechanism. While the study relied on secondary data due to the COVID-19 pandemic, utilizing primary data could lead to a more comprehensive evaluation of the topic.

15.5 Conclusion

The cost–benefit analysis makes it evident that the expenses incurred for environmental management measures are not a loss but rather a long-term profit. Therefore, it is recommended that the EDFC incorporate such measures into all its future projects. In comparison to other modes of transportation, the Indian Railways (IR) already emits fewer greenhouse gases (GHGs) and is environmentally cleaner. By further investing in efficient technologies, such as solar panels on station buildings, solar panels along DFC walls, rainwater harvesting on station premises, and quieter trains, the Dedicated Freight Corridor (DFC) model for freight transportation can alleviate congestion, save time for passenger trains, adopt environmentally sustainable practices, and reduce harmful gas emissions. This study's pivotal finding underscores the substantial environmental benefits and low-carbon characteristics of major transportation infrastructure projects, such as the Western DFC. This aligns with global commitments to achieve climate-related milestones. It is important to emphasize that transportation investments are long-term assets, necessitating a focus on long-term goals to maximize social welfare benefits.

Finally, the case study highlights the symbiotic relationship between transportation and development. Effective growth plans require robust transportation infrastructure, and, conversely, regional economic development relies on transportation infrastructure to reach its full potential. The DFC decision is a crucial element of transportation investment planning, with the success of the Delhi–Mumbai Industrial Corridor (DMIC) approach influencing implementation outcomes. Therefore, integrating low-carbon, sustainable measures across multiple facets is essential to maximize social welfare benefits.

References

Acharya A, Gahlaut U, Sharma K, Sharma SK, Vishwakarma PN, Phanden RK (2020) Crashworthiness analysis of a thin-walled structure in the frontal part of automotive chassis. Int J Veh Struct Syst 12. https://doi.org/10.4273/ijvss.12.5.06

Adeoye AO, Quadri RO, Lawal OS (2022) Assessment of biofuel potential of tenera palm kernel shell via fixed bed pyrolysis and thermal characterization. Results Surf Interfaces 9:100091. https://doi.org/10.1016/j.rsurfi.2022.100091

Adeoye AO, Quadri RO, Lawal OS (2023a) Wet synthesis, characterization of goethite nanoparticles and its application in catalytic pyrolysis of palm kernel shell in TGA. Results Surf Interfaces 11:100118. https://doi.org/10.1016/j.rsurfi.2023.100118

Adeoye AO, Quadri RO, Lawal OS (2023b) Biomass waste pyrolysis in the context of Nigeria's oil and gas industry: an overview. Int J Ambient Energy 44:2525–2541. https://doi.org/10.1080/01430750.2023.2256325

Adeoye AO, Lawal OS, Quadri RO, Malomo D, Aliyu MT, Dang GE, Emojevu EO, Maikato MJ, Yahaya MG, Omonije OO, Edidem VG, Abubakar YK, Offor OF, Sochima EH, Peter BE, Hikon BN (2023) Sustainable energy via thermochemical and biochemical conversion of biomass wastes for biofuel production. In: Transportation energy and dynamics. energy, environment, and sustainability. Springer, Singapore, pp 245–306. https://doi.org/10.1007/978-981-99-2150-8_11

Agarwal KM, Mohapatra S, Sharma P, Sharma S, Bhatia D, Mishra A (2020) Study and overview of the novel corona virus disease (COVID-19). Sensors Int 1:100037. https://doi.org/10.1016/j.sintl.2020.100037

Agrahari R, Mohanty S, Vishwakarma K, Nayak SK, Samantaray D, Mohapatra S (2021b) Update vision on COVID-19: Structure, immune pathogenesis, treatment and safety assessment. Sensors Int 2:100073. https://doi.org/10.1016/j.sintl.2020.100073

Agrahari R, Sarraf G, Joshi NC, Mohapatra S, Varma A (2021) Insight of biopolymers and applications of polyhydroxyalkanoates. In Microbial polymers, pp 177–191. Springer Singapore, Singapore (2021). https://doi.org/10.1007/978-981-16-0045-6_8

Avesh M, Srivastava R (2012) Modeling simulation and control of active suspension system in Matlab Simulink environment. In 2012 Students conference on engineering and systems, pp 1–6. IEEE. https://doi.org/10.1109/SCES.2012.6199124

Avesh M, Srivastava R (2019) Parametric optimization to design a passenger car suspension system for better dynamic performance. Eur Transp—Trasp Eur

Avesh M, Srivastava R, Chandmal Sharma R, Sharma N (2019) Effective design of active suspension system using fuzzy logic control approach. Int J Veh Struct Syst 11. https://doi.org/10.4273/ijvss.11.5.13

Avesh M, Srivastava R, Sharma RC (2019b) Parametric optimization for the design of passenger vehicle suspension system with the application of genetic algorithm. Int J Veh Struct Syst 11. https://doi.org/10.4273/ijvss.11.2.08

Avesh M, Srivastava R (2020) Passenger car active suspension system model for better dynamic characteristics. Natl Acad Sci Lett 43:37–41. https://doi.org/10.1007/s40009-019-00807-z

Bandekar D, Chouhan OP, Mohapatra S, Hazra M, Hazra S, Biswas S (2017) Putative protein VC0395_0300 from Vibrio cholerae is a diguanylate cyclase with a role in biofilm formation. Microbiol Res 202:61–70. https://doi.org/10.1016/j.micres.2017.05.003

Bhardawaj S, Chandmal Sharma R, Kumar Sharma S (2019a) Development and advancement in the wheel-rail rolling contact mechanics. IOP Conf Ser Mater Sci Eng 691:012034. https://doi.org/10.1088/1757-899X/691/1/012034

Bhardawaj S, Sharma RC, Sharma SK (2020b) Development in the modeling of rail vehicle system for the analysis of lateral stability. Mater Today Proc 25:610–619. https://doi.org/10.1016/j.matpr.2019.07.376

Bhardawaj S, Sharma RC, Sharma SK (2020c) Analysis of frontal car crash characteristics using ANSYS. Mater Today Proc 25:898–902. https://doi.org/10.1016/j.matpr.2019.12.358

Bhardawaj S, Chandmal Sharma R, Kumar Sharma S (2019) A survey of railway track modelling. Int J Veh Struct Syst 11. https://doi.org/10.4273/ijvss.11.5.08

Bhardawaj S, Sharma RC, Sharma SK (2020a) Ride analysis of track-vehicle-human body interaction subjected to random excitation. J Chin Soc Mech Eng Trans Chinese Inst Eng Ser C/ Chung-Kuo Chi Hsueh K. Ch'eng Hsuebo Pao 41:229–238

Bhattarai S, Mhamdi A, Hossain I, Raoui Y, Pandey R, Madan J, Bouazizi A, Maiti M, Gogoi D, Sharma A (2022) A detailed review of perovskite solar cells: Introduction, working principle, modelling, fabrication techniques, future challenges. Micro Nanostruct 172:207450. https://doi.org/10.1016/j.micrna.2022.207450

Biswas N, Saha S, Biswas BK, Chowdhury M, Rahaman A, Junghare V, Mohapatra S, Hazra S, Zangrando E, Roy Choudhury R, Roy Choudhury C (2021) The DNA- and protein-binding properties and cytotoxicity of a new copper(II) hydrazone Schiff base complex. J Coord Chem 74:1482–1504. https://doi.org/10.1080/00958972.2021.1913128

Chatterjee A, Nirwan S, Mohapatra S, Sharma P, Agnihotri A, Shrivastava N (2022) Biochemical aspects of pathogenic variability in white rust infected Indian mustard. Mycologia 114:757–768. https://doi.org/10.1080/00275514.2022.2060007

COVID-19: Analysing the legal nuances of the lockdown order. Indian J Forensic Med Toxicol (2020). https://doi.org/10.37506/ijfmt.v14i4.11553

Das S, Swapnarekha H, Vimal S (2023) Integration of blockchain technology with renewable energy for sustainable development: issues, challenges and future direction. Presented at the (2023). https://doi.org/10.1007/978-981-99-3734-9_48

Dedicated Freight Corridor Projects (2020)

Devadarshini D, Mohapatra S, Pati S, Maity S, Rath CC, Jena PK, Samantaray D (2022) Evaluation of PHAs production by mixed bacterial culture under submerged fermentation. Biologia (bratisl). 78:895–902. https://doi.org/10.1007/s11756-022-01302-5

Gopala Rao L, Sharma RC, Satyanarayana V, Palli S (2022) Stochastic optimal preview control response of a quarter-car nonlinear suspension model using spectral decomposition method. Noise Vib Worldw 53:225–232. https://doi.org/10.1177/09574565221093234

Goswami B, Rathi A, Sayeed S, Das P, Sharma RC, Sharma SK (2019) Optimization design for aerodynamic elements of Indian locomotive of passenger train. In: Advances in engineering design. Lecture notes in mechanical engineering. Springer, Singapore, pp 663–673. https://doi.org/10.1007/978-981-13-6469-3_61

Hossain I, Velkin VI, Shcheklein SE (2017) The study of passive vibration dampers in pipelines using PIV-methodology for single phase flow. In WIT transactions on ecology and the environment, pp 565–570. https://doi.org/10.2495/ESUS170521

Kanth Choppara R, Chandmal Sharma R, Kumar Sharma S, Gupta T (2019) Aero dynamic cross wind analysis of locomotive. IOP Conf Ser Mater Sci Eng 691:012035. https://doi.org/10.1088/1757-899X/691/1/012035

Kohli I, Joshi NC, Mohapatra S, Varma A (2020) Extremophile – an adaptive strategy for extreme conditions and applications. Curr Genomics 21:96–110. https://doi.org/10.2174/1389202921666200401105908

Kohli I, Mohapatra S, Kumar P, Goel A, Varma A, Joshi NC (2022) Role of bacterial endophytes in the promotion of plant growth. In: Bacterial endophytes for sustainable agriculture and environmental management, pp 235–253. Springer Singapore, Singapore (2022). https://doi.org/10.1007/978-981-16-4497-9_12

Maity S, Das S, Mohapatra S, Tripathi AD, Akthar J, Pati S, Pattnaik S, Samantaray DP (2020) Growth associated polyhydroxybutyrate production by the novel Zobellellae tiwanensis strain DD5 from banana peels under submerged fermentation. Int J Biol Macromol 153:461–469. https://doi.org/10.1016/j.ijbiomac.2020.03.004

Misra S, Mohanty D, Mohapatra S (2022) Food-based probiotics: Functional dietary ingredients. In Probiotics, pp 257–275. Elsevier. https://doi.org/10.1016/B978-0-323-85170-1.00006-3

Mohanty DP, Mohapatra S, Misra S, Sahu PS (2016) Milk derived bioactive peptides and their impact on human health—A review. Saudi J Biol Sci 23:577–583. https://doi.org/10.1016/j.sjbs.2015.06.005

Mohapatra S, Rath SN, Pradhan SK, Samantaray DP, Rath CC (2016b) Secondary structural models (16S rRNA) of polyhydroxyalkanoates producing bacillus species isolated from different rhizospheric soil: Phylogenetics and chemical analysis. Int J Bioautomation 20:329–338

Mohapatra S, Mishra R, Roy P, Yadav KL, Satapathi S (2017b) Systematic investigation and in vitro biocompatibility studies on implantable magnetic nanocomposites for hyperthermia treatment of osteoarthritic knee joints. J Mater Sci 52:9262–9268. https://doi.org/10.1007/s10853-017-1136-0

Mohapatra S, Sarkar B, Samantaray DP, Daware A, Maity S, Pattnaik S, Bhattacharjee S (2017c) Bioconversion of fish solid waste into PHB using *Bacillus subtilis* based submerged fermentation process. Environ Technol 38:3201–3208. https://doi.org/10.1080/09593330.2017.1291759

Mohapatra S, Maity S, Dash HR, Das S, Pattnaik S, Rath CC, Samantaray D (2017d) Bacillus and biopolymer: prospects and challenges. Biochem Biophys Reports 12:206–213. https://doi.org/10.1016/j.bbrep.2017.10.001

Mohapatra S, Pattnaik S, Maity S, Mohapatra S, Sharma S, Akhtar J, Pati S, Samantaray DP, Varma A (2020) Comparative analysis of PHAs production by *Bacillus megaterium* OUAT 016 under submerged and solid-state fermentation. Saudi J Biol Sci 27:1242–1250. https://doi.org/10.1016/j.sjbs.2020.02.001

Mohapatra S, Samantaray DP, Samantaray SM (2015) Study on polyhydroxyalkanoates production using rhizospheric soil bacterial isolates of sweet potato. Indian J Sci Technol 8:57. https://doi.org/10.17485/ijst/2015/v8iS7/64027

Mohapatra S, Samantaray DP, Samantaray SM, Mishra BB, Das S, Majumdar S, Pradhan SK, Rath SN, Rath CC, Akthar J, Achary KG (2016) Structural and thermal characterization of PHAs produced by Lysinibacillus sp. through submerged fermentation process. Int J Biol Macromol 93:1161–1167. https://doi.org/10.1016/j.ijbiomac.2016.09.077

Mohapatra S, Mohanta PR, Sarkar B, Daware A, Kumar C, Samantaray DP (2017a) Production of polyhydroxyalkanoates (PHAs) by bacillus strain isolated from waste water and its biochemical characterization. Proc Natl Acad Sci India Sect B Biol Sci 87:459–466. https://doi.org/10.1007/s40011-015-0626-6

Mohapatra S, Vishwakarma K, Joshi NC, Maity S, Kumar R, Ramchander M, Pattnaik S, Samantaray DP (2021) A review on PHAs: the future biopolymer. In: Environmental and agricultural microbiology, pp 83–100. Wiley. https://doi.org/10.1002/9781119525899.ch5

Mohapatra S, Mohapatra S, Han H, Ariza-Montes A, López-Martín M del C (2022) Climate change and vulnerability of agribusiness: assessment of climate change impact on agricultural productivity. Front Psychol 13. https://doi.org/10.3389/fpsyg.2022.955622

Mohd A, Srivastava R (2019) Parametric optimization and experimental validation for nonlinear characteristics of passenger car suspension system. Period Polytech Transp Eng 49:103–113. https://doi.org/10.3311/PPtr.12999

Nayak J, Swapnarekha H, Routray AR, Nayak SR, Behera HS (2022) Big data in education: present and future. In Computational intelligence in data mining. smart innovation, systems and technologies, 281. Springer, Singapore, pp 721–739. https://doi.org/10.1007/978-981-16-9447-9_54

Nibedita S, Swati P, Pattnaik M, Mohapatra S (2021) Methane emission and strategies for mitigation in livestock. In: Environmental and agricultural microbiology, pp 257–274. Wiley. https://doi.org/10.1002/9781119525899.ch12

Palli S, Dontikurti R, Sharma RC, Sharma N (2021) Analysis of dynamic response of a railway locomotive using ANSYS. Int J Veh Struct Syst 13. https://doi.org/10.4273/ijvss.13.2.21

Palli S, Duppala A, Sharma RC, Gopala Rao LVV (2022) Dynamic simulation of automotive vehicle suspension using MATLAB simulink. Int J Veh Struct Syst 14. https://doi.org/10.4273/ijvss.14.3.04

Palli S, Koona R (2015) Analyses of dynamic response of a railway bogie. Int J Veh Noise Vib 11:103. https://doi.org/10.1504/IJVNV.2015.070015

PanduRanga Vital T, Lakshmi BG, Swapna Rekha H, DhanaLakshmi M (2019) Student performance analysis with using statistical and cluster studies. In Soft computing in data analytics. Advances in intelligent systems and computing, 758. Springer, Singapore, pp 743–757. https://doi.org/10.1007/978-981-13-0514-6_71

Pati S, Maity S, Dash A, Jema S, Mohapatra S, Das S, Samantaray DP (2020) Biocompatible PHB production from Bacillus species under submerged and solid-state fermentation and extraction through different downstream processing. Curr Microbiol 77:1203–1209. https://doi.org/10.1007/s00284-020-01922-7

Pati S, Mohanty MK, Mohapatra S, Samantaray D (2022) Bioethanol production by *Enterobacter hormaechei* through carbon monoxide-rich syngas fermentation. Int J Energy Res 46:20096–20106. https://doi.org/10.1002/er.8678

Pattnaik S, Dash D, Mohapatra S, Pattnaik M, Marandi AK, Das S, Samantaray DP (2020) Corrigendum to "Improvement of rice plant productivity by native Cr(VI) reducing and plant growth promoting soil bacteria Enterobacter cloacae" [Chemosphere 240 (2020) 124895]. Chemosphere 238:125011. https://doi.org/10.1016/j.chemosphere.2019.125011

Pattnaik S, Mohapatra S, Pati S, Dash D, Devadarshini D, Tanaya K, Mishra BB, Samantaray D (2022) Microbial bioremediation of Cr(VI)-contaminated soil for sustainable agriculture. In: Microbial biodegradation and bioremediation, pp 395–407. Elsevier (2022). https://doi.org/10.1016/B978-0-323-85455-9.00008-4

Pundir S, Upadhyay S, Kumar N, Chandra Joshi N, Priya R, Ahmad Mir R, Hossain I, Pandey OP (2023a) Interconnected CuO nanoplates as a highly efficient electrocatalyst for oxygen evolution reaction. Mater Lett 336:133921. https://doi.org/10.1016/j.matlet.2023.133921

Pundir S, Upadhyay S, Priya R, Kumar N, Chetana S, Hossain I, Joshi NC, Pandey OP (2023b) Synthesis of 1D β-MnO2 for high-performance supercapacitor application. J Solid State Electrochem 27:531–538. https://doi.org/10.1007/s10008-022-05347-z

Samantaray D, Mohapatra S, Mishra BB (2014) Microbial bioremediation of industrial effluents. In: Microbial biodegradation and bioremediation, pp 325–339. Elsevier. https://doi.org/10.1016/B978-0-12-800021-2.00014-5

Sarkar MS, Segu H, Bhaskar JV, Jakher R, Mohapatra S, Shalini K, Shivaji S, Reddy PA (2018) Ecological preferences of large carnivores in remote, high-altitude protected areas: insights from Buxa Tiger Reserve, India. Oryx 52:66–77. https://doi.org/10.1017/S0030605317000060

Satyanarayana VSV, Sharma RC, Sateesh B, Gopala Rao LVV, Mohan Rao N, Palli S (2022) Optimum response of a nonlinear passive vehicle suspension system under random road excitations. SAE Int J Commer Veh 16:02-16-01-0004. https://doi.org/10.4271/02-16-01-0004

Sharma SK, Kumar A (2017) Ride performance of a high speed rail vehicle using controlled semi active suspension system. Smart Mater Struct 26:055026. https://doi.org/10.1088/1361-665X/aa68f7

Sharma SK, Lee J (2020b) Design and development of smart semi active suspension for nonlinear rail vehicle vibration reduction. Int J Struct Stab Dyn 20:2050120. https://doi.org/10.1142/S0219455420501205

Sharma SK, Lee J (2021) Crashworthiness analysis for structural stability and dynamics. Int J Struct Stab Dyn 21:2150039. https://doi.org/10.1142/S0219455421500395

Sharma RC, Palli S (2016) Analysis of creep force and its sensitivity on stability and vertical-lateral ride for railway vehicle. Int J Veh Noise Vib 12:60. https://doi.org/10.1504/IJVNV.2016.077474

Sharma RC, Sharma SK, Sharma N, Sharma S (2020d) Analysis of ride and stability of an ICF railway coach. Int J Veh Noise Vib 16:127. https://doi.org/10.1504/IJVNV.2020.117820

Sharma RC, Sharma S, Sharma SK, Sharma N (2020e) Analysis of generalized force and its influence on ride and stability of railway vehicle. Noise Vib Worldw 51:95–109. https://doi.org/10.1177/0957456520923125

Sharma SK, Sharma RC, Lee J (2021c) Effect of rail vehicle-track coupled dynamics on fatigue failure of coil spring in a suspension system. Appl Sci 11:2650. https://doi.org/10.3390/app11062650

Sharma SK, Mohapatra S, Sharma RC, Alturjman S, Altrjman C, Mostarda L, Stephan T (2022a) Retrofitting existing buildings to improve energy performance. Sustainability 14:666. https://doi.org/10.3390/su14020666

Sharma SK, Sharma RC, Lee J, Jang H-L (2022b) Numerical and experimental analysis of DVA on the flexible-rigid rail vehicle carbody resonant vibration. Sensors 22:1922. https://doi.org/10.3390/s22051922

Sharma SK, Lee J, Jang H-L (2022c) Mathematical modeling and simulation of suspended equipment impact on car body modes. Machines 10:192. https://doi.org/10.3390/machines10030192

Sharma SK, Sharma RC, Lee J (2022d) In situ and experimental analysis of longitudinal load on carbody fatigue life using nonlinear damage accumulation. Int J Damage Mech 31:605–622. https://doi.org/10.1177/10567895211046043

Sharma RC, Palli S, Sharma SK (2023b) Ride analysis of railway vehicle considering rigidity and flexibility of the carbody. J Chinese Inst Eng 46:355–366. https://doi.org/10.1080/02533839.2023.2194918

Sharma SK, Kumar A (2018) Ride comfort of a higher speed rail vehicle using a magnetorheological suspension system. Proc Inst Mech Eng Part K. J Multi-body Dyn 232:32–48. https://doi.org/10.1177/1464419317706873

Sharma SK, Lee J (2020a) Finite element analysis of a fishplate rail joint in extreme environment condition. Int J Veh Struct Syst 12. https://doi.org/10.4273/ijvss.12.5.03

Sharma SK, Sharma RC (2019) Pothole detection and warning system for Indian roads. In Advances in interdisciplinary engineering. Lecture notes in mechanical engineering. Springer, Singapore, pp 511–519. https://doi.org/10.1007/978-981-13-6577-5_48

Sharma S, Sharma RC, Sharma SK, Sharma N, Palli S, Bhardawaj S (2020) Vibration isolation of the quarter car model of road vehicle system using dynamic vibration absorber. Int J Veh Struct Syst 12. https://doi.org/10.4273/ijvss.12.5.05

Sharma RC, Sharma SK, Palli S (2020b) Linear and non-linear stability analysis of a constrained railway wheelaxle. Int J Veh Struct Syst 12. https://doi.org/10.4273/ijvss.12.2.04

Sharma SK, Sharma RC, Sharma N (2020c) Combined multi-body-system and finite element analysis of a rail locomotive crashworthiness. Int J Veh Struct Syst 12. https://doi.org/10.4273/ijvss.12.4.15

Sharma RC, Palli S, Sharma N, Sharma SK (2021) Ride behaviour of a four-wheel vehicle using H infinity semi-active suspension control under deterministic and random inputs. Int J Veh Struct Syst 13. https://doi.org/10.4273/ijvss.13.2.18

Sharma RC, Sharma S, Sharma N, Sharma SK (2021) Linear and nonlinear analysis of ride and stability of a three-wheeled vehicle subjected to random and bump inputs using bond graph and simulink methodology. SAE Int J Commer Veh 14:02-15-01-0001. https://doi.org/10.4271/02-15-01-0001

Sharma RC, Sharma S, Sharma SK, Sharma N, Singh G (2021d) Analysis of bio-dynamic model of seated human subject and optimization of the passenger ride comfort for three-wheel vehicle using random search technique. Proc Inst Mech Eng Part K J Multi-body Dyn 235:106–121. https://doi.org/10.1177/1464419320983711

Sharma RC, Palli S, Bhardawaj S (2023) A review of wheel-rail dynamic behavior on curved tracks. Int J Veh Struct Syst 15. https://doi.org/10.4273/ijvss.15.1.04

Sinha AK, Sengupta A, Gandhi H, Bansal P, Agarwal KM, Sharma SK, Sharma RC, Sharma SK (2019) Performance enhancement of an all-terrain vehicle by optimizing steering, powertrain and brakes. In Advances in engineering design. lecture notes in mechanical engineering. Springer, Singapore, pp 207–215. https://doi.org/10.1007/978-981-13-6469-3_19

Swapna Rekha H, Nayak J, Behera HS (2020) Pi-sigma neural network: survey of a decade progress. In: Computational intelligence in pattern recognition. Advances in intelligent systems and computing, 1120. Springer, Singapore, pp 429–441. https://doi.org/10.1007/978-981-15-2449-3_37

Chapter 16
Comparative Analysis of Weir Plan Forms and Their Discharge Characteristics

Abhinav Singh, Keerat Kumar Gupta, and Sanjeev Kumar

16.1 Introduction

Weirs, structures used in hydraulic engineering to control water flow, are integral components of various water management systems. They find applications in diverse fields such as irrigation, flood control, hydroelectric power generation, and environmental conservation. Understanding the discharge characteristics of weirs and how different plan forms impact their performance is vital for ensuring efficient water resource management and sustainable environmental practices (Machiels et al. 2011; Anderson and Tullis 2011; Anderson 2011; Leite Ribeiro et al. 2012). Throughout history, the development and optimization of weir structures have been of great importance to engineers and water resource managers (Sharma and Kumar 2017; Sharma and Kumar 2018). This introduction sets the stage for the exploration of the complexities and nuances of weir design by elucidating the primary objectives of the study.

A. Singh (✉) · K. K. Gupta · S. Kumar
Department of Civil Engineering, Graphic Era (Deemed to Be University), Dehradun 248002, India
e-mail: amu.abhinav@gmail.com

K. K. Gupta
e-mail: kkguptafce@geu.ac.in

S. Kumar
e-mail: er.sanjeevkr@gmail.com

S. K. Sharma et al. (eds.), *Dynamics of Transportation Ecosystem, Modeling, and Control*, Energy, Environment, and Sustainability,
https://doi.org/10.1007/978-981-97-0437-8_16

16.1.1 Background and Significance

Weirs have been employed for centuries to harness the power of water for various purposes, from ancient water mills to modern-day hydroelectric power stations. Over time, engineers and scientists have striven to refine weir designs, seeking to maximize their efficiency and minimize environmental impacts. This background section also underscores the critical role that water resources play in sustaining ecosystems, agriculture, and human populations, underscoring the importance of effective weir design. Moreover, the introduction emphasizes the environmental implications of weir construction and operation (Sangsefidi et al. 2021; Bhukya et al. 2022; Anderson and Tullis 2012; Kabiri-Samani and Javaheri 2012; Mehboudi et al. 2016). Weirs can alter natural flow patterns, affect sediment transport, and influence aquatic habitats. Consequently, their design must be approached with careful consideration of environmental consequences. This section highlights the significance of balancing hydraulic efficiency with ecological sustainability.

16.1.2 Objectives of the Study

The primary goal is to conduct a comprehensive analysis of different weir plan forms, including rectangular, trapezoidal, triangular, and compound shapes. The study aims to elucidate how each plan form influences discharge characteristics, flow patterns, and hydraulic efficiency. Additionally, it seeks to identify the advantages, limitations, and practical considerations associated with each plan form. A key objective is to present a comparative analysis of these weir plan forms, enabling engineers, researchers, and policymakers to make informed decisions based on project objectives, site conditions, and environmental considerations. The study will examine the impact of weir geometry, crest shape, submergence, and other design parameters on discharge characteristics, facilitating a deeper understanding of weir behavior. Furthermore, this section of the introduction outlines the scope and structure of the manuscript, providing readers with a roadmap for what to expect in subsequent sections. It establishes the manuscript's intent to synthesize existing knowledge and contribute valuable insights to the field of hydraulic engineering and water resource management.

16.2 Weir Plan Forms

In this section, we delve into the various plan forms of weirs, which serve as the foundational structures in hydraulic engineering. Weirs come in different shapes and configurations, each with its unique characteristics and suitability for specific applications. Understanding these plan forms is crucial for making informed decisions in weir design and management. Figure 16.1 shows different Weirs.

16.2.1 Rectangular Weirs

Rectangular weirs are one of the most common and straightforward plan forms. They consist of a straight, vertical upstream face and a flat crest (Norouzi 2018). Engineers often use rectangular weirs in open-channel flow measurement due to their simplicity. The discharge over a rectangular weir can be calculated using Eq. (16.1).

$$Q = CLH^{\frac{3}{2}} \tag{16.1}$$

where Q is the discharge (flow rate), C is the discharge coefficient is the length of the weir crest, and H is the head (the height of water above the weir crest). Table 16.1 provides a summary of the key characteristics and considerations for rectangular weirs.

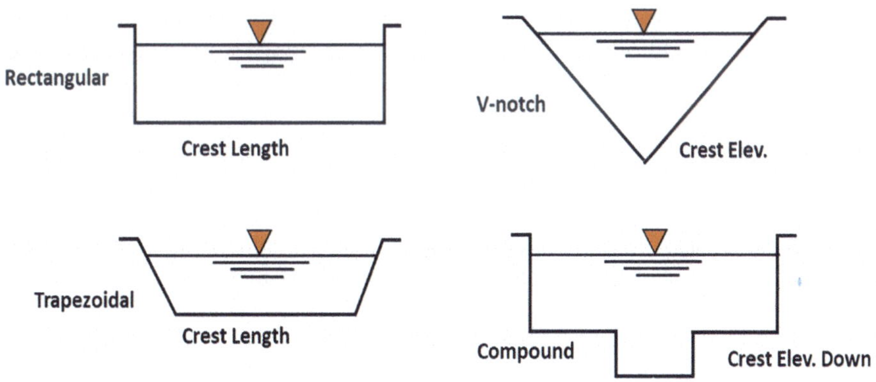

Fig. 16.1 Schematic of different Weirs

Table 16.1 Characteristics of rectangular weirs

Characteristic	Description
Shape	Straight, vertical upstream face with a flat crest
Common applications	Flow measurement, small-scale water diversion
Discharge formula	$Q = CLH^{\frac{3}{2}}$
Advantages	Simplicity, ease of construction, accurate flow measurement
Limitations	Limited applicability to certain flow conditions, potential for cavitation at high velocities

Table 16.2 Characteristics of trapezoidal weirs

Characteristic	Description
Shape	Inclined upstream face resembling a trapezoid
Common applications	Flow regulation, sediment control, spillways
Discharge formula	$Q = K \bullet H^{\frac{3}{2}}$
Advantages	Improved flow control, versatility in various flow conditions
Limitations	More complex design and construction, potential for sediment accumulation at the base

16.2.2 Trapezoidal Weirs

Trapezoidal weirs feature an inclined upstream face, resembling a trapezoid in shape. This design allows for more controlled flow characteristics and improved performance in certain conditions. The discharge over a trapezoidal weir can be calculated using Eq. (16.2).

$$Q = K.H^{\frac{3}{2}} \tag{16.2}$$

where:

Q is the discharge, K is the weir coefficient, and H is the head over the weir. Table 16.2 provides an overview of the key features and considerations for trapezoidal weirs.

16.2.3 Triangular Weirs

Triangular weirs are characterized by an upstream face that forms a triangle. These weirs offer advantages in terms of flow control and are particularly useful for measuring low flows. The discharge over a triangular weir can be calculated using Eq. (16.3).

Table 16.3 Characteristics of Triangular Weirs

Characteristic	Description
Shape	Upstream face forms a triangle
Common applications	Low-flow measurement, precision flow control
Discharge formula	$Q = 2.49 \bullet L \bullet H^{\frac{3}{2}}$
Advantages	Accurate measurement of low flows, minimal maintenance required
Limitations	Limited to low-flow applications, not suitable for high-flow or sediment-laden conditions

$$Q = 2.49 \bullet L \bullet H^{\frac{3}{2}} \tag{16.3}$$

where: Q is the discharge, L is the length of the weir crest, and H is the head over the weir. Table 16.3 summarizes the primary characteristics and considerations for triangular weirs.

16.2.4 Compound Weirs

Compound weirs combine multiple weir shapes in a single structure, allowing for enhanced control over a wide range of flow conditions. These weirs are versatile and find applications in various hydraulic engineering scenarios. The discharge over a compound weir can be determined by combining the discharge formulas for the individual weir sections. Let's consider a compound weir consisting of a trapezoidal section followed by a triangular section. The total discharge can be calculated using Eq. (16.4).

$$Q_{total} = Q_{trapezoidal} + Q_{tringular} \tag{16.4}$$

The coefficients and dimensions for each section must be appropriately accounted for in the calculations. Table 16.4 outlines the key characteristics and considerations for compound weirs.

Table 16.4 Characteristics of Compound Weirs

Characteristic	Description
Shape	Combination of different weir shapes
Common applications	Versatile flow regulation, adaptable to varying flow conditions
Discharge calculation	Sum of discharges over individual sections
Advantages	Enhanced flow control, adaptability to changing flow conditions
Limitations	Increased complexity in design and analysis, careful consideration of section interactions

Table 16.4 has provided an in-depth exploration of various weir plan forms, including rectangular, trapezoidal, triangular, and compound weirs. Each plan form has its unique characteristics, advantages, and limitations, making it essential for engineers and water resource managers to carefully select the appropriate weir design based on project requirements and site-specific conditions. The accompanying tables and mathematical formulas offer valuable reference material for weir design and analysis.

16.3 Discharge Characteristics

The fundamental principles governing discharge characteristics are discussed, along with a better understanding of how different factors influence discharge rates and flow patterns, which is essential for designing effective weir structures. Additionally, mathematical formulas and tables are introduced to illustrate key concepts.

16.3.1 Factors Influencing Discharge

The discharge over a weir is influenced by several factors, including the weir's geometry, the height of water over the weir crest (head), and the shape of the flow approaching the weir. One of the critical factors is the weir coefficient (C or K), which varies depending on the weir plan form and the flow conditions.

The choice of the appropriate weir coefficient is crucial for accurate discharge calculations. Engineers often rely on empirical formulas to determine these coefficients for different weir shapes and flow conditions. Table 16.5 provides typical values of weir coefficients for common weir plan forms.

The appropriate coefficient should be selected based on the weir's geometry and the type of flow (submerged or free-flowing). Submerged flow occurs when the water level upstream of the weir is higher than the weir crest, while free-flowing flow occurs when the water level is lower.

Table 16.5 Typical weir coefficients for common plan forms

Weir plan form	Weir coefficient (C or K) range
Rectangular weir	0.59–0.61
Trapezoidal weir	0.60–0.65
Triangular weir	2.49
Compound weir (Varies)	Depends on specific configurations

Table 16.6 Example hydraulic efficiency values for weir plan forms

Weir plan form	Typical hydraulic efficiency (E) range
Rectangular weir	0.85–0.95
Trapezoidal weir	0.90–0.97
Triangular weir	0.95–0.99
Compound weir (Varies)	Depends on specific configurations

16.3.2 Flow Patterns and Hydraulic Efficiency

Understanding flow patterns over a weir is essential for optimizing hydraulic efficiency and avoiding issues such as cavitation and turbulence. Flow patterns are influenced by the shape of the weir crest and the approach velocity of the water.

In rectangular weirs, the flow generally converges smoothly over the crest, producing a stable, well-defined flow pattern. Trapezoidal weirs, with their inclined upstream faces, create a more gradual transition of flow and are effective at reducing turbulence. Triangular weirs are often used for precise low-flow measurements due to their ability to maintain laminar flow patterns.

The hydraulic efficiency of a weir is another critical aspect. It is often characterized by the ratio of actual discharge to theoretical discharge. The actual discharge (Q) is determined through measurement or calculation, while the theoretical discharge is calculated using the appropriate discharge formula for the specific weir plan form.

The hydraulic efficiency can be calculated using Eq. (16.5).

$$E = \frac{Q}{Q_{theory}} \tag{16.5}$$

A higher hydraulic efficiency indicates that the weir is operating closer to its theoretical capacity. Table 16.6 provides an example of hydraulic efficiency values for different weir plan forms.

16.3.3 Design Parameters and Submergence Effects

Weir design involves considering various parameters beyond just the weir plan form. These parameters include the length of the weir crest (L), the head (H), and the approach velocity (V). Engineers must also account for the potential effects of submergence, where the water level upstream of the weir is significantly higher than the weir crest (Ghanbari and Heidarnejad 2020; Crookston et al. 2018; Karimi et al. 2018; Safarrazavi Zadeh et al. 2021).

The effects of submergence can alter the flow characteristics and require adjustments in the discharge calculations. For submerged flow, a correction factor (P) is introduced to the discharge formula can be calculated using Eq. (16.6).

Table 16.7 Submergence correction factor values

Submergence ratio	Submergence correction factor
0.2	0.95
0.4	0.89
0.6	0.82
0.8	0.74
1.0 (Free Flow)	1.00

$$Q = CLH^{\frac{3}{2}} \bullet P \tag{16.6}$$

The correction factor (P) is a function of the submergence ratio. Table 16.7 illustrates the submergence correction factor values for various submergence ratios.

Engineers and designers must carefully consider submergence effects when selecting weir plan forms and dimensions to ensure accurate discharge estimation in real-world scenarios (Karimi et al. 2019; Saghari et al. 2019; Guo et al. 2019; Kadia et al. 2023; Dabling and Tullis 2012). The factors influencing discharge characteristics in weir structures. It discussed the significance of weir coefficients, flow patterns, hydraulic efficiency, and design parameters. Additionally, mathematical formulas and tables were introduced to aid in the understanding and application of these principles in weir design and analysis. Understanding these factors is essential for engineers and water resource managers aiming to optimize weir performance and ensure accurate flow measurement and control (Torre-Gómez et al. 2022).

16.4 Comparative Analysis

A comprehensive comparative analysis of different weir plan forms, with a particular emphasis on their practical applications and field performance, is presented. By dissecting the advantages, limitations, and unique considerations associated with each weir plan form, engineers, researchers, and decision-makers gain critical insights needed to make well-informed choices in weir design and implementation.

16.4.1 Advantages and Limitations of Different Weir Plan Forms

In the hydraulic engineering, the choice of a weir plan form is akin to selecting the perfect tool for a specific task. Each plan form offers its own set of advantages and limitations, making them suitable for distinct applications. The following Table 16.8 provides a concise summary of these aspects, allowing for a quick comparison among common weir plan forms.

Table 16.8 Advantages and limitations of common weir plan forms

Weir plan form	Advantages	Limitations
Rectangular weir	–Simplicity in design and construction	–Limited applicability to certain flow conditions
	–Accurate flow measurement	–Potential for cavitation at high velocities
Trapezoidal weir	–Improved flow control and versatility in various flow conditions	–More complex design and construction
	–Effective sediment control	–Potential for sediment accumulation at the base
Triangular weir	–Accurate measurement of low flows	–Limited to low-flow applications
	–Minimal maintenance required	–Not suitable for high-flow or sediment-laden conditions
Compound weir	–Enhanced flow control and adaptability to changing flow conditions	–Increased complexity in design and analysis
	–Versatility in addressing varied hydraulic scenarios	–Careful consideration of section interactions required

These insights illuminate the diverse strengths and limitations of each plan form. Rectangular weirs, known for their simplicity and precision in flow measurement, are best suited for scenarios where these attributes are paramount. Trapezoidal weirs, with their advanced flow control capabilities, shine in applications where sediment management is vital. Triangular weirs, while confined to low-flow scenarios, excel in precise measurement tasks. Compound weirs offer versatility across a broad spectrum of hydraulic conditions, but their implementation requires careful consideration due to their increased complexity.

16.4.2 Case Studies: Weir Performance in Practical Applications

To further underscore the practical relevance of different weir plan forms, this section presents a series of case studies, each demonstrating the remarkable performance of a specific weir plan form in real-world hydraulic engineering scenarios.

Case Study 1: Rectangular Weir in Flow Measurement.

In this case study, a rectangular weir takes center stage as a crucial tool for accurate flow measurement in a small irrigation canal. The straightforward design and exceptional accuracy of rectangular weirs make them indispensable in applications where precision is paramount. Engineers employ the venerable Francis formula to

Table 16.9 Case study 1—rectangular weir in flow measurement

Parameter	Results or findings
Weir length	2 m
Head (H)	0.6 m
Discharge (Q)	2.5 cubic meters per second (m^3/s)
Accuracy	±2%
Application	Small irrigation canal
Significance	Precise flow measurement for irrigation

calculate discharge with precision, and the results are a testament to the effectiveness of rectangular weirs in providing precise flow measurements. The data obtained from this case study (See table 16.9) are invaluable for efficient water distribution in irrigation systems, ensuring that crops receive the appropriate amount of water, thereby optimizing agricultural productivity and water resource management.

Case Study 2: Trapezoidal Weir for Sediment Control

This captivating case study shines a spotlight on the practical applications of trapezoidal weirs in a river system prone to sediment transport. The inclined upstream face of the trapezoidal weir plays a pivotal role in promoting controlled flow patterns, thereby reducing the likelihood of sediment accumulation downstream (See Table 16.10). By maintaining a gradual transition of flow, the weir effectively prevents sediment buildup, contributing to improved water quality and ecosystem health. The results of this case study are a testament to the engineering finesse that trapezoidal weirs bring to sediment control scenarios, paving the way for cleaner waterways and healthier aquatic ecosystems.

Case Study 3: Triangular Weir for Low-Flow Monitoring

In this case study, the remarkable precision of triangular weirs is showcased in a water quality assessment project. Triangular weirs, known for their inherent accuracy in measuring low flows, are deployed to provide vital data in an environmental monitoring and research context. With minimal maintenance requirements, triangular weirs deliver dependable measurements of low-flow conditions, essential for understanding and safeguarding environmental health (See Table 16.11). The results of this case study underscore the indispensable role of triangular weirs in applications that demand precise low-flow monitoring, setting the stage for more informed decisions in environmental preservation.

Table 16.10 Case study 2—trapezoidal weir for sediment control

Parameter	Results or findings
Weir dimensions	4 m (length), 30° (slope)
Sediment reduction	90% reduction in downstream sediment
Application	River prone to sediment transport
Significance	Improved water quality and ecosystem health

Table 16.11 Case study 3—triangular weir for low-flow monitoring

Parameter	Results or findings
Weir length	1.5 m
Head (H)	0.2 m
Discharge (Q)	0.15 cubic meters per second (m^3/s)
Accuracy	±1%
Application	Water quality assessment project
Significance	Precise measurement of low-flow conditions for environmental monitoring

This table presents specific data and findings from Case Study 3, showcasing the accuracy and suitability of a triangular weir in the context of low-flow monitoring for environmental assessments.

16.5 Conclusion

The manuscript synthesizes the key findings from the comparative analysis of different weir plan forms and their discharge characteristics. The insights gained from the previous sections are synthesized to provide a holistic view of the implications for weir design, water resource management, and future research. Additionally, tables are introduced to summarize essential conclusions and recommendations.

16.5.1 Key Findings

Throughout this manuscript, an in-depth exploration of various weir plan forms has been conducted, shedding light on their individual characteristics, advantages, and limitations. The analysis of discharge characteristics has revealed the critical role played by factors such as weir geometry, head, and approach velocity. Practical case studies have demonstrated the real-world applicability of different weir plan forms, showcasing their performance in diverse hydraulic engineering scenarios. The key findings from this manuscript can be summarized in Table 16.12.

16.5.2 Implications for Weir Design and Water Resource Management

The findings from this manuscript have several implications for weir design and water resource management are given in Table 16.13.

Table 16.12 Key findings

Finding	Implication
Weir plan forms offer unique advantages and limitations	Engineers should carefully select the appropriate plan form based on project objectives and hydraulic conditions
Discharge accuracy is influenced by weir coefficients	Accurate discharge estimation requires selecting the appropriate coefficient for the specific weir shape and flow conditions
Flow patterns and hydraulic efficiency vary among weir plan forms	Understanding flow patterns is essential for optimizing weir performance and minimizing issues such as cavitation and turbulence
Weir design parameters and submergence effects must be considered	Engineers must account for parameters like weir length, head, and submergence effects to ensure accurate discharge calculations
Practical case studies demonstrate the real-world effectiveness of weir plan forms	Weir plan forms can significantly impact sediment control, flow measurement, and low-flow monitoring in various applications

Table 16.13 Implications for Weir Design and Water Resource Management

Implication	Action required
Weir plan form selection	Engineers and decision-makers should carefully evaluate project objectives and hydraulic conditions when choosing a weir plan form
Weir coefficient calibration	Proper calibration of weir coefficients is essential for accurate flow measurement and control
Flow pattern optimization	Engineers should aim to optimize flow patterns over weirs to enhance hydraulic efficiency and minimize potential issues
Submergence considerations	Submergence effects must be considered in weir design to ensure accurate discharge calculations
Environmental impact assessment	The environmental consequences of weir construction and operation should be thoroughly assessed to minimize negative effects on ecosystems

16.5.3 Future Directions and Research Needs

While this manuscript provides valuable insights into weir plan forms and their discharge characteristics, there remain avenues for future research and exploration are given in Table 16.14.

Table 16.14 Future research directions

Research Area	Key questions and considerations
Enhanced weir coefficient models	Are there more accurate models for weir coefficients that account for a wider range of flow conditions and weir shapes?
Flow control innovations	Can innovatively design approaches further improve flow control and sediment management in weir systems?
Environmental impact mitigation	How can we further minimize the environmental impact of weir construction and operation?
Emerging weir plan forms	Are there new weir plan forms or modifications that offer advantages in specific applications?

16.5.4 Conclusion and Recommendations

This manuscript has provided a comprehensive analysis of different weir plan forms and their discharge characteristics. The findings emphasize the importance of selecting the appropriate plan form based on project objectives and hydraulic conditions. Proper calibration of weir coefficients, optimization of flow patterns, and consideration of submergence effects are essential for accurate weir design and flow measurement. To enhance water resource management and minimize environmental impact, it is crucial to conduct thorough assessments of weir projects. Future research can focus on developing more accurate weir coefficient models, innovative flow control approaches, and strategies for mitigating environmental impact. In light of the research presented in this manuscript, the following recommendations are made and given in Table 16.15.

This manuscript's findings and recommendations provide valuable guidance for engineers, researchers, and decision-makers in the field of hydraulic engineering,

Table 16.15 Recommendations

Recommendation	Action steps
Weir plan form selection	Evaluate project objectives and hydraulic conditions thoroughly when selecting a weir plan form
Weir coefficient calibration	Ensure proper calibration of weir coefficients for accurate flow measurement and control
Flow pattern optimization	Strive to optimize flow patterns over weirs to enhance hydraulic efficiency
Submergence considerations	Consider submergence effects in weir design to ensure accurate discharge calculations
Environmental impact assessment	Conduct comprehensive assessments of weir projects to minimize environmental impact
Future research focus	Explore new weir coefficient models, innovative flow control approaches, and environmental impact mitigation strategies

contributing to the sustainable management of water resources and the continued advancement of weir design and operation.

References

Anderson RM (2011) *Piano Key Weir Head Discharge Relationships*; Utah State University: Logan. UT, USA

Anderson RM, Tullis BP (2012) Comparison of piano key and rectangular labyrinth weir hydraulics. J Hydraul Eng 138:358–361

Anderson RM, Tullis B (2011) Influence of piano key weir geometry on discharge. In Labyrinth and piano key weirs. CRC Press, Boca Raton, FL, USA, pp 75–80

Bhukya RK, Pandey M, Valyrakis M, Michalis P (2022) Discharge estimation over piano key weirs: a review of recent developments. Water 14:3029

Crookston BM, Anderson RM, Tullis BP (2018) Free-flow discharge estimation method for Piano Key weir geometries. J Hydro-Environ Res 19:160–167

Dabling MR, Tullis BP (2012) Piano key weir submergence in channel applications. J Hydraul Eng 138:661–666

Ghanbari R, Heidarnejad M (2020) Experimental and numerical analysis of flow hydraulics in triangular and rectangular piano key weirs. Water Sci 34:32–38

Guo X, Liu Z, Wang T, Fu H, Li J, Xia Q, Guo Y (2019) Discharge capacity evaluation and hydraulic design of a piano key weir. Water Supply 19:871–878

Kabiri-Samani A, Javaheri A (2012) Discharge coefficients for free and submerged flow over Piano Key weirs. J Hydraul Res 50:114–120

Kadia S, Pummer E, Kumar B, Rüther N, Ahmad Z (2023) A reformed empirical equation for the discharge coefficient of free-flowing type—A piano key weirs. J Irrig Drain Eng 149:04023003

Karimi M, Ghazizadeh MJ, Saneie M, Attari J (2019) Flow characteristics over asymmetric triangular labyrinth side weirs. Flow Meas Instrum 68:101574

Karimi M, Attari J, Saneie M, Jalili Ghazizadeh MR (2018) Side weir flow characteristics: comparison of piano key, labyrinth, and linear types. J Hydraul Eng 144:04018075

Leite Ribeiro M, Pfister M, Boillat JL, Schleiss A, Laugier F (2012) Piano key weirs as efficient spillway structure. In: Proceedings of the 24th Congress of CIGB–ICOLD, Kyoto, Japan, pp 176–186

Machiels O, Erpicum S, Dewals BJ, Archambeau P, Pirotton M (2011) Experimental observation of flow characteristics over a Piano Key Weir. J Hydraul Res 49:359–366

Mehboudi A, Attari J, Hosseini SA (2016) Experimental study of discharge coefficient for trapezoidal piano key weirs. Flow Meas Instrum 50:65–72

Norouzi B (2018) Experimental and numerical study on hydraulics of the piano key weirs with modified keys. J Dam Hydroelectr Powerpl 5:36–47

Safarrazavi Zadeh M, Esmaeili Varaki M, Biabani R (2021) Experimental study on flow over sinusoidal and semicircular labyrinth weirs. ISH J Hydraul Eng 27:304–313

Saghari A, Saneie M, Hosseini K (2019) Experimental study of one-and two-cycle trapezoidal piano-key side weirs in a curved channel. Water Supply 19:1597–1603

Sangsefidi Y, Tavakol-Davani H, Ghodsian M, Mehraein M, Zarei R (2021) Hydrodynamics and free-flow characteristics of piano key weirs with different plan shapes. Water 13:2108

Sharma SK, Kumar A (2017) Ride performance of a high speed rail vehicle using controlled semi active suspension system. Smart Mater Struct 26:055026. https://doi.org/10.1088/1361-665X/aa68f7

Sharma SK, Kumar A (2018) Ride comfort of a higher speed rail vehicle using a magnetorheological suspension system. Proc Inst Mech Eng Part K J Multi-body Dyn 232:32–48. https://doi.org/10.1177/1464419317706873

Torre-Gómez G, Erpicum S, Pugliese F, Giugni M (2022) Hydraulic behavior assessment of Type A and Type B piano key weirs from experimental and numerical results. Environ Sci Proc 21:84

Chapter 17
Warehousing and Supply Chain Management of Dangerous Goods

Lakshya Bhardwaj, Aditi Aiswarya, and Abhilasha Saksena

17.1 Introduction

It is difficult to conceive a modern world without things having some quantity of chemicals. There are several consumer items created from hazardous ingredients in the hygiene sector alone, including laundry detergents, dishwashing detergents, soaps, shampoos, and almost all modern home chemicals. Furthermore, chemical influence affects more than just home items. Unfortunately, consumers are also exposed to high levels of synthetic tastes, food-grade colorants, chemicals, and stabilizers. Even though domestic chemical use is fairly minimal when compared to industrial scale, it is apparent that chemicals are created, transported, stored, and consumed all over the world (Adeoye et al. 2023a).

Every day, numerous radioactive, explosive, or hazardous chemicals are engaged in the lives of many people living in space, ranging from industrial cities with dense populations to small communities along the transitional route of various sorts. Road traffic, in particular (Sharma et al. 2020a). These are compounds that, by accident, can cause immense damage to the environment and human safety due to many dangers during manufacture, transit, use, or storage. The increasing need for dangerous products transportation is driven by the usage of hazardous materials in a variety of areas, a growth in transportation operations, and technological advancements. Millions of tons of dangerous commodities are moved by numerous forms of transportation each year. Because of the ubiquity of hazardous chemicals, transportation of hazardous materials is a severe challenge in developed countries (Mohapatra et al. 2016a).

As a result, persons transporting hazardous products have more accidents than other modes of transportation i.e. Rail, road, air, ship etc (Badgujar et al. 2024; Kokkinos et al. 2024; Raju Rallabandi et al. 2024; Rallabandi et al. 2024; Sharma

L. Bhardwaj · A. Aiswarya · A. Saksena (✉)
School of Technology, Gati Shakti Vishwavidyalaya, Vadodara 390004, India
e-mail: asaksena1978@yahoo.com

S. K. Sharma et al. (eds.), *Dynamics of Transportation Ecosystem, Modeling, and Control*, Energy, Environment, and Sustainability,
https://doi.org/10.1007/978-981-97-0437-8_17

et al. 2024a, b, c, d, e, f, g; Upadhyay and Sharma Vikram 2024a, b). Accidents do happen during the transportation of hazardous chemicals, even if they happen in a unique fashion. The heightened danger of hazardous material transport events has boosted awareness in a variety of areas, including business, government, and academia (Gopala Rao et al. 2022). Much study has been conducted in the latter field to assess and decrease risks in the transport of dangerous commodities, where risk is an indicator of the chance and severity of loss of a crucial organ. Hazardous materials may be exposed as a result of probable undesirable occurrences (Mohapatra et al. 2016b).

The usage of hazardous materials in numerous sectors, as well as the rise in transportation operations as technology advances, has increased the necessity to carry harmful items. Millions of tons of dangerous commodities are moved by numerous forms of transportation each year. Special laws have been developed all across the world to prevent any hazard from these modes of transportation. The goal of these standards is to develop a safe procedure for transporting dangerous items, which includes delivery, storage and packing, sorting, labeling, and handling. Simultaneously, these guidelines instruct us on what to do in an emergency. The basic purpose of all of these guidelines is to remove possible hazards before they cause an accident. As a result, the preventative strategy mainly focuses on logistics for risky items. In this process, the regulatory approach is the backup plan. The second reason for the relevance of the regulatory approach is that making changes or taking action would not remove environmental pollution or compensate for potential loss of life. As a result, a preventative and prediction strategy based on global research on risky goods, road transport hazards and payload/route selection techniques has been developed. They created an innovative, globally applicable page-oriented framework. Because of the ubiquity of these compounds, the transportation of hazardous materials (hazmat) is a severe concern in industrialized countries. Living without hazardous chemicals is inconceivable for most people in developed nations. Unfortunately, most hazardous products are not utilized at the place of manufacture and must be transported long distances.

17.2 Hazardous Goods

A hazardous product is any item or material that might unreasonably jeopardize people's health, safety, or property during commercial transportation. Identifying dangerous commodities is the first step in reducing the dangers caused by a commodity through adequate packaging, handling, communication, and storage. Because of the dangers involved, any mistake, misunderstanding, or incorrect documentation might have major consequences and be devastating for human life on land or onboard a ship. Goods that are classified as hazardous cargo under the IMDG Code due to their flammability, corrosiveness, poisonousness, or other qualities.

17.2.1 Classes of Dangerous Goods

Class 1: Explosive Materials Subclass

Explosives with the potential for large-scale explosions are among the materials and items that represent a danger of mass explosion. (A mass explosion is one that hits the whole load very instantaneously.) have a significant projection risk but no mass explosion risk. comprises explosives with a predicted danger but no risk of a large-scale explosion have a fire risk, a little blast risk, a minor projection risk, or both, but no mass explosion risk, consists of explosives that pose a little fire danger, a minor blast risk, a slight projection risk, or both, but pose no threat of a large-scale explosion. They emit a huge quantity of radiant heat as they burn or burn one after the other. Items and materials pose no major risk. This category includes chemicals and products that provide only a minor risk of ignite or start during shipment. Substances that are very insensitive and pose a mass explosion risk. Substances have a mass explosion danger but are so insensitive that there is very little chance of initiation or transition from burning to detonation under typical transport conditions. Extremely insensitive items that do not provide a threat of a large-scale explosion, as well as extremely insensitive goods that do not pose a threat of a large-scale explosion.

Class 2: Gases

Volatile Gases having a flammable range of at least 12% regardless of the lower flammable limit and gases that at 20 °C and a standard pressure of 101.3 kPa are ignitable when in a combination of 13% or less by volume with air (Kohli et al. 2022). Gases that are not poisonous or combustible Pressurized gases, liquefied gases, pressurised cryogenic gases, compressed gases in a solution, and oxidising gases are all included in this category (Sharma and Kumar 2017). Gases that do not fall under Divisions 2.1 or 2 but exert an absolute pressure in the packaging of 280 kPa (40.6 psia) or more at 20 °C (68 °F) are considered non-flammable and non-toxic. Hazardous Gases known to be extremely dangerous or corrosive, such as testes, are assumed to be poisonous to humans or corrosive since their LC50 value is equal to or less than 5,000 ml/ m^3 (ppm).

Class 3: Flammable Liquids

Substances that fulfill the criteria for being deemed liquids under subparagraph (a) of the ADR/RID/IMDG Code 1.2.1 definition of "liquids" include those having flash temperatures less than 60 °C and vapour pressures less than 300 kPa (3 bars) at 50 °C and 101.3 kPa at 20 °C, respectively (Sharma and Kumar 2018).

Class 4: Flammable Solids

Solid desensitized explosives, self-reactive chemicals, and polymerizing substances are examples of flammable solids. Self-reactive liquids or solids include desensitized solid explosives, quickly combustible solids, self-reactive chemicals, and polymerizing substances ((Sharma et al. 2021a; Bhardawaj et al. 2020a)). Materials that burn spontaneously. Pyrophoric compounds are chemicals that burn spontaneously. They are substances that heat up or ignite quickly when they come into touch with air within five minutes. Items containing chemicals that emit flammable gasses when exposed to water, as well as similar substances. These substances have the ability to react with water and produce explosive combinations with air.

Class 5: Oxidizing Substances and Organic Peroxides

Oxidizing agents are substances that, while not necessarily combustible, begin or help in the burning of other materials, frequently by providing oxygen. Peroxides derived from organic sources. Organic peroxides are susceptible to exothermic breakdown at high or low temperatures. Heat, friction, impact, or contact with contaminants (such as acids, heavy metal compounds, or amines) can all cause decomposition.

Class 6: Toxic and Infectious Substances

Dangerous compounds that are known to be hazardous to human health at low levels or to have a single or quick lethal effect if consumed, breathed, or come into touch with the skin. Contaminated substances Infectious substances are pathogen-containing chemicals that are known or expected to be present. Pathogens are microorganisms that may cause disease in people and animals. Bacteria, viruses, rickettsia, parasites, fungi, and other things such as prions are examples of microorganisms (Bandekar et al. 2017).

Class 7: Radioactive Material

Radioactive material is defined as any substance containing radionuclides that exceeds the values specified in paragraphs 2.7.2.2.1 to 2.7.2.2.6 of the IMDG Code in terms of both activity concentration and total activity (Hossain et al. 2017; Mohapatra et al. 2017a).

Class 8: Corrosive Substances.

Compounds that, when touched, may irrevocably hurt the skin or, in the case of a leak, will substantially harm or even destroy other products or modes of transportation.

Class 9: Miscellaneous Dangerous

Objects and substances and products that pose a risk not addressed by other groups during transportation.

17.2.2 Rules and Regulations Governing the Dangerous Goods: Sea – IMDG Code and Others

The International Maritime Dangerous products (IMDG) Code is "the standard guide to all aspects of handling dangerous goods and marine pollutants in sea transport," according to its own definition (Sharma and Sharma 2019). Three of its primary purposes are to protect human lives, avoid maritime pollution, and facilitate the free circulation of dangerous items. The International Maritime Organization (IMO) developed it, and it is revised every two years to apply to all cargo-carrying ships. It appears to apply to the transportation of 3,500 packed risky items. There is one "Primary Class" and one or more "Subsidiary Risks" for a medicine having several hazards.

Air—DGR and ICAO Technical Instructions

According to the International Air Transport Association (IATA), the Dangerous products Regulations (DGR) are "the only norm recognized by airlines for carrying dangerous goods by air." It specifies the shipper's/operator's responsibilities, transit and storage capacity, and forbidden items (Swapna Rekha et al. 1120). It also contains mechanisms for incident reporting, security, and training. The International Civil Aviation Organization's (ICAO) Technical Instructions are a second set of standards designed to guarantee that risky cargo is delivered by airlines without jeopardizing the aircraft or its passengers.

Road—Motor Vehicles Rules and Hazardous Substances Rules

Unlike air and sea travel, which are controlled by international standards, dangerous goods transit by road is governed by the laws of a variety of countries, including India. In India, these are known as the Hazardous Substances (Classification, Packaging, and Labeling) Rules and the Central Motor Vehicles Rules ((Sinha et al. 2019; Goswami et al. 2019)). Central Motor Vehicle Rules: These laws, which are implemented by the Ministry of Road Transport and Highways, relate to all aspects of motor vehicle transportation. Some, however, are exclusively concerned with destructive goods: Vehicles transporting hazardous goods must be equipped with a spark arrester and a tachograph (a fire safety device) that measures driving behavior such as speed and distance. The cargo's class, UN number, and PSN must be prominently posted on "emergency information panels" carried by the truck. The shipper's cargo information must be correct, and the vehicle owner must disclose this information to the driver in writing. The written information must be kept at all times in the driver's cabin. Drivers must adhere to regulations about fire and explosion prevention to keep the cargo safe. Driver must follow the path laid forth by the owner and shipper of the vehicle. Driver must be trained to handle the dangers of moving hazardous cargo. The Hazardous Substances Rules, issued by the Ministry of Environment and Forests, regulate categorization, packaging, labelling, testing, certification, transport, and recordkeeping. Imports, radioactive materials, and pharmaceutical and cosmetic finished goods intended for consumer consumption are not included. Imported goods are excluded since they very definitely conform to the IMDG Code and DGR. Radioactive material is governed by the Atomic Energy Regulatory Board's safety code. National limitations are superseded by the European Agreement Concerning the International Carriage of Dangerous Goods by Road (ADR), which is followed by a bloc of predominantly European governments.

Rail—IRCA Red Tariff and Concor Rules

The Red Tariff of the Indian Railways Conference Association (IRCA) oversees rail booking and delivery of hazardous products in India. Among its regulations are the placement of warning signs to waggon doors and the use of unique lead seals for waggons hauling explosives, gases, flammables, and oxidizers. In recent years, significant research efforts have advanced our understanding of the aerodynamic optimization of rail vehicles, a crucial aspect in enhancing the safety and efficiency of warehousing and supply chain management of dangerous goods (Palli et al. 2018; Sharma et al. 2017; Sharma and Sharma 2018; Sharma and Kumar 2018a, b; Sharma and Sharma 2018a, b). Furthermore, the Container Corporation of India (Concor), a public sector subsidiary under the Ministry of Railways, establishes criteria for the booking of dangerous items in containers. The Regulations Governing the International Carriage of Dangerous Goods by Rail (RID), signed by 45 contracting governments from Europe, Asia, and North Africa, is a comparable international treaty to the ADR. India is not a party to it. The recent advancements in rail vehicle system modeling, structural analysis, and performance evaluation have been extensively studied in various contexts, including crashworthiness, lateral stability, critical speed analysis, and multi-objective optimization (Goyal et al. 2019; Sharma 2019; Sharma et al. 2019; Bhardawaj et al. 2020; Lee and Sharma 2020; Bhardawaj et al. 2021; Choi et al. 2021; Sharma and Sharma 2021). These investigations hold significant implications for the safe handling and effective supply chain management of dangerous goods within warehousing and transportation systems (Bhardawaj et al. 2021; Choi et al. 2021; Lee et al. 2021; Sharma and Sharma 2021).

Barge—ADN and Others

The Inland Waterway Authority of India supervises and enforces hazardous material transportation by barge in India, including the National Waterways, Safety of Navigation, and Shipping Regulations. The mechanisms established in the Explosives Act are applied to explosives under these rules (Sharma et al. 2022b). The European Agreement Concerning the International Carriage of Dangerous Goods by Inland Waterways (ADN) was established on a global scale to ensure high levels of security in the transportation of hazardous cargo on inland waterways, promote global trade in dangerous goods, and prevent pollution during such movement. The ADN is represented by delegates from eighteen different nations.

17.2.3 Warehousing of Dangerous Goods

United Nations Recommendations of Warehousing

The United Nations provides recommendations on how to establish the ideal structure, manage hazards, and apply control mechanisms in "Chemical Warehousing: The Storage of Packaged Dangerous Substances," a 2009 document. The paragraph accurately describes the best sort of warehousing. The warehouse's physical layout must correspond to the following requirements, according to the 2009 document "Chemical Warehousing: The Storage of Packaged Dangerous Substances." The warehouse must be one-story with no basements or cellars. Everything must be made of fire-resistant materials, and the interior areas must be split into various sections separated by fire-resistant partition walls. Electrical equipment should be placed outdoors and protected from explosion, and each section should have its own entrance and fire extinguisher. It is also critical to allow quick access to the warehouse, especially in the event of an emergency, by constructing many exits, because an accident might render one of them inoperable (Mohanty et al. 2016; Samantaray et al. 2014; Pattnaik et al. 2022).

The storage area should be arranged in such a way that handling procedures are simplified. Containers should be stored safely and openly so that any leaky container is instantly recognized, removed, and dealt with correctly. The stacks must not obstruct ventilation or make escape difficult in an emergency. Depending on the type of commodities stored in the warehouse, the most appropriate equipment should be utilized. Containers must be adequately secured to avoid unintentional movement, which might damage the containers and create accidents. If the aisles are narrow, fork-lift trucks should be used. Any vehicle involved in the transportation of hazardous materials should be parked properly, especially during loading and unloading. To avoid this, the distances between aisles should be carefully calculated. Stack sizes should be kept to a minimum to avoid congestion of lower levels and to ensure pile integrity. It is critical that no unauthorized persons enter the warehouse in order to avoid vandalism. Only individuals with the appropriate level of authority should be able to enter the storage area, which should be locked.

17.2.4 Types of Warehousing

Rack Storage

Racks are used in small warehouses where individual containers or cargo are stored or retrieved by hand. Racks and walk-through racks are also used to order smaller packaging pieces. For safety concerns, chemical handling overhead height is banned. Steps and ladders or storage retrieval equipment must be used with higher racks. Fragile packages must be secured to avoid falling. Small trays, specialty shelf inserts, upturned shelf edges, and spacing away from the shelf edge can all assist.

Block Storage and Line Storage

Pallets are placed directly on the floor in block warehouses rather than on shelves. Because access is confined to the front pallets, block storage is only ideal for large numbers of similar items. Aisles divide rows of blocks that are organized in tight rows. Pallets and stack-nest containers may not be piled higher than the permissible carrying, vertical, and stacking loads.

Storage in Cabinets

Small quantities can be stored safely in chemical cabinets. For the storage of non-flammable hazardous compounds that do not come under no. 5 of TRGS 510, a chemical cabinet (up to roughly 200 kg or 200 l) without additional fire protection features is suitable. If the risk assessment demonstrates that they are suitable for storage in these cabinets, hazardous substances that fall under TRGS 510 No. 5 may also be kept there. It is important to remember that lockable cabinets must be used to store acutely poisonous liquids and solids (formerly exceedingly dangerous and toxic chemicals). Similar to shelves, precautions must be made to stop containers from falling here.

17.3 Different Stages of Shipping Process:

17.3.1 Packaging

Because dangerous commodities endanger human life and property, careful packing is critical. Here are some recommendations for safeguarding your shipment: Use proper packing. The material used to pack dangerous items is assessed for durability and compliance with norms and regulations. To avoid damage and leakage, cargo must be properly packed, cushioned, and secured. Any cargo that will be exposed to water, moisture, or heat must be packed in airtight, windtight, and watertight containers. When integrated with conventional cargo during shipment, hazardous goods should be located near the container doors for easy evacuation in an emergency.

17.3.2 Labelling

Labels, markers, and placards are required by law for dangerous commodities. They are the first to detect the dangerous nature of the cargo and contain crucial information on how it should be handled. They should be readable, well-placed, and free of previous labels and marks.

Label: It quickly recognizes the cargo and conveys its dangers. According to the UN Model Regulations, a label is applied to the packaging or overpack ("an enclosure used by a single consignor to contain one or more packages and to form one unit for the convenience of handling and stowage during transport"). It should be near the UN number and PSN, not on a corner where it may collapse.

Placard: Another common hazardous identification, it resembles a label but is bigger and more durable. It is used on containers, cylinders, lorries, and other modes of transportation.

Marking: It is a mixture of the UN number, PSN, weight, specifications, precautions, and emergency response directions that is placed on the exterior container. It guarantees that cargo is handled securely. Markings, unlike labels and placards, do not have a defined style, color, or size. Some cargo (liquid hazardous products, radioactive stuff, and toxic items) necessitates the use of particular markings.

17.3.3 Documentation

Documentation is just as crucial as packaging and labeling. Incorrect information can cause the carrier to make incorrect handling, segregation, and stowage decisions, which can have devastating repercussions. Documents for dangerous products varies depending on the manner of transit, but typically contain the following fundamental information as PSN, Hazard class, and UN number, Exporter and importer names and addresses, Product weight and quantity, Packaging quantity and type (cartons, drums, containers).

17.3.4 Segregation

Moving hazardous cargo also necessitates understanding which risky commodities may and cannot be kept and carried together. The IMDG Code defines segregation as the process of separating two or more substances that are mutually incompatible. Some frequent segregation don'ts are as Store various types of explosives separately, Store flammable compounds away from oxidizing agents, a strong concentrated acid should not be stored with a strong alkali.

However, by adhering to segregation requirements, certain classes of dangerous items can be loaded and carried alongside another class of dangerous goods. An oxidizing agent, for example, can be loaded with different oxidizers as well as non-flammable and dangerous gases. It cannot, however, be used with flammables (solids, liquids, and gases) or corrosives.

17.4 Case Study

17.4.1 Tianjin Explosion:

According to government sources, a series of explosions at the Port of Tianjin in Tianjin, northern China, on August 12, 2015, killed 173 people and wounded hundreds more. The explosions happened at a container storage depot in Tianjin's Binhai New Area. The initial two explosions happened 33 s apart. The second explosion was much greater, involving the detonation of around 800 tonnes of ammonium nitrate (equal to 256 tones TNT). The fires started by the first explosions raged uncontrollably all weekend, culminating in eight further explosions on August 15.

The accident was caused by a jar of dried nitrocellulose. The container warmed, and the wetting substances within evaporated, resulting in spontaneous combustion. The dry nitrocellulose fire spread to another dangerous chemical kept in the warehouse: ammonium nitrate fertilizer. The fertilizer exploded due to the fire. Of course, with so many deadly chemicals nearby, another explosion is unavoidable. It is critical to emphasize that what occurred in that warehouse was not a natural calamity. Finally, the explosion in Tianjin, China, was fully preventable owing to human error. The incident was eventually blamed on far over 100 persons. Some were careless, while others took an active part in placing others in risk. The warehouse operator, Tianjin Ruihai International Logistics, was unlawfully keeping massive volumes of dangerous goods while following lax safety standards (Nibedita et al. 2021; Biswas et al. 2021; Sharma et al. 2022c).

17.4.2 *Oversight is Critical for Maintaining Safety*

To begin, this catastrophe illustrates the need of having oversight from government agencies or other recognized groups that develop laws and ensure firms follow them in order to prevent tragedies like Tianjin. While these restrictions may appear to make your job more difficult at times or to place unnecessary strain on your organization, the fact is that they are completely necessary. The monitoring was missing in Tianjin, and the outcome was disastrous. The Occupational Safety and Health Administration, the Environmental Protection Agency, the Department of Transportation, and the Department of Homeland Security all have an impact on the storage, handling, and transportation of hazardous materials in the United States. Safe processes begin with acknowledging the importance of these authorities' regulations.

17.4.3 *Workers Must Be Aware of Hazards and Have Hazmat Training*

Hazardous materials don't only get a special label, though that is part of it. Workers need training on how to handle, store and ship these goods safely. Not only must they Employers must also have a HazCom plan written down, which outlines how they will go about informing their employees of hazards, maintain safety data sheets and train their employees. To ensure total transparency, something that was clearly lacking in the case of the Tianjin warehouse, employers must maintain a chemical inventory list, which identifies every hazardous chemical in the workplace, how much is there and its location.

17.4.4 *Facilities Containing Hazmat Need Emergency Prevention and Response Plans*

Finally, companies that work with hazardous materials should prepare for an emergency. Thinking about the possibility of an emergency should start with a prevention strategy, but it shouldn't stop there. An emergency response plan is also crucial. Consider the most efficient and effective ways employees should respond in the event of a hazardous material igniting or spilling, for example. You can even contact your local first responders to apprise them of any unique hazards in your facility. That way, they'll be ready to help if it's ever necessary.

17.4.5 Aqaba Toxic Gas Leak

A poisonous gas leak occurred on June 27, 2022, in the Port of Aqaba, Jordan, when a container carrying 25 tons of chlorine dropped off a crane onto a parked ship and exploded. At least thirteen individuals were murdered and another 265 were injured in the tragedy (Sharma et al. 2020b).

Jordan's lone seaport lies at the head of the Gulf of Aqaba, off the Red Sea, close the Israeli border. One of six pressurized chemical storage containers was being loaded aboard the container ship for transportation to Djibouti by a crane. The cabling system on the crane failed, and one container holding about 25 tonnes (55,000 lb) of chlorine fell onto the ship and fractured, causing the chemical to escape from the container. As people fled, a cloud of brilliant yellow gas swept around the harbor. Humans are poisoned by chlorine gas. When inhaled and mixed with moisture within the human body, it produces hypochlorous and hydrochloric acid, both of which can produce oxygen free radicals that break down cell walls in the pulmonary system, causing irritation under mild exposure but potentially leading to pulmonary edema, acute respiratory distress syndrome, chronic respiratory problems, and death. The port was promptly evacuated, and first responders worked to provide medical assistance to dock employees who were injured. The nearest residential area is around 15 miles away, and Nidal Majali, the local tourist and environment commissioner, told state media that the impacts of the gas leak were limited and that there was no risk to public health.

17.4.6 Chlorine Leak on Mumbai Port Trust's Sewri Yard

In India, chemical incidents involving dangerous substances are not uncommon. On the morning of July 14, 2010, a chlorine leak was detected at the Haji Bunder hazardous cargo storage of the Mumbai Port Trust (MPT), Sewri, affecting around 120 persons in the area, including students, laborers, port personnel, and fire fighters, 70 of whom were deemed critical (Sharma et al. 2021b). On one side, Sewri Yard is near to LBS College of Advance Lifetime Studies and Research, while on the other side, there is a large amount of sea and unoccupied ground. The leak was discovered at 3 a.m. Many students from the LBS College dormitory, which is within 100 m from the source of the leak, were impacted.

Observations and Lesson Learnt

According to the Manufacture, Storage, and Import of Hazardous Chemicals (MSIHC) rules, 1989, under the Environment (Protection) Act 1986, the imported cylinders lying in the open should have been accepted for delivery as empties only if the residual chlorine was fully evacuated, either at the exporting country's facility or here in Mumbai as created in the vicinity of the port, or with assistance taken from a chlorine manufacturing or user industry having facilities. The storage room should have included suction pipes and suction hoods linked to a neutralizing tank that was in continuous operation. Cylinders must have been imported with certification from a competent authority on the residual life as well as the current state of soundness, as determined by Non-Destructive Testing (NDT) such as hydraulic testing, thickness surveys, and radiography of the welded joints, and approved by the Chief Controller of Explosives (CCE).

17.4.7 Beirut Explosion

On the evening of August 4th, 2020 (around 15:00 GMT), the world witnessed one of the deadliest explosions in history. The fire and subsequent blasts were initially witnessed and reported by civilians of the Lebanese republic and were published as videos and posts in various social media websites, after which news reports began to arrive (Pundir et al. 2023a). According to sources, the big explosion happened in a warehouse that was holding ammonium nitrate as well as dozens of bags of fireworks (The Guardian, 2020). A minor blow followed by a larger bomb happened in a succession, resulting in blast and shock waves stretching across diverse locations surrounding the port. According to news and outcome reports, the fireworks that caught fire were found close to the confiscated shipment of ammonium nitrate, which was seized in 2014 from the Rhosus (cargo ship) and stored for approximately 6 years, containing approximately 2750 tonnes of ammonium nitrate (Das et al. 2023; Agrahari et al. 2021a; Sharma et al. 2022d; Maity et al. 2020; Misra et al. 2022; Sharma and Lee 2020a; Mohapatra et al. 2017b). The original report of the explosion revealed a fire with dense grey smoke followed by flames visible from the facility accompanied by flashes and sparks, which were assumed to be caused by the fireworks. The first explosion was accompanied by a flurry of white and grey smoke and increasingly fast sparks. After 30 s, there was a huge explosion, a fireball, and red and orange gases emanating from the plant. The second explosion produced a spherical blast wave, which was later discovered to be a white mushroom cloud erupting from the source.

17.5 Factors in the Incident

17.5.1 Improper Storage Practices

The ammonium nitrate was a known hazard, but delays in removing the seized cargo exposed the port and the public to unnecessary risk. The cargo was intermingled with a variety of other incompatible materials, including fireworks and other combustible and hazardous materials (Kohli et al. 2020; Devadarshini et al. 2022; Sharma et al. 2021c; Sarkar et al. 2018). The AN was allowed to leak from bags and merge, resulting in a larger exposed mass and contamination.

17.5.2 Inadequate Fire Protection and Emergency Response

A possible reason is insufficient hot work during door repairs. The structure lacked an automated sprinkler system, and the fireworks triggered the explosion. The firemen were completely oblivious of the dangers they were in. Following the initial fire, neither the terminal nor the city was evacuated.

17.5.3 Onsite Impacts

Multiple crew fatalities and injuries, including ten city firefighters. All port operations buildings on the west side of the port are being demolished. Almost all warehouse space has been lost (Palli et al. 2022; Bhardawaj et al. 2020b, 2019a; Sharma and Lee 2020b; Acharya et al. 2020). There were eight ships declared lost. Perimeter security and adjacent structures are no longer available. Gantry cranes have been damaged. Container port shutdown (10 days of reduced operation).

17.5.4 Offsite Impacts

Disruption to the port, companies, the city, and the nation's economy. Estimated losses of $15 billion or more, with unclear effects on future GDP. There were around 200 fatalities, over 6,500 injuries, and over 300,000 people were displaced. The city's principal grain silos, 17 hospitals, and 178 schools were all damaged. The sitting administration was brought down as a result (Sharma and Lee 2021; Pundir et al. 2023b; COVID-19 2020; Pattnaik et al. 2020; Mohapatra et al. 2020).

17.5.5 Operational Impacts and Resilience

The site was not completely destroyed (because to its distance from the hanger, the cargo port suffered only minor damage). Despite broken gantry cranes, demand could be satisfied despite lower demand owing to COVID-19 and overall economic difficulties. There were significant gaps in other activities, necessitating a move to other ports or modes of transportation. The incident emphasizes the need of risk assessment, emergency preparedness, resiliency, and sustainability planning (Sharma et al. 2020c; Mohapatra et al. 2022, 2017c; Adeoye et al. 2023b; Pati et al. 2022).

17.6 Data Analysis

Visakhapatnam (Vizag), Kakkinada, Chennai, Cochin, Kandla, Mumbai, and Nhava Sheva are the major seaports in India for chemical imports. Based on import statistics from the last four years, it is clear that the quantity of hazardous chemicals imported is growing year by year. However, the overall number of imports in 2007–08 is 3.25 lacs MT, which is 23.24% smaller than the previous year number (Pati et al. 2020; Chatterjee et al. 2022; Adeoye et al. 2023c; Nayak et al. 2022; Mohapatra et al. 2017d). Ammonia has been imported in greater quantities for the previous three years in a row. In 2007–08, it again ranked top, accounting for 31.2% of total import volume, followed by Styrene (15.67%), Phosphoric acid (13.58%), and Methanol (12.95%). It has been discovered that 117 hazardous substances were imported via 12 sea ports and 2 air ports. In terms of volume, Vizakhapatnam port received 2.022 million MT of chemicals in 2007–08, accounting for 62.14% of total import volume. Mumbai port got the most chemicals (30 in total), followed by Chennai (26 in total) (Adeoye et al. 2022; Sharma et al. 2020d, 2020e; Bhardawaj et al. 2020c; Palli et al. 2021).

Over the years, there have been significant developments in the importation of chemicals into India. In the fiscal year 2003–04, 25 different types of chemicals were imported, with specific chemicals predominantly arriving through distinct ports (Sharma et al. 2021d; Palli and Koona 2015; Kanth Choppara et al. 2019; Mohapatra et al. 2015; Bhardawaj et al. 2019b). Methanol, for instance, was exclusively imported via the Visakhapatnam port, constituting 58.64% of the total import quantity.

In 2004–05, industries expanded their chemical imports, utilizing six different ports to bring in 46 types of dangerous chemicals. Visakhapatnam continued to be the top port for chemical imports, followed by Kandla, Kakinada, Chennai, and Nhava Sheva. Ammonia was a major import, primarily arriving through Visakhapatnam, accounting for 42.24% of the total import quantity.

By 2005–06, the variety of chemicals imported increased to 55 types through various ports. Visakhapatnam remained the leading port for chemical imports, mainly of ammonia, contributing to 35.3% of the total imports. Chennai, despite receiving the greatest variety of chemicals (39 in total), only accounted for approximately 17.5% in terms of quantity (Mohapatra et al. 2021; Bhattarai et al. 2022; Agrahari et al. 2021b; Agarwal et al. 2020; PanduRanga Vital et al. 2019).

In the subsequent year, 2006–07, a staggering 90 different types of chemicals were detected as being smuggled into India through eight different ports. Visakhapatnam emerged as the primary port for chemical imports, receiving 83.23% of the total import volume, including ammonia, which constituted 55% of all chemical imports. Chennai port had the most types of chemicals, with 53 in total, but its overall import percentage decreased to 1.92 from the previous year. Additionally, Kakinada port received a substantial amount of naphthalene, making it the highest hazardous chemical import at a port. (Sharma et al. 2023a, 2023b; Satyanarayana et al. 2022; Sharma and Palli 2016).

In 2007–08, the import landscape further evolved, with 117 hazardous substances arriving through 12 sea ports and 2 air ports, although air imports remained minimal at around 3.26 MTs. Visakhapatnam continued to dominate with 62.14% of the total import volume, including exclusive imports of ammonia, which had also been imported through Kakinada port during the year. Mumbai and Chennai ports led in terms of the number of chemicals received, with 30 and 26 types, respectively. Kakinada received 12,500 MTs of chemicals, New Mangalore and Ennore also emerged as significant ports in chemical imports, while imports through other ports like Kolkata, Cochin, Pipavav, and ICD Delhi remained negligible (Sarkar et al. 2018; Palli et al. 2022; Bhardawaj et al. 2020b, 2019a).

17.7 Conclusion

In conclusion, the warehousing and supply chain management of dangerous goods is a complex and critical aspect of modern logistics operations. Throughout this exploration, we have delved into the intricacies and challenges associated with handling hazardous materials, emphasizing the paramount importance of safety, compliance, and efficiency. The primary goal of managing dangerous goods within the supply chain is to mitigate risks to human health, the environment, and property. From the moment hazardous materials enter the warehouse to their distribution and eventual end-use, strict adherence to regulations and best practices is imperative. This involves proper classification, packaging, labeling, and documentation, as well as thorough training for personnel involved at every stage of the process. Furthermore, technological advancements have significantly impacted how dangerous goods are managed in the supply chain. Innovations such as real-time tracking, sensor-based monitoring, and data analytics have enhanced visibility and control, allowing for proactive risk management and quicker response to potential incidents. Automation has streamlined processes, minimized human error and ensured consistency in compliance.

Sustainability is an emerging consideration in the management of dangerous goods. As the global focus on environmental responsibility intensifies, industries are pressured to adopt greener practices in the supply chain. From sustainable packaging to emissions reduction strategies during transportation, a holistic approach to sustainability is becoming a key driver in shaping the future of dangerous goods management. Collaboration among stakeholders is a linchpin in the successful management of dangerous goods. Close cooperation between manufacturers, logistics providers, regulatory bodies, and emergency responders is essential to harmonize efforts and respond effectively to any unforeseen events. Communication, information sharing, and joint training exercises play pivotal roles in building a resilient and responsive supply chain ecosystem. In conclusion, the warehousing and supply chain management of dangerous goods is an intricate dance of regulations, technology, sustainability, and collaboration. It requires meticulous attention to detail, continuous improvement, and adaptability to evolving regulations and industry standards. Striking the balance between efficiency and safety remains a perpetual challenge, but it is one that must be met to ensure the well-being of individuals, communities, and the environment. Through rigorous adherence to best practices, embracing innovation, and fostering collaboration, the management of dangerous goods can continue to evolve into a safer, more sustainable, and resilient discipline within the broader field of supply chain management.

References

Acharya A, Gahlaut U, Sharma K, Sharma SK, Vishwakarma PN, Phanden RK (2020) Crashworthiness analysis of a thin-walled structure in the frontal part of automotive chassis. Int J Veh Struct Syst 12. https://doi.org/10.4273/ijvss.12.5.06

Adeoye AO, Quadri RO, Lawal OS (2022) Assessment of biofuel potential of tenera palm kernel shell via fixed bed pyrolysis and thermal characterization. Results Surf Interfaces 9:100091. https://doi.org/10.1016/j.rsurfi.2022.100091

Adeoye AO, Quadri RO, Lawal OS (2023a) Wet synthesis, characterization of goethite nanoparticles and its application in catalytic pyrolysis of palm kernel shell in TGA. Results Surf Interfaces 11:100118. https://doi.org/10.1016/j.rsurfi.2023.100118

Adeoye AO, Quadri RO, Lawal OS (2023b) Biomass waste pyrolysis in the context of Nigeria's oil and gas industry: an overview. Int J Ambient Energy 44:2525–2541. https://doi.org/10.1080/01430750.2023.2256325

Adeoye AO, Lawal OS, Quadri RO, Malomo D, Aliyu MT, Dang GE, Emojevu EO, Maikato MJ, Yahaya MG, Omonije OO, Edidem VG, Abubakar YK, Offor OF, Sochima EH, Peter BE, Hikon BN (2023) Sustainable energy via thermochemical and biochemical conversion of biomass wastes for biofuel production. In: Transportation energy and dynamics. energy, environment, and sustainability. Springer, Singapore, pp 245–306. https://doi.org/10.1007/978-981-99-2150-8_11

Agarwal KM, Mohapatra S, Sharma P, Sharma S, Bhatia D, Mishra A (2020) Study and overview of the novel corona virus disease (COVID-19). Sensors Int 1:100037. https://doi.org/10.1016/j.sintl.2020.100037

Agrahari R, Mohanty S, Vishwakarma K, Nayak SK, Samantaray D, Mohapatra S (2021b) Update vision on COVID-19: Structure, immune pathogenesis, treatment and safety assessment. Sensors Int 2:100073. https://doi.org/10.1016/j.sintl.2020.100073

Agrahari R, Sarraf G, Joshi NC, Mohapatra S, Varma A (2021) Insight of biopolymers and applications of polyhydroxyalkanoates. In Microbial polymers, pp 177–191. Springer Singapore, Singapore (2021). https://doi.org/10.1007/978-981-16-0045-6_8

Badgujar P, Sharma S, Rahul D, Sharma SK, Upadhyay RK (2024) Feasibility study of Mumbai-Pune Hyperloop corridor: challenges and implementation. Res Transport Business Manage 52:101089

Bandekar D, Chouhan OP, Mohapatra S, Hazra M, Hazra S, Biswas S (2017) Putative protein VC0395_0300 from Vibrio cholerae is a diguanylate cyclase with a role in biofilm formation. Microbiol Res 202:61–70. https://doi.org/10.1016/j.micres.2017.05.003

Bhardawaj S, Chandmal Sharma R, Kumar Sharma S (2019a) Development and advancement in the wheel-rail rolling contact mechanics. IOP Conf Ser Mater Sci Eng 691:012034. https://doi.org/10.1088/1757-899X/691/1/012034

Bhardawaj S, Sharma RC, Sharma SK (2020b) Development in the modeling of rail vehicle system for the analysis of lateral stability. Mater Today Proc 25:610–619. https://doi.org/10.1016/j.matpr.2019.07.376

Bhardawaj S, Sharma RC, Sharma SK (2020c) Analysis of frontal car crash characteristics using ANSYS. Mater Today Proc 25:898–902. https://doi.org/10.1016/j.matpr.2019.12.358

Bhardawaj S, Chandmal Sharma R, Kumar Sharma S (2019) A survey of railway track modelling. Int J Veh Struct Syst 11. https://doi.org/10.4273/ijvss.11.5.08

Bhardawaj S, Sharma RC, Sharma SK (2020a) Ride analysis of track-vehicle-human body interaction subjected to random excitation. J Chin Soc Mech Eng Trans Chinese Inst Eng Ser C/ Chung-Kuo Chi Hsueh K. Ch'eng Hsuebo Pao 41:229–238

Bhardawaj S, Sharma RC, Sharma SK (2020) Development in the modeling of rail vehicle system for the analysis of lateral stability. Mater Today: Proc 25:610–619. https://doi.org/10.1016/j.matpr.2019.07.376

Bhardawaj S, Sharma RC, Sharma SK, Sharma N (2021) On the planning and construction of railway curved track. Int J Veh Struct Syst 13(2):151–159. https://doi.org/10.4273/ijvss.13.2.04

Bhattarai S, Mhamdi A, Hossain I, Raoui Y, Pandey R, Madan J, Bouazizi A, Maiti M, Gogoi D, Sharma A (2022) A detailed review of perovskite solar cells: Introduction, working principle, modelling, fabrication techniques, future challenges. Micro Nanostruct 172:207450. https://doi.org/10.1016/j.micrna.2022.207450

Biswas N, Saha S, Biswas BK, Chowdhury M, Rahaman A, Junghare V, Mohapatra S, Hazra S, Zangrando E, Roy Choudhury R, Roy Choudhury C (2021) The DNA- and protein-binding properties and cytotoxicity of a new copper(II) hydrazone Schiff base complex. J Coord Chem 74:1482–1504. https://doi.org/10.1080/00958972.2021.1913128

Chatterjee A, Nirwan S, Mohapatra S, Sharma P, Agnihotri A, Shrivastava N (2022) Biochemical aspects of pathogenic variability in white rust infected Indian mustard. Mycologia 114:757–768. https://doi.org/10.1080/00275514.2022.2060007

Choi S, Lee J, Sharma SK (2021) A study on the performance evaluation of hydraulic tank injectors. In: Joshi P, Gupta SS, Shukla AK, Gautam SS (eds) Advances in engineering design. Lecture Notes in Mechanical Engineering, pp 183–190, https://doi.org/10.1007/978-981-33-4684-019

COVID-19: Analysing the legal nuances of the lockdown order. Indian J Forensic Med Toxicol (2020). https://doi.org/10.37506/ijfmt.v14i4.11553

Das S, Swapnarekha H, Vimal S (2023) Integration of blockchain technology with renewable energy for sustainable development: issues, challenges and future direction. Presented at the (2023). https://doi.org/10.1007/978-981-99-3734-9_48

Devadarshini D, Mohapatra S, Pati S, Maity S, Rath CC, Jena PK, Samantaray D (2022) Evaluation of PHAs production by mixed bacterial culture under submerged fermentation. Biologia (bratisl). 78:895–902. https://doi.org/10.1007/s11756-022-01302-5

Gopala Rao L, Sharma RC, Satyanarayana V, Palli S (2022) Stochastic optimal preview control response of a quarter-car nonlinear suspension model using spectral decomposition method. Noise Vib Worldw 53:225–232. https://doi.org/10.1177/09574565221093234

Goswami B, Rathi A, Sayeed S, Das P, Sharma RC, Sharma SK (2019) Optimization design for aerodynamic elements of Indian locomotive of passenger train. In: Advances in engineering design. Lecture notes in mechanical engineering. Springer, Singapore, pp 663–673. https://doi.org/10.1007/978-981-13-6469-3_61

Goyal S, Anand CS, Sharma SK, Sharma RC (2019) Crashworthiness analysis of foam filled star shape polygon of thin-walled structure. Thin-Walled Struct 144:106312. https://doi.org/10.1016/j.tws.2019.106312

Hossain I, Velkin VI, Shcheklein SE (2017) The study of passive vibration dampers in pipelines using PIV-methodology for single phase flow. In WIT transactions on ecology and the environment, pp 565–570. https://doi.org/10.2495/ESUS170521

Kanth Choppara R, Chandmal Sharma R, Kumar Sharma S, Gupta T (2019) Aero dynamic cross wind analysis of locomotive. IOP Conf Ser Mater Sci Eng 691:012035. https://doi.org/10.1088/1757-899X/691/1/012035

Kohli I, Joshi NC, Mohapatra S, Varma A (2020) Extremophile – an adaptive strategy for extreme conditions and applications. Curr Genomics 21:96–110. https://doi.org/10.2174/1389202921666200401105908

Kohli I, Mohapatra S, Kumar P, Goel A, Varma A, Joshi NC (2022) Role of bacterial endophytes in the promotion of plant growth. In: Bacterial endophytes for sustainable agriculture and environmental management, pp 235–253. Springer Singapore, Singapore (2022). https://doi.org/10.1007/978-981-16-4497-9_12

Kokkinos NC, Emmanouilidou E, Sharma SK (2024) Waste-to-biofuel production for the transportation sector. In: Intelligent transportation system and advanced technology, pp 99–122, Springer

Lee J, Han J, Sharma SK (2021) Structural analysis on the separated and integrated differential gear case for the weight reduction. Adv Eng Design: Select Proc FLAME 2020:175

Maity S, Das S, Mohapatra S, Tripathi AD, Akthar J, Pati S, Pattnaik S, Samantaray DP (2020) Growth associated polyhydroxybutyrate production by the novel Zobellellae tiwanensis strain DD5 from banana peels under submerged fermentation. Int J Biol Macromol 153:461–469. https://doi.org/10.1016/j.ijbiomac.2020.03.004

Misra S, Mohanty D, Mohapatra S (2022) Food-based probiotics: Functional dietary ingredients. In Probiotics, pp 257–275. Elsevier. https://doi.org/10.1016/B978-0-323-85170-1.00006-3

Mohanty DP, Mohapatra S, Misra S, Sahu PS (2016) Milk derived bioactive peptides and their impact on human health—A review. Saudi J Biol Sci 23:577–583. https://doi.org/10.1016/j.sjbs.2015.06.005

Mohapatra S, Rath SN, Pradhan SK, Samantaray DP, Rath CC (2016b) Secondary structural models (16S rRNA) of polyhydroxyalkanoates producing bacillus species isolated from different rhizospheric soil: Phylogenetics and chemical analysis. Int J Bioautomation 20:329–338

Mohapatra S, Mishra R, Roy P, Yadav KL, Satapathi S (2017b) Systematic investigation and in vitro biocompatibility studies on implantable magnetic nanocomposites for hyperthermia treatment of osteoarthritic knee joints. J Mater Sci 52:9262–9268. https://doi.org/10.1007/s10853-017-1136-0

Mohapatra S, Sarkar B, Samantaray DP, Daware A, Maity S, Pattnaik S, Bhattacharjee S (2017c) Bioconversion of fish solid waste into PHB using *Bacillus subtilis* based submerged fermentation process. Environ Technol 38:3201–3208. https://doi.org/10.1080/09593330.2017.1291759

Mohapatra S, Maity S, Dash HR, Das S, Pattnaik S, Rath CC, Samantaray D (2017d) Bacillus and biopolymer: prospects and challenges. Biochem Biophys Reports 12:206–213. https://doi.org/10.1016/j.bbrep.2017.10.001

Mohapatra S, Pattnaik S, Maity S, Mohapatra S, Sharma S, Akhtar J, Pati S, Samantaray DP, Varma A (2020) Comparative analysis of PHAs production by *Bacillus megaterium* OUAT 016 under submerged and solid-state fermentation. Saudi J Biol Sci 27:1242–1250. https://doi.org/10.1016/j.sjbs.2020.02.001

Mohapatra S, Samantaray DP, Samantaray SM (2015) Study on polyhydroxyalkanoates production using rhizospheric soil bacterial isolates of sweet potato. Indian J Sci Technol 8:57. https://doi.org/10.17485/ijst/2015/v8iS7/64027

Mohapatra S, Samantaray DP, Samantaray SM, Mishra BB, Das S, Majumdar S, Pradhan SK, Rath SN, Rath CC, Akthar J, Achary KG (2016) Structural and thermal characterization of PHAs produced by Lysinibacillus sp. through submerged fermentation process. Int J Biol Macromol 93:1161–1167. https://doi.org/10.1016/j.ijbiomac.2016.09.077

Mohapatra S, Mohanta PR, Sarkar B, Daware A, Kumar C, Samantaray DP (2017a) Production of polyhydroxyalkanoates (PHAs) by bacillus strain isolated from waste water and its biochemical characterization. Proc Natl Acad Sci India Sect B Biol Sci 87:459–466. https://doi.org/10.1007/s40011-015-0626-6

Mohapatra S, Vishwakarma K, Joshi NC, Maity S, Kumar R, Ramchander M, Pattnaik S, Samantaray DP (2021) A review on PHAs: the future biopolymer. In: Environmental and agricultural microbiology, pp 83–100. Wiley. https://doi.org/10.1002/9781119525899.ch5

Mohapatra S, Mohapatra S, Han H, Ariza-Montes A, López-Martín M del C (2022) Climate change and vulnerability of agribusiness: assessment of climate change impact on agricultural productivity. Front Psychol 13. https://doi.org/10.3389/fpsyg.2022.955622

Nayak J, Swapnarekha H, Routray AR, Nayak SR, Behera HS (2022) Big data in education: present and future. In Computational intelligence in data mining. smart innovation, systems and technologies, 281. Springer, Singapore, pp 721–739. https://doi.org/10.1007/978-981-16-9447-9_54

Nibedita S, Swati P, Pattnaik M, Mohapatra S (2021) Methane emission and strategies for mitigation in livestock. In: Environmental and agricultural microbiology, pp 257–274. Wiley. https://doi.org/10.1002/9781119525899.ch12

Palli S, Koona R, Sharma SK, Sharma RC (2018) A review on dynamic analysis of rail vehicle coach. Int J Veh Struct Syst 10(3):204–211. https://doi.org/10.4273/ijvss.10.3.10

Palli S, Dontikurti R, Sharma RC, Sharma N (2021) Analysis of dynamic response of a railway locomotive using ANSYS. Int J Veh Struct Syst 13. https://doi.org/10.4273/ijvss.13.2.21

Palli S, Duppala A, Sharma RC, Gopala Rao LVV (2022) Dynamic simulation of automotive vehicle suspension using MATLAB simulink. Int J Veh Struct Syst 14. https://doi.org/10.4273/ijvss.14.3.04

Palli S, Koona R (2015) Analyses of dynamic response of a railway bogie. Int J Veh Noise Vib 11:103. https://doi.org/10.1504/IJVNV.2015.070015

PanduRanga Vital T, Lakshmi BG, Swapna Rekha H, DhanaLakshmi M (2019) Student performance analysis with using statistical and cluster studies. In Soft computing in data analytics. Advances in intelligent systems and computing, 758. Springer, Singapore, pp 743–757. https://doi.org/10.1007/978-981-13-0514-6_71

Pati S, Maity S, Dash A, Jema S, Mohapatra S, Das S, Samantaray DP (2020) Biocompatible PHB production from Bacillus species under submerged and solid-state fermentation and extraction through different downstream processing. Curr Microbiol 77:1203–1209. https://doi.org/10.1007/s00284-020-01922-7

Pati S, Mohanty MK, Mohapatra S, Samantaray D (2022) Bioethanol production by *Enterobacter hormaechei* through carbon monoxide-rich syngas fermentation. Int J Energy Res 46:20096–20106. https://doi.org/10.1002/er.8678

Pattnaik S, Dash D, Mohapatra S, Pattnaik M, Marandi AK, Das S, Samantaray DP (2020) Corrigendum to "Improvement of rice plant productivity by native Cr(VI) reducing and plant growth promoting soil bacteria Enterobacter cloacae" [Chemosphere 240 (2020) 124895]. Chemosphere 238:125011. https://doi.org/10.1016/j.chemosphere.2019.125011

Pattnaik S, Mohapatra S, Pati S, Dash D, Devadarshini D, Tanaya K, Mishra BB, Samantaray D (2022) Microbial bioremediation of Cr(VI)-contaminated soil for sustainable agriculture. In: Microbial biodegradation and bioremediation, pp 395–407. Elsevier (2022). https://doi.org/10.1016/B978-0-323-85455-9.00008-4

Pundir S, Upadhyay S, Kumar N, Chandra Joshi N, Priya R, Ahmad Mir R, Hossain I, Pandey OP (2023a) Interconnected CuO nanoplates as a highly efficient electrocatalyst for oxygen evolution reaction. Mater Lett 336:133921. https://doi.org/10.1016/j.matlet.2023.133921

Pundir S, Upadhyay S, Priya R, Kumar N, Chetana S, Hossain I, Joshi NC, Pandey OP (2023b) Synthesis of 1D β-MnO2 for high-performance supercapacitor application. J Solid State Electrochem 27:531–538. https://doi.org/10.1007/s10008-022-05347-z

Rallabandi SR, Srinivas L, Palli S, Sharma RC, Sharma N, Sharma A, Sharma SK (2024) A contrastive characterization of pure Mg and AZ91D alloy based on the testing of mechanical, corrosion, wear, and erosion properties. Eng Res Exp 6(1):015017

Raju Rallabandi S, Pilla DP, Dowluru S, Palli S, Sharma N, Sharma SK, Sharma RC (2024) Critical evaluation of epoxy-hemp-pineapple-palm fiber composites using hybrid AHM-TOPSIS technique for sustainable structural applications. J Chin Inst Eng 1–12

Samantaray D, Mohapatra S, Mishra BB (2014) Microbial bioremediation of industrial effluents. In: Microbial biodegradation and bioremediation, pp 325–339. Elsevier. https://doi.org/10.1016/B978-0-12-800021-2.00014-5

Sarkar MS, Segu H, Bhaskar JV, Jakher R, Mohapatra S, Shalini K, Shivaji S, Reddy PA (2018) Ecological preferences of large carnivores in remote, high-altitude protected areas: insights from Buxa Tiger Reserve, India. Oryx 52:66–77. https://doi.org/10.1017/S0030605317000060

Satyanarayana VSV, Sharma RC, Sateesh B, Gopala Rao LVV, Mohan Rao N, Palli S (2022) Optimum response of a nonlinear passive vehicle suspension system under random road excitations. SAE Int J Commer Veh 16:02-16-01-0004. https://doi.org/10.4271/02-16-01-0004
Sharma SK, Sharma RC, Kumar A, Palli S (2015) Challenges in rail vehicle-track modeling and simulation. Int J Veh Struct Syst 7(1):1–9. https://doi.org/10.4273/ijvss.7.1.01
Sharma RC, Palli S, Sharma SK, Roy M (2017) Modernization of railway track with composite sleepers. Int J Veh Struct Syst 9(5):321–329. https://doi.org/10.4273/ijvss.9.5.10
Sharma SK, Kumar A (2017) Impact of electric locomotive traction of the passenger vehicle Ride quality in longitudinal train dynamics in the context of Indian railways. Mech Ind 18(2): 222. https://doi.org/10.1051/meca/2016047
Sharma RC, Sharma SK, Palli S (2018) Rail vehicle modelling and simulation using lagrangian method. Int J Veh Struct Syst 10(3):188–194. https://doi.org/10.4273/ijvss.10.3.0
Sharma SK, Kumar A (2018) Disturbance rejection and force-tracking controller of nonlinear lateral vibrations in passenger rail vehicle using magnetorheological fluid damper. J Intel Mater Syst Struct 29(2):279–297. https://doi.org/10.1177/1045389X17721051
Sharma RC, Sharma SK (2018b) Sensitivity analysis of three-wheel vehicle's suspension parameters influencing ride behavior. Noise Vibrat Worldwide 49(7–8):272–280. https://doi.org/10.1177/0957456518796846
Sharma SK, Kumar A (2018b) Impact of longitudinal train dynamics on train operations: a simulation-based study. J Vibrat Eng Technol 6(3):197–203. https://doi.org/10.1007/s42417-018-0033-4
Sharma SK, Sharma RC (2018c) An investigation of a locomotive structural crashworthiness using finite element simulation. SAE Int J Commer Veh 11(4):235–244. https://doi.org/10.4271/02-11-04-0019
Sharma SK, Sharma RC (2018b) Simulation of quarter-car model with magnetorheological dampers for ride quality improvement. Int J Veh Struct Syst 10(3):169–173. https://doi.org/10.4273/ijvss.10.3.03
Sharma SK (2019) Multibody analysis of longitudinal train dynamics on the passenger ride performance due to brake application. Proc Inst Mech Eng Part K: J Multi-Body Dynam 233(2):266–279. https://doi.org/10.1177/1464419318788775
Sharma SK, Saini U, Kumar A (2019) Semi-active control to reduce lateral vibration of passenger rail vehicle using disturbance rejection and continuous state damper controllers. J Vibrat Eng Technol 7(2):117–129. https://doi.org/10.1007/s42417-019-00088-2
Sharma SK, Sharma RC (2021) Multi-objective design optimization of locomotive nose. sae technical paper, 10. https://doi.org/10.4271/2021-01-5053
Sharma RC, Palli S, Dharmana L, Sharma SK, Sharma A, Kharola A, Sharma N (2024a) Simulation of transient dynamic response of a locomotive considering its flexibility and rigidity. Noise Vibrat Worldwide 55(1–2):33–47
Sharma RC, Palli S, Duppala A, Bhardawaj S, Sharma SK, Sharma N, Sharma A (2024b) Rigid body analysis of ride behaviour of roadway vehicle using lagrangian dynamics. Int J Veh Struct Syst (IJVSS) 16(1)
Sharma RC, Sharma N, Palli S, Sharma SK, Joshi S, Rallabandi SR, Sharma A (2024c) Dynamic behavior of three-wheel vehicle with inerter-based dynamic absorber suspension system. Noise Vibrat Worldwide 09574565241243398
Sharma SK, Kumar N, Avesh M, Sharma RC, Siddiqui MIH, Lee J (2024d) Navigating tranquillity with h∞ controller to mitigate ship propeller shaft vibration. J Vibrat Eng Technol 1–11
Sharma SK, Sharma RC, Lee J (2024e) Modelling and simulation of MR damper characterization and uncertainty assessment in marine engine vibration isolation with FLPID control. Int J Model Simulat 1–11
Sharma SK, Sharma RC, Lee J (2024f) Propelling precision of longitudinal vibration mitigation in ship propeller shafts through advanced nonlinear intelligent semi-active control leveraging adaptive neuro-fuzzy inference system with linear quadratic regulator. J Vibrat Control. https://doi.org/10.1077/5463241244836

Sharma SK, Sharma RC, Upadhyay RK, Lee J (2024g) Enhanced ship engine vibration reduction using magnetorheological dampers and Adaptive Neuro-Fuzzy Control System. Ships Offshore Struct 1–17, https://doi.org/10.1080/17445302.2024.2335449

Sharma SK, Kumar A (2017) Ride performance of a high speed rail vehicle using controlled semi active suspension system. Smart Mater Struct 26:055026. https://doi.org/10.1088/1361-665X/aa68f7

Sharma SK, Lee J (2020b) Design and development of smart semi active suspension for nonlinear rail vehicle vibration reduction. Int J Struct Stab Dyn 20:2050120. https://doi.org/10.1142/S0219455420501205

Sharma SK, Lee J (2021) Crashworthiness analysis for structural stability and dynamics. Int J Struct Stab Dyn 21:2150039. https://doi.org/10.1142/S0219455421500395

Sharma RC, Palli S (2016) Analysis of creep force and its sensitivity on stability and vertical-lateral ride for railway vehicle. Int J Veh Noise Vib 12:60. https://doi.org/10.1504/IJVNV.2016.077474

Sharma RC, Sharma SK, Sharma N, Sharma S (2020d) Analysis of ride and stability of an ICF railway coach. Int J Veh Noise Vib 16:127. https://doi.org/10.1504/IJVNV.2020.117820

Sharma RC, Sharma S, Sharma SK, Sharma N (2020e) Analysis of generalized force and its influence on ride and stability of railway vehicle. Noise Vib Worldw 51:95–109. https://doi.org/10.1177/0957456520923125

Sharma SK, Sharma RC, Lee J (2021c) Effect of rail vehicle-track coupled dynamics on fatigue failure of coil spring in a suspension system. Appl Sci 11:2650. https://doi.org/10.3390/app11062650

Sharma SK, Mohapatra S, Sharma RC, Alturjman S, Altrjman C, Mostarda L, Stephan T (2022a) Retrofitting existing buildings to improve energy performance. Sustainability 14:666. https://doi.org/10.3390/su14020666

Sharma SK, Sharma RC, Lee J, Jang H-L (2022b) Numerical and experimental analysis of DVA on the flexible-rigid rail vehicle carbody resonant vibration. Sensors 22:1922. https://doi.org/10.3390/s22051922

Sharma SK, Lee J, Jang H-L (2022c) Mathematical modeling and simulation of suspended equipment impact on car body modes. Machines 10:192. https://doi.org/10.3390/machines10030192

Sharma SK, Sharma RC, Lee J (2022d) In situ and experimental analysis of longitudinal load on carbody fatigue life using nonlinear damage accumulation. Int J Damage Mech 31:605–622. https://doi.org/10.1177/10567895211046043

Sharma RC, Palli S, Sharma SK (2023b) Ride analysis of railway vehicle considering rigidity and flexibility of the carbody. J Chinese Inst Eng 46:355–366. https://doi.org/10.1080/02533839.2023.2194918

Sharma SK, Kumar A (2018) Ride comfort of a higher speed rail vehicle using a magnetorheological suspension system. Proc Inst Mech Eng Part K. J Multi-body Dyn 232:32–48. https://doi.org/10.1177/1464419317706873

Sharma SK, Lee J (2020a) Finite element analysis of a fishplate rail joint in extreme environment condition. Int J Veh Struct Syst 12. https://doi.org/10.4273/ijvss.12.5.03

Sharma SK, Sharma RC (2019) Pothole detection and warning system for Indian roads. In Advances in interdisciplinary engineering. Lecture notes in mechanical engineering. Springer, Singapore, pp 511–519. https://doi.org/10.1007/978-981-13-6577-5_48

Sharma S, Sharma RC, Sharma SK, Sharma N, Palli S, Bhardawaj S (2020) Vibration isolation of the quarter car model of road vehicle system using dynamic vibration absorber. Int J Veh Struct Syst 12. https://doi.org/10.4273/ijvss.12.5.05

Sharma RC, Sharma SK, Palli S (2020b) Linear and non-linear stability analysis of a constrained railway wheelaxle. Int J Veh Struct Syst 12. https://doi.org/10.4273/ijvss.12.2.04

Sharma SK, Sharma RC, Sharma N (2020c) Combined multi-body-system and finite element analysis of a rail locomotive crashworthiness. Int J Veh Struct Syst 12. https://doi.org/10.4273/ijvss.12.4.15

Sharma RC, Palli S, Sharma N, Sharma SK (2021) Ride behaviour of a four-wheel vehicle using H infinity semi-active suspension control under deterministic and random inputs. Int J Veh Struct Syst 13. https://doi.org/10.4273/ijvss.13.2.18

Sharma RC, Sharma S, Sharma N, Sharma SK (2021) Linear and nonlinear analysis of ride and stability of a three-wheeled vehicle subjected to random and bump inputs using bond graph and simulink methodology. SAE Int J Commer Veh 14:02-15-01-0001. https://doi.org/10.4271/02-15-01-0001

Sharma RC, Sharma S, Sharma SK, Sharma N, Singh G (2021d) Analysis of bio-dynamic model of seated human subject and optimization of the passenger ride comfort for three-wheel vehicle using random search technique. Proc Inst Mech Eng Part K J Multi-body Dyn 235:106–121. https://doi.org/10.1177/1464419320983711

Sharma RC, Palli S, Bhardawaj S (2023) A review of wheel-rail dynamic behavior on curved tracks. Int J Veh Struct Syst 15. https://doi.org/10.4273/ijvss.15.1.04

Sinha AK, Sengupta A, Gandhi H, Bansal P, Agarwal KM, Sharma SK, Sharma RC, Sharma SK (2019) Performance enhancement of an all-terrain vehicle by optimizing steering, powertrain and brakes. In Advances in engineering design . lecture notes in mechanical engineering. Springer, Singapore, pp 207–215. https://doi.org/10.1007/978-981-13-6469-3_19

Swapna Rekha H, Nayak J, Behera HS (2020) Pi-sigma neural network: survey of a decade progress. In: Computational intelligence in pattern recognition. Advances in intelligent systems and computing, 1120. Springer, Singapore, pp 429–441. https://doi.org/10.1007/978-981-15-2449-3_37

Upadhyay RK, Sharma SK, Kumar V (Eds) (2024a) Intelligent transportation system and advanced technology. Springer Nature Singapore. https://doi.org/10.1007/978-981-97-0515-3

Upadhyay RK, Sharma SK, Kumar V (2024b) Introduction to intelligent transportation system and advanced technology. In: Intelligent transportation system and advanced technology. Energy, Environment, and Sustainability. pp 3–6. Springer, Singapore. https://doi.org/10.1007/978-981-97-0515-3_1

The manufacturer's authorised representative in the EU is Springer Nature Customer Service Centre GmbH, Europaplatz 3, 69115 Heidelberg, Germany. If you have any concerns regarding our products, please contact ProductSafety@springernature.com

Printed and bound by CPI Group (UK) Ltd, Croydon, CR0 4YY
01/10/2026
02225863-0012